# Membrane Transport in Plants

# Annual Plant Reviews

A series for researchers and postgraduates in the plant sciences. Each volume in this series focuses on a theme of topical importance and emphasis is placed on rapid publication.

# Membrane Transport in Plants

Edited by

MICHAEL R. BLATT
Regius Professor of Botany
Laboratory of Plant Physiology and Biophysics
IBLS - Plant Sciences Bower Building
University of Glasgow
UK

**Blackwell**
Publishing

**CRC Press**

© 2004 by Blackwell Publishing Ltd

Editorial Offices:
Blackwell Publishing Ltd, 9600 Garsington Road, Oxford OX4 2DQ, UK
  *Tel:* +44 (0)1865 776868
Blackwell Publishing Asia Pty Ltd, 550 Swanston Street, Carlton, Victoria 3053, Australia
  *Tel:* +61 (0)3 8359 1011

ISBN 1-4051-1803-2
ISSN 1460-1494

Published in the USA and Canada (only) by CRC Press LLC, 2000 Corporate Blvd., N.W.,
Boca Raton, FL 33431, USA
Orders from the USA and Canada (only) to CRC Press LLC

USA and Canada only:
ISBN 0-8493-2351-7
ISSN 1097-7570

First published 2004

Agr
GH
509
.M46x
2004

Library of Congress Cataloging-in-Publication Data:
A catalog record for this title is available from the Library of Congress

British Library Cataloguing-in-Publication Data:
A catalogue record for this title is available from the British Library

Set in 10/12 pt Times New Roman by TechBooks, New Delhi, India
Printed and bound in Great Britain by MPG Books Ltd, Bodmin, Cornwall

For further information on Blackwell Publishing, visit our website:
www.blackwellpublishing.com

# Contents

# 3   Structure, function and regulation of primary H$^+$ and Ca$^{2+}$ pumps

**ROSA L. LÓPEZ-MARQUÉS, MORTEN SCHIØTT, MIA KYED JAKOBSEN and MICHAEL G. PALMGREN**

## 6 Voltage-gated ion channels
### INGO DREYER, BERND MÜLLER-RÖBER
### and BARBARA KÖHLER

**150**

## 9   Ca$^{2+}$ and pH as integrating signals in transport control      252
TATIANA N. BIBIKOVA, SARAH M. ASSMANN
and SIMON GILROY

## 10   Vesicle traffic and plasma membrane transport      279
ANNETTE C. HURST, GERHARD THIEL
and ULRIKE HOMANN

# List of contributors

**Dr Anna Amtmann**  Laboratory of Plant Physiology and Biophysics, Plant Sciences, Bower Building, IBLS - Plant Sciences, University of Glasgow, Glasgow G12 8QQ, UK

**Dr Patrick Armengaud**  Laboratory of Plant Physiology and Biophysics, Plant Sciences, Bower Building, IBLS, University of Glasgow, Glasgow G12 8QQ, UK

**Prof. Sarah Assmann**  Biology Department, 208 Mueller Laboratory, Penn. State University, University Park, PA 16802-5301, USA

**Dr Tatiana N. Bibikova**  Biology Department, 208 Mueller Laboratory, Penn. State University, University Park, PA 16802-5301, USA

**Dr Yann Boursiac**  Biochimie et Physiologie Moleculaire des Plantes, Agro-M/INRA/CNRS/UM2 UMR 5004, 2 Place Viala, F34060 Montpellier Cedex 1, France

**Prof. Michael Blatt**  Laboratory of Plant Physiology and Biophysics, Bower Building, IBLS, Plant Sciences University of Glasgow, Glasgow G12 8QQ, UK

**Prof. Dan Bush**  Department of Biology, Colorado State University, Fort Collins, Colorado 80253, USA

**Dr Peter Dominy**  Plant Sciences, Bower Building, IBLS, University of Glasgow, Glasgow G12 8QQ, UK

**Dr Ingo Dreyer**  Universität Potsdam, Institut für Biochemie und Biologie, Abt. Molekularbiologie, Karl-Liebknecht-Str. 24/25, Haus 20, D-14476 Golm, Germany

**Dr Simon Gilroy**                Biology Department, 208 Mueller Laboratory,
                                   Penn. State University, University Park, PA
                                   16802-5301, USA

**Dr Malcolm Hawkesford**          Crop Performance and Improvement Division,
                                   Rothamsted Research, Harpenden,
                                   Hertfordshire AL5 2JQ, UK

**Dr Adrian Hills**                Laboratory of Plant Physiology and
                                   Biophysics, Plant Sciences, Bower Building,
                                   IBLS, University of Glasgow, Glasgow G12
                                   8QQ, UK

**Dr Ulrike Homann**               Botanisches Institut, TU-Darmstadt,
                                   Schnittspahnstrasse 3, D64287 Darmstadt,
                                   Germany

**Dr Annette C. Hurst**           Botanisches Institut, TU-Darmstadt,
                                   Schnittspahnstrasse 3, D64287 Darmstadt,
                                   Germany

**Dr Mia Kyed Jakobsen**           Department of Plant Biology, Royal Veterinary
                                   and Agricultural University, Thorvaldsenvej
                                   40, DK1871 Frederiksberg C, Copenhagen,
                                   Denmark

**Dr Barbara Köhler**              Universität Potsdam, Institut für Biochemie
                                   und Biologie, Abt. Molekularbiologie,
                                   Karl-Liebknecht-Str. 24/25, Haus 20, D-14476
                                   Golm, Germany

**Dr Rosa L. López-Marqués**       Instituto de Bioquímica Vegetal y Fotosíntesis,
                                   Universidad de Sevilla-CSIC, Sevilla, Spain

**Dr Frans Maathuis**              Biology Department, Area 9, University of
                                   York, York YO10 5DD, UK

**Dr Christophe Maurel**           Biochimie et Physiologie Moleculaire des
                                   Plantes, Agro-M/INRA/CNRS/UM2 UMR
                                   5004, 2 Place Viala, F34060 Montpellier Cedex
                                   1, France

**Dr Tony Miller**                 Crop Performance and Improvement Division,
                                   Rothamsted Research, Harpenden,
                                   Hertfordshire AL5 2JQ, UK

**Dr Bernd Müller-Röber**    Universität Potsdam, Institut für Biochemie und Biologie, Abt. Molekularbiologie, Karl-Liebknecht-Str. 24/25, Haus 20, D-14476 Golm, Germany

**Prof. Mickey Palmgren**    Department of Plant Biology, Royal Veterinary and Agricultural University, Thorvaldsenvej 40, DK1871 Frederiksberg C, Copenhagen, Denmark

**Dr Susan Rosser**    Plant Sciences, Bower Building, IBLS, University of Glasgow, Glasgow G12 8QQ, UK

**Dr Morten Schiøtt**    Department of Plant Biology, Royal Veterinary and Agricultural University, Thorvaldsenvej 40, DK1871 Frederiksberg C, Copenhagen, Denmark

**Dr Gerhard Thiel**    Botanisches Institut, TU-Darmstadt, Schnittspahnstrasse 3, D64287 Darmstadt, Germany

**Dr Lionel Verdoucq**    Biochimie et Physiologie Moleculaire des Plantes, Agro-M/INRA/CNRS/UM2 UMR 5004, 2 Place Viala, F34060 Montpellier Cedex 1, France

**Dr Vadim Volkov**    Laboratory of Plant Physiology and Biophysics, Plant Sciences, Bower Building, IBLS, University of Glasgow, Glasgow G12 8QQ, UK

**Dr Clare Vander Willigen**    Department of Molecular and Cellular Biology, University of Cape Town, Private Bag, Rondebosch 7701, South Africa

# Preface

Nutrition lies at the heart of physiological studies in plants, but to define a modern understanding of the field of membrane transport in this context alone neglects both the wealth of its scientific inheritance and the breadth of its impact today. A brief historical review serves to illustrate this point. As early as the seventeenth and eighteenth centuries, naturalists were absorbed by the question 'from where do plants derive their mass?' Van Helmont's conclusion that plants come from water alone (van Helmont, 1648, translated in Howe, 1965) and Woodward's (Woodward, 1699; see Thomas, 1955) experiments suggesting that plants arose from soil were followed by the observations of Stephen Hales and Joseph Priestley identifying the importance of gas exchange to plant growth. However, it was von Liebig (1840), demonstrating the need for inorganic minerals in the soil around the root, who led the way to considerations of their uptake. The studies of Brezeale (1906), Hasselbring (1914) and Stiles and Kidd (1919) showed that plants could regulate solute intake independent of transpiration. Thus, by the early 1900s, selective mineral nutrition was recognised as an important characteristic of the growing plant.

The period around 1900 also marked other new ideas that would define our present thinking about transport in cells. Nernst (1888,1889) and Planck (1890a,b) introduced what is now known as the Nernst–Planck Electrodiffusion Equation, combining Fick's and Ohm's Laws, from which is derived the Nernst Equation equating chemical and electrical driving forces across a membrane at equilibrium. Not long before, Pfeffer (1877) had proposed the existence of a delimiting and semi-permeable cell membrane or skin ('Plasmahaut') from his studies of osmosis and plant movements. Finally, Bernstein (1902) postulated the cell membrane and its ionic permeability as the basis for 'bioelectricity' in cells.

These and related ideas on the nature of membranes soon became foci for much experimental work with large and easily manipulated cells, including invertebrate eggs, muscle, squid axon and the giant algae *Chara* and *Nitella*. By the late 1930s, Cole and Curtis (1938, 1939) had demonstrated, both in squid axon and *Nitella*, that action potentials were accompanied by very large changes in membrane conductance. They correctly surmised the underlying transmembrane flux of ions, but were puzzled by the fact that the voltage would overshoot zero and reverse sign by tens of millivolts. If the membrane became transiently permeable to all ions, the voltage should simply approach zero. Clearly, the problem related to the selectivity for permeation among ions present on either side of the membrane. Its solution, by Hodgkin, Huxley and Katz (Hodgkin & Katz, 1949; Hodgkin *et al.*, 1952), led to a recognition of the ionic basis of the membrane potential and the existence of discrete and specialised ionic conductances, which later would come to be known as ion channels.

Another major piece was added to this conceptual jigsaw puzzle in the 1960s, although its acceptance was slow to come in many circles. Hodgkin, Katz, Cole and others had worked with giant cells and syncitia that would survive cutting, permeabilisation and other manipulations necessary to gain direct access to the cytoplasmic side of the membrane. Microcapillary impalement techniques had been introduced by plant electrophysiologists from the early 1920s (see Umrath, 1930, 1932, and Osterhout, 1931). As these techniques attracted wider attention (see Ling & Gerard, 1949, and Chapter 1), resting potentials of animal cells were found, almost without exception, to lie close to $-40$ to $-50$ mV, consistent with the passive diffusion of $K^+$ from the cell and additional minor contributions from $Na^+$ and $Cl^-$ conductances (Goldman, 1943; Hodgkin & Katz, 1949). Radiotracer studies, introduced after the Second World War, supported this idea, suggesting that the energetic input to maintaining these ionic gradients was achieved through a background exchange of $Na^+$ and $K^+$ that was all but electrically silent (see Tanford, 1983). Indeed, the electrogenic nature of the $Na^+/K^+$-ATPase in animals would remain a matter of debate for almost three decades, coming to an end only with the studies of Gadsby and DeWeer in the mid-1980s (DeWeer $et$ $al.$, 1988).

The idea that the membrane potential arose solely from passive ionic diffusion was first challenged by Slayman (1965) in his work with $Neurospora$, and subsequently by Spanswick $et$ $al.$ (1970) and Higinbotham $et$ $al.$ (1967) in their studies of $Nitella$, pea and oat mesophyll cells. Slayman showed that the resting membrane potential of the fungus $Neurospora$ was typically around $-200$ mV – far more negative than could be explained by diffusion of any of the major ions present – and that this potential relaxed to approximately the diffusion potential expected for $K^+$ when respiration was suppressed. Kitasato (1968) noted that the resting membrane potential of $Nitella$ was strictly dependent on the concentration of $H^+$ in the medium. Against the backdrop of the new ideas of chemiosmosis (see Mitchell, 1969), transport in plants and fungi was soon recognised to be powered by an ATPase with a significant proportion of the energy output used to maintain the large transmembrane voltage as well as a pH gradient. Changes in membrane potential with the uptake of uncharged nutrients such as glucose (Slayman & Slayman, 1974; Schwab & Komor, 1978) were strong indications that transport in plant and fungal cells was coupled to the $H^+$ electrochemical gradient. A truly chemiosmotic coupling to this gradient – of high-affinity $K^+$ uptake affecting both membrane potential and cytosolic pH – was first demonstrated in $Neurospora$ (Rodriguez-Navarro $et$ $al.$, 1986; Blatt & Slayman, 1987; Blatt $et$ $al.$, 1987). A decade later, sufficient evidence for a similar mechanism of $K^+$ transport in $Arabidopsis$ (Maathuis & Sanders, 1994) closed discussions of a $H^+/K^+$-ATPase in plants analogous to the $Na^+/K^+$-ATPase of animals. The 1960s and 1970s also marked a growing awareness of the roles of $H^+$ and, even more, of $Ca^{2+}$ and their transport in cell signalling. Lionel Jaffe's studies of fucoid 'eggs' and especially his introduction of the vibrating reed electrode (Jaffe, 1968; Robinson & Jaffe, 1975; Jaffe, 1981) brought the fields of cell development and transport into close juxtaposition. Thus, the period from the 1950s through the 1980s largely defined the compass of transport across membranes, both plant and animal, and established its importance for cell signalling and related processes; but it yielded fewer clues to the identities of the proteins responsible or the physical mechanisms for transport.

These last twenty years, we have witnessed another major leap in understanding of membranes and transport. More than ever before, these developments are cross-disciplinary in their origins, arising on one side from advancements in electronics, leading to so-called patch clamp methods for single-channel (Hamill *et al.*, 1981) and capacitance (Travis & Wightman, 1998) recording, through to chemistry and microscopy that have engendered the techniques of high-resolution imaging *in vivo* of ion concentrations (McCormack & Cobbold, 1991; Angleson & Betz, 1997; Roy *et al.*, 1999) and even proteins (Stephens & Allan, 2003; Lippincott-Schwartz & Patterson, 2003). The biophysical and molecular details of ion transport across membranes has become closely interwoven into our understanding of plant cell signalling and intercellular communication, of plant-pathogen and symbiotic inter-actions, and of membrane biogenesis, polarity and development, to name just a few topics of intense research interest today.

The revolution in molecular genetics has had a still greater impact. Even before the complete sequencing of the *Arabidopsis* genome in 2000 (Arabidopsis Genome Initiative, 2000), a significant number of transporters had been cloned, including $H^+$-ATPases, $H^+$-driven ion pumps and ion and water channels. In general, these transporters have proved to be members of much larger gene families, some of which are discussed in this volume. Thus, one challenge for the next generation of researchers will be to establish the functions of many members of these families and to associate the proteins with development, signalling and nutrition of the plant as a whole. Other transport functions still lack obvious molecular identities, among these are the several $Ca^{2+}$ channels that play central roles in various cell-signalling events. In most cases, these obstacles will be overcome through the now well-established methods of molecular genetics, (electro)physiological analysis *in vivo* and after heterologous expression. A still bigger challenge lies in understanding the integration of transport across family boundaries for nutrition and homeostasis, for example, of $K^+$. This information, in turn, will inform new initiatives to utilise plants in resolving some of the dilemmas that face our present society.

It would be impossible to cover all aspects of membrane transport in plants in a volume of this kind without reducing its content to the most trivial of descriptions. Instead, the chapters that follow offer a selection of the most important and hotly pursued topics in the field. The first two chapters cover approaches to membrane transport analysis, emphasizing concepts, techniques and tools for electrophysiol-ogy. Topics of Chapters 3–8 divide along boundaries of pumps, coupled transporters and channels; the addition of a chapter on water channels highlights this rapidly ex-panding and, until recently, highly controversial topic. Chapters 9 and 10 address issues of $Ca^{2+}$ and $H^+$ signalling, and of membrane traffic that increasingly attracts attention of researchers in plant development as well as in membrane transport.

Finally, Chapters 11 and 12 take a post-genomic look at the problems of under-standing the integration of transport mechanisms and its relevance to inorganic nutrition and phytoremediation. An overriding theme throughout is the extent to which the research on membrane transport now informs the fields of plant cell biology and physiology, and is itself enriched in return. It is my hope that these pages will enthuse, guide and inform advanced students and researchers unfamiliar with membrane transport in plants, and will serve as a useful resource for those who are.

Michael R. Blatt
(citations will be found in Chapter 1)

# 1 Concepts and techniques in plant membrane physiology

Michael R. Blatt

## 1.1 Introduction

The vast majority of known biological transport processes carry electrical charge associated with the transported solute(s) and therefore affect the voltage across membranes. Other membrane-related activities, such as the exo- and endocytotic traffic of vesicles, also influence the electrical capacitance and, at least transiently therefore, the voltage across membranes. Indeed, (trans)membrane voltage or 'membrane potential' is the simplest of parameters to record across biological membranes. It is also one of the most commonly cited, although its significance is frequently misunderstood. Accounting for membrane voltage is essential, but is only part of any quantitative description of transport that moves charge across membranes.

From an enzymologist's viewpoint, membrane transporters are proteins that facilitate the conversion of substrates to products, thereby consuming or releasing free energy. It happens only that these changes in free energy relate to the electrochemical states of the substrates and products on the two sides of the membrane, rather than to any chemical bonding. Substrates and products are chemically identical, whether the transport mechanism entails coupling to $H^+$ or passive diffusion through an ion channel. (Obvious exceptions to this rule are the ATPases and pyrophosphatases that utilise the energy of the phosphate bond to drive movement of ions across membranes. However, none of these phosphate compounds are transported themselves in these instances.) Indeed, membrane physiologists make use of conventional reaction-kinetic paradigms, incorporating ion and other substrate concentrations, cyclic binding and debinding steps, to give the same mathematical substance to transport as with any other enzyme kinetic process.

Of course, there are important differences that set membrane transport apart from other fields of enzyme biochemistry. One relates to advantages for the electrophysiologist in the speed, simplicity and tremendous sensitivity afforded by recording membrane current. Current and chemical flux are related stoichiometrically to the charge on the ion substrates that are carried by any one transporter. So the current – the charge moved per second – through a transporter gives a direct measure of enzyme turnover (transport) rate in proportion with the charge stoichiometry for transport. Transmembrane currents on the order of a few picoamperes are easily recorded in whole-cell studies, and currents on the order of 0.5 pA and less are frequently resolved in measurements from single channels. Considering that an ion channel may open for 3–5 ms, these currents are equivalent to the movement of no more than a few thousand ions through a single functional protein unit. Furthermore, a full spectrum of transport rates can be obtained in a matter of seconds in many cases.

Why, then, is membrane voltage important? On one hand, ion transport will affect charge distribution across the membrane and, hence, its voltage. A change in voltage is a sure sign of a change in charge transport activity. On the other hand, charges that move across a membrane necessarily do so through the transmembrane electric field (at the macroscopic level, the membrane voltage) and therefore will be influenced both in their rate of movement as well as direction. The essence of these statements is obvious, but their significance is less often appreciated. Herein lies another important difference that sets transport analysis apart. Membrane voltage can be expected to affect the kinetic characteristics of transport in much the same way as does ion (substrate or product) concentration on either side of the membrane. In a very real sense, membrane voltage is the 'electrical' substrate for all charge-carrying transporters. At the same time, charge transport activity necessarily affects membrane voltage and, thereby, the activity of all processes transporting charge across the same membrane. Clearly, in this situation a satisfactory kinetic description requires that the membrane voltage be controlled experimentally. It is for this reason that the 'voltage clamp' remains one of the most powerful tools in membrane biology, now more than 50 years after its inception (Cole, 1949; Hodgkin & Katz, 1949; Hodgkin *et al.*, 1952).

This chapter reviews some of the essential concepts behind our present understanding of membrane transport in plants and the techniques that have given them experimental substance. A compendium of techniques in the field could easily fill several volumes (and has!). Three important topics are not discussed here. Techniques for single-channel analysis using artificial planar bilayers are very similar to those of the patch clamp, and the reader will find excellent source material in Chris Miller's book *Ion Channel Reconstitution* (1986) (see also Rudy & Iverson, 1992; Conn, 1998). Also not discussed are methods for recording ion fluxes around cells using the vibrating reed electrode developed by Lionel Jaffe (Jaffe *et al.*, 1974) and variants that incorporate ion-selective probes (Shabala *et al.*, 1997). These are non-invasive methods, but are restricted by the lack of access to membrane voltage. Finally, methods for measurement and control of intracellular ion concentrations, especially of the second messengers $Ca^{2+}$ and $H^+$ (see Chapter 9), are the subject of several volumes in themselves (McCormack & Cobbold, 1991; Nuccitelli, 1994). Techniques for microinjection and cell dialysis discussed in this chapter are relevant to these topics. Practical information on expressing membrane transport proteins in heterologous systems will be found in volumes by Gould (1994) and Conn (1998). Methods for analysis of heterologously expressed transporters, however, are essentially identical to those for their study *in planta*. For readers interested in pursuing research on their own, especially using the electrophysiological techniques discussed here, an introduction to the essential tools will be found in the following chapter. Excellent practical guides to methods and data analysis will be found also in the Plymouth Workshop Handbook entitled *Microelectrode Techniques*

(Standen *et al.*, 1987). Purves' volume (Purves, 1981) is also well worth a look, especially his chapters dealing with microinjection and troubleshooting, and the author's own chapter in *Methods in Plant Biochemistry* (Blatt, 1991) offers a number of tips for intracellular microelectrode recording from plants.

## 1.2 Plant membrane transport

A fundamental characteristic of transporters that operate across a common biological membrane is that they all share common intermediates (substrates and products) of ion and other solute concentrations on the two sides of the membrane. For transporters that move a net charge across the membrane, the membrane voltage is also a common intermediate. In these respects, the transport 'network' of a membrane is unlike other enzyme pathways that describe a sequential series of reaction steps. In the latter, each reaction step acts on a set of substrates yielding products that, in turn, feed the next step in the chain; any one set of substrate/product intermediates is generally common only to two links in the chain of reactions. This difference affects both the way that individual membrane transport processes interact and the experimental approaches available for their analysis.

Consider the plant plasma membrane shown schematically in Fig. 1.1. With few exceptions (McCulloch *et al.*, 1990; Maathuis *et al.*, 1996; GarciaSanchez *et al.*, 2000), this membrane is energised by a family of $H^+$-ATPases (see Chapter 3) that drive $H^+$ out of the cell, generating an electrochemical gradient for $H^+$ ($\Delta\mu_H$) directed inward across the membrane. One $H^+$ is transported for each molecule of ATP hydrolysed (Blatt, 1987a; Blatt *et al.*, 1990) and this energetic input sustains gradients of 2–3 pH units and $-150$ to $-200$ mV (inside negative) across the membrane in many instances. The output of the $H^+$-ATPase is defined by the sum of the electrical and chemical gradients, plus the energy released by ATP hydrolysis. Thus, at equilibrium the voltage generated by the pump

$$E_P = \frac{RT}{zF} \left\{ \ln K_{ATP} + \ln \frac{[ATP]_i}{[ADP]_i[P_i]_i} + z \ln \frac{[H^+]_i}{[H^+]_o} \right\} \quad (1.1)$$

where $K_{ATP}$ is the equilibrium constant for ATP under standard conditions, $[P_i]_i$ is the cytosolic concentration of inorganic phosphate, and $R$, $T$, $z$ and $F$ have their usual meanings. For respiring plant cells, this relationship predicts an equilibrium voltage for the $H^+$-ATPase around $-500$ mV in the absence of a $[H^+]$ difference between inside and outside the cell. In other words, with 1 $H^+$ transported per ATP hydrolysed ($z = +1$), the pump will stall against the 'back pressure' of this voltage across the membrane. The fact that the membrane voltages of plants and fungi generally do not exceed $-300$ mV, even at an external pH near 7 (see, for example, Slayman, 1965; Beilby, 1984; Blatt, 1987b; Blatt *et al.*, 1987), is not surprising. The output of the pump is used to do work in moving other solutes across the membrane, with the effect that a constant circuit of current and $H^+$ pass through the pump and return via these other transport mechanisms.

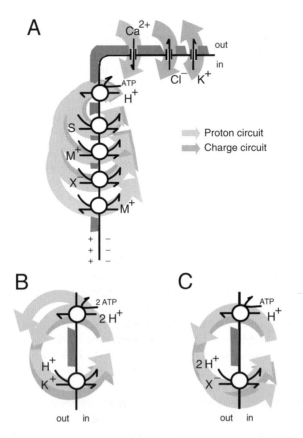

**Fig. 1.1** An overview of the proton and charge circuit of the plant plasma membrane. (A) The energy of ATP hydrolysis is used to drive $H^+$ out of the cell, generating a $\Delta\mu_H$ of membrane voltage and $[H^+]$ gradients directed back into the cell. This so-called proton-motive force energises $H^+$-coupled transport for uncharged (S) and charged ($M^+$ and $X^-$) solutes. Coupled transport thus contributes return pathways for $H^+$ flux and, with the exception of electroneutral ion exchange (e.g. $H^+$-$M^+$ exchange), charge movement back across the membrane. Ion channels for $Ca^{2+}$, $Cl^-$ and $K^+$ (above) contribute to the charge circuit, but not the $H^+$ circuit across the membrane. (B) $H^+$-coupled $K^+$ uptake in *Neurospora* (Rodriguez-Navarro *et al.*, 1986; Blatt *et al.*, 1987; Blatt & Slayman, 1987) requires export of 2 $H^+$ to balance charge. The overall effect is a 1:1 exchange of $H^+$ export with $K^+$ uptake and an overall rise in cytosolic pH. (C) Transport of many anions requires coupling with 2 $H^+$ to overcome the opposing electrical barrier of moving a negatively charged ion into the (inside-negative) cell. Charge balance via the $H^+$-ATPase in this case must result in a net fall in cytosolic pH.

Transport of other solutes, when otherwise energetically unfavourable, generally proceeds by coupling flux of the solute to $H^+$ return across the membrane. Coupling ratios $n$(solute):$m$($H^+$) are such that forward transport is often characterised by a net, cytoplasm-directed current (hence, completing the electrical circuit) and, when the membrane voltage is not clamped, by a depolarising voltage response. Classic examples of $H^+$-coupled transport of this kind include the glucose transporters in

*Neurospora* (Slayman & Slayman, 1974; Hansen & Slayman, 1978) and *Chlorella* (Schwab and Komor, 1978), high-affinity transport of $K^+$ in *Neurospora* (Rodriguez-Navarro *et al.*, 1986; Blatt *et al.*, 1987; Blatt & Slayman, 1987), $Cl^-$ transport in *Chara* (Beilby & Walker, 1981), $NO_3^-$ uptake in *Arabidopsis* and barley (Glass *et al.*, 1992; Meharg & Blatt, 1995) and amino acid transport in *Neurospora* and *Riccia* (Felle, 1981; Sanders *et al.*, 1983) (see Chapters 4 and 5). These transporters draw on both the voltage and concentration difference for $H^+$ that make up the electrochemical potential for the ion, defined at equilibrium by the equation

$$E_{HS} = \frac{RT}{(n+m)F} \left\{ n \ln \frac{[S]_o}{[S]_i} + m \ln \frac{[H^+]_o}{[H^+]_i} \right\} \tag{1.2}$$

where S is the solute.

Important exceptions to this rule are typified by extrusion of $Na^+$ that operates in electroneutral exchange with $H^+$ in *Chara* (Clint & MacRobbie, 1987) and across the tonoplast of many plants (Blumwald & Poole, 1987; Quintero *et al.*, 2000). These transporters draw on the $H^+$ electrochemical potential but depend, in the first instance, only on the $H^+$ concentration difference across the membrane. They neither respond to nor affect membrane voltage and, because transport is via electroneutral exchange of two ions, their activities cannot be measured as membrane current. However, all $H^+$-coupled transport, whether electrically silent or not, facilitates $H^+$ return across the membrane, and so contributes to $H^+$ cycling across the membrane – literally, the $H^+$ circuit – between the cytosol and the extracellular space (see Fig. 1.1A).

One important consequence of this interplay between $\Delta\mu_H$-generating and -utilising pathways is that the balance between different $H^+$ return pathways can affect pH on both sides of the membrane. For example, in $K^+$-starved *Neurospora* (Blatt *et al.*, 1987) the energy output of the $H^+$-ATPase is utilised almost entirely to drive $K^+$ uptake through high-affinity, $H^+$-coupled transport. Because physical laws dictate that charge balance must be maintained, current carried by $H^+$ and $K^+$ entry – i.e. 2(+) – is balanced by export of 2 $H^+$, with the overall effect of a one-to-one exchange of $H^+$ for $K^+$ (Fig. 1.1B). Indeed, $H^+$-selective electrode measurements demonstrated a roughly stoichiometric rise in cytosolic pH with $K^+$ uptake (Blatt & Slayman, 1987). It is very probable that the same applies also to $H^+$-coupled transport of $K^+$ plants (Newman *et al.*, 1987; Maathuis & Sanders, 1994). By contrast, anion uptake is often coupled to 2 $H^+$ to provide sufficient driving force (Meharg & Blatt, 1995; Blatt *et al.*, 1997), but charge balance in this case requires the export of only one $H^+$ by the $H^+$-ATPase (Fig. 1.1C). So $H^+$-coupled anion transport can be expected to lead to a net fall in cytosolic pH.

Another consequence of $\Delta\mu_H$ coupling – the commonality of $H^+$, and especially of the membrane voltage – is that block current through one half of the circuit immediately affects the activity of the other half. Thus, for the same $H^+$-ATPase/ $H^+$-coupled $K^+$ transport circuit, poisoning the $H^+$-ATPase blocks $K^+$ uptake (Blatt *et al.*, 1987; Maathuis & Sanders, 1994), and eliminating current through the

$H^+$-coupled $K^+$ transporter by removing $K^+$ outside immediately suppresses cytosolic alkalinisation and current output through the $H^+$-ATPase (Blatt & Slayman, 1987). These examples also serve to illustrate the fundamental importance of the voltage clamp to transport analysis: In effect it introduces an experimentally controlled pathway for current that bypasses the membrane and, hence, can substitute for the $H^+$-ATPase when its output is blocked by metabolic poisons, and for current return when $H^+$-coupled $K^+$ transport is eliminated by the absence of extracellular $K^+$.

Ion channels also contribute to the electrical circuit of the membrane, albeit not directly to transmembrane $H^+$ flux (Fig. 1.1A). The plant plasma membrane is known to harbour a large number of channels that are permeant to different ions, including $K^+$, $Cl^-$ and $Ca^{2+}$ (see Chapters 6 and 7). Movement of each ionic species through these channels is driven solely by the prevailing electrochemical gradient for the ion X that is permeant as defined at equilibrium by the Nernst equation

$$E_X = \frac{RT}{zF} \left\{ \ln \frac{[X^\pm]_o}{[X^\pm]_i} \right\} \tag{1.3}$$

Unlike ion pumps and coupled transporters that show high selectivities for the transported species – for example, selectivities $>500{:}1$ for $K^+$ over $Rb^+$ among high-affinity $H^+$-coupled $K^+$ transporters (Rodriguez-Navarro et al., 1986; Blatt et al., 1987; Rubio et al., 2000) – ion channels are poor at discriminating between ions. $Ca^{2+}$ channels are equally (if not more) permeable to $Ba^{2+}$ (Gelli & Blumwald, 1997; Hamilton et al., 2000, 2001) and $Cl^-$ channels generally carry $NO_3^-$ more readily than $Cl^-$ (Tyerman & Findlay, 1989; Tyerman, 1992). Among the most exacting, many $K^+$ channels show selectivities between $K^+$, $Rb^+$ and $Na^+$ of approximately 10:7:1 (Very & Sentenac, 2003). Compared with ion pumps and coupled transporters, ion channels also carry much greater current and therefore can be major contributors to charge (but not $H^+$) return despite their very low protein densities in most membranes. Not surprisingly, the activities of virtually all ion channels are tightly regulated through various gating mechanisms (see Chapters 5, 6, 9 and 11). Many plasma membrane $K^+$ channels, for example, become active (open) only at very negative membrane voltages at which the driving force for $K^+$ flux is directed inward across the membrane. These 'inward-rectifying' $K^+$ channels (so called by analogy with diodes of electrical circuits) thus are self-limiting and balance the current of $H^+$-extruding ATPases with $K^+$ entry into the cell only at these voltages.

As with coupled transport, factors that affect the membrane voltage will influence $K^+$ flux through the channel. However, for voltage-gated channels changes in voltage influence the current not only by altering the driving force on $K^+$ movement but also by affecting the activity of the channel – the proportion of time it is open – through its voltage-sensitive gating mechanism. Again, the importance of the voltage clamp lies in its introducing an experimentally controlled pathway for current passage that bypasses the membrane and, hence, enables direct control of the membrane voltage and a measure of membrane current. It has other advantages as well. In particular, because the response times of voltage clamp circuits are commonly 1000-fold shorter than the protein conformational changes of channel gating, the voltage clamp permits the experimenter to 'see' the development of channel current as channels open and close, and therefore to identify channel activity from the background of other transporters (see Section 5.2 of this chapter).

## 1.3   Intracellular recording and the voltage clamp

Following the studies of Ling and Gerard (1949), the term *microelectrode* has come to mean a glass capillary drawn to a microscopic point at one end and filled with an electrically conductive solution. In its use for intracellular recording the micro-electrode is in fact no electrode at all, but serves as a micro salt bridge between the cytosol and the true electrode. Strictly speaking, the electrode is formed by the junction between the electrolyte of this salt bridge and the electrical recording circuit, commonly formed by a silver wire coated with AgCl precipitate. Metal–electrolyte junctions – also known as half-cells – inevitably are associated with a voltage difference, or junction potential, and can be very sensitive to the electrolyte composition. So it is important to separate this junction from any variations introduced by the cell in salt composition or concentration. Often a AgCl-coated silver wire is inserted in the back of a KCl-filled microelectrode, but the metal–electrolyte junction can also be incorporated within a holder filled with concentrated KCl solution. To compensate for the junction potential *per se*, it is common practice to use an equivalent junction as a reference to electrically 'earth' or 'ground' the bath; because the orientations of the two junctions oppose one another in the recording circuit (Fig. 1.2), the voltages generated at the two junctions are effectively cancelled.

For reproducibility, microelectrodes are commonly formed on a microelectrode puller (Purves, 1981; Blatt, 1991). A length of glass capillary tubing is pulled by heating the central region of the tubing until soft, and then rapidly drawing out the softened glass until it separates to give two fine tips. The best microelectrode pullers are based on the original horizontal design of Alexander and Nastuk (1953) and apply a weak 'sensing' pull while heating the glass before the final hard pull that draws out the tip. In our laboratory, multi-barrelled microelectrodes are routinely made with a horizontal puller of this design (Narashige Instruments, model PD5) modified to take up to four capillaries at a time. Thus, for double-barrelled electrodes, two capillaries are mounted, heated without pulling and twisted about their axis 360° before cooling. Thereafter, a normal pull cycle is initiated. With proper adjustment, tip diameters of 0.05–0.08 $\mu$m are routinely formed this way from borosilicate capillary.

By far the most common electrolyte used to fill microelectrodes is 3 M KCl, but almost any salt solution will do. The reasons for using a concentrated KCl solution are twofold, related (i) to the need to minimise electrical resistance between the recording circuit and the cell when measuring rapid ($<10$ ms) transients, and (ii) to uncertainties in the recorded voltage introduced by the liquid junction between the cytosol and the electrolyte in the microelectrode (see Tasaki & Singer, 1968, but also Blatt & Slayman, 1983). The disadvantage of concentrated KCl filling

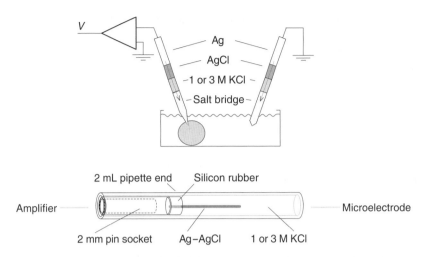

**Fig. 1.2** A simple circuit for measuring membrane voltage (*above*), drawn to emphasise the opposing orientation of the half-cell junctions (indicated by dotted arrows). On its own, a single Ag–AgCl|Cl⁻ junction has a potential of approx. 0.22 V with 1 M KCl. Note that the amplifier earth (ground) is linked to the bath half-cell junction and salt bridge. Half-cell junctions in our laboratory (*below*) are constructed from a clean Ag wire soldered to a 2 mm pin socket (to match the 2 mm pin input to the amplifier), a piece of silicon rubber and a 3–4-cm length of a 2 mL Sterilin plastic serological pipette. The silicon rubber is pressed into the pipette end, the wire pushed through the silicon rubber, and the socket then pressed into the pipette with the silicon and wire in front. The pipette is then back-filled with $NaHClO_4$ solution, which reacts with the Ag wire, coating it with AgCl. Finally, the $NaHClO_4$ solution is replaced with 1 M KCl. Filled microelectrodes are fixed in the open end of the assembly with dental impression compound so that a liquid junction is formed between the filling solution and the 1 M KCl in the half-cell.

solutions is that salt leakage from the microelectrode into the cytosol can substantially alter the transport properties of cells (Blatt & Slayman, 1983; Blatt, 1987b). Diffusional leakage rates are on the order of 10–30 fmol KCl/s from a typical, 3 M KCl-filled microelectrode. For cells with cytosolic volumes of even several tens of picoliters, the effect can be to raise cytosolic ion concentrations several fold in a matter of minutes, especially of Cl⁻, which effectively 'short circuits' the membrane as it activates Cl⁻ channels and diffuses out of the cell. The author and others have made use of microelectrode holders that incorporate the Ag–AgCl|KCl metal–electrolyte junction to side-step this problem (see Fig. 1.2; also Blatt, 1991). The advantage of such a holder is that the Ag–AgCl|KCl junction is isolated from any changes in salt composition outside and the microelectrode can be filled with physiologically compatible solutions, even with electrolytes without Cl⁻ (Blatt & Slayman, 1983; Blatt, 1987b; Lew, 1991, 1998; Roelfsema & Prins, 1997). There are other considerations, too, relating to the reversibility of the Ag–AgCl|KCl junction, microelectrode tip potentials and capacitance for which separating the Ag–AgCl|KCl junction offers advantages (see Purves, 1981; Blatt, 1991).

Intracellular recording is surprisingly easy to achieve in most plant cells. A stable read-out of membrane voltage can be obtained from virtually any tissue, if a suitably high-impedance amplifier is used and the tissue and microelectrode are isolated from mechanical vibrations and electrical disturbance (see Chapter 2). It is worth noting that intracellular recordings from plant cells are not generally degraded by 'leakage current' between the glass of the microelectrode and the membrane, contrary to some notions. For example, recordings from guard cells typically give input resistances of 1–5 G$\Omega$, and in some circumstances, in excess of 20 G$\Omega$ (Blatt, 1987b; Thiel et al., 1992; Roelfsema & Prins, 1997). These values are directly comparable to typical tight-seal resistances of patch pipettes (below) and, indeed, are significantly 'tighter' considering that the input resistance of intracellular measurements *includes* the conductance of the entire cell membrane. Another common fallacy is that deep impalement will place the electrode tip within the vacuole. The author's experience (cf. Blatt & Slayman, 1983; Blatt & Armstrong, 1993), and that of others (Beilby & Blatt, 1986; Felle, 1987, 1988; Miller et al., 1990), is that successful impalements always leave the tip in the cytosol, a conclusion that has been verified, for example, through ion-selective microelectrode measurements. The fact that advancing a microelectrode into a tissue can give a positive-directed (depolarising) shift in voltage – as if entering the vacuole – is easily explained by mechanical damage to the membrane seal once it forms around the microelectrode.

The requirements for voltage clamp recordings are not substantially different except that an additional connection to the cell is needed to input the clamp current. Most often (though not always; see Forestier et al.,1998) this connection is afforded by impaling the cell with a second microelectrode or with a double-barrelled microelectrode. The essence of this approach, often referred to as the *two-electrode voltage clamp* (Fig. 1.3), is an electronic negative feedback loop between the cell and a high-gain, differential amplifier, the voltage clamp. The membrane voltage is recorded via the first electrode (barrel) and read-out from one amplifier (A in Fig. 1.3). The voltage clamp (C in Fig. 1.3) compares this signal with a command voltage, inverts the difference, and generates an output proportional to this difference. Finally, the clamp output is fed into a second, high-impedance amplifier (B in Fig. 1.3) that then converts the signal to a current and pumps it back into the cell via the second microelectrode (barrel). In effect, the clamp drives the membrane voltage to the command voltage, as any deviation from the command voltage engenders a compensating change in current pumped back into the cell. In turn, the clamp current (or, more precisely, its inverse) is equal to the current across the membrane at the same voltage.

Although the two-electrode voltage clamp originated with the studies of Cole (1949) and Hodgkin et al. (1952), it has since taken many different forms depending on the particular characteristics of the cells (Purves, 1981; Standen et al., 1987; Blatt, 1991). A particular requirement for voltage clamp recording from plant cells relates to the physical geometry and spread of the clamp current within the cell and tissue. The efficacy of the voltage clamp depends on measuring current carried by a population of transporters, all of which experience the same voltage across the membrane. Because the plant plasma membrane usually comprises a high resistance compared with the cytosol, even when current is delivered from a point source – the

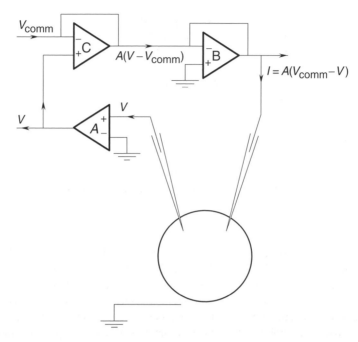

**Fig. 1.3** A common two-electrode voltage clamp circuit. Membrane voltage ($V$) is measured using one microelectrode (barrel) and amplifier A. The output from amplifier A is fed to high-gain differential amplifier C that compares $V$ against a command voltage $V_{comm}$. The high-gain difference [$=A(V - V_{comm})$] is inverted by amplifier B that drives a current proportional to $V_{comm} - V$ back into the cell. Thus, any deviation of $V$ from $V_{comm}$ is met with an opposing charge movement into the cell via the second microelectrode. By the same token, this clamp current is equal to the inverse of the current generated by the cell across the membrane.

microelectrode tip – it is effective for controlling the membrane voltage of roughly isodiametric cells such as guard cells (Blatt, 1987a), short root hairs (Lew, 1991) and ellipsoidal algae (Coleman & Findlay, 1985). However, other cell geometries do not lend themselves to this approach and require correction for current dissipation along the length of the cell (Meharg *et al.*, 1994; Meharg & Blatt, 1995). For cylindrical cells, uniform current spread is best achieved with an axial wire inserted along the length of the cell (Findlay & Hope, 1964; Beilby & Coster, 1979). Obviously, this approach is suitable only for large and robust cells such as giant algae.

A still greater problem is cell-to-cell coupling through plasmodesmata. Mathematical corrections for current spread are generally possible only for spatially regular (ideally homogeneous) tissues and their application demands that the electrically conductive properties of the tissue be determined experimentally (see Jack *et al.*, 1983; Smith, 1984). Because changes in membrane transport – and, hence, membrane conductance – will also affect current spread, a two-electrode (point-source) voltage clamp will not 'see' the conductance change without this correction. For this single reason, more than any other, voltage clamp studies of intact plants over the decades have been limited to single-cell preparations, and tissues from which single cells can be isolated from their neighbours (or are so *in situ*, as is the case for guard cells). By contrast, the fact that current injected into plant cells must pass across (or around) the tonoplast should be of little concern. With few exceptions (Beilby & Bisson, 1999), its conductance is generally an order of magnitude (or more) greater than that of the plasma membrane (Bentrup *et al.*, 1986; Tester *et al.*, 1987; Allen & Sanders, 1997).

## 1.4 Patch clamp

The patch clamp was developed as a variant of early extracellular (so-called loose seal or suction pipette) recording methods (see Byerly & Hagiwara, 1982, and references therein) and was introduced by Neher and Sakman in the mid-1970s (Neher & Sakmann, 1976). Techniques for obtaining high-resistance recordings necessary to resolve single ion channels have changed very little from those described in the now classic paper on the method by Hamill *et al.* (1981), and a wealth of detail can be found in the volume by Rudy and Iverson (1992) and Conn (1998) as well as in recent articles (see, for example, White *et al.*, 1999). Patch clamp recording is particularly suited to analysis of ion channels and, as a technique, comes into its own in measurements of ion flux through single channels. However, in the whole-cell mode (below) it also finds application for measurement of current through ion pumps and coupled transporters (Lohse & Hedrich, 1992; Findlay *et al.*, 1994; Gambale *et al.*, 1994; Weiser & Bentrup, 1994).

A simplified electrical circuit for the patch clamp is shown in Fig. 1.4. Unlike the two-electrode voltage clamp (Fig. 1.3), the signal from the patch electrode does

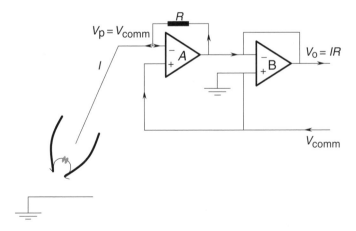

**Fig. 1.4** The patch clamp circuit. The patch electrode is connected directly to the low-impedance input (also known as the summing or reference input) of the clamp amplifier A. The command voltage $V_{comm}$ is fed into the high-impedance input. Output from the clamp is fed back across resistance $R$ and ensures that the patch electrode is always driven to $V_{comm}$. Any deviation from this value generates a compensating current across $R$. The output from amplifier A is conditioned across amplifier B to give an output scaled by $R$ and proportional to the clamp current $I$.

not pass through a voltage follower in front of the clamp itself but, instead, is fed directly into the (low-impedance) input of a high-gain, current-to-voltage converter (A in Fig. 1.4). The command voltage is supplied to the (high-impedance) follower input of the converter. The elegance of this circuit is that it effectively always drives the pipette voltage (the command signal), and, with appropriate conditioning, its voltage output is also directly proportional to the pipette current.

Many of the electrical and mechanical considerations important for intracellular recordings also apply to work with the patch clamp (see also Chapter 2). Like the intracellular microelectrode, the patch pipette is manufactured by drawing out glass capillary tubing to form a fine tip, and several microelectrode pullers are available commercially, designed specifically for this purpose. Patch pipettes are commonly pulled in two, sometimes three, stages, with relatively little force applied to the softened glass. The result is to give a rapid taper of the shank to the tip, which minimises the overall resistance of the pipette, and to produce a tip with a diameter typically of 0.5–2 μm. Although originally common practice (Hamill *et al.*, 1981), many electrophysiologists no longer 'polish' the pipette tip (normally achieved by bringing it in close proximity to a resistive coil which is heated by passing current), but use the pipettes directly as pulled (Böhle & Benndorf, 1994). For satisfactory recordings of single-channel currents, however, it is usually necessary to minimise the distributed capacitance of the pipette when in solution by coating as much of the shank as possible with a thick layer of a water-insoluble, resistive material such as an encapsulating compound (e.g. Sylgard) or even paraffin wax (Hamilton *et al.*, 2000). Once coated, patch pipettes are filled with solution that often roughly matches the physiological electrolyte on one side of the membrane or the other. Patch amplifier suppliers generally provide standardised pipette holders that permit the user to apply a gentle suction (or pressure) to the back of the pipette and incorporate a AgCl-coated silver wire as the half-cell junction when inserted into the pipette shaft. Thus, to ensure electrical reversibility of the junction, pipette-filling solutions commonly include a few millimolar Cl$^-$, at least in the shaft back from the tip of the pipette.

Patch clamp recording generally involves one or more of the three standard configurations shown in Fig. 1.5. Commonly the pipette tip is placed against the cell membrane and a gentle suction is applied to the pipette to seal a small patch of membrane over the pipette tip. The quality of the seal is checked by applying a small voltage step to the clamp voltage and monitoring the clamp current until a seal resistance of 1–10 GΩ is obtained. So-called gigaseals in the cell-attached configuration can be used to record single-channel current in the patch. The overall conductance of the rest of the cell membrane is generally hundreds, even thousands of times greater than that of the patch; so the principle resistance to charge flow is the patch itself and changes in clamp current with channel opening/closing events within the patch are easily resolved. An advantage of cell-attached recording is that the contents of the cell remain intact, a factor of particular importance if ion channel activity requires essential cytosolic factors (Philippar *et al.*, 1999). However, the approach suffers because the voltage clamped is that between the inside of the pipette and the bath. In other words, the clamp voltage is *added* to the voltage

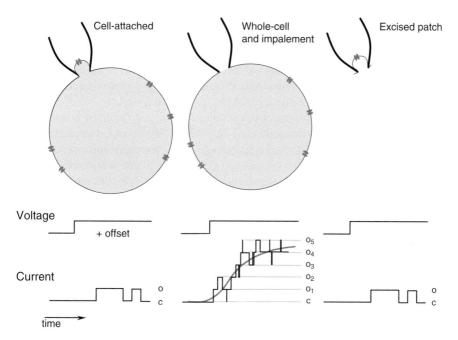

**Fig. 1.5** Three modes of recording using patch clamp methods. Cell-attached recordings (*left*) are obtained after gentle suction to seal a small area of membrane in the pipette tip. Note that current passing across this membrane patch also passes in series across the rest of the membrane of the cell to the (earthed) bath. Because the clamp will reference to the total voltage difference to the bath, any voltage generated by the cell itself will sum with the clamp voltage. Thus, it is not possible to know the precise value of the voltage across the cell-attached patch. The patch, nonetheless, presents the principal resistive barrier to current. So, single-channel events can be resolved (*below*). Whole-cell recordings (*centre*) give direct electrical and bulk solute access to the cytosol (or, for example, to the lumen of the vacuole when patching the tonoplast). Current recorded in this case corresponds to charge movement across the entire membrane and the results are essentially identical to those obtained using a two-electrode voltage clamp with intracellular microelectrodes (*below*, idealised activation kinetics showing the equivalent process as a sum of individual channel opening and closing events). Excised patch recordings (*right*) are achieved by physically pulling away from a patched cell to leave a small patch of membrane in the pipette tip. Excision from the cell-attached configuration gives inside-out patches, and excision from the whole-cell configuration gives outside-out patches. In each case, the entire voltage difference to the bath occurs across the patch, ensuring that channels in the membrane 'see' the entire clamp voltage. As in the cell-attached configuration, changes in conductance – channel opening and closing events – draw compensating changes in clamp current and are faithfully reported as such.

generated by the sum of transport activities across the membrane as a whole. As a result, the true voltage across the membrane patch can never be known accurately unless concurrent intracellular data can be obtained (see Rudy & Iverson, 1992).

The whole-cell configuration is achieved by applying a short burst of suction to break through the membrane patch under the pipette tip. In principle, the data obtained in this configuration is identical to that of intracellular recording methods and the currents recorded relate to the ensemble of transporters in the membrane as a whole. One important difference is that the whole-cell configuration results in direct access to the cytosol through a pathway that has a much lower resistance than intracellular microelectrodes to bulk diffusion of inorganic and organic compounds, including soluble proteins. As a result, whole-cell patch clamp recordings also *dialyse* the cell against the contents of the patch pipette and can lead to near-complete exchange of ions and small compounds within a few seconds (Pusch & Neher, 1988; Tang *et al.*, 1992). An advantage of such 'pipette dialysis' is that concentrations of these materials inside the cell can be brought under experimental control, but a disadvantage is that ion channel activity often decays within a few seconds after breaking through the patch. The 'rundown' of channel activity in many instances has proven a good indicator of channel regulation by protein (de)phosphorylation (Hoshi, 1995; Schmidt *et al.*, 1995; Tang & Hoshi, 1999; Horvath *et al.*, 2002; Köhler & Blatt, 2002).

Finally, the membrane patch can also be excised from the whole cell, in which case the ion transport, gating and regulatory properties of a single, functional protein or protein complex are available for direct experimentation with complete control of the voltage across the membrane. The manner of patch excision – whether the patch electrode is pulled off the cell before or after achieving the whole-cell configuration – determines the final orientation of the membrane patch, inside exposed to the bath (inside out) or to the solution in the pipette (outside out). Thus, it is common to carry a recording through two or all three configurations, beginning with cell-attached measurement to confirm channel activity before progressing to the whole-cell and/or excised patch measurements.

Some differences to intracellular recording are worth bearing in mind. Amplifiers used for patch clamp measurements, unlike those used with intracellular microelectrodes, incorporate the clamp within the primary recording circuit so that the membrane is permanently driven to a command signal, either current or voltage. The 'current clamp' mode approximates the free-running voltage of intracellular recordings by clamping the current across the membrane to zero. Another important difference is that patch pipettes are commonly filled with near-physiological salt solutions and, hence, compensating for junction potentials and (unknown) changes in these potentials can become a significant concern (Barry & Lynch, 1991; Amtmann & Sanders, 1997). Finally, direct access to the membrane surface is an absolute prerequisite of patch clamp recording. For virtually all plant cells, access to the membrane requires that the cell wall be removed to release the protoplast, either by enzyme digestion (Schroeder *et al.*, 1987; Elzenga *et al.*, 1991; Ivashikina *et al.*, 2001) or by mechanical means (Stoeckel & Takeda, 1989; Findlay *et al.*, 1994; Pottosin & Andjus, 1994; Henriksen *et al.*, 1996; Taylor *et al.*, 1996). Working with protoplasts in the absence of the cell wall or turgor pressure can have implications for interpreting the data (Clint, 1985; McCulloch & Beilby, 1997; Brudern & Thiel, 1999). Most important, however, the single-channel methods of the patch clamp permit the experimenter to dissect the component contributions of channel conductance and gating, which, in whole-cell patch and intracellular measurements, must be derived indirectly (see below).

In the last decade, the patch clamp has also been adapted to recording the vesicle trafficking events of exo- and endocytosis at the plasma membrane. The specific capacitance of the lipid bilayer is remarkably stable at a value around 1 $\mu F/cm^2$. Thus, changes in membrane capacitance can be used to monitor the balance between exo- and endocytosis and, in some circumstances, reveal single vesicle fusion events. The technique relies on imposing a small sinusoidal waveform to the clamp voltage while monitoring the phase lag associated with the capacitative component of the clamp current (Gillis, 1995). Again, work with plants requires that protoplasts are isolated and the patch pipette used in cell-attached or whole-cell mode. Recent work from several plant tissues has confirmed many of the early observations from electron microscopy, and has also yielded information about $Ca^{2+}$-dependent and -independent fusion of vesicles (Homann & Thiel, 1999; Sutter *et al.*, 2000; Thiel *et al.*, 2000; Weise *et al.*, 2000) and its relation to bulk movement of ion channels (Homann & Thiel, 2002; Blatt & Thiel, 2003; see also Chapter 10). No doubt, further studies will show that the specificity of traffic among different ion channels is important for channel regulation (Pratelli *et al.*, 2004).

## 1.5  Separating and analysing membrane currents

Voltage clamp experiments yield raw currents that, in whole-cell patch and intra-cellular recordings from plants, represent the flux of ions through an ensemble of ion channels, pumps and $H^+$-coupled transporters. Thus, the first step in characterising any one transport process is to isolate the corresponding current. Only single-channel records differentiate individual channel currents directly, and these methods are outlined separately. Current subtraction is by far the most common method used to identify the component of total membrane current carried by any one transporter. Details of this approach depends on whether the data are derived as isopotential or as steady-state measurements, and on whether or not the analysis must take account of ion channel gating. The following are thumbnail sketches of several common techniques used in current analysis.

### 1.5.1  Steady-state current

ATP-driven ion pumps and coupled transporters generally show no changes in activity during the short periods of voltage clamp measurements other than those imposed by the clamp voltage. It is usual in these instances to record steady-state current over the widest range of voltages possible without damage to the membrane, for most plant cells roughly between $-250$ and $+100$ mV. The clamp is driven either through a slow ramp or, preferably, stepwise across this voltage range. In the latter case, a continuous record of current provides verification that steady state is reached by the end of each step. A plot of the steady-state current against clamp voltage then gives a current–voltage (*IV*) characteristic that defines the sum of charge transport across the membrane. In principle, to identify the current carried by a specific transporter requires only that the measurement be

repeated once the transport activity is eliminated, either by pharmacological inhibition or by removing substrate(s). Subtracting currents recorded in the second
measurement from total current at the same voltages then yields the *IV* [or, more
precisely, the difference *IV* (d*IV*)] curve for the transport process. Elimination or
substitution experiments also yield information about the probable nature of the
substrate(s) or ion(s) carried by the transporter. Applications of this approach include studies of the $H^+$-ATPase, coupled amino acid transport and the $H^+$–$K^+$
cotransporter of *Neurospora* (Gradmann *et al.*, 1978; Hansen *et al.*, 1981; Sanders
*et al.*, 1983; Blatt *et al.*, 1987; Blatt & Slayman, 1987), the $H^+$-ATPase of *Vicia* guard cells (Blatt, 1987a; Lohse & Hedrich, 1992), $H^+$-coupled transport of
$K^+$ and $NO_3^-$ in *Arabidopsis* (Maathuis & Sanders, 1994; Meharg & Blatt, 1995),
for the vacuolar $H^+$-ATPase of beet (Davies *et al.*, 1996) and the plasma membrane $H^+$-ATPase (see Fig. 1.6) and $H^+$–$Cl^-$ cotransporter of *Chara* (Beilby &
Walker, 1981; Blatt *et al.*, 1990).

   Why analyse the steady-state *IV* curve of a transporter? The *IV* curve of any charge
transporter includes kinetic information for transport as a function of the voltage
'substrate', much as a conventional substrate concentration vs enzyme velocity plot
reports the kinetic characteristics of a soluble enzyme. It also defines the equilibrium
voltage – empirically, the reversal voltage at which the *IV* curve crosses the voltage
axis – that describes the thermodynamic limit or *stalling point* for net transport in any
one direction. Thus, increases in electrical driving force away from the equilibrium
voltage increase transport 'turnover' (current) until, eventually, the current saturates
with respect to the voltage. Of course, transport can be driven in either direction
across a membrane, at least in principle, and so a kinetic analysis must also take
account of negative voltages and currents. Such characteristics are accommodated
by kinetic carrier models for transport in which a carrier molecule cycles between
the two sides of the membrane (Läuger & Stark, 1970; Hansen *et al.*, 1981; Sanders
*et al.*, 1984; Blatt, 1986; Gradmann *et al.*, 1987; Wierzbicki *et al.*, 1990; Segel,
1993). In these mathematical models, substrate binding and debinding reactions
are generally assumed to occur at the membrane faces and voltage contributes to
movement of the charged carrier or carrier–substrate complex across the membrane.
Even with the more complex kinetic systems, often a simplified form of the carrier
cycle can be adopted for specific limiting conditions of substrate concentration and
voltage (Sanders *et al.*, 1984). Indeed, the simplest carrier model in this case is one
that takes account of membrane voltage only, all other experimental variables (e.g.
ion concentrations) being constant, so that (see Fig. 1.6)

$$I = zFN \frac{k_{io}\kappa_{oi} - k_{oi}\kappa_{io}}{k_{io} + \kappa_{oi} + k_{oi} + \kappa_{io}} \tag{1.4}$$

where $k_{io}, \kappa_{oi}, k_{oi}$ and $\kappa_{io}$ are the forward and reverse (pseudo-) reaction rate constants
and incorporate membrane voltage and substrate binding and debinding, $N$ is the
number of carriers and $z$ and $F$ have their usual meanings (see Hansen *et al.*,
1981, for details). Thus, from the experimentalist's point of view, the challenge is

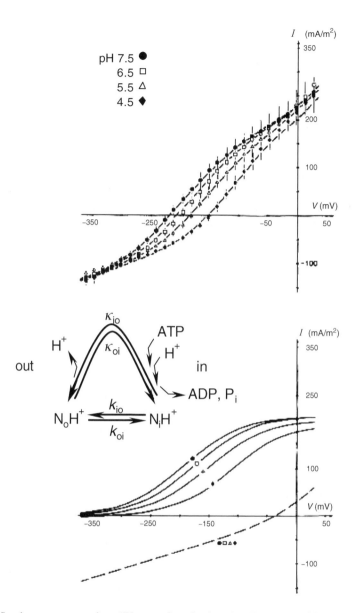

**Fig. 1.6** Steady-state current–voltage (*IV*) curves from the giant alga *Chara* at four different external pH values (*above*) after Beilby (1984). The *IV* curves were fitted jointly to a simplified, two-state model for the $H^+$-ATPase (*inset, below*) summed with an ionic diffusion component described by the Goldman–Hodgkin–Katz function (Hodgkin & Katz, 1949) to approximate the sum of all other transport activities in the membrane (after Blatt *et al.*, 1990). Fitting was by a least-squares parametric algorithm (Marquardt, 1963). The results (*below*) indicated a positive-going shift in the *IV* curve for the pump (solid lines) against a constant background of other transport currents (dashed line). The fitted lines (above) are obtained by summing the fitted $H^+$-ATPase current at each pH with the background current. The two-state model (*inset*) subsumes $H^+$ binding and debinding steps and ATP hydrolysis within the rate constants $\kappa_{io}$ and $\kappa_{oi}$. Membrane voltage is incorporated explicitly in the rate constants $k_{io}$ and $k_{oi}$ (see Hansen *et al.*, 1981).

to describe the kinetic parameters by fitting such limiting-condition data sets to an appropriately constrained model (see, for example, Fig. 1.6). This approach applies equally well to the permeation characteristics of ion channels (Hille & Schwarz, 1978; Läuger, 1980; Gradmann *et al.*, 1987) in the open state (White *et al.*, 2000; White & Davenport, 2002). Ion channel gating is normally treated separately.

A word of caution is in order here. Removing substrate(s) – including pharmacological treatments such as metabolic blockade with cyanide to suppress ATP availability for $H^+$-ATPases – will affect both kinetic *and* thermodynamic properties of transport. In other words, these manipulations affect the equilibrium voltage for transport and therefore the 'null point' for net transport in any one direction across the membrane. Subtractions in this case will not reflect the true transporter current near equilibrium and, at voltages that drive ion flux in the physiologically reverse direction, can be very misleading. An analytical treatment of the problem – and its solution – is available (Blatt, 1986), and good examples and discussion of current subtraction will be found in several publications (Blatt *et al.*, 1987, 1990; McCulloch *et al.*, 1990).

### 1.5.2 Current relaxations and ion channel gating

Most ion channels show little evidence of current saturation with respect to voltage within the normal physiological range, and so ensemble channel current can be approximated by Ohm's law, where the current ($I$) is related to the ensemble conductance ($g$) and the voltage difference from the equilibrium voltage of ion X, $E_X$, as

$$I = g(V - E_X) \tag{1.5}$$

Unlike pumps and carriers, ion channel gating affects $g$ over time-scales of a few milliseconds to tens of seconds. For ion channels that are gated by voltage – including many $K^+$, $Cl^-$ and $Ca^{2+}$ channels at the plant plasma membrane and tonoplast (see Chapter 6) – changes in clamp voltage therefore affect both the driving force for permeation (i.e. $V - E_X$) and the ensemble activity of (number of open) channels ($g$) at steady state. Three common voltage clamp protocols are used in these instances as outlined below: (1) to identify the steady-state current carried by one type of channel, (2) to determine the ionic species carrying the current and (3) to separate the effects of voltage on gating and on permeation. Note that these approaches must be combined with pharmacological treatments in some instances to eliminate other currents showing similar relaxation characteristics that might contribute to (and thereby confuse) instantaneous and/or steady-state current measurements (see, for example, Grabov *et al.*, 1997; Pei *et al.*, 1997).

1)  To identify steady-state current the membrane is stepped from a conditioning (or 'holding') voltage at which the channels are nominally closed (inactive) to

a test voltage at which the channels may open (activate) in what is sometimes called a 'two-step' protocol. Both conditioning voltage and test voltage steps are of sufficient duration to ensure that the current achieves steady state. The process is repeated for a range of different test voltages and the current during each test clamp step recorded continuously at a suitably high frequency (see Chapter 2). After the initial capacitance transient of the voltage step itself, voltages that open (activate) the channels are marked by a current that develops in amplitude over time as more and more channels open, relaxing to a new time-averaged steady state between the numbers of closed and open channels. The initial, so-called 'instantaneous' current at each voltage (discounting the capacitance transient) thus corresponds to the background of currents through other transporters, while the final steady-state current corresponds to the sum of these currents plus the new steady-state channel current. Subtracting the instantaneous current from the steady-state current at any one voltage yields the steady-state current carried by the channels at that voltage (see Fig. 1.7; also Blatt, 1990). This $IV$ curve represents the product of current carried by all the channels when open and their probability of being open at each voltage (see Section 5.3).

2)   To determine the ionic species carried by the current a so-called 'tail current' protocol is used. The membrane is stepped from a conditioning voltage at which the channels are open to a test voltage at which the channels may close (deactivate). The process is repeated for a range of different test voltages both positive and negative of the equilibrium voltage for the ion anticipated to carry the current. Current during each test clamp step is recorded continuously at a suitably high frequency. After the initial capacitance transient of the voltage step itself, voltages that favour channel closure are marked by a current that relaxes in amplitude over time as more and more channels deactivate until a new time-averaged steady state is achieved between the numbers of closed and open channels. Significantly, the *sign* (positive- vs negative-going) of the relaxations reverses on test voltage clamp steps on either side of the current reversal voltage, $E_{rev}$. Thus, subtracting the steady-state currents at the end of the test voltage steps from the instantaneous currents at the start of the steps yields a d$IV$ curve that crosses the voltage axis at $E_{rev}$ (see Fig. 1.8). A comparison of $E_{rev}$ with the predicted $E_X$ values for ions that might permeate is then normally used to determine the ion selectivity for the channel.

It is important here to distinguish between $E_{rev}$ and $E_X$. By far the most commonly used method for determining the ion selectivity of a channel relies on the so-called constant-field theory of Goldman (1943) and Hodgkin and Katz (1949), which, in turn, is derived from the Nernst–Planck electrodiffusion equation (Nernst, 1888, 1889; Planck, 1890a,b). Constant-field theory makes two important and simplifying assumptions: (i) that the electrical field acting on an ion crossing the membrane is constant and linear, and (ii) that individual ions do not interact with other ions that may also cross the membrane.

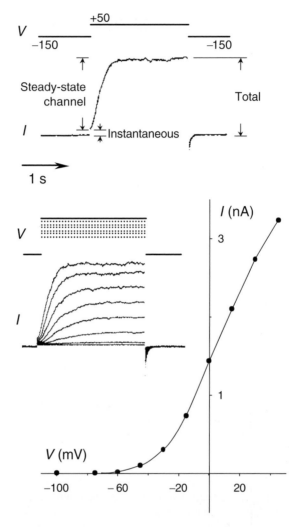

**Fig. 1.7** Preliminary analysis of a voltage-gated ion channel recorded by whole-cell patch clamp or two-electrode voltage clamp. Data are of the outward-rectifying $K^+$ channel of *Vicia* guard cells. The schematic and trace (*above*) show a single clamp cycle of steps from $-150$ to $+50$ mV and back to $-150$ mV and the current record (sampling frequency: 1 kHz; time-scale: below). Stepping to $+50$ mV uncovers, after a short capacitance spike, an instantaneous jump in current, followed by a slow rise in current amplitude that reaches a new steady state within 1 s. On return to $-150$ mV, a sizable inward (downward) current is visible that subsequently relaxes to the original steady-state level of the start of the cycle as the channels close. Steady-state channel current at $+50$ mV is described by the slow-relaxing current component of the total current at this voltage. Repeating this cycle of clamp steps with test voltages between $-100$ and $+50$ mV gives the currents and steady-state channel *IV* curve shown (*below*). Pharmacological treatments with the $K^+$-channel-blocker tetraethylammonium chloride (not shown) eliminate the slow-relaxing component, confirming its origin with the $K^+$ channels.

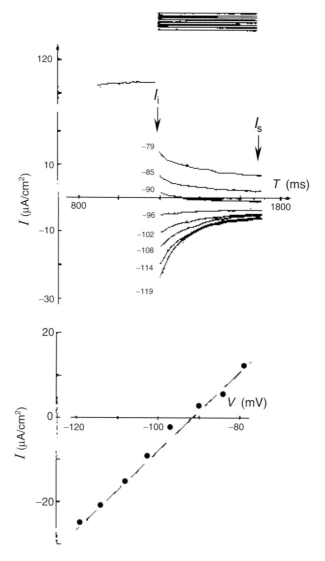

**Fig. 1.8** Tail-current analysis for a voltage-gated ion channel recorded by whole-cell patch clamp or two-electrode voltage clamp. Data are of the outward-rectifying K$^+$ channel of *Vicia* guard cells and were recorded with 3 mM K$^+$ in the bath. The K$^+$ equilibrium voltage $E_K$ was approx. −95 mV. After activating the channels at +50 mV, steps to voltages between −79 and −90 mV gave current traces that relaxed slowly to move negative values (downward) whereas steps to voltages between −102 and −119 mV gave traces that relaxed to more positive values (upward). In each case, the current relaxes to a new steady state that includes a much smaller K$^+$-channel component. At voltages negative of $E_K$, the K$^+$ channels add a large inward component to the total current and this decays as the channels close down. Note that the trace is virtually unaffected at −96 mV, very close to $E_K$. A plot of the current difference $I_i − I_s$ at each voltage is shown below and crosses the voltage axis at −94 mV.

These assumptions lead to the Goldman–Hodgkin–Katz voltage equation that is widely used to define relative ionic permeabilities of channels. For the monovalent ions $K^+$, $Na^+$ and $Cl^-$, the equation predicts the zero-current voltage

$$E_{rev} = \frac{RT}{F} \ln \left\{ \frac{P_K[K^+]_o}{P_K[K^+]_i} + \frac{P_{Na}[Na^+]_o}{P_{Na}[Na^+]_i} + \frac{P_{Cl}[Cl^-]_i}{P_{Cl}[Cl^-]_o} \right\} \quad (1.6)$$

where the permeabilities for each ion are defined by the constants $P_K$, $P_{Na}$ and $P_{Cl}$. Equation 1.6 can be used to calculate permeability ratios – for example, the relative permeability for $K^+$ vs $Na^+$, $P_K/P_{Na}$ – provided that the ion concentrations are known. With only one permeant ion, Equation 1.6 reduces to the Nernst equation (Equation 1.3). If only one ionic species is present on either side of the membrane, Equation 1.6 reduces to the so-called bi-ionic form

$$E_{rev} = \frac{RT}{F} \ln \left\{ \frac{P_K[K^+]_o}{P_{Na}[Na^+]_i} \right\} \quad (1.7)$$

for example, with only $K^+$ outside and $Na^+$ inside.

This latter form of the Goldman–Hodgkin–Katz equation is useful for single-channel recordings (see below) in which the composition of solutions on both sides of the membrane can be controlled experimentally. For intracellular recordings (and for some whole-cell measurements), however, the ion concentrations are generally not precisely defined. In these circumstances, it is best practice to *change* ion concentrations outside and determine the change in $E_{rev}$ that results. By assuming that the internal ion concentrations remain constant (a reasonable assumption in most experiments), the permeability ratio can be determined from the change in $E_{rev}$. For example, exposing a cell to $K^+$ to find

---

**Fig. 1.9** Using a two-step protocol to separate the effects of voltage on ion permeation and gating. Data are for the outward-rectifying $K^+$ channel of *Vicia* guard cells (after Blatt & Gradmann, 1997). The activity of the $K^+$ channels is assayed at the end of conditioning voltage steps using the instantaneous current at the start of test steps to six voltages between $-100$ and $+50$ mV (*arrows on left*). Repeating this protocol with different conditioning voltages yields a family of *IV* curves (*centre*; conditioning voltage as indicated on right of each curve) that show an increase in conductance (slope) with positive-going values for the conditioning voltage. Thus, the *IV* curves show the effect of voltage on permeation, while the dependence of the conductance on the conditioning step shows the effect of voltage on gating. A plot of this conductance as a function of conditioning voltage (*right*) shows a steep voltage dependence ($\sim$ e-fold rise per 12 mV) and a sensitivity to extracellular $[K^+]$. Plotted on a linear scale (*inset, right*), the voltage giving half-maximal conductance (marked by arrows) shifts approx. $+52$ mV per $[K^+]$ decade. Adapted from Blatt and Gradmann (1997).

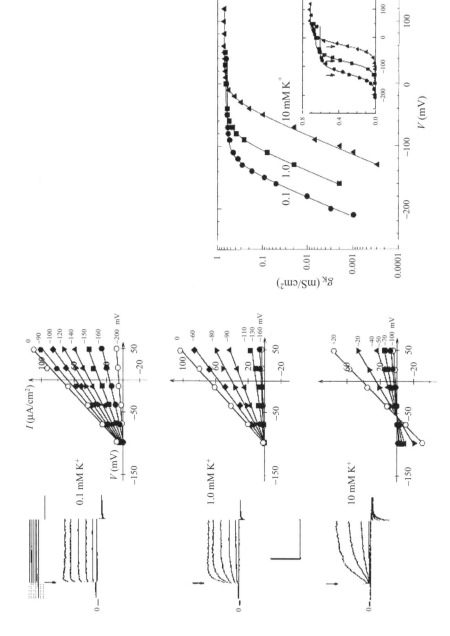

**Fig. 1.9**

$E_{rev,K}$, and then $Na^+$ to find $E_{rev,Na}$, gives

$$\Delta E_{rev} = E_{rev,K} - E_{rev,Na} = \frac{RT}{F} \ln \left\{ \frac{P_K[K^+]_o}{P_{Na}[Na^+]_o} \right\} \qquad (1.8)$$

Much the same approach can be used with mixed solutions. An added advantage of changing solutions – for example, varying $[K^+]$ against a constant background of $[Na^+]$ – is that errors introduced by (unknown) liquid junction potentials can be eliminated.

The Goldman–Hodkin–Katz formalism also predicts well-behaved patterns in conductance and current rectification that are responsible for these changes in $E_{rev}$ (see, for example, Hille, 2001). In principle, it is important to confirm that current through the open channels – or the instantaneous current recorded for any fixed proportion of open channels – conforms to this pattern. In many cases it does not (see Figs. 1.9 and 1.10; also White *et al.*, 2000). Even so, it is remarkable how well Equation 1.6 does as an empirical description and as a tool for defining permeability ratios and ion selectivity. In such cases, however, it is important to bear in mind that the selectivities calculated will be weakly voltage- and/or concentration-dependent.

3) To separate the effects of voltage on gating and permeation a two-step protocol similar to that in point 2 is used with test voltages chosen over a suitable range of values. Recordings are repeated successively with the conditioning voltage set to different values, including those at which the channels activate. For each conditioning voltage, instantaneous currents at the start of the test voltage steps are plotted. These instantaneous *IV* curves each represents the current through a fixed proportion of channels in the open state at the corresponding conditioning voltage. Thus, the *slope* of the instantaneous *IV* curve defines the conductance ($g$) prevailing at the conditioning voltage just before the test voltage steps, and is therefore independent of the effect of voltage on permeation *per se*. This analysis is much simplified if the instantaneous *IV* plots are roughly linear in accordance with Equation 1.5. (They should therefore also intersect at the equilibrium voltage for the permeant ion.) The conductance is plotted against conditioning voltage (see Fig. 1.9) and may be normalised to the maximum conductance to give the equivalent of a plot of relative channel open probability vs voltage (see Section 5.3).

Important information about channel gating is also to be found in the kinetics of current relaxations. For voltage-gated ion channels, the steady-state current at any one voltage reflects the distribution in the channel population between channels in the open (conductive) state and those in closed (non-conductive) states. When the membrane is stepped to a new voltage, the propensity of individual channels to reside in one or more states is affected. Consequently, the closed/open distribution between channels in the population also changes, as is marked by relaxation of the

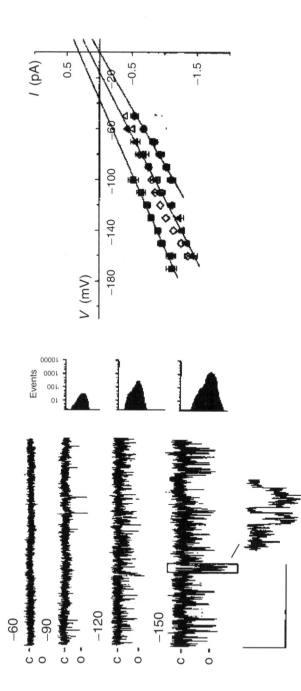

**Fig. 1.10** Using outside-out, single-channel measurements to identify the voltage-gating and permeation characteristics of an ion channel. Data are for the inward-rectifying Ca$^{2+}$ channel of *Vicia* guard cells (after Hamilton *et al.*, 2000). Steady-state traces at four voltages (*left*) show a dramatic increase in the number of channel opening events as the voltage is driven negative to −120 and −150 mV. Point-amplitude plots (*centre*) indicate the number of openings during a 5-s time period as well as the distribution of open amplitudes (vertical scale). Scale (*below*): 2 pA, vertical; 1 s, horizontal. Mean open- and closed-channel levels were determined from these distributions and are indicated on the left of each trace. A section of the trace at −150 mV is expanded to show two opening events. The amplitude of the open channel current depends on electrochemical driving force for the permeant ion. The plot (*right*) shows that the current amplitudes extrapolate to the voltage axis at voltages predicted for a Ca$^{2+}$ (Ba$^{2+}$)-selective channel. Measurements were carried out in solutions inside/outside containing 30 mM Ba$^{2+}$/30 mM Ba$^{2+}$ (●), 30 mM Ba$^{2+}$/10 mM Ba$^{2+}$ (△), 30 mM Ba$^{2+}$/2 mM Ba$^{2+}$ (■), and 30 mM Ca$^{2+}$/10 mM Ca$^{2+}$ (◇).

current to a new steady state. For a channel that can exist in one of only two states, either open ($O_1$) or closed ($C_2$)

$$C_2 \underset{k_{12}}{\overset{k_{21}}{\rightleftharpoons}} O_1 \tag{1.9}$$

relaxation between steady states will follow single-exponential kinetics with a time constant, $\tau = 1/(k_{21} + k_{12})$, where $k_{21}$ and $k_{12}$ are the reaction rate constants for the transitions in the two directions. Thus, effects of voltage (and of other factors) on the reaction rates between the two states can be obtained directly from current measurements. Often, however, current relaxations follow more complex kinetics from which additional closed (and sometimes open) states can be inferred. Each new state, and the associated transitions between states, introduces additional exponential components to the current kinetics. For example, the current relaxations on depolarising voltage steps in Fig. 1.9 (see also Fig. 1.7) show a pronounced sigmoidicity in 10 mM $K^+$ and, thus, can be accommodated only with the sum of at least two exponential components. So the simplest explanation in this case is that the channels can reside in one of three states, two closed and one open, as

$$C_3 \underset{k_{32}}{\overset{k_{23}}{\rightleftharpoons}} C_2 \underset{k_{21}}{\overset{k_{12}}{\rightleftharpoons}} O_1 \tag{1.10}$$

Here the sigmoid relaxations on transition to the open state $O_1$ must arise because a significant proportion of channels reside in closed state $C_3$ at the conditioning voltage and because, on stepping the voltage positive, their exit to closed state $C_2$ (reaction rate $k_{32}$) is slower than their exit from $C_2$, either back to $C_3$ or to $O_1$ (reaction rates $k_{23}$ and $k_{21}$). In other words, channels accumulate in the open state only slowly, because this accumulation must be 'bled off' the cycling of channels between the two closed states. An explicit mathematical solution for current relaxation is defined by the sum of two exponential components

$$I = I_s + I_1 e^{-\lambda_1 t} + I_2 e^{-\lambda_2 t} \tag{1.11}$$

where $I_s$ is a steady-state component, $I_1$ and $I_2$ are the amplitudes of the two components dependent on time $t$ and the relaxation time constants $\lambda_1$ and $\lambda_2$. The relaxation time constants, in turn, are complex functions of the reaction constants, where

$$\lambda_{1,2} = [a \pm (a^2 - 4b)^{1/2}]/2 \tag{1.12}$$
$$a = k_{23} + k_{32} + k_{12} + k_{21} \tag{1.13}$$
$$b = k_{23}k_{21} + k_{32}k_{12} + k_{12}k_{23} \tag{1.14}$$

In this case, finding values for the individual reaction rate constants $k_{23}$, $k_{32}$, $k_{12}$ and $k_{21}$ and their dependence on membrane voltage (and, in this case, on external $K^+$ concentration) requires fitting equation 1.11 to sets of current relaxations such as shown in Fig. 1.9 (see Blatt & Gradmann, 1997).

It is worth noting that important information about channel gating can be obtained without the need for extensive mathematical analysis. For a channel that exists in only two states, closed and open (Equation 1.9), a lower limit on the number of internal charges moved to open the gate can be calculated from the limiting slope of the conductance–voltage curve (Fig. 1.9). In this case, the fraction of open channels, or relative conductance

$$g_{rel} = \frac{1}{1 + e^{-\delta z F(V - V_{1/2})/RT}} \tag{1.15}$$

Here $V$ is the clamp voltage, $V_{1/2}$ defines the voltage giving half-maximal conductance, and $\delta$ is the apparent charge moved in gating or the 'voltage sensitivity' coefficient (the limiting slope of the conductance–voltage curve; see Fig. 1.9; also Hille, 2001). Note that even a simple channel that gates between two states does not show an 'activation threshold' at which the conductance–voltage curve actually *reaches* zero. This conception should be avoided in favour of the parameters $V_{1/2}$ and $\delta$. Equation 1.15 is a useful approximation also for channels that show complex gating behaviours and does remarkably well as a predictive tool in these circumstances (see Blatt & Gradmann, 1997).

### 1.5.3 Analysing single-channel current

The ensemble channel current of the whole cell arises from the sum of (more or less) random opening and closing events of all the channels in the membrane. Thus, the total membrane current can be described by an expanded form of Equation 1.5 as

$$I = N \gamma_X P_o (V - E_X) \tag{1.16}$$

to take account of the component contributions to the ensemble conductance $g$. Here $N$ is the number of functional channel units in the membrane, $\gamma_X$ is the conductance of a single channel for ion X, and $P_o$ is the mean open probability of a single channel. $N$ and $\gamma_X$ are effectively scalars, while $P_o$ reflects the activity of the channel and, thus, is most commonly of experimental interest for understanding channel gating and its regulation. For voltage-gated channels, it is a function of the membrane voltage, and may also depend on other factors, including (de)phosphorylation state and the (de)binding of regulatory ligands.

One of the principal advantages of single-channel recording is that it distinguishes between changes in $N$, $\gamma_X$ and $P_o$ that result from experimental treatments, for example with pharmacological agents. Changes in $N$ and $\gamma_X$, in particular, cannot be distinguished by whole-cell and intracellular recording. Single-channel records, by contrast, give a direct measure of current $i_X$ passing through one functional channel unit and hence of $\gamma_X$ [$=i_X/(V - E_X)$ if approximated by Ohm's law]. Parameters $N$ and $P_o$ can be obtained through appropriate analysis. Disadvantages of the single-channel approach are that obtaining satisfactory recordings is often technically challenging (especially in plants), that identifying the one channel of interest among other single-channel events can be difficult, and that proper analysis of records is extremely laborious and often confounded by unaccounted fluctuations in $P_o$. In fact, whole-cell measurements can identify changes in $P_o$, for example as it affects the voltage sensitivity of gating. So it is no surprise that following the early imperatives for single-channel documentation in the 1980s and 1990s, many laboratories have reverted to whole-cell measurements for most purposes, finding it far more productive to rely on the cell to 'integrate' channel activity over the surface of the membrane. Nonetheless, there are circumstances in which patch excision and single-channel measurement offers important benefits, for example in analysing specific signalling and/or regulatory factors that act from the cytosolic side of the membrane (Booij *et al.*, 1999; Pottosin *et al.*, 1999; Tang & Hoshi, 1999; Hamilton *et al.*, 2000). Here are a few thumbnail descriptions for the most common analytical techniques:

1) To determine single-channel conductance and ion selectivity, current data at each voltage are plotted in so-called 'all-points' or 'point-amplitude' histograms (see Fig. 1.10 and, for example, Hoshi, 1995; Hamilton *et al.*, 2000) to determine the sign and mean amplitudes of the single events. The resulting mean open-channel currents are plotted against voltage to derive the open-channel *IV* curve and conductance (slope). Ion selectivity is determined as for intracellular and whole-cell recordings (above).

2) Several approaches are available to identify $N$. If $\gamma_X$ and $P_o$ are known for a given voltage, then $N$ can be derived directly from Equation 1.16. A statistical analysis of channel frequency per patch can also be informative. However, it is useful to compare methods, especially as many channels occur in clusters at membrane surfaces and can therefore complicate the latter approach.

3) To measure steady-state $P_o$ from the record of a single channel, opening and closing events are logged against time from records at one voltage, usually using rapid software algorithms (see Chapter 2 and also, for example, Sachs *et al.*, 1982, Rudy & Iverson, 1992, and references therein). $P_o$ at the given voltage is determined as the fraction of time spent in the open state averaged over a suitable time period. (For well-behaved data, $P_o$ can also be obtained by fitting the all-points histogram to a set of Gaussian distribution functions; the fraction of the total area under the peak corresponding to the open channel then corresponds to $P_o$.) Such analysis also yields important kinetic information on the rates of transitions between closed and open states, usually derived as the mean closed and mean open lifetimes of the channel (see Colquhoun, 1987; Colquhoun & Sigworth, 1995). Channel lifetime analysis offers an alternative and, arguably, more immediate approach to finding values for the rates of transition between closed and open states of a channel and their dependence on membrane voltage or other factors.

## 1.6    Microinjection and perfusion

It is often desirable to introduce compounds into cells during electrophysiological recordings. A wide variety of methods have been adapted to this end, for use with both intracellular microelectrodes and patch pipettes (Purves, 1981; Rudy & Iverson, 1992), and deserve brief mention here. These techniques have important applications for fluorescent dye loading and measurements of cytosolic $Ca^{2+}$ and $H^+$ (Malho et al., 1995; Taylor et al., 1996; Grabov & Blatt, 1997; Garcia-Mata et al., 2003) among others, but have also been used for 'in vivo biochemistry', to introduce into the single cell antibodies (Lin & Yang, 1997; Carroll et al., 1998), second messengers (Blatt, 1991; Leckie et al., 1998; Lemtiri-Chlieh et al., 2000) and functional proteins (Moutinho et al., 1998; Booij et al., 1999; Leyman et al., 1999).

Pressure injection into intact plant cells with intracellular microelectrodes can prove difficult, especially with small cells, although this is generally the only option for loading cells with material such as dextran-coupled fluorescent dyes (Feijo et al., 1999; Franklin-Tong et al., 2002). To introduce small, charged molecules ($<1000$–$2000$ molecular weight) an electrophoretic approach is a simple alternative. Current driven through the microelectrode will be carried by charged molecules in solution and, with a judicious choice of solute concentration, the current can be used alternately to withhold and to inject these charged molecules. Electrophoretic injections can be achieved with little more than a battery and resistor in series (Gilroy et al., 1991; McAinsh et al., 1995; Wu et al., 1997), but this approach leads to extreme swings in voltage – with consequent effects on cell signalling intermediates such as cytosolic $Ca^{2+}$ – as the injection current must pass across the plasma membrane to complete the circuit. For this reason, it is far better to use the voltage clamp to 'sink' the injection current (Grabov & Blatt, 1998; Garcia-Mata et al., 2003) or a current pump to balance current source and sink between microelectrodes (Blatt et al., 1987) within the cell so that membrane voltage is not affected during injection. Because some leakage always occurs from microelectrodes, it is also possible to introduce material into cells simply relying on diffusion. Large molecules will diffuse slowly, but in many instances only very low concentrations are needed. Leyman et al. (1999) used this approach successfully to load guard cells both with a SNARE peptide of 29 kDa and with the Clostridium botulinum endopeptidase BotN/C ($\sim$50 kDa).

The patch pipette offers a pathway with much lower resistance to diffusional flow between the electrolyte and the cytosol (Neher & Eckert, 1988; Pusch & Neher, 1988) and has been used extensively for experimental manipulation of cytosolic composition (Homann & Tester, 1997; Carroll et al., 1998; Leckie et al., 1998; Booij et al., 1999; Lemtiri-Chlieh et al., 2000; Köhler & Blatt, 2002). It is also possible to change solutions within the pipette (and cytosol), provided that rapid changes in composition are not required and, hence, that stable recordings can be maintained for periods of many minutes or even tens of minutes. Neher and Eckert (1988), Tang et al. (1992) and Maathuis et al. (1997) describe variations on the theme of introducing a fine capillary tubing to within 100–300 $\mu$m of the tip. The capillary is fed by an external reservoir adjacent to the patch pipette and solution flow is driven into the region of the tip by either a small positive or negative pressure. The effect on the cell is of flow dialysis and, with positive pressure feed, several capillary lines can be positioned behind the tip and engaged independently.

## 1.7  Radiotracer flux analysis

Of the tools available to the transport physiologist, radiotracer techniques are some of the least invasive. By their nature, they are readily adapted to work with whole tissues (Hirsch *et al.*, 1998) and isolated cells (MacRobbie, 1995a) as well as membrane vesicles and organelles (Bewell *et al.*, 1999). The requirements for radiotracer flux measurements are minimal. Knowledge of the transported substrate(s) and availability of a suitable radioactive tracer is sufficient to begin work. An advantage of using inorganic ions as radiotracers is that they are not chemically altered within the tissue, and so problems of identifying radioactive metabolic by-products is not an issue. Another advantage is that ion concentrations can be determined in the steady state. A limitation of these techniques is that they do not provide information about membrane voltage or changes in voltage that can affect fluxes. Nonetheless, radio-tracer techniques were central to Gaffey and Mullins' demonstration of $K^+$ and $Cl^-$ fluxes that drive the *Chara* action potential (Gaffey & Mullins, 1958), to Fischer's evidence of roles for the same ions in stomatal movements (Fischer & Hsiao, 1968), and to Robinson and Jaffe's proof of a polarised $Ca^{2+}$ flux during polar development of fucoid 'eggs' (Robinson & Jaffe, 1975). In recent years, radiotracer techniques have found application in demonstrating roles for $K^+$ channels in mineral nutrition (Hirsch *et al.*, 1998) and the interaction between $NO_3^-$ and $NH_4^+$ assimilation in maize roots (Amancio & Santos, 1992), among others. Significantly, flux analysis is the only technique available that bridges the gap between the *in vivo* situation and patch clamp studies of isolated endomembranes such as the tonoplast. In this regard, MacRobbie's studies (MacRobbie, 1995a,b, 2000) are notable, providing important evidence for parallel actions of abscisic acid at the plasma membrane and tonoplast during stomatal closure.

The design of experiments depends on material – tissue, cells or membrane fractions – and whether it is important to determine unidirectional or bidirectional flux. However, in principle the approaches are similar. Measurements are generally carried out using a small quantity of the radiotracer to 'spike' a test solution with sufficient tracer to monitor changes in its content within the solution and material, for example with a scintillation counter. Uptake (unidirectional influx) by cells and tissues is often measured by sampling the bathing solution at intervals to determine the time course of its disappearance as the tracer is taken up. However, tracer content in the tissue can also be monitored, especially in cell suspensions from which aliquots can be taken at intervals and the cells separated from the bath solution by filtration or centrifugation before scintillation counting. Measuring efflux requires that the cells or tissues are first loaded to a stationary state in which the tracer is equilibrated with the radiolabelled solution outside. Measurements are started by transferring the cells or tissue to fresh, unlabelled solution. Transfer to fresh solution is repeated at intervals and the radiotracer accumulated during each interval is determined from the used solutions. At the end of the experiment, radiotracer remaining in the tissue or cells is also determined.

Strictly speaking, radiotracer uptake and efflux will show kinetics with exponential characteristics. For uptake this occurs because, as radiotracer begins to accumulate in the cells or tissue, its rate of efflux approaches the rate of influx. Therefore, uptake measurements usually emphasise data from earlier in the experiment – normally showing a quasi-linear rate of uptake – on the assumption that tracer efflux will not have time to build up in the cells or tissue and be lost by efflux back into the bath. The rate of efflux also decays with exponential characteristics but, because the external bath is constantly replaced with fresh (unlabelled) solution, this occurs as radiotracer is gradually depleted. For cells and tissues, tracer efflux commonly shows two or three (or more) phases and is best described mathematically as a sum of exponential components. This behaviour is expected of a series of compartments that communicate (and exchange radiotracer) with each other. Mac-Robbie has used efflux analysis extensively for characterising the transport of $K^+$, $Cl^-$ and $Ca^{2+}$ in guard cells and their response to hormonal and environmental stimuli (MacRobbie, 1981, 1983, 1984, 1989, 1995a; Clint & MacRobbie, 1984). Her results have shown, for example, that $K^+$ (using $^{86}Rb^+$ as a radiotracer) exchanges across the plasma membrane within the first 30–60 min and that subsequent efflux is dominated by tracer remaining in the vacuole. Treatments with abscisic acid had differential effects on $K^+$ ($^{86}Rb^+$) flux across the two membranes, depending on the hormone concentration used (MacRobbie, 1995a,b), suggesting that abscisic acid may have two, independent sites of action in guard cells.

## 1.8 Conclusion

Many of the techniques for measuring the electrical and ion transport characteristics of biological membranes are now a century old, and more (see Osterhout, 1931). Their endurance is testimony to the facility they offer in gaining an understanding of these structures and the proteins that are found in them. In recent decades new tools – including those of microelectronics, molecular genetics and fluorescence imaging – have added further dimensions to the exploration of plant membrane biology. Now, more than ever, the biophysical and molecular details of ion transport across membranes have become closely interwoven into our understanding of cellular signalling, development and homeostasis. Without a doubt, membrane physiology and the electrophysiologist's 'tool chest' have grown from the arcane to become central elements of research in plant cell biology.

### Acknowledgements

I am grateful to Drs A. Amtmann and A. Hills for comments on the manuscript, and to the many colleagues and friends who have contributed to this wealth of knowledge. Funding for this work over the years has come from the Royal Society, the Gatsby Charitable Foundation and the UK Research Councils.

### References

Alexander, J. & Nastuk, W.L. (1953) An instrument for the production of microelectrodes used in electrophysiological studies, *Rev. Sci. Instrum.*, **24**, 528–531.

Allen, G. & Sanders, D. (1997) Vacuolar ion channels in higher plants, *Adv. Bot. Res.*, **25**, 217–252.

Amancio, S. & Santos, H. (1992) Nitrate and ammonium assimilation by roots of maize (*Zea mays* L.) seedlings as investigated by in vivo $^{15}$N-NMR, *J. Exp. Bot.*, **43**, 633–639.

Amtmann, A. & Sanders, D. (1997) A unified procedure for the correction of liquid junction potentials in patch clamp experiments on endo- and plasma membranes, *J. Exp. Bot.* **48**, 361–364.

Angleson, J.K. & Betz, W.J. (1997) Monitoring secretion in real time: capacitance, amperometry and fluorescence compared, *Trends Neurosci.*, **20**, 281–287.

Arabidopsis Genome Initiative (2000) Analysis of the genome sequence of the flowering plant *Arabidopsis thaliana, Nature* **408**, 796–815.

Barry, P.H. & Lynch, J.W. (1991) Liquid junction potentials and small cell effects in patch-clamp analysis, *J. Membr. Biol.*, **121**, 101–117.

Beilby, M. & Walker, N.A. (1981) Chloride transport in *Chara*, I: Kinetics and current–voltage curves for a probable proton symport, *J. Exp. Bot.*, **32**, 43–54.

Beilby, M.J. (1984) Current–voltage characteristics of the proton pump at *Chara* plasmalemma, I: pH dependence, *J. Membr. Biol.*, **81**, 113–126.

Beilby, M.J. & Bisson, M.A. (1999) Transport systems of *Ventricaria ventricosa: I/V* analysis of both membranes in series as a function of [K$^+$](o), *J. Membr. Biol.*, **171**, 63–73.

Beilby, M.J. & Blatt, M.R. (1986) Simultaneous measurements of cytoplasmic K$^+$ concentration and plasma membrane electrical parameters in single membrane samples of *Chara corallina, Plant Physiol.*, **82**, 417–422.

Beilby, M.J. & Coster, H.G.L. (1979) The action potential in *Chara corallina*, II: Two activation–inactivation transients in voltage clamps of the plasmalemma, *Aust. J. Plant Physiol.*, **6**, 323–325.

Bentrup, F.-W., Gogarten-Boekels, M., Hoffmann, B., Gogarten, J. & Baumann, C. (1986) ATP-dependent acidification and tonoplast hyperpolarization in isolated vacuoles from green suspension cells of *Chenopodium rubrum, Proc. Natl. Acad. Sci. U.S.A.*, **83**, 2431–2433.

Bernstein, J. (1902) Untersuchungen zur Thermodynamik der bioelektrischen Ströme, *Pflugers Arch.*, **92**, 521–562.

Bewell, M.A., Maathuis, F.J.M., Allen, G.J. & Sanders, D. (1999) Calcium-induced calcium release mediated by a voltage-activated cation channel in vacuolar vesicles from red beet, *FEBS Lett.*, **458**, 41–44.

Blatt, M.R. (1986) Interpretation of steady-state current–voltage curves: consequences and implications of current subtraction in transport studies, *J. Membr. Biol.*, **92**, 91–110.

Blatt, M.R. (1987a) Electrical characteristics of stomatal guard cells: the contribution of ATP-dependent, 'electrogenic' transport revealed by current–voltage and difference-current–voltage analysis, *J. Membr. Biol.*, **98**, 257–274.

Blatt, M.R. (1987b) Electrical characteristics of stomatal guard cells: the ionic basis of the membrane potential and the consequence of potassium chloride leakage from microelectrodes, *Planta*, **170**, 272–287.

Blatt, M.R. (1990) Potassium channel currents in intact stomatal guard cells: rapid enhancement by abscisic acid, *Planta*, **180**, 445–455.

Blatt, M.R. (1991) A primer in plant electrophysiological methods, in *Methods in Plant Biochemistry* (ed. K. Hostettmann), Academic Press, London, pp. 281–321.

Blatt, M.R. & Armstrong, F. (1993) K$^+$ channels of stomatal guard cells: abscisic acid-evoked control of the outward rectifier mediated by cytoplasmic pH, *Planta*, **191**, 330–341.

Blatt, M.R., Beilby, M.J. & Tester, M. (1990) Voltage dependence of the *Chara* proton pump revealed by current–voltage measurement during rapid metabolic blockade with cyanide, *J. Membr. Biol.*, **114**, 205–223.

Blatt, M.R. & Gradmann, D. (1997) K$^+$-sensitive gating of the K$^+$ outward rectifier in *Vicia* guard cells, *J. Membr. Biol.*, **158**, 241–256.

Blatt, M.R., Maurousset, L. & Meharg, A.A. (1997) High-affinity NO$_3^-$-H$^+$ cotransport in the fungus *Neurospora*: induction and control by pH and membrane voltage, *J. Membr. Biol.*, **160**, 59–76.

Blatt, M.R., Rodriguez-Navarro, A. & Slayman, C.L. (1987) Potassium–proton symport in *Neurospora*: kinetic control by pH and membrane potential, *J. Membr. Biol.*, **98**, 169–189.

Blatt, M.R. & Slayman, C.L. (1983) KCl leakage from microelectrodes and its impact on the membrane parameters of a nonexcitable cell, *J. Membr. Biol.*, **72**, 223–234.

Blatt, M.R. & Slayman, C.L. (1987) Role of 'active' potassium transport in the regulation of cytoplasmic pH by nonanimal cells, *Proc. Natl. Acad. Sci. U.S.A.*, **84**, 2737–2741.

Blatt, M.R. & Thiel, G. (2003) SNARE components and mechanisms of exocytosis in plants, in *The Golgi Apparatus and the Plant Secretory Pathway* (ed. D.G. Robinson), Blackwell Publishing Oxford, pp. 208–237.

Blumwald, E. & Poole, R.J. (1987) Salt tolerance in suspension cultures of sugar beet: induction of $Na^+/H^+$ antiport activity at the tonoplast by growth in salt, *Plant Physiol.*, **83**, 884–887.

Böhle, T. & Benndorf, K. (1994) Facilitated giga-seal formation with a just originated glass surface, *Pflugers Arch.*, **427**, 487–491.

Booij, P.P., Roberts, M.R., Vogelzang, S.A., Kraayenhof, R. & deBoer, A.H. (1999) 14-3-3 proteins double the number of outward-rectifying $K^+$ channels available for activation in tomato cells, *Plant J.*, **20**, 673–683.

Brezeale, J.F. (1906) The relation of sodium to potassium in soil and solution cultures, *J. Am. Chem. Soc.*, **28**, 1013–1025.

Brudern, A. & Thiel, G. (1999) Effect of cell-wall-digesting enzymes on physiological state and competence of maize coleoptile cells, *Protoplasma*, **209**, 246–255.

Byerly, L. & Hagiwara, S. (1982) Calcium currents in internally perfused nerve cell bodies of *Limnea stagnalis*, *J. Physiol. London*, **322**, 503–528.

Carroll, A.D., Moyen, C., VanKesteren, P., Tooke, F., Battey, N.H. & Brownlee, C. (1998) $Ca^{2+}$, annexins, and GTP modulate exocytosis from maize root cap protoplasts, *Plant Cell*, **10**, 1267–1276.

Clint, G.M. (1985) The investigation of stomatal ionic relations using guard cell protoplasts, *J. Exp. Bot.*, **36**, 1726–1738.

Clint, G.M. & MacRobbie, E.A.C. (1984) Effects of fusicoccin in 'isolated' guard cells of *Commelina communis*, *J. Exp. Bot.*, **35**, 180–192.

Clint, G.M. & MacRobbie, E.A.C. (1987) Sodium efflux from perfused giant algal cells, *Planta*, **171**, 247–253.

Cole, K.S. (1949) Dynamic electrical characteristics of the squid axon membrane, *Arch. Sci. Physiol.*, **3**, 253–257.

Cole, K.S. & Curtis, H.J. (1938) Electrical impedance of *Nitella* during activity, *J. Gen. Physiol.*, **22**, 37–64.

Cole, K.S. & Curtis, H.J. (1939) Electrical impedance of the squid giant axon during activity, *J. Gen. Physiol.*, **22**, 649–670.

Coleman, H.A. & Findlay, G.P. (1985) Ion channels in the membrane of *Chara inflata*, *J. Membr. Biol.*, **83**, 109–118.

Colquhoun, D. (1987) Practical analysis of single channel records, in *Microelectrode Techniques* (eds N.B. Standen, P.T.A. Gray & M.J. Whitaker), Company of Biologists, Cambridge, UK, pp. 83–104.

Colquhoun, D. & Sigworth, F.J. (1995) Practical analysis of single channel records, in *Single Channel Recording* (eds B. Sakmann & E. Neher), Plenum Press, New York, pp. 397–482.

Conn, P.M. (ed.) (1998) *Methods in Enzymology, Vol. 298: Ion channels, Part B*, Academic Press, New York, pp. 1–805.

Davies, J.M., Sanders, D. & Gradmann, D. (1996) Reaction-kinetics of the vacuolar $H^+$-pumping ATPase in *Beta vulgaris*, *J. Membr. Biol.*, **150**, 231–241.

DeWeer, P., Gadsby, D.C. & Rakowski, R.F. (1988) Voltage dependence of the Na–K pump, *Annu. Rev. Physiol.*, **50**, 225–241.

Elzenga, J.T.M., Keller, C.P. & Van Volkenburgh, E. (1991) Patch clamping protoplasts from vascular plants, *Plant Physiol.*, **97**, 1573–1575.

Feijo, J.A., Sainhas, J., Hackett, G.R., Kunkel, J.G. & Hepler, P.K. (1999) Growing pollen tubes possess a constitutive alkaline band in the clear zone & a growth-dependent acidic tip, *J. Cell Biol.*, **144**, 483–496.

Felle, H. (1981) Stereospecificity and electrogenicity of amino acid transport in *Riccia fluitans*, *Planta*, **152**, 505–512.

Felle, H. (1987) Proton transport and pH control in *Sinapis alba* root hairs: a study carried out with double-barrelled pH microelectrodes, *J. Exp. Bot.*, **340**, 354.

Felle, H. (1988) Cytoplasmic free calcium in *Riccia fluitans* L. and *Zea mays* L.: interaction of $Ca^{2+}$ and pH? *Planta*, **176**, 248–255.

Findlay, G.P. & Hope, A.B. (1964) Ionic relations of cells of *Chara australis*, VII: The separate electrical characteristics of the plasmalemma and tonoplast, *Aust. J. Biol. Sci.*, **17**, 62–77.

Findlay, G.P., Tyerman, S.D., Garrill, A. & Skerrett, M. (1994) Pump and $K^+$ inward rectifiers in the plasmalemma of wheat root protoplasts, *J. Membr. Biol.*, **139**, 103–116.

Fischer, R. & Hsiao, T. (1968) Stomatal opening in isolated epidermal strips of *Vicia faba*, II: Response to KCl concentration and role of potassium absorption, *Plant Physiol.*, **43**, 1953–1958.

Forestier, C., Bouteau, F., Leonhardt, N. & Vavasseur, A. (1998) Pharmacological properties of slow anion currents in intact guard cells of *Arabidopsis*. Application of the discontinuous single-electrode voltage-clamp to different species, *Pflugers Arch.*, **436**, 920–927.

Franklin-Tong, V.E., Holdaway-Clarke, T.L., Straatman, K.R., Kunke, J.G. & Hepler, P.K. (2002) Involvement of extracellular calcium influx in the self-incompatibility response of *Papaver rhoeas*, *Plant J.*, **29**, 333–345.

Gaffey, C.T. & Mullins, L.J. (1958) Ion fluxes during the action potential in *Chara*, *J. Physiol. London*, **144**, 505–524.

Gambale, F., Kolb, H.A., Cantu, A.M. & Hedrich, R. (1994) The voltage-dependent $H^+$-ATPase of the sugar-beet vacuole is reversible, *Eur. Biophys. J.*, **22**, 399–403.

Garcia-Mata, C., Gay, R., Sokolovski, S., Hills, A., Lamattina, L. & Blatt, M.R. (2003) Nitric oxide regulates $K^+$ and $Cl^-$ channels in guard cells through a subset of abscisic acid-evoked signaling pathways, *Proc. Natl. Acad. Sci. U.S.A.*, **100**, 11116–11121.

GarciaSanchez, M.J., Jaime, M.P., Ramos, A., Sanders, D. & Fernandez, J.A. (2000) Sodium-dependent nitrate transport at the plasma membrane of leaf cells of the marine higher plant *Zostera marina* L., *Plant Physiol.*, **122**, 879–885.

Gelli, A. & Blumwald, E. (1997) Hyperpolarization-activated $Ca^{2+}$-permeable channels in the plasma membrane of tomato cells, *J. Membr. Biol.*, **155**, 35–45.

Gillis, K.D. (1995) Techniques for membrane capacitance measurements, in *Single-Channel Recording* (eds B. Sakmann & E. Neher), Plenum Press, New York, pp. 155–198.

Gilroy, S., Fricker, M.D., Read, N.D. & Trewavas, A.J. (1991) Role of calcium in signal transduction of *Commelina* guard cells, *Plant Cell*, **3**, 333–344.

Glass, A.D.M., Shaff, J.E. & Kochian, L.V. (1992) Studies of the uptake of nitrate in barley, 4: Electrophysiology, *Plant Physiol.*, **99**, 456–463.

Goldman, D.E. (1943) Potential, impedance and rectification in membranes, *J. Gen. Physiol.*, **27**, 37–60.

Gould, G.W. (1994) *Membrane Protein Expression Systems*, Portland Press, London, pp. 1–306.

Grabov, A. & Blatt, M.R. (1997) Parallel control of the inward-rectifier $K^+$ channel by cytosolic-free $Ca^{2+}$ and pH in *Vicia* guard cells, *Planta*, **201**, 84–95.

Grabov, A. & Blatt, M.R. (1998) Membrane voltage initiates $Ca^{2+}$ waves and potentiates $Ca^{2+}$ increases with abscisic acid in stomatal guard cells, *Proc. Natl. Acad. Sci. U.S.A.*, **95**, 4778–4783.

Grabov, A., Leung, J., Giraudat, J. & Blatt, M.R. (1997) Alteration of anion channel kinetics in wild-type and *abi1-1* transgenic *Nicotiana benthamiana* guard cells by abscisic acid, *Plant J.*, **12**, 203–213.

Gradmann, D., Hansen, U.-P., Long, W., Slayman, C.L. & Warnke, J. (1978) Current–voltage relationships for the plasma membrane and its principle electrogenic pump in *Neurospora crassa*, I: Steady-state conditions, *J. Membr. Biol.*, **29**, 333–367.

Gradmann, D., Kleiber, H.-G., & Hansen, U.-P. (1987) Reaction kinetic parameters for ion transport from steady-state current–voltage curves, *Biophys. J.*, **51**, 569–585.

Hamill, O.P., Marty, A., Neher, E., Sakmann, B. & Sigworth, F.J. (1981) Improved patch-clamp techniques for high-resolution current recording from cells and cell-free membrane patches, *Pfluegers Arch. Eur. J. Physiol.*, **391**, 85–100.

Hamilton, D.W.A., Hills, A. & Blatt, M.R. (2001) Extracellular $Ba^{2+}$ and voltage interact to gate $Ca^{2+}$ channels at the plasma membrane of stomatal guard cells, *FEBS Lett.*, **491**, 99–103.

Hamilton, D.W.A., Hills, A., Kohler, B. & Blatt, M.R. (2000) $Ca^{2+}$ channels at the plasma membrane of stomatal guard cells are activated by hyperpolarization and abscisic acid, *Proc. Natl. Acad. Sci. U.S.A.*, **97**, 4967–4972.

Hansen, U.-P., Gradmann, D., Sanders, D. & Slayman, C.L. (1981) Interpretation of current–voltage relationships for 'active' ion transport systems, I: Steady-state reaction-kinetic analysis of class I mechanisms, *J. Membr. Biol.*, **63**, 165–190.

Hansen, U.-P. & Slayman, C.L. (1978) Current–voltage relationships for a clearly electrogenic cotransport system, in *Membrane Transport Processes* (ed. J.F. Hoffman), Raven Press, New York, pp. 141–154.

Hasselbring, H. (1914) The relation between the transpiration stream and the absorption of salts, *Bot. Gazette*, **57**, 72–73.

Henriksen, G.H., Taylor, A.R., Brownlee, C. & Assmann, S.M. (1996) Laser microsurgery of higher-plant cell walls permits patch clamp access, *Plant Physiol.*, **110**, 1063–1068.

Higinbotham, N., Etherton, B. & Foster, R.J. (1967) Mineral ion contents and cell transmembrane electropotentials of pea and oat seedling tissue, *Plant Physiol.*, **42**, 37–46.

Hille, B. (2001) *Ionic Channels of Excitable Membranes*, Sinauer Press, Sunderland, MA, pp. 1–813.

Hille, B. & Schwarz, W. (1978) Potassium channels as multi-ion single-file pores, *J. Gen. Physiol.*, **72**, 409–442.

Hirsch, R.E., Lewis, B.D., Spalding, E.P. & Sussman, M.R. (1998) A role for the AKT1 potassium channel in plant nutrition, *Science*, **280**, 918–921.

Hodgkin, A.L., Huxley, A.F. & Katz, B. (1952) Measurements of current–voltage relations in the membrane of the giant axon of *Loligo*, *J. Physiol. London*, **116**, 424–448.

Hodgkin, A.L. & Katz, B. (1949) The effect of sodium ions on the electrical activity of the giant axon of the squid, *J. Physiol. London*, **108**, 37–77.

Homann, U. & Tester, M. (1997) $Ca^{2+}$-independent and $Ca^{2+}$/GTP-binding protein-controlled exocytosis in a plant cell, *Proc. Natl. Acad. Sci. U.S.A.*, **94**, 6565–6570.

Homann, U. & Thiel, G. (1999) Unitary exocytotic and endocytotic events in guard cell protoplasts during osmotically driven volume changes, *FEBS Lett.*, **460**, 495–499.

Homann, U. & Thiel, G. (2002) The number of $K^+$ channels in the plasma membrane of guard cell protoplasts changes in parallel with the surface area, *Proc. Natl. Acad. Sci. U.S.A.*, **99**, 10215–10220.

Horvath, F., Erdei, L., Wodala, B., Homann, U. & Thiel, G. (2002) $K^+$ outward rectifying channels as targets of phosphatase inhibitor deltamethrin in *Vicia faba* guard cells, *J. Plant Physiol.*, **159**, 1097–1103.

Hoshi, T. (1995) Regulation of voltage-dependence of the KAT1 channel by intracellular factors, *J. Gen. Physiol.*, **105**, 309–328.

Howe, H.M. (1965) A root of Helmont's tree, *ISIS*, **56**, 408–419.

Ivashikina, N., Becker, D., Ache, P., Meyerhoff, O., Felle, H.H. & Hedrich, R. (2001) $K^+$ channel profile and electrical properties of *Arabidopsis* root hairs, *FEBS Lett.*, **508**, 463–469.

Jack, J.J.B., Noble, D. & Tsien, R.W. (1983) *Electric Current Flow in Excitable Cells*, Clarendon Press, Oxford.

Jaffe, L.F. (1968) Localization in the developing *Fucus* egg and the general role of localizing currents, *Adv. Morphog.*, **7**, 295–328.

Jaffe, L.F. (1981) Calcium explosions as triggers of development, *Ann. N. Y. Acad. Sci.*, **339**, 86–101.

Jaffe, L.F., Robinson, K.R. & Nuccitelli, R. (1974) Transcellular currents and ion fluxes through developing fucoid eggs, in *Membrane Transport in Plants* (eds U. Zimmermann & J. Dainty), Springer, New York, pp. 226–233.

Kitasato, H. (1968) The influence of $H^+$ on the membrane potential and ion fluxes of *Nitella clavata*, *J. Gen. Physiol.*, **52**, 60–87.

Köhler, B. & Blatt, M.R. (2002) Protein phosphorylation activates the guard cell $Ca^{2+}$ channel and is a prerequisite for gating by abscisic acid, *Plant J.*, **32**, 185–194.

Läuger, P. (1980) Kinetic properties of ion carriers and channels, *J. Membr. Biol.*, **57**, 163–178.

Läuger, P. & Stark, G. (1970) Kinetics of carrier-mediated ion transport across lipid bilayer membranes, *Biochim. Biophys. Acta*, **211**, 458–466.

Leckie, C.P., McAinsh, M.R., Allen, G.J., Sanders, D. & Hetherington, A.M. (1998) Abscisic acid-induced stomatal closure mediated by cyclic ADP-ribose, *Proc. Natl. Acad. Sci. U.S.A.*, **95**, 15837–15842.

Lemtiri-Chlieh, F., MacRobbie, E.A.C. & Brearley, C.A. (2000) Inositol hexakisphosphate is a physiological signal regulating the $K^+$-inward rectifying conductance in guard cells, *Proc. Natl. Acad. Sci. U.S.A.*, **97**, 8687–8692.

Lew, R.R. (1991) Electrogenic transport properties of growing *Arabidopsis* root hairs the plasma membrane proton pump and potassium channels, *Plant Physiol.*, **97**, 1527–1534.

Lew, R.R. (1998) Immediate and steady state extracellular ionic fluxes of growing *Arabidopsis thaliana* root hairs under hyperosmotic and hypoosmotic conditions, *Physiol. Plant.*, **104**, 397–404.

Leyman, B., Geelen, D., Quintero, F.J. & Blatt, M.R. (1999) A tobacco syntaxin with a role in hormonal control of guard cell ion channels, *Science*, **283**, 537–540.

Lin, Y.K. & Yang, Z.B. (1997) Inhibition of pollen tube elongation by microinjected anti-Rop1Ps antibodies suggests a crucial role for Rho-type GTPases in the control of tip growth, *Plant Cell*, **9**, 1647–1659.

Ling, G. & Gerard, R.W. (1949) The normal membrane potential of frog sartorius fibers, *J. Cell. Comp. Physiol.*, **34**, 383–396.

Lippincott-Schwartz, J. & Patterson, G.H. (2003) Development and use of fluorescent protein markers in living cells, *Science*, **300**, 87–91.

Lohse, G. & Hedrich, R. (1992) Characterization of the plasma-membrane $H^+$-ATPase from *Vicia faba* guard cells, *Planta*, **188**, 206–214.

Maathuis, F. & Sanders, D. (1994) Mechanism of high-affinity potassium uptake in roots of *Arabidopsis thaliana*, *Proc. Natl. Acad. Sci. U.S.A.*, **91**, 9272–9276.

Maathuis, F.J.M., Taylor, A.R., Assmann, S.M. & Sanders, D. (1997) Seal-promoting solutions and pipette perfusion for patch clamping plant cells, *Plant J.*, **11**, 891–896.

Maathuis, F.J.M., Verlin, D., Smith, F.A., Sanders, D., Fernandez, J.A. & Walker, N.A. (1996) The physiological relevance of $Na^+$-coupled $K^+$-transport, *Plant Physiol.*, **112**, 1609–1616.

MacRobbie, E.A.C. (1981) Ion fluxes in 'isolated' guard cells of *Commelina communis* L., *J. Exp. Bot.*, **32**, 545–562.

MacRobbie, E.A.C. (1983) Effects of light/dark on cation fluxes in guard cells of *Commelina communis* L., *J. Exp. Bot.*, **34**, 1695–1710.

MacRobbie, E.A.C. (1984) Effects of light/dark on anion fluxes in isolated guard cells of *Commelina communis*, *J. Exp. Bot.*, **35**, 707–726.

MacRobbie, E.A.C. (1989) Calcium influx at the plasmalemma of isolated guard cells of *Commelina communis* effects of abscisic acid, *Planta*, **178**, 231–241.

MacRobbie, E.A.C. (1995a) ABA-induced ion efflux in stomatal guard-cells – multiple actions of ABA inside and outside the cell, *Plant J.*, **7**, 565–576.

MacRobbie, E.A.C. (1995b) Effects of ABA on $^{86}Rb^+$ fluxes at plasmalemma and tonoplast of stomatal guard cells, *Plant J.*, **7**, 835–843.

MacRobbie, E.A.C. (2000) ABA activates multiple $Ca^{2+}$ fluxes in stomatal guard cells, triggering vacuolar $K^+(Rb^+)$ release, *Proc. Natl. Acad. Sci. U.S.A.*, **97**, 12361–12368.

Malho, R., Read, N.D., Trewavas, A.J. & Pais, M.S. (1995) Calcium channel activity during pollen tube growth and reorientation, *Plant Cell*, **7**, 1173–1184.

Marquardt, D. (1963) An algorithm for least-squares estimation of nonlinear parameters, *J. Soc. Ind. Appl. Math.*, **11**, 431–441.

McAinsh, M.R., Webb, A.A.R., Taylor, J.E. & Hetherington, A.M. (1995) Stimulus-induced oscillations in guard cell cytosolic-free calcium, *Plant Cell*, **7**, 1207–1219.

McCormack, J.G. & Cobbold, P.H. (1991) *Cellular Calcium*, Vol. 1, Oxford University, Oxford, pp. 1–418.

McCulloch, S.R. & Beilby, M.J. (1997) The electrophysiology of plasmolysed cells of *Chara australis*, *J. Exp. Bot.*, **48**, 1383–1392.

McCulloch, S.R., Beilby, M.J. & Walker, N.A. (1990) Transport of potassium in *Chara australis*, II: Kinetics of a symport with sodium, *J. Membr. Biol.*, **115**, 129–143.

Meharg, A.A. & Blatt, M.R. (1995) Nitrate transport in root hairs of *Arabidopsis thaliana*: kinetic control by membrane voltage and pH, *J. Membr. Biol.*, **145**, 49–66.

Meharg, A.A., Maurousset, L. & Blatt, M.R. (1994) Cable correction of membrane currents recorded from root hairs of *Arabidopsis thaliana* L., *J. Exp. Bot.*, **45**, 1–6.

Miller, A.J., Vogg, G. & Sanders, D. (1990) Cytosolic calcium homeostasis in fungi: roles of plasma membrane transport and intracellular sequestration of calcium, *Proc. Natl. Acad. Sci. U.S.A.*, **87**, 9348–9352.

Miller, C. (1986) *Ion Channel Reconstitution*, Plenum Press, New York, pp. 1–283.

Mitchell, P. (1969) Chemiosmotic coupling and energy transduction, *Theor. Exp. Biophys.*, **2**, 159–216.

Moutinho, A., Love, J., Trewavas, A.J. & Malho, R. (1998) Distribution of calmodulin protein and mRNA in growing pollen tubes, *Sex. Plant Reprod.*, **11**, 131–139.

Neher, E. & Eckert, R. (1988) Fast patch-pipette internal perfusion with minimum solution flow, in *Calcium and Ion Channel Modulation* (eds A. Grinnell, D. Armstrong & M.B. Jackson), Plenum Press, New York, pp. 371–377.

Neher, E. & Sakmann, B. (1976) Single-channel currents recorded from the membrane of denervated frog muscle fibres, *Nature*, **260**, 779–802.

Nernst, W. (1888) Zur Kinetik der in Lösung befindlichen Körper: Theorie der Diffusion, *Z. Phys. Chem.*, **3**, 613–637.

Nernst, W. (1889) Die electromotorische Wirksamkeit der Ionen, *Z. Phys. Chem.*, **4**, 129–181.

Newman, I.A., Kochian, L.V., Grusak, M.A. & Lucas, W.J. (1987) Fluxes of $H^+$ and $K^+$ in corn roots, *Plant Physiol.*, **84**, 1177–1184.

Nuccitelli, R. (1994) *A Practical Guide to the Study of Calcium in Living Cells*, Academic Press, London, pp. 1–368.

Osterhout, W.J.V. (1931) Physiological studies of single plant cells, *Biol. Rev.*, **6**, 369–411.

Pei, Z.M., Kuchitsu, K., Ward, J.M., Schwarz, M. & Schroeder, J.I. (1997) Differential abscisic acid regulation of guard cell slow anion channels in *Arabidopsis* wild-type and *abi1* and *abi2* mutants, *Plant Cell*, **9**, 409–423.

Pfeffer, W. (1877) *Osmotische Untersuchungen*, Wilhelm Engelmann, Leipzig, pp. 1–236.

Philippar, K., Fuchs, I., Luthen, H., *et al.* (1999) Auxin-induced $K^+$ channel expression represents an essential step in coleoptile growth and gravitropism, *Proc. Natl. Acad. Sci. U.S.A.*, **96**, 12186–12191.

Planck, M. (1890a) Über die Erregung von Elektricität und Wärme in Elektrolyten, *Ann. Phys. Chem.*, **39**, 161–186.

Planck, M. (1890b) Über die Potentialdifferenz zwischen zwei verdünnten Lösungen binärer Elektrolyte, *Ann. Phys. Chem.*, **40**, 561–578.

Pottosin, I.I. & Andjus, P.R. (1994) Depolarization-activated $K^+$ channel in *Chara* droplets, *Plant Physiol.*, **106**, 313–319.

Pottosin, I.I., Dobrovinskaya, O.R. & Muniz, J. (1999) Cooperative block of the plant endomembrane ion channel by ruthenium red, *Biophysical J.*, **77**, 1973–1979.

Pratelli, R., Sutter, J.-U. & Blatt, M.R. (2004) A new catch to the SNARE, *Trends Plant Sci.*, **9**, 187–195.

Purves, J.D. (1981) *Microelectrode Methods for Intracellular Recording and Ionophoresis*, Vol. 1, Academic Press, London, pp. 1–146.

Pusch, M. & Neher, E. (1988) Rates of diffusional exchange between small cells and a measuring patch pipette, *Pfluegers Arch. Eur. J. Physiol.*, **411**, 204–214.

Quintero, F.J., Blatt, M.R. & Pardo, J.M. (2000) Functional conservation between yeast and plant endosomal $Na^+/H^+$ antiporters, *FEBS Lett.*, **471**, 224–228.

Robinson, K.R. & Jaffe, L.F. (1975) Polarizing fucoid eggs drive a calcium current through themselves, *Science*, **187**, 70–72.

Rodriguez-Navarro, A., Blatt, M.R. & Slayman, C.L. (1986) A potassium–proton symport in *Neurospora crassa*, *J. Gen. Physiol.*, **87**, 649–674.

Roelfsema, M.R.G. & Prins, H.B.A. (1997) Ion channels in guard cells of *Arabidopsis thaliana* (L.) Heynh, *Planta*, **202**, 18–27.

Roy, S.J., HoldawayClarke, T.L., Hackett, G.R., Kunkel, J.G., Lord, E.M. & Hepler, P.K. (1999) Uncoupling secretion and tip growth in lily pollen tubes: evidence for the role of calcium in exocytosis, *Plant J.*, **19**, 379–386.

Rubio, F., Santa-Maria, G.E. & Rodriguez-Navarro, A. (2000) Cloning of *Arabidopsis* and barley cDNAs encoding HAK potassium transporters in root and shoot cells, *Physiol. Plant.*, **109**, 34–43.

Rudy, B. & Iverson, L.E. (eds) (1992) *Methods in Enzymology, Vol. 207: Ion Channels*, Academic Press, New York, pp. 1–917.

Sachs, F., Neil, J. & Barkakati, N. (1982) The automated analysis of data from single ionic channels, *Pfluegers Arch. Eur. J. Physiol.*, **395**, 331–340.

Sanders, D., Hansen, U.-P., Gradmann, D. & Slayman, C.L. (1984) Generalized kinetic analysis of ion-driven cotransport systems: a unified interpretation of selective ionic effects on Michaelis parameters, *J. Membr. Biol.*, **77**, 123–152.

Sanders, D., Slayman, C.L. & Pall, M. (1983) Stoichiometry of $H^+$/amino-acid cotransport in *Neurospora crassa* revealed by current–voltage analysis, *Biochim. Biophys. Acta*, **735**, 67–76.

Schmidt, C., Schelle, I., Liao, Y.J. & Schroeder, J.I. (1995) Strong regulation of slow anion channels and abscisic acid signaling in guard cells by phosphorylation and dephosphorylation events, *Proc. Natl. Acad. Sci. U.S.A.*, **92**, 9535–9539.

Schroeder, J.I., Raschke, K. & Neher, E. (1987) Voltage dependence of $K^+$ channels in guard-cell protoplasts, *Proc. Natl. Acad. Sci. U.S.A.*, **84**, 4108–4112.

Schwab, W. & Komor, E. (1978) A possible mechanistic role of the membrane potential in proton–sugar cotransport of *Chlorella*, *FEBS Lett.*, **87**, 157–160.

Segel, I.H. (1993) *Enzyme Kinetics*, Wiley-Interscience, New York, pp. 1–957.

Shabala, S.N., Newman, I.A. & Morris, J. (1997) Oscillations in $H^+$ and $Ca^{2+}$ ion fluxes around the elongation region of corn roots and effects of external pH, *Plant Physiol.*, **113**, 111–118.

Slayman, C.L. (1965) Electrical properties of *Neurospora crassa*. Respiration and the intracellular potential, *J. Gen. Physiol.*, **49**, 93–116.

Slayman, C.L. & Slayman, C.W. (1974) Depolarization of the plasma membrane of *Neurospora* during active transport of glucose: evidence for a proton-dependent cotransport system, *Proc. Natl. Acad. Sci. U.S.A.*, **71**, 1935–1939.

Smith, J.R. (1984) The electrical properties of plant cell membranes, II: Distortion of non-linear current–voltage characteristics induced by the cable properties of *Chara*, *Aust. J. Plant Physiol.*, **11**, 211–224.

Spanswick, R.M. (1970) Electrophysiological techniques and the magnitudes of the membrane potentials and resistances of *Nitella translucens*, *J. Exp. Bot.*, **21**, 617–627.

Standen, N.B., Gray, P.T.A. & Whitaker, M.J. (1987) *Microelectrode Techniques*, Company of Biologists, Cambridge, UK, pp. 1–253.

Stephens, D.J. & Allan, V.J. (2003) Light microscopy techniques for live cell imaging, *Science*, **300**, 82–86.

Stiles, W. & Kidd, F. (1919) The influence of external concentration on the position of the equilibrium attained in the intake of salts by plant cells, *Proc. R. Soc. Lond. B Biol. Sci.*, **90**, 448–470.

Stoeckel, H. & Takeda, K. (1989) Calcium-activated voltage-dependent non-selective cation currents in endosperm plasma membrane from higher plants, *Proc. R. Soc. Lond. B Biol. Sci.*, **237**, 213–231.

Sutter, J.U., Homann, U. & Thiel, G. (2000) $Ca^{2+}$-stimulated exocytosis in maize coleoptile cells, *Plant Cell*, **12**, 1127–1136.

Tanford, C. (1983) Mechanism of free energy coupling in active transport, *Annu. Rev. Biochem.*, **52**, 379–409.

Tang, J.M., Wang, J. & Eisenberg, R.S. (1992) Perfusing patch pipettes, in *Methods in Enzymology, Vol. 207: Ion Channels* (eds B. Rudy & L.E. Iverson), Academic Press, New York, pp. 176–181.

Tang, X.D. & Hoshi, T. (1999) Rundown of the hyperpolarization-activated KAT1 channel involves slowing of the opening transitions regulated by phosphorylation, *Biophys. J.*, **76**, 3089–3098.

Tasaki, I. & Singer, I. (1968) Some problems involved in electric measurements of biological systems, *Proc. N. Y. Acad. Sci.*, **148**, 36–53.

Taylor, A.R., Manison, N.F.H., Fernandez, C., Wood, J. & Brownlee, C. (1996) Spatial organization of calcium signaling involved in cell volume control in the *Fucus* rhizoid, *Plant Cell*, **8**, 2015–2031.

Tester, M., Beilby, M.J. & Shimmen, T. (1987) Electrical characteristics of the tonoplast of *Chara corallina*: a study using permeabilized cells, *Plant Cell Physiol.*, **28**, 1555–1568.

Thiel, G., MacRobbie, E.A.C. & Blatt, M.R. (1992) Membrane transport in stomatal guard cells: the importance of voltage control, *J. Membr. Biol.*, **126**, 1–18.

Thiel, G., Sutter, J.U. & Homann, U. (2000) $Ca^{2+}$-sensitive and $Ca^{2+}$-insensitive exocytosis in maize coleoptile protoplasts, *Pflugers Arch.*, **439**, R152–R153.

Thomas, H.H. (1955) Experimental plant biology in pre-Linnean times, *Bull. Br. Soc. Hist. Sci.*, **2**, 15–22.

Travis, E.R. & Wightman, R.M. (1998) Spatio-temporal resolution of exocytosis from individual cells, *Annu. Rev. Biophys. Biomol. Struct.*, **27**, 77–103.

Tyerman, S.D. (1992) Anion channels in plants, *Annu. Rev. Plant Physiol. Plant Mol. Biol.*, **43**, 351–373.

Tyerman, S.D. & Findlay, G.P. (1989) Current–voltage curves of single chloride channels which coexist with two types of potassium channel in the tonoplast of *Chara corallina*, *J. Exp. Bot.*, **40**, 105–118.

Umrath, K. (1930) Untersuchungen über Plasma und Plasmaströmung an Characean, IV: Potentialmessungen an *Nitella mucronata* mit besonderer Berücksichtigung der Erregungserscheinungen, *Protoplasma*, **9**, 576–597.

Umrath, K. (1932) Die Bildung von Plasmalemma (Plasmahaut) bei *Nitella mucronata*, *Protoplasma*, **16**, 173–188.

Very, A.A. & Sentenac, H. (2003) Molecular mechanisms and regulation of $K^+$ transport in higher plants, *Annu. Rev. Plant Biol.*, **54**, 575–603.

von Liebig, J. (1840) *Die Chemie in ihrer Anwendung auf Agrikultur und Physiologie*, Wilhelm Engelmann, Leipzig, pp. 1–835.

Weise, R., Kreft, M., Zorec, R., Homann, U. & Thiel, G. (2000) Transient and permanent fusion of vesicles in *Zea mays* coleoptile protoplasts measured in the cell-attached configuration, *J. Membr. Biol.*, **174**, 15–20.

Weiser, T. & Bentrup, F.W. (1994) The chaotropic anions thiocyanate and nitrate inhibit the electric current through the tonoplast ATPase of isolated vacuoles from suspension cells of *Chenopodium rubrum*, *Physiol. Plant.*, **91**, 17–22.

White, P.J., Biskup, B., Elzenga, J.T.M., *et al.* (1999) Advanced patch-clamp techniques and single-channel analysis, *J. Exp. Bot.*, **50**, 1037–1054.

White, P.J. & Davenport, R.J. (2002) The voltage-independent cation channel in the plasma membrane of wheat roots is permeable to divalent cations and may be involved in cytosolic $Ca^{2+}$ homeostasis, *Plant Physiol.*, **130**, 1386–1395.

White, P.J., Pineros, M., Tester, M. & Ridout, M.S. (2000) Cation permeability and selectivity of a root plasma membrane calcium channel, *J. Membr. Biol.*, **174**, 71–83.

Wierzbicki, W., Berteloot, A. & Roy, G. (1990) Pre-steady-state kinetics and carrier-mediated transport: a theoretical analysis, *J. Membr. Biol.*, **117**, 11–27.

Wu, Y., Kuzma, J., Marechal, E., *et al.* (1997) Abscisic acid signaling through cyclic ADP-ribose in plants, *Science*, **278**, 2126–2130.

# 2 Electrophysiology equipment and software

Adrian Hills and Vadim Volkov

## 2.1 Introduction

There is a wide range of methods available for the study of ion transport across biological membranes. Early, simple measurements of pH and ion concentrations resulted in striking insights, such as the chemiosmotic theory (Mitchell, 1961) and, for plant nutrition, the discovery of low- and high-affinity systems for potassium uptake (Epstein *et al.*, 1963). Modifications to these techniques (radioactive tracers, fluorescent probes, etc.) have enabled better temporal and spatial resolution of such ion transport mechanisms. However, some of the most important breakthroughs have come with the advent of the voltage clamp (Cole, 1949; Cole & Moore, 1960) and, subsequently, with the development of the patch clamp (Neher & Sakmann, 1976; Hamill *et al.*, 1981). The theory and applicability of the voltage clamp are discussed in the previous chapter; in this chapter, we discuss some important practical aspects of the techniques involved and present an overview of some of the equipment and computer software available to researchers in this field.

The traditional, two-electrode voltage clamp (TEVC) technique is widely used for measuring whole-cell currents (that is, the total current flowing through all active ion channels on the cell surface). Although, in many fields, the patch clamp is the dominant tool for electrophysiological investigations, there are important areas in which TEVC is far more applicable. In plant science, as well as in mammalian and pharmacological research, impalements of *Xenopus* oocytes expressing particular ion channels or transporters of interest (Gurdon *et al.*, 1971; Miller & Zhou, 2000), combined with the site-directed mutagenesis of these proteins, provide an extremely powerful, yet readily accessible, means of elucidating the relationships between structure and function in ion transport proteins. In work with plants, impalement methods are particularly relevant also to analysing ion channel and transport characteristics in the intact cell, for example of stomatal guard cells, bounded by the cell wall and attached to the leaf epidermis (Blatt *et al.*, 1987; Thiel *et al.*, 1992), and, more recently, to the study of guard cell electrophysiology in intact organisms (Roelfsema *et al.*, 2001). (The fact that guard cells are electrically isolated from the rest of the leaf's symplast makes them especially suitable for such studies [see Chapter 1].)

The various patch clamp configurations now available (see Chapter 1) have considerably widened the horizons for electrophysiological research in the plant sciences. The cell-attached mode is most useful in situations where loss of cell contents results in a rundown and loss of channel activity (Philippar *et al.*, 1999). Nevertheless, this mode is subject to uncertainties about the clamp voltage, since the monitored voltage is the sum of the potential difference across the patch and that across the membrane as a whole. Exposing the patch to agonists or antagonists can also be difficult if the compounds must reach the 'outer' surface of the membrane. The whole-cell patch mode allows the recording of currents similar in nature to those available in TEVC experiments. One advantage of the whole-cell patch mode is that it can be used with tissues for which impalement is difficult or impossible; indeed, one consequence of isolating protoplasts for whole-cell patch measurements is that problems of symplastic coupling are eliminated. The inside-out configuration is particularly suitable for the study of intracellular messengers (e.g. Köhler & Blatt, 2002) and the outside-out configuration is extensively used in the study of ligand-gated channels, where various concentrations of agonists can be applied and removed (e.g. Maathuis & Sanders, 1995).

## 2.2   Voltage clamp protocols

The design of voltage clamp protocols depends mostly on the nature of the currents and ion channels being investigated. Fast currents, which appear virtually as soon as the membrane potential is changed, can be mediated by several types of ion channel, such as cyclic nucleotide gated channels (CNGCs) (Maathuis & Sanders, 2001) and glutamate receptors (Demidchik *et al.*, 2002). Slower, time-dependent currents reach their maximum (steady-state) value only some tens, hundreds or even thousands of milliseconds after an applied change in membrane potential. There are large differences in activation kinetics among ion channels in plants: for example, so-called slow anion channels in guard cells are among the slowest, with half-activation times $(t_{1/2})$ in the order of 5–30 s (Schroeder & Keller, 1992); potassium channels are usually considerably faster, with $t_{1/2}$ significantly less than 1 s (Blatt & Gradmann, 1997). The following are a few examples of common protocols used to assess the kinetic and conductance characteristics of ion channels and other transporters (see also Chapter 1).

### 2.2.1   Voltage stepping protocols

The simplest form of voltage clamp protocol comprises a series of consecutive 'pulses', changing the cell membrane potential from a fixed, holding potential (at which the channels under investigation are nominally inactive or closed) to an increasingly different test potentials, at which the channels open (see Fig. 2.1). An important feature of many step protocols is that voltage steps are held long enough to ensure that the current reaches steady state (see Chapter 1). More complex protocols can be devised along these lines to identify multiple channels with differing voltage- and/or time-dependent characteristics (see Fig. 2.2).

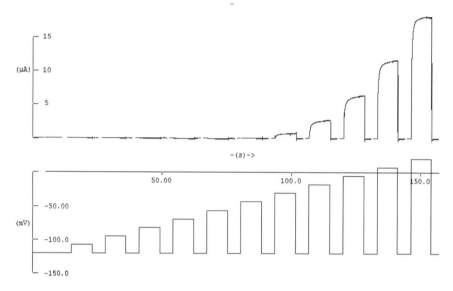

**Fig. 2.1** The current response (upper trace) of the *Saccharomyces cerevisiae* 'TOK1' channel, expressed in *Xenopus laevis* oocytes, to a single-step voltage clamp protocol (lower trace). The holding potential is −120 mV, and the channel activity is tested at potentials ranging from −120 to +20 mV (in steps of just under 13 mV). The channel activates around −30 mV, with the current increasing thereafter. Data courtesy of Dr Ingela Johansson, Laboratory of Plant Physiology and Biophysics, IBLS, University of Glasgow, Glasgow, UK.

## 2.2.2 *Voltage ramp protocols*

Step protocols are ideal for studying the kinetics of voltage-activated ion channels and even their steady-state voltage dependence. However, each cycle of clamp pulses takes time (11 s in the two-step example shown in Fig. 2.2, giving a total run time of 1.5 min), during which cells may deteriorate, or otherwise become irresponsive. Voltage ramp protocols offer a more rapid approach to measuring membrane current, if only the steady-state current–voltage characteristics are important to know (see Fig. 2.3). The use of a ramp takes advantage of the fact that kinetic transitions of many ion channels (and other transporters) will be essentially complete – independent of changes in voltage – over part of the voltage range covered by the ramp. Thus, a significant time saving can be realised. Note that, in such cases, it is vital that the ramp be slow enough to ensure true steady-state conditions across the entire voltage range. (A discussion of some of the aspects of the analysis of such current–voltage relationships can be found in Section 2.4.3.)

There are problems inherent in the voltage ramp approach, the most significant of which concerns the separation of genuine, transport currents from capacitive transients. All cell membranes behave as electrical capacitors; thus, whenever the clamp potential is changed, a rapid flow of current is required to charge this capacitor. In voltage stepping protocols, these short-lived currents manifest as narrow current

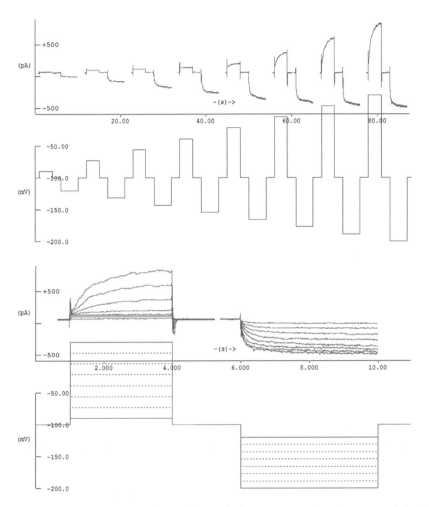

**Fig. 2.2** The response of an impaled *Vicia faba* guard cell to a two-step voltage clamp protocol, showing the induced whole-cell currents of outward- and inward-rectifying $K^+$ channels to, respectively, depolarisation and hyperpolarisation of the membrane. The upper diagram shows the clamped potential and elicited current for the entire protocol in 'real time'; the lower diagram presents the same data in the more traditional format, with each consecutive cycle of steps overlaid in time. Data kindly supplied by Dr Sergei Sokolovski, Laboratory of Plant Physiology and Biophysics, IBLS, University of Glasgow, Glasgow, UK.

'spikes' immediately following any change in potential (as can be seen in Fig. 2.2) and compensation for their contribution to total measured currents is straightforward; indeed, at the time of steady state, they are simply irrelevant. However, in a ramp protocol, the membrane potential is changing continuously, and so there is always a (frequently unknown) capacitive component to the measured currents.

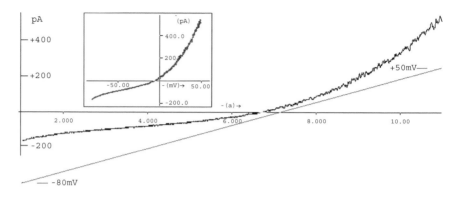

**Fig. 2.3** The $K^+$ outward current (black line) of an impaled *Vicia faba* guard cell, elicited by a voltage ramp running from $-80$ to $+50$ mV over 10 s (grey line). The inset shows the corresponding *IV* curve. Data kindly provided by Prof. Teresa Hernandez-Sotomayor, CICY, Mexico.

### 2.2.3 'Tail current' protocols

Tail current protocols are commonly used to determine the selectivity for different ions of the channel of interest. For voltage-gated ion channels, ion selectivity is very often determined by comparing the current reversal potential, $E_{rev}$, with the predicted ionic equilibrium calculated using the Nernst equation (see Chapter 1). Determining $E_{rev}$ requires some prior knowledge of the kinetic properties of the ion channel of interest before designing the tail current protocol. Although similar to the protocols of Figs. 2.1 and 2.2, where the holding potential is such that the channel is inactive (closed), tail current measurements start with a voltage at which the channels are fully active (open) and step, successively, to levels more positive and more negative of an initial estimate for the reversal potential. The actual value can be interpolated from the relaxations of the current 'tails', which reverse sign either side of $E_{rev}$ (see Fig. 2.4 and Chapter 1). It is worth noting that a protocol such as this can only be used for channels that show a significantly reduced level of activity in the steady state at or near their reversal potential.

### 2.2.4 Time-variant protocols

In the protocols described in Section 2.2.1, the only variation between successive 'cycles' is the test (activation) potential. However, there are occasions on which, rather than changing the voltage to which the clamp should step, it is desirable to change the time at which to make the step. A good example is the protocol shown in Fig. 2.5, for measuring the slow deactivation kinetics of the *Saccharomyces cerevisiae* outward-rectifying TOK1 $K^+$ channel expressed in *Xenopus* oocytes (Vergani & Blatt, 1999).

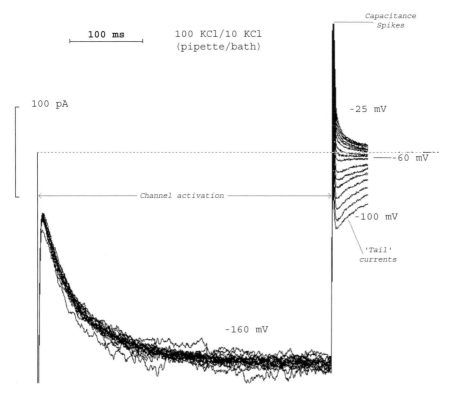

**Fig. 2.4** Recorded current traces illustrating a two-step protocol used to measure the reversal potential of an inward-rectifying channel from the tail currents of its deactivation; the current, from a *Hordeum vulgare* protoplast, is activated by holding at −160 mV for 500 ms (labelled 'channel activation'). The voltage is then increased in a test step to −100 mV, deactivating the inward rectifier (labelled 'tail currents'); repeating these steps with gradual (5 mV) increases in the test voltage steps gives a point around −60 mV at which no current relaxation is seen during deactivation. Here, the value of −60 mV agrees well with the high selectivity for potassium over chloride and other ions in the pipette and bath solutions. The data shown are from a patch clamp experiment in whole-cell mode.

The TOK1 channel shows gating that is consistent with a single open state of the channel that communicates with a set of three serial closed states (Vergani *et al.*, 1998)

$$C_4 \longleftrightarrow C_3 \longleftrightarrow C_2 \longleftrightarrow O_1$$

where the transitions between $C_2$ and $O_1$ are strongly voltage dependent and so rapid as to make their kinetics unresolved in whole-cell measurements. Thus, the slower, time-dependent activation is associated with the $C_4 \leftrightarrow C_3 \leftrightarrow C_2$ transitions. However, once activated by a positive voltage step, returning to negative voltages that close the channel gives an almost instantaneous loss of current as the channels enter $C_2$ from $O_1$. In order to access the relaxation times from $C_2$ to the other closed states, a time-variant, two-step protocol can be used to measure the proportion of channels in the $C_2$ state. The protocol begins with a conditioning step to $+50$ mV to fully activate the channels and to establish the zero-time current maximum. The voltage is then stepped to $-120$ mV to close the channels; at successively longer intervals thereafter, short steps are made to $+50$ mV, to determine the amplitude of the near-instantaneous current (Fig. 2.5, arrows), corresponding to the number of channels remaining in the immediately activatable $C_2$ state. Normalised to the maximum current amplitude and plotted against time, these data then give the time course for transitions from $C_2$ into the more distal $C_3$ and $C_4$ states (see Vergani et al., 1998).

### 2.2.5   Extended single-channel recording

In patch clamp experiments recording the activity of single channels, it is often desirable to clamp at a particular voltage for a long period of time in order to calculate the open probability of the channels of interest, and to compare such values

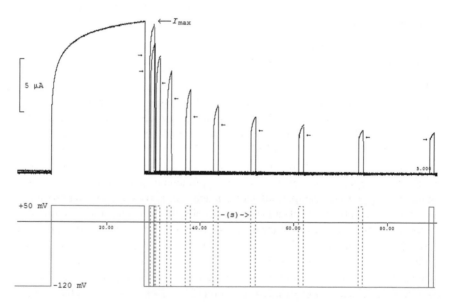

**Fig. 2.5** A three-step, 'time-variant' protocol. The clamp command voltage is the grey line below (holding potential $= -120$ mV, activation potential $= +50$ mV), with consecutive cycles overlaid; values for the first and last (of 10) cycles are shown as solid lines, with the intermediate cycles as dashed lines. The elicited (outward $K^+$) currents are shown as black lines above (capacitance 'spikes' have been removed for the sake of clarity). The small arrows show the 'instantaneous' reactivation levels that, when normalised against the $I_{max}$ value, give the time course for the slow transitions between different closed states (see text). Data kindly provided by Dr Ingela Johansson, Laboratory of Plant Physiology and Biophysics, IBLS, University of Glasgow, Glasgow, UK.

at different potentials and ion/agonist concentrations. In these cases, a predefined protocol is not necessarily the best strategy; rather, it is more usual to collect current data continuously (sometimes recording to a digital tape recorder rather than directly to computer hard disk), with manual control of the clamped potential. Even so, voltage ramp protocols are useful to the patch clamper for estimating the reversal potential of active channels (e.g. Amtmann *et al.*, 1997).

## 2.3 Equipment and hardware

The principal demands on any hardware (and software) used in an electrophysiology laboratory are sharply determined by the requirements of the type(s) of experiment being undertaken. For example, to study the conductivities of single channels (typically several picosiemens), high-precision measurement of currents less than 1 pA and a consequent low background noise (see Sections 2.3.1, 2.4.2 and 2.5.4) are essential, as is good temporal resolution, requiring sampling rates of around 15–20 kHz (Islas & Sigworth, 1999). In the plant kingdom, examples of such channels include CNGCs (Maathuis & Sanders, 2001; see Chapters 7 and 11). For whole-cell patch clamps and TEVC work, where current levels are much higher, lower precision will suffice, and sampling rates seldom need exceed 5 kHz.

### 2.3.1 The working environment

Whether using TEVC or patch clamp techniques, the experimental working environment will be similar, and there are several essential requirements for successful electrophysiology. First, the cell and the buffer in which it is suspended must be accessible with, and under visual control of a microscope, and the capillary or micropipette (see Section 2.3.2) should be attached to a suitable micromanipulator device, such as those supplied by the Sutter Instrument (http://www.sutter.com) and Narishige Scientific Instruments (http://www.narishige.co.jp/main.htm). The choice of a micromanipulator is especially important both in patch clamp work and for single-cell impalements of cells such as root hairs and guard cells, where a precision of a few tenths of a micrometre is necessary. Indeed, good seals, like good impalements, depend critically on truly perpendicular contact between the micropipette and the membrane surface (Vogelzang & Blom-Zandstra, 1998). Micromanipulators are available with manual (knobs or joysticks) or remote (via a computer) control; micromanipulators have been described in more detail by the editor (Blatt, 1991). Because of the delicate nature of the operations required for impaling or patching, all equipment should be mounted on a vibration-free isolation table, such as those manufactured by Melles-Griot (http://www.mellesgriot.com/products/tablesandbreadboards/tablesale.asp).

An inverted microscope is not essential, but can offer some advantages for voltage clamp work, as these microscopes generally permit more open access to the tissue. Visual control of microcapillaries or micropipettes in all cases is imperative when attempting an impalement or patch seal. A magnification of ×400 is sufficient for most applications, although it is extremely useful to have lower powers available: the low magnification can be used for selecting a protoplast and moving the micropipette into its vicinity but the higher power is required for making the contact with the plasma membrane and establishing a high-quality seal.

Second, and perhaps most important, the entire assembly (cell, bath, microscope, vibration-isolation table, alternate bath solution bottles, amplifier headstages, etc.) must be earthed (grounded) and shielded from electrical noise, such as that emitted by the ubiquitous fluorescent tube, computer video monitors, and radio and TV signals. It is standard to surround the workbench with an earthed Faraday cage – a chamber made of either fine metal mesh or solid plates, the interior being accessible through either hinged doors or a copper-mesh-backed roller blind at the front; the finer details of noise reduction techniques have been well described elsewhere (Morrison, 1986). Typical systems, used in our laboratory, are shown in Fig. 2.6.

While it is possible to remove (almost completely) external electrical interference from electrophysiological signals, other random fluctuations, inherent in the system, are more difficult to deal with. The most common of such problem is Johnson noise – random fluctuations in voltage measurements caused by the thermal motion of electrons (in a metal conductor) or ions (in a solution or cell membrane) (DeFelice, 1981). Various electronic means are available to remove such noise, the

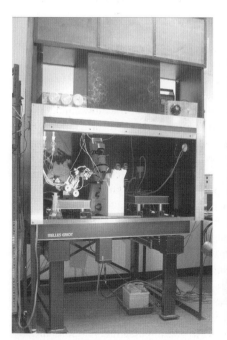
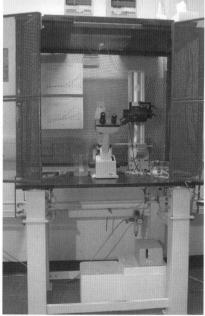

**Fig. 2.6** Representative electrophysiology workbenches. The rig of the left-hand picture is configured for TEVC whole-cell impalement of stomatal guard cells with concurrent fluorescence imaging of cytosolic ion concentrations, and that of right-hand picture is primarily designed for patch clamp experiments.

most common of which are the Bessel filter and Butterworth filter, many models of which are commercially available (see Section 2.3.3). However, some noise will inevitably penetrate such analogue filters, and so it is important that any software chosen provides techniques or tools for post-acquisition digital filtering of data (see Sections 2.4.2 and 2.5.4).

### 2.3.2 Capillaries and micropipettes

For TEVC whole-cell impalements, the microelectrode assemblies serve to transfer current between the voltage clamp and/or amplifier and the cytosol. To do so, they must be physically sharp (in order to penetrate the cell wall and membrane with minimal collateral damage to the cell) and have a non-conducting outer surface (in order to form a high-resistance seal between the capillary glass and the repaired puncture). The most common system used in electrophysiology is the silver/silver chloride (Ag/AgCl) electrode (see Chapter 1). In our laboratory, we make microelectrodes for impalements from borosilicate capillary glass with a triangular cross section, a wall thickness of 0.25 mm and an internal diameter of 1.25 mm (Dial Glassworks, Stourbridge, West Midlands, UK), using a two-stage PD5 horizontal puller (Narishige, http://www.narishige.co.jp/main.htm). The two 'barrels' are clamped into the puller, heated for 20 s and twisted slowly through 360°; after cooling for 30 s, the capillaries are then pulled slowly apart (over 30–45 s), giving two sets of two-electrode assemblies (see also Blatt, 1991).

For the patch clamp, the glass micropipette is possibly the most crucial element when working with plant cells, protoplasts or vacuoles; formation of a good, high-resistance seal between the glass tip and the membrane is vital. In the cell-attached, inside-in and outside-out configurations, a seal resistance of at least 2–3 GΩ is required (the higher the better) and, for ion channels of particularly low conductance, only a seal of 10 GΩ or more will allow sufficiently low-noise recording. In whole-cell mode, lower resistances (~1 GΩ, depending on the membrane's total surface area) suffice, as the currents involved are very much higher (as with TEVC impalement experiments). Removing a patch by pipette suction becomes more difficult as the seal strengthens, and often, this is performed before the resistance reaches 1 MΩ, allowing the resistance to build up afterwards.

The formation of the seal depends on the characteristics of the glass, the shape of the pipette tip and, especially, the membrane properties. Soft soda glass, with a melting point around 800°C, gives higher Johnson noise (see Section 2.3.1), because of its lower resistivity. Coating the pipette tips with a non-conductive material, such as Sylgard (Penner, 1995) or even paraffin wax (Hamilton et al., 2000), can greatly increase the bulk resistance of the pipette and reduce the distributed capacitance near the tip by increasing the physical separation between solutions on either side of the pipette wall. Hard borosilicate glass, with a melting point around 1200°C, is preferable in plant electrophysiology, with its higher resistivity and consequently lower inherent Johnson noise; however, coating with Sylgard or another non-conductive material is still important to reduce capacitance. For impalement work with microelectrodes, coating, e.g. with paraffin wax, can also be important to improve electrode response times (Blatt, 1991).

The issue of which type of glass to use is still very much an open question. Some experiments with animal cells seem to show no difference in the quality of measurements, and borosilicate glass is generally assumed to be more efficient in the formation of a gigohm seal with artificial membranes (Penner, 1995), but other authors have reported efficient sealing to plant protoplasts only with soda glass (Keunecke *et al.*, 1997). In this latter case, seal formation was extremely slow (taking up to 1 h). However, a major problem with soda glass is that it releases cations (such as $Ba^{2+}$), which can poison the ion channels (Copollo *et al.*, 1991); internal coating of the pipette tip can solve this problem. In our laboratory, we mainly use borosilicate glass, such as Kimax-51 capillaries (Kimble-Kontes, http://www.kimble-kontes.com). These have an outer diameter of 1.5–1.8 mm and a wall thickness of 0.2 mm. These capillary blanks lack an internal glass fibre, often included in microelectrode capillary tubing to aid filling, which can interfere with patch formation. When making micropipettes (see the discussion in Chapter 1), a useful tip is to heat the ends of the capillary before pulling, to exclude sharp surfaces; this helps to avoid scratching the electrode wire and the holder.

It is noteworthy that formation of a high-resistance seal in patch clamp work is much more difficult to achieve with the plasma membranes of plant cells than in the case of mammalian cells; also, it is simpler with the tonoplast than with the plasma membrane. Numerous factors impact on patch sealing, including the physiological state of the plant, the procedure used for removal of the cell wall, and the composition of the bath solution (sometimes a high concentration of $Ca^{2+}$ can increase sealing efficiency) (Vogelzang & Blom-Zandstra, 1998). When forming a whole-cell patch, a suction pulse is more effective for the plasma membrane but, in our experience, an electric pulse is usually more efficient for the tonoplast. In fact, obtaining high-quality seals reproducibly is something of a 'black art': periods of bad seal formation can occur, lasting for several weeks; these are not generally related to the quality of micropipettes or chemicals used, but rather to the physiological state of the plants, their plasma membranes and their cell walls. An interesting observation is that activity of $K^+$-selective, large-conductance channels in guard cell plasma membranes was detected when isolating protoplasts with laser microsurgery, but not when using traditional enzymatic digestion of the cell wall (Miedema *et al.*, 1999).

## 2.3.3   Electronics

The essential electronics required for voltage clamp work are (i) the voltage clamp itself, (ii) a high-gain, high-fidelity amplifier/ammeter and (iii) some form of active analogue filter unit. Several well-established companies now supply excellent hardware in this respect, some of the more popular of which are summarised here:

HEKA Elektronik GmbH (http://www.heka.com) has long supplied the EPC series of patch clamp amplifiers, which have built-in active filter units. The earlier models (EPC7 and EPC8) have only manual control, but the latest in the range (the EPC10) is fully computer controlled.

Axon Instruments (http://www.axon.com/cn_Neuroscience.html) produces a number of voltage clamp modules with integrated amplifiers; the most notable of these are the Axoclamp-2B (particularly suitable for *Xenopus* oocyte TEVC work), the Axopatch 200B (ideal for single-channel patch clamp studies) and the GeneClamp 500B (an excellent, general-purpose module, providing both TEVC and patch clamp functions).

A-M Systems, Inc.® (http://www.a-msystems.com/physiology/Instruments) manufactures a number of low-cost patch clamp (and other) amplifiers and associated electronic equipment.

Kemo® Limited (http://www.kemo.com) specialises in high-performance laboratory filters; it produces a large range of models, of varying cost and performance. The theory (Martin, 1991) and design (Horowitz & Hill, 1980) of active analogue filters are outside the scope of this chapter; briefly, the most commonly used today are the four-pole and eight-pole *Butterworth* and *Bessel* filters.

We (http://www.gla.ac.uk/ibls/BMB/mrb/lppbh.htm) also manufacture our own voltage clamps and amplifiers, and have done so for nearly two decades. Of particular note here is that our equipment caters especially for the plant scientist, whereas most of that commercially produced is more oriented towards mammalian and medical studies. One issue arising from this is the effective range of voltage clamp command potentials – for example, the Axon systems generally offer the facility to clamp voltage over the range ±200 mV but plant cell work frequently involves clamping at more negative potentials, even as low as −300 mV, and seldom more positive than +100 mV. Potentials below −230 mV have been recorded in the root cells of *Arabidopsis thaliana*, which may play an important role in channel-mediated potassium uptake from solutions with low potassium concentrations (Hirsch *et al.*, 1998). The membrane potentials of guard cells can exceed −300 mV in some circumstances (Blatt *et al.*, 1987; Thiel *et al.*, 1992), and voltages as low as −350 mV have been recorded in *Chara corallina* (Pesacreta & Lucas, 1984). In such circumstances, it is clearly important to be able to clamp at voltages both above and below these values.

### 2.3.4 Data acquisition and control boards

In order for a (digital) computer to control and record (analogue) current and voltage signals, special hardware is required, which requires basic circuitry for digital-to-analogue (DAC; output) and analogue-to-digital (ADC; input) conversion. To maintain sampling at precisely determined and often very high frequencies, this hardware should include its own sampling clock, which triggers each input conversion, and the ability to transfer large amounts of data directly to the computer's memory, without having to depend on the system's central processor unit. Most modern data acquisition boards implement this latter process, known as direct memory access (DMA), although there are some notable exceptions, as discussed below. Another requirement is an on-board, user-programmable clock that can (directly or indirectly) trigger DAC (output) conversions, so controlling the time-dependent clamp command voltage when running a protocol.

There is a vast range of suitable hardware available, varying greatly in cost and precision – one generally being proportional to the other. In many cases, the choice of board depends on the other hardware and, more frequently, the software in use (see Section 2.4), as many commercial packages are specifically designed to operate only with vendor supplied or supported boards. However, with customised systems, there are three important factors to bear in mind when choosing the type of data acquisition and control hardware.

First, the data precision of the board should be considered. Most available boards have an analogue input range selectable from a maximum of $\pm10$ V to as low as $\pm1$ V (or even less). For whole-cell impalement of guard cells, the input signal amplifier would typically be set to 100 mV/V for monitoring the voltage ($V$) electrode and 1 nA/V for the current ($I$) electrode; an input range of $\pm5$ V would thus give suitable working ranges of $\pm500$ mV ($V$ electrode) and $\pm5000$ pA ($I$ electrode). (By comparison, for whole-cell *Xenopus* oocyte work a gain of 40 $\mu$A/V would be more appropriate, giving a working current range of $\pm100$ $\mu$A.) Digitised signals are represented as 2-byte binary integers, with the data range (precision) varying between boards. The maximum precision (full 16-bit) divides the input range into 65 536 ($2^{16}$) intervals – giving a resolution (for the parameters we have just outlined) of 15 $\mu$V ($V$) and 150 fA ($I$); lower cost boards have only 12-bit precision, dividing the input range into 4096 ($2^{12}$) intervals – the corresponding resolution is 0.25 mV ($V$) and 2.5 pA ($I$). For recording whole-cell currents, this lower precision is sufficient but, for single-channel work, the higher precision is essential.

Second, the maximum sampling rate of the board must meet the requirements of the experiments to be undertaken: generally, the rate at which data are sampled should be an order of magnitude greater than the most rapid transitions under investigation. Today, almost all available boards can reliably collect (and transfer) data at rates in excess of 100 kHz (100 000 sample per second), which is certainly sufficient for TEVC, whole-cell and many single-channel patch clamp investigations. Some circumstances, however, may demand higher sampling rates – in such cases, the speed of data storage (writing to hard disk) should also be considered (a typical modern PC can reliably store data to disk at rates of 200–500 kHz).

Finally, one should consider the versatility of the board and, if considering developing your own software (a task not to be undertaken lightly), then the ease of programming is important. For example, a board that works well for patch clamp experiments with one software package may be entirely unsuitable for TEVC work, or with different programs.

### 2.3.4.1 *Scientific Solutions' LabMaster® Boards*
Since 1981, Scientific Solutions (http://www.labmaster.com) has been manufacturing high-fidelity laboratory interfaces for PCs. The LabMaster series of boards are ideally suited for electrophysiological applications, although (surprisingly) none

of the commercially available packages supports them (Section 2.4). However, the hardware (Section 2.3.3) and software (Section 2.5) developed in our laboratory are largely based on the LabMaster DMA, which now serves as the type-defining ADC/DAC board for our other hardware interfaces (see Section 2.5.6). Unfortunately, although this board (and the higher precision LabMaster ADEX) is still perfectly well suited for many voltage clamp experiments, it is built using the ISA specification. The ISA interface (between the host computer and the attached hardware) is fast becoming obsolete on new PCs. The newer LabMaster PRO, with the more modern PCI interface, is the natural successor, and this shows great promise as a replacement in the next generation of systems.

### 2.3.4.2  Instrutech Corporation's ITC Interfaces
Instrutech Corporation (http://www.instrutech.com) manufactures the ITC-16 (now no longer produced), ITC-18 and ITC-1600 interfaces, which are the primary choices of ADC/DAC board for the HEKA Pulse software (Section 2.4.4.1). However, while the boards are extremely reliable and efficient, they lack the versatility of the Lab-Master series, and those discussed below, lacking DMA, an on-board event clock and independent control of analogue input and output.

### 2.3.4.3  Axon Instruments' DigiData Systems
Axon Instruments (http://www.axon.com) initially relied on a LabMaster board as the core of their DigiData 1200 system, and their subsequent models, the DigiData 1200A and 1200B, have many features in common with the LabMaster ADEX; they are supported (unsurprisingly) by all Axon software (Section 2.4.4.2) and by HEKA software. However, as with the LabMaster DMA/ADEX, they have the ISA interface and are not suitable for newer PCs. Their successor, the DigiData 1320, uses an external SCSI (small computer systems interface) connexion to the host PC, which may very well suffer a fate similar to the (internal) ISA interface in the not-too-distant future.

### 2.3.4.4  National Instruments' Cards
The National Instruments Corporation (http://digital.ni.com) has been producing a wide range of data acquisition hardware for many years. Currently, its PCI 'E' and 'S' series boards are the most suitable for electrophysiological applications. It provides a large choice of boards, at varying cost and performance specifications. Both 12-bit and 16-bit analogue input and output are available, with sampling rates up to a maximum of 1.25 MHz.

### 2.3.4.5  Data Translation
Data Translation (http://www.datatranslation.com) is probably the largest manufacturer of data acquisition hardware, and it produces an almost incredible range of suitable boards. The most appropriate for electrophysiology are the DT-300 and DT-3000 series, both of which utilise the PCI interface. As with the LabMaster boards, however, its products are *not* supported by commercially available packages (but see Section 2.5.6).

### 2.3.5 Choosing a computer

Low-cost PCs of sufficient power for experimental electrophysiology are now readily available from many manufacturers at relatively low cost. Two main 'flavours' exist: the IBM-compatible PC (generally running one of the Microsoft Windows® operating systems) and the Apple Macintosh™ (with its own operating system). With the exception of the HEKA Pulse software (Section 2.4.4.1), commercial and other software are available almost exclusively for the PC/Windows combination. Other important considerations that heavily favour this model are its inherent compatibility with data-acquisition hardware, and the fact that many manufacturers do not supply device drivers for the Macintosh system.

## 2.4 Computer software

### 2.4.1 Basic requirements

For TEVC and whole-cell patch clamp experiments, it is vital for software to support the easy creation and modification of multi-step and voltage ramp protocols and to allow specification of when data collection should start and stop, the sampling frequency, and from which analogue input channels data should be collected. All these parameters need to be immediately accessible to the user and easily adjusted 'on the fly' during recording sessions. Although it is possible to create relatively complex protocols in all of the software packages reviewed below (Section 2.4.4), we have yet to find any that matches the protocol editor of the Henry II application (Section 2.5.2.1), for its ease of use, its accessibility during recording sessions and its lack of limitations in protocol design.

During the active run, the progress of the clamp and elicited currents must be visible to the electrophysiologist in 'real time' (i.e. as it happens), allowing the possibility of aborting the operation in certain circumstances (such as cell or equipment failure). On successful completion of the run, the collected data should be stored in a filing system with an orderly naming/numbering convention; an extremely useful feature is the option to either accept or reject data after (immediate) visual inspection, thus removing unnecessary strain on (frequently overloaded) computer hard disks.

### 2.4.2 Signal conditioning

Despite the precautions that can be taken to reduce interference (Section 2.3.1), it is inevitable that some noise will be present in the majority of recorded signals. Data collection and/or analysis software should include some means of reducing this. As a minimum, basic signal smoothing (Gaussian filtering) should be provided; however, more elaborate filters, based on the Fast Fourier Transform (FFT), are far more powerful. Such FFT digital filters include the high-pass and low-pass filters (respectively removing all signal variations with frequencies below or above a specified cut-off), the band-pass filter (allowing signals only within a specified frequency range) and the notch filter (the logical converse of the band-pass, especially useful for removing 'regular' noise such as 50/60-Hz mains-induced or 75-Hz video-monitor-induced interference). Whatever software system is in use, the N-Pro V2 application (Section 2.5.4) is a useful addition in this respect.

### 2.4.3  Data analysis tools

#### 2.4.3.1  IV analysis

One of the most elementary, yet elucidative, tools for the analysis of all charge transport processes, including voltage-activated ion-channel currents, is the current vs voltage ($IV$) curve (see also Chapter 1). For ion channels, for example, the current, $I$, flowing through a membrane (or single channel) can be approximated by

$$I = G(V - E_{rev}) \tag{2.1}$$

where $V$ is the membrane potential, $E_{rev}$ is the reversal potential (see Section 2.2.3) and $G$ is the conductance (of either the single channel or all channels in the membrane). For a channel selective for a single ion (X), $E_{rev}$ is numerically equal to the equilibrium potential, $E_X$, which is related to the intracellular, $[X]_{in}$, and extracellular, $[X]_{out}$, concentrations of that ion, and its charge ($z$) by the Nernst equation

$$E_X = \frac{RT}{zF} \ln \frac{[X]_{out}}{[X]_{in}} \tag{2.2}$$

Far from being a constant value, the conductance $G$ is generally both voltage- *and* time-dependent, and so values measured at equilibrium are, in reality, the sum of both the steady-state ($I_\infty$) and instantaneous ($I_0$) currents (see Section 2.2); so, when performing $IV$ analysis, the data plotted should, wherever possible, be $[I_\infty - I_0]$ vs $V$ (see Fig. 2.7). In many (if not most) studies of voltage-activated channels, a key element is the determination of how $G$ varies with $V$ (see Chapter 1). The $IV$ curve for the data of Fig. 2.1 is shown in Fig. 2.8a.

Although the $IV$ curve can provide startling visual insights into the behaviour of voltage-activated channels, a more direct representation of the underlying voltage sensitivity is provided by the conductance vs voltage ($GV$) curve, shown in Fig. 2.8b. Current data can be readily converted to conductances by a simple rearrangement of Equation 2.1

$$G = I/(V - E_{rev}) \tag{2.3}$$

where the value of $E_{rev}$ can be estimated either by interpolation of the $IV$ curve (as the point at which the current changes sign) or approximated by assuming $E_{rev} = E_X$ from Equation 2.2 (where $[X]_{out}$ is experimentally controlled and $[X]_{in}$

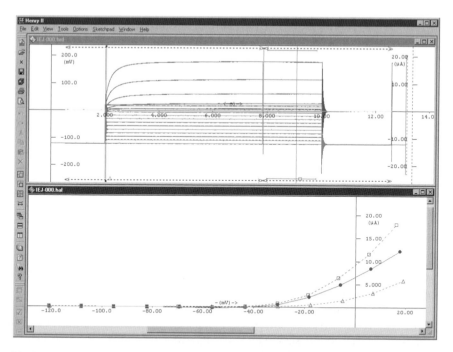

**Fig. 2.7** Voltage-induced activation of the *Saccharomyces cerevisiae* TOK1 channel expressed in *Xeno-pus* oocytes (current and voltage in upper window) and the corresponding $IV$ analysis (lower window). The *instantaneous* currents ($I_0$) for each voltage are shown as open triangles, the *steady-state* ($I_\infty$) currents as open squares, and the *difference* ($I_\infty - I_0$) as solid circles. The data shown are as in Fig. 2.1.

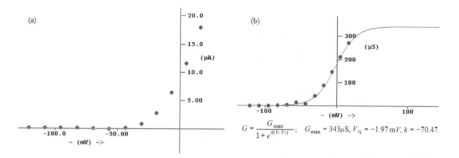

**Fig. 2.8** (a) The current vs voltage ($IV$) curve and (b) conductance vs voltage ($GV$) curve for the *Saccharomyces cerevisiae* TOK1 channel data of Fig. 2.1. The smooth line is the least-squares fitted Boltzmann sigmoid function, with the parameters as shown; the value of $-70.47$ for $k$ corresponds to a gating charge of $-1.86$ (see Chapter 1 and Section 2.4.3.2 of this chapter for an explanation).

can be estimated from physiological experience); the $GV$ curve of Fig. 2.7b was plotted using a value of $-50$ mV for $E_{rev}$.

### 2.4.3.2  Curve fitting

Another essential analytical tool, especially for investigating the kinetics of TEVC and whole-cell patch data, is the ability to fit a variety of equations to current time courses and $IV$ curves. For an excellent introduction to the principles of fitting equations to electrophysiological data, see Dempster (1993) and references therein. Some of the more common functions used in fitting curves to electrophysiological data are outlined below

$$I = A_{\infty} + A_1 e^{-t/\tau_1} + \cdots + A_n e^{-t/\tau_n} \tag{2.4}$$

Equation 2.4, the sum of $n$ exponentials relaxing to a constant value, is of widespread applicability in electrophysiological studies directed toward understanding the kinetics of transport and of gating (see also Chapter 1). Examples include the inwardly and outwardly rectified ion currents flowing through voltage-gated membrane channels, such as the data of Figs. 2.1, 2.2 and 2.5.

When fitting such exponential curves to data, there are several important points to be considered. First, the functions are statistically ill-defined – that is to say, if the range of the experimental data is insufficiently extensive, then the calculated parameters are likely to be largely inaccurate: generally, data should span the time domain from at (or near) zero to at least two or three 'half-life' $(1/\tau)$ values. Second, unless there is a sound biological reason for doing otherwise, the number of sums $(n)$ should be kept to a minimum – never use two exponentials when one will suffice, or three when two will give equally good approximations to the data. If two or more $\tau$ values are very similar, or if any of the $A$ parameters are very small compared to the others, then it is likely that too many terms are being used.

$$I = G_{max}(V - E_{rev})/[1 + e^{k(V - V_{1/2})}] \tag{2.5}$$

$$G = G_{max}/[1 + e^{k(V - V_{1/2})}] \tag{2.6}$$

The Boltzmann sigmoid functions are important in the interpretation of current vs voltage (Equation 2.5) and conductance vs voltage (Equation 2.6) analyses. Here the $k$ parameter in these formulae corresponds to the values normally expressed as $\delta Fz/RT$, where $F$, $R$ and $T$ have their conventional meanings, $z$ is the ionic charge (valency) and $\delta$ is the apparent 'gating charge' or 'voltage sensitivity coefficient' for the channel. It is important to remember that least-squares curve-fitting algorithms are particularly prone to finding 'false minima' for these functions, especially when using unsuitable initial estimates of the $V_{1/2}$ and $E_{rev}$ parameters.

Commercially available software invariably provides some support for curve fitting, but the ease of use, robustness and versatility of the implementations varies greatly between packages (see Sections 2.4.4 and 2.5.2.3).

### 2.4.3.3   Single-channel analysis

The analysis of single-channel data from patch clamp recordings is very much more complex. The determination of conductance levels and open probabilities using current–amplitude histograms – perhaps the simplest analyses – are supported by most major applications. However, event (open-closed state transition) detection and dwell-time histograms still involve largely state-of-the-art mathematical and statistical techniques. Among the most powerful of these, hidden Markov modelling (Fredkin & Rice, 1992; Venkataramanan & Sigworth, 2002), is not (yet) implemented in any of the major commercial packages. (By the time of publication, these techniques should be available in Henry II's EP Suite – see Section 2.5.5.) Even with these statistical tools, it is sometimes necessary to manually determine transition events by visual inspection, although this can be extremely labour-intensive; this is frequently the case when channel conductances are of similar magnitude to noise levels, or when open times are very small.

### 2.4.3.4   Data export

Although a variety of primary data analysis tools, as discussed above, should be available within any electrophysiological software, it is often desirable (and sometimes necessary) to perform more customised analyses with third-party statistical or mathematical software. Once analysed, it is generally desirable to produce charts, tables or graphs of publication quality. Thus, it is essential for any software package to be able to export binary data to a form readable by other, word-processing, mathematical and graphical software, such as Microsoft Excel™, PowerPoint™ and SigmaPlot™.

### 2.4.4   Commercially available software

#### 2.4.4.1   Pulse+PulseFit (HEKA Elektronik GmbH)

The Pulse+PulseFit application, together with various utilities, forms a widely used, robust and scientifically powerful electrophysiological control and analysis system. The software is available for the Apple Macintosh and Microsoft Windows operating systems although, for the latter, the user interface is far from standard (see Fig. 2.9).

   We have used the Pulse+PulseFit program extensively for many years, running a variety of protocols (which are easy to change during an experiment); in this time, we have *never* experienced a 'crash', or any other problem resulting in a loss of data – a remarkably unusual feature among Windows applications, as many readers will know only too well!

   A major drawback to this software is the seemingly unnecessary complexity of the user interface and data output structures and the complete lack of online help; in view of such complexities, this seems to us to be rather a pity. However, in compensation, the package ships with a comprehensive, printed user manual, and the company provides useful technical support, via their Web site. The user interface is itself difficult to master, comprising a seemingly endless combination of windows

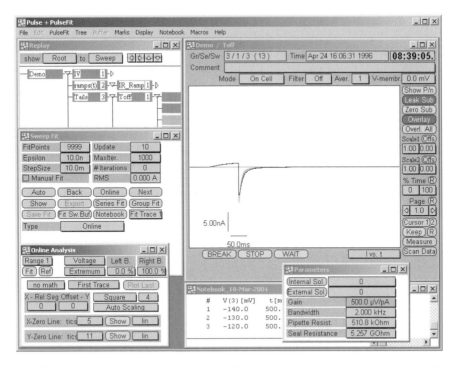

**Fig. 2.9** A typical display from the 'Pulse+PulseFit' application, illustrating the non-standard style of the user interface under the Microsoft Windows operating system and the large number of 'button' controls and windows.

and buttons; even after many months of practice, colleagues are often not well versed in the use of the software.

As with almost all software, the Pulse+PulseFit program stores its binary data in a unique (and extremely complex) format, not readily accessible to third-party or user-developed programs. However, a plug-in module is available for the SigmaPlot™ application that enables the binary data files to be directly imported; also, programs in Henry II's EP Suite (Section 2.5) can read/write HEKA binary files.

### 2.4.4.2 The pClamp Suite (Axon Instruments)

The ClampEx and ClampFit programs (version 9.0 at the time of writing) are the latest in a long line of programs originally developed in academia (Kegel *et al.*, 1985). These two applications together comprise an extremely comprehensive data acquisition and analysis system.

The applications' user interfaces are user-friendly, compatible with most Windows applications and easy to learn in a short time. However, the programs are designed primarily for 'standard' applications; for example, the protocol editor offers less opportunities for unrestricted changes than that of the HEKA Pulse+PulseFit suite. It is also very poor in provision for 'on the fly' adjustment during sessions, a factor that can cause loss of much useful recording time. Equally important, the software is not particularly versatile with respect to hardware manufactured by companies other than Axon, and the marketing strategy clearly concentrates on mammalian electrophysiology and standardised, high-throughput pharmacological assays.

In 1993, Axon Instruments published an invaluable guide to the basic principles of practical electrophysiology, 'The Axon Guide', which can now be downloaded (as a PDF document) from their Web site (http://www.axon.com/mr_Axon_Guide.html). This guide covers, in some detail, many of the topics overviewed herein, such as the construction of microelectrodes and voltage clamp theory.

A distinct advantage of the pClamp software is the format of its data files. A large number of third-party analysis programs support this Axon Binary File format, including SigmaPlot (http://www.spssscience.com), Data Access (http://www.bruxton.com) and Henry II's EP Suite (see Section 2.5).

### 2.4.4.3   Other commercial packages

Cambridge Electronic Designs (http://www.ced.co.uk) seems to have shifted the emphasis from its Vclamp and Patch applications to its new, all-embracing Spike2 software; this latter software caters for such diverse fields as electrophysiology, general neuroscience, sleep studies (*sic*) and behavioural research. The data acquisition software supports only its own devices (the CED-1401 series).

Bruxton® Corporation (http://www.bruxton.com) produces the Acquire and TAC programs that, respectively, provide for data acquisition and analysis of single-channel patch clamp experiments. The software supports ADC/DAC boards provided by Instrutech (Section 2.3.4.2), Axon (Section 2.3.4.3) and National Instruments (Section 2.3.4.4).

### 2.4.4.4   Whole Cell Patch

Although this software, written by John Dempster at the University of Strathclyde, is not, strictly speaking, commercial (it is freely downloadable to academic users from his Web site, http://www.strath.ac.uk/Departments/PhysPharm/ses.htm), it has been extensively used in the academic community for some two decades. Even if your system is based on one of the commercially produced packages, this software is a useful addition, and works with a variety of hardware systems. Dr Dempster has also written an excellent guide to the principles involved in the analysis of electrophysiological data (Dempster, 1993).

## 2.5   Henry II's EP Suite

### 2.5.1   Overview

In Section 2.4.4, several of the more widely used electrophysiological applications have been briefly reviewed, and there are many others available. Increasingly, commercial software packages are becoming 'closed systems', implicitly linked with the manufacturers' hardware, and freeware programs, like Whole Cell Patch, written by scientists rather than professional programmers, are suffering from the ever more complex environment of the Windows operating systems. Our laboratory has been committed, for some years now, to developing flexible and high-quality software catering specifically for, and with the feedback of, research-active electrophysiologists with interests in plant science. The remainder of this chapter discusses some of the more important aspects of the fruits of this project, Henry II's EP Suite.

The software runs under Windows 98$^{TM}$, Windows ME$^{TM}$, Windows 2000$^{TM}$ and Windows XP$^{TM}$, and is designed specifically for flexibility and speed of use during experimentation. The component applications take full advantage of the 'point, click and drag' aspects of the graphical user interface (GUI) for designing voltage and current clamp protocols, and for selecting acquired data for analysis. Utilities include user-defined analysis envelopes, which provide all-point or time-averaged $IV$ plots in real time during data acquisition, a variety of post-acquisition data-selection, analysis and curve-fitting tools, digital filtering and export of data to a number of standard file formats. Extensive online help is provided within the suite, but students in our laboratory, research visitors and other colleagues using the software outside the laboratory have found the software to be intuitive and easy to use.

The suite comprises four main applications and a modular library of drivers, providing the interface between the host computer and the attached clamp control and data acquisition hardware. Each of these five components is discussed in detail in the following sections. The suite maintains a 'common interface' between its applications, providing a broad spectrum of user-selectable options, such as display colours, fonts and axes. Perhaps more importantly, the software maintains the 'standard' expected of professional Windows applications, such as dynamic menus, graphical toolbars, easily memorised keyboard 'shortcuts' and efficient use of the mouse for viewing and zooming data sets.

Another important feature, common to all component applications, is the sketch-pad. This is, essentially, a bitmap window, on which any active view – whatever that may be – can be copied or overlaid, thus enabling the visual comparison of, say, the $IV$ response of a cell to a given stimulus before and after changing its environmental conditions.

Henry II's EP Suite and the stand-alone components are available for free download from the laboratory Web site (http://www.gla.ac.uk/ibls/BMB/mrb/lppbh.htm). There are typically three to five updates posted on the Web site per year, with each release incorporating new features and/or support for additional hardware. Many recent developments have originated from requests and/or feedback from colleagues within our own laboratory and those from other establishments. Comments and requests are welcome through the Web site and are dealt with expediently.

### 2.5.2 The Henry II application

This is the 'core' application of the suite, providing tools for designing and running voltage (or current) clamp experiments, interactive communication with a variety of

ADC/DAC converter boards installed on the computer, efficient data-collection, run-time clamp monitoring and many analysis and data-selection tools (see Fig. 2.10).

### 2.5.2.1   The protocol editor

Henry II's protocol editor allows the design of voltage or current clamp experiments using an intuitive GUI, whereby clamp values, times, sampling parameters and digital output signals can be defined and changed using the mouse in a 'point-and-press' and 'drag-and-drop' system. When protocol file is opened or created, a view window like the one shown in Fig. 2.11 is displayed; this is a *Protocol Edit Window*. (The colours used for the various components, the background, and the axes' fonts are the program's defaults; they can be changed on a per-user basis; also, the protocol shown is rather more complex than would normally be used, in order to demonstrate the available features.)

Each of the components can be 'selected' by *clicking* the mouse at (or near) the particular element in question. When selected, a component can be modified by *dragging* the mouse, deleted, copied into the program's clipboard or cut (a

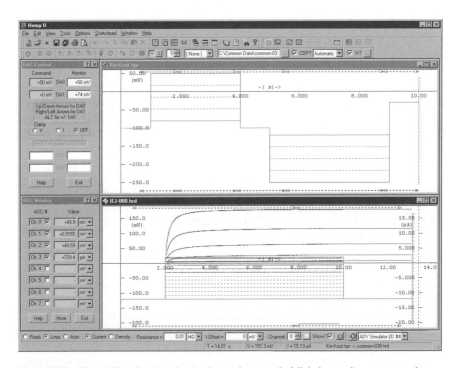

**Fig. 2.10** The 'Henry II' application, showing interactive control of digital-to-analogue output and monitoring of analogue-to-digital input (left-hand windows), the protocol editor (upper-right-hand window) and a pre-recorded data set (lower-right-hand window), to which a set of curves (each the sum of two exponentials) has been fitted.

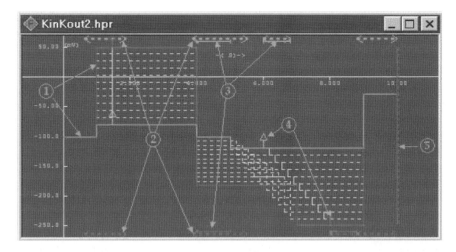

**Fig. 2.11** A protocol editor window, showing the various elements of a protocol (not all of which need be present): (1) The clamp trace for the first (green) and last (red) cycles of the protocol defines the voltage (or current for current clamp mode) values and time course over which the clamp will be driven; the calculated traces for the intermediate cycles are shown as dashed yellow lines. (2) Sampling windows for the first cycle (top of window) and last cycle (bottom of window) define the time periods for sampling and sampling frequencies to be used. (3) Analysis envelopes within sample windows (an optional component) are used primarily for the pre-selection of data within sample windows for run-time and post-run analysis. (4) Triggers/signals (shown for the first and last cycles and expanded to cycles between) permit optional time points at which a specified byte value is sent to the digital output port on the ADC/DAC board for communication and synchronisation with other equipment. (5) The protocol end-marker – this is not actually part of the protocol; rather, it is an editor tool that can be used to 'stretch' or 'shrink' the entire protocol (and all its component elements) along the time axis. Some examples of protocols generated with this tool are shown in Figs. 2.1, 2.2, 2.5 and 2.10. This figure is reproduced in the colour plate section, see Plate 5.

combination of copy followed by delete). The parameters for a sampling window or trigger/signal can be specified by *double-clicking* the element in the protocol editor; this opens a pop-up dialogue box, wherein all relevant parameters (sampling frequency, number of channels, etc.) can be entered textually. Double-clicking other elements in the protocol – clamp trace 'holds' (horizontal lines), 'steps' (vertical lines) and 'hooks' (corners) – causes an appropriate pop-up dialogue box to appear; all these dialogues allow the specification of the times (and $V$ or $I$ values for clamp-trace components) for the first and last cycles. Double-clicking the end-marker opens a protocol options dialogue box, wherein one can specify details such as how to step between cycles and where to save collected data. When a component has been cut or copied into the clipboard, it can then be pasted (inserted) into the protocol at a specified insertion point; new components can be inserted at any point in the protocol, and elements can be copied from one protocol to another. An important feature of the protocol editor is the ability to undo changes (mistakes), in much the same way as in typical word-processor applications.

### 2.5.2.2   Run-time monitoring and analysis

When running a protocol, the progress is displayed in a run monitor window; a run-time $IV$ plot (all-point or time-averaged) can also be shown. Optionally, the final data display (either voltage and current vs time or current vs voltage) can be added to the sketchpad, thus comparing the data from consecutive protocol runs (see Fig. 2.12). The arrangement of these three 'run-time' windows is fully customisable to the user's preferences.

On completion of a protocol 'run', the full set of collected data is displayed, at which point there is an option to accept or reject the data; on acceptance, the data are stored, in binary format, to a file with a name comprising a user-specified 'root' and an automatically appended, numerical suffix, for ease of future reference. This data file contains a complete copy of the original protocol; at any time in the future, the program can *extract* this protocol into the protocol editor, where it can be modified, saved as is, or rerun – particularly useful for repeating an experiment when the original protocol file is not available.

### 2.5.2.3   Post-acquisition data analysis

The Henry II application provides a wide range of tools for analysing data sets, including a variety of $IV$ analyses (plotting the command clamp potential against either current or conductance), data normalisation (whereby amplitude differences in elicited current traces can be eliminated, thus making it easier to find kinetic

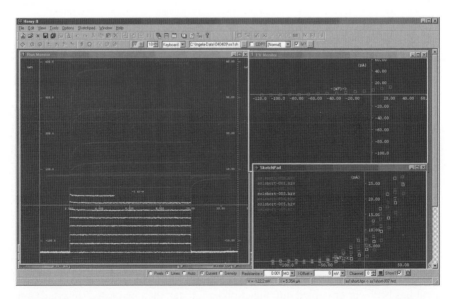

**Fig. 2.12** The 'run-time' display, showing the 'in-progress' clamp monitor (left-hand window: voltage protocol is shown in background grey, recorded voltage in yellow, recorded current in blue), the real-time $IV$ analysis for this data set (upper-right-hand window) and the sketchpad, on which the $IV$ plots of previous voltage clamp data sets are shown (colour coded to the overlaid filename list). This figure is reproduced in the colour plate section, see Plate 6.

differences) and efficient and robust curve fitting. Data can be readily exported for use by third-party statistical and mathematical applications, and if the computer has either Microsoft Excel or SigmaPlot installed, then the program can write data directly (and rapidly) to the file formats required for these applications.

One of the most useful tools for the analysis of the kinetic properties of ion channels is the ability to fit curves to measured data. Although many software packages provide this function, the curve-fitting tools of the Henry II application are remarkably easy to use, yet versatile, mathematically robust and extremely efficient (for example, on a 450 MHz Pentium III computer, fitting the sum of three exponentials jointly to the combined data sets shown in Fig. 2.13, comprising some 100 000 data points, takes only a few seconds).

The Henry II application also provides comprehensive, yet easy to use, *IV* analysis tools (see Section 2.4.3). Data can be plotted during a protocol run or post-acquisition, as all-point 'scatter' diagrams or window-averages, and there is

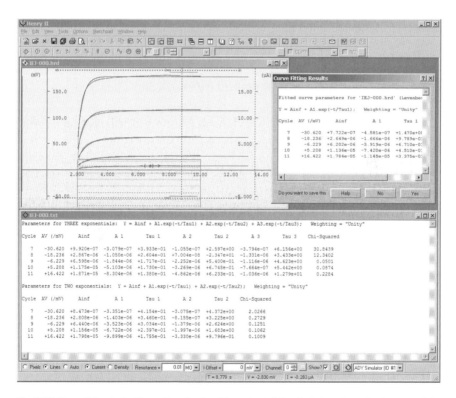

**Fig. 2.13** Curve fitting in the Henry II application. The upper-left-hand window shows the data set and the curve fit to the recorded currents currently under review (a single exponential function, fitted parameters shown in the active dialogue box to the upper-right-hand box). The lower (text file) window shows the results from previous curve fits, using the sum of two and three exponentials, indicating that the most suitable fit is the sum of two exponentials.

a dynamic, cursor-controlled mode for automatic subtraction of instantaneous currents from steady-state values (see Fig. 2.8). Once an $IV$ analysis has been created, a variety of further analytical tools are available, including curve fitting, data selection and de-selection, conductance–voltage plots, and export of data to a variety of third-party formats.

### 2.5.3 The Vicar V2 virtual chart recorder

The Vicar V2 (Fig. 2.14) application provides the means to record analogue input data over a long period. Designed specifically to remove the need for a physical chart recorder, the program also provides some additional features, such as the ability to send simple commands to the voltage clamp hardware and inserting predefined comments into the record with the press of a key (other strings can be inserted via a pop-up dialogue box).

Unlike some other 'chart recorder' software, Vicar V2 can continue recording when a voltage clamp protocol is active (in the Henry II application); in such cases, the names of both the protocol file and the stored data file are included automatically on the chart record (see Fig. 2.15).

### 2.5.4 Noise reduction and removal with N-Pro V2

This stand-alone program allows the analysis and/or removal of noise in or from electrophysiological data recorded in a wide variety of file formats. The application includes traditional digital filtering (notch and band-pass) and smoothing (Gaussian) techniques, as well as the less-used (in electrophysiology) Savitsky–Golay smoothing (Savitsky & Golay, 1964) – which, unlike Gaussian smoothing, preserves peak width and sharpness – and a novel 'noise profiling' technique, especially useful for removing semi-regular mains-induced noise from patch clamp records (see Fig. 2.16). These tools have proved particularly useful to colleagues working with

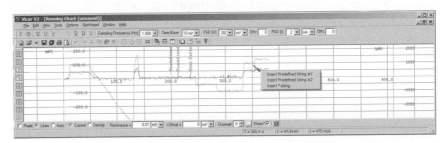

**Fig. 2.14** The Vicar V2 application during recording. Note the vertical toolbar to the left of the picture, whereby the input channels to be monitored can be quickly selected, and the horizontal toolbar under the Tools menu, whereby up to eight predefined comments can be written to the record with a single function key stroke. The pop-up menu shown slightly to the right of centre is activated by clicking the mouse at any point in the record. This figure is reproduced in the colour plate section, see Plate 7.

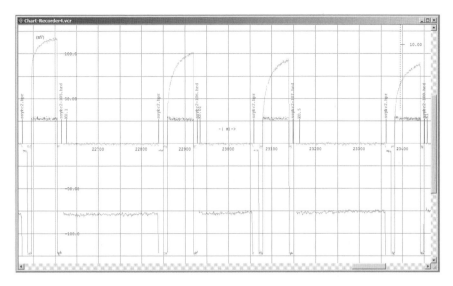

**Fig. 2.15** A section of a chart recorded with the Vicar V2 application, showing how the program can record data collected while running a protocol in Henry II. At the start of each run, the chart displays the name of the protocol file (ssykc2.hpr) and, on completion, the name of the saved data file (e.g. ssykc2-107.hrd for the run to the right of centre). At any stage, the user can insert comments, just as if writing on a chart-paper record (e.g. the 'K0.1', 'K0.01', 'K0.5' remarks, indicating a change in the potassium concentration of the bath solution). Data from two analogue input channels are displayed – the *V* electrode (red) and the *I* electrode (green). Data courtesy of Dr Ingela Johansson, Laboratory of Plant Physiology and Biophysics, IBLS, University of Glasgow, Glasgow, UK. This figure is reproduced in the colour plate section, see Plate 8.

other software on single-channel currents (Hamilton *et al.*, 2000; Köhler *et al.*, 2003).

At the time of writing, the program can read (and save) data in the following third-party file formats: Axon Binary Files, Bruxton Exchange Format, HEKA Pulse files and Whole Cell Patch (Strathclyde Electrophysiology Software), and there is limited support for files in the Cambridge File System (as used by CED software). If required, support for additional file formats can be added on request.

### 2.5.5   *The Pandora! application*

This new program, which at the time of writing is still very much in the developmental stages, provides a variety of analytical and modelling tools for the electrophysiologist. For example, the software allows the creation of 'model' *I V* curves (see Fig. 2.17) using a variety of established formulae (e.g. Amtmann & Sanders, 1999).

Currently under development is a collection of tools for the analysis of data from single-channel recordings. This includes traditional methods, such as current–amplitude histograms, open-channel probabilities and dwell-time histograms, as

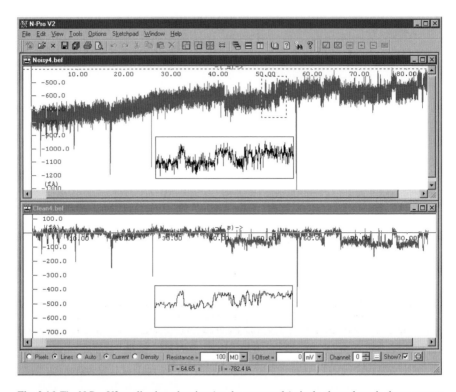

**Fig. 2.16** The N-Pro V2 application, showing (on the same scale) single-channel patch clamp current data, both the raw current recorded (upper window) and the same data after baseline adjustment, noise-profiling and Savitsky–Golay smoothing (lower window). The inset shows the area within the dotted box of the raw data 'zoomed-in', and demonstrates both the efficiency of the combined noise-removal tools, and their ability to preserve transition sharpness. Data courtesy of Dr Barbara Köhler, University of Potsdam, Golm-Potsdam, Germany.

well as more 'state-of-the-art' techniques, such as hidden Markov modelling and other statistical approaches to the 'restoration' of current traces.

### 2.5.6 The Y-Science ADC/DAC board drivers

The Microsoft Windows family are highly complex operating systems in which, unlike earlier, simpler environments like MS-DOS, the application programmer does not have direct access to low-level, hardware-related code. Briefly, all programs running on the computer fall into one of two categories: User mode applications include the explorer, word processors and the 'front-end' electrophysiology software. Kernel mode code, such as the operating system itself and the many 'device drivers' (the interfaces between the operating system and hardware attached to the computer, such as the hard disk(s), the keyboard and the clamp-control/data-acquisition

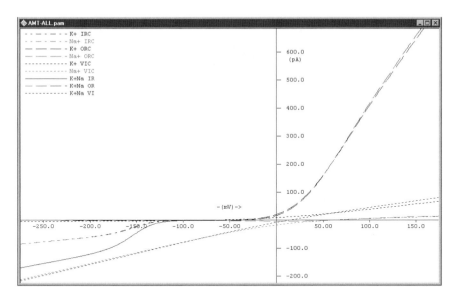

**Fig. 2.17** A collection of $I/V$ model currents produced in the Pandora! application. Three types of channel (inward-rectifying, outward-rectifying and voltage-independent) are shown for both $Na^+$ and $K^+$, as are the combined currents for both ions. The intra- and extracellular ion concentrations chosen were representative of the natural environment; these values, and the channel parameters, were from Amtmann and Sanders (1999).

hardware), is not directly accessible to the applications programmer. All communication between a user mode application and the ADC/DAC board must thus pass via its device driver.

Hardware manufacturers provide device drivers for their boards, together with an application programmer's interface (API) library, enabling software developers to access the hardware (indirectly) from their programs. These drivers and APIs are generally efficient and comprehensive, but there is no recognised coding standard; so, for software to support a variety of hardware types, a large amount of code is required for each device. A more significant issue with Windows is that it is a multi-tasking operating system, where many different processes are running concurrently; this makes it extremely difficult for user mode applications to perform operations in real time.

To address these problems, the Y-Science device driver system was developed, which currently supports most of the hardware reviewed in Section 2.3.4 (and new modules are added when required). This system presents an effectively hardware-independent interface to the user mode application, thus making it possible to add support for new devices to existing software, without the need for extra development, in much the same way as a new printer can be added to a computer. (All printer drivers use the same communications protocol, and so an application need not be aware of the physical details of the printer in order to use it.)

More importantly, the Y-Science drivers are specifically designed for use with electrophysiology applications, and the issue of 'real-time' control is resolved. Before running a protocol, information for the entire run is loaded into the driver's own, kernel mode memory, whence it can be accessed immediately, as and when required. The driver runs at a priority higher than any available to user mode applications, and events (such as a change in the clamp command potential or the start of a sampling window) are triggered by high-priority hardware interrupts, ensuring that they actually execute within a few microseconds of the requested time.

An easy to use API is provided, for those who wish to develop their own software using the Y-Science drivers, and is well documented within the online help of Henry II's EP Suite.

## References

Amtmann, A., Laurie, S., Leigh, R. & Sanders, D. (1997) Multiple inward channels provide flexibility in $Na^+/K^+$ discrimination at the plasma membrane of barley suspension culture cells, *J. Exp. Bot.*, **48**, 481–497.

Amtmann, A. & Sanders, D. (1999) Mechanisms of $Na^+$ uptake by plants, *Adv. Bot. Res.*, **29**, 75–112.

Blatt, M.R. (1991) A primer in plant electrophysiological methods, *Meth. Plant Biochem.*, **6**, 281–321.

Blatt, M.R. & Gradmann, D. (1997) $K^+$-sensitive gating of the $K^+$ outward rectifier in *Vicia* guard cells, *J. Membr. Biol.*, **158**, 241–256.

Blatt, M.R., Rodriguez-Navarro, A. & Slayman, C.L. (1987) Potassium–proton symport in Neurospora: kinetic control by pH and membrane potential, *J. Membr. Biol.*, **98**, 169–189.

Cole, K.S. (1949) Dynamic electrical characteristics of the squid axon membrane, *Arch. Sci. Physiol.*, **3**, 253–258.

Cole, K.S. & Moore, J.W. (1960) Ionic current measurements in the squid giant axon membrane, *J. Gen. Physiol.*, **44**, 123–167.

Copollo, J., Simon, S.B., Segal, Y., *et al.* (1991) $Ba^{2+}$ release from soda glass modifies single maxi $K^+$ channel activity in patch clamp experiments, *Biophys. J.*, **60**, 931–941.

DeFelice, I.J. (1981) *Introduction to Membrane Noise*, Plenum Press, New York.

Demidchik, V., Davenport, R.J. & Tester, M. (2002) Nonselective cation channels in plants, *Annu. Rev Plant Biol.*, **53**, 67–107.

Dempster, J. (1993) *Computer Analysis of Electrophysiological Signals*, Academic Press, New York.

Epstein, E., Rains, D.W. & Elzam, O.E. (1963) Resolution of dual mechanisms of potassium absorption by barley roots, *Proc. Natl. Acad. Sci. U.S.A.*, **49**, 684–692.

Fredkin, D.R. & Rice, J.A. (1992) Bayesian restoration of single-channel patch clamp recordings, *Biometrics*, **48**, 427–448.

Gurdon, J.B., Lane, C.D., Woodland, H.R. & Marbaix, G. (1971) Use of frog eggs and oocytes for the study of messenger RNA and its transition in living cells, *Nature*, **233**, 177–182.

Hamill, O.P., Marty, A., Neher, E., Sakmann, B. & Sigworth, F.J. (1981) Improved patch clamp techniques for high-resolution current recording from cells and cell-free membrane patches, *Pflugers Arch.*, **391**, 85–110.

Hamilton, D.W.A., Hills A., Köhler, B. & Blatt, M.R (2000) $Ca^{2+}$ channels at the plasma membrane of stomatal guard cells are activated by hyperpolarization and abscisic acid, *Proc. Natl. Acad. Sci. U.S.A.*, **97**, 4967–4072.

Hirsch, R.E., Lewis, B.D., Spalding, E.P. & Sussman, M.R. (1998) A role for the AKT1 potassium channel in plant nutrition, *Science*, **280**, 918–921.

Horowitz, P. & Hill, W. (1980) *The Art of Electronics*, Cambridge University Press, Cambridge, UK.

Islas, L.D. & Sigworth, F.J. (1999) Voltage sensitivity and gating charge in Shaker and Shab family potassium channels, *J. Gen. Physiol.*, **114**, 723–742.

Kegel, D.R., Wolf, B.D., Sheridan, R.E. & Lester, H.A. (1985) Software for electrophysiological experiments with a personal computer, *J. Neurosci. Meth.*, **12**, 317–330.

Keunecke, M., Sutter, J.-U., Sattelmacher, B. & Hansen, U.P. (1997) Isolation and patch clamp measurements of xylem contact cells for the study of their role in the exchange between apoplast and symplast of leaves, *Plant Soil*, **196**, 239–244.

Köhler, B. & Blatt, M.R. (2002) Protein phosphorylation activates the guard cell $Ca^{2+}$ channel and is a prerequisite for gating by abscisic acid, *Plant J.*, **32**, 185–194.

Köhler, B., Hills, A. & Blatt, M.R. (2003) Control of guard cell ion channels by hydrogen peroxide and abscisic acid indicates their action through alternate signalling pathways, *Plant Physiol.*, **131**, 385–388.

Maathuis, F.J. & Sanders, D. (1995) Contrasting roles in ion transport of two $K^+$-channel types in root cells of *Arabidopsis thaliana*, *Planta*, **197**, 456–464.

Maathuis, F.J. & Sanders, D. (2001) Sodium uptake in *Arabidopsis* roots is regulated by cyclic nucleotides, *Plant Physiol.*, **127**, 1617–1625.

Martin, J.D. (1991) *Signals and Processes – A Foundation Course*, Pitman, London.

Miedema, H., Henriksen, G.H. & Assmann, S.M. (1999) A laser microsurgical method of cell wall removal allows detection of large-conductance channels in the guard cell plasma membrane, *Protoplasma*, **209**, 58–67.

Miller, A.J. & Zhou, J.J. (2000) *Xenopus* oocytes as an expression system for plant transporters, *Biochem. Biophys. Acta*, **1465**, 343–358.

Mitchell, M. (1961) Coupling of phosporylation to electron and hydrogen transfer by a chemi-osmotic type of mechanism, *Nature*, **191**, 144–148.

Morrison, R. (1986) *Grounding and Shielding Techniques in Instrumentation*, John Wiley & Sons, New York.

Neher, E. & Sakmann, B. (1976) Single channel currents recorded from membrane of denervated frog muscle fibres, *Nature*, **260**, 799–802.

Penner, R. (1995) A practical guide to patch clamping, in *Single-Channel Recording*, 2nd edn (eds B. Sakmann & E. Neher), Plenum Press, New York, pp. 3–30.

Pesacreta, T.C. & Lucas, W.J. (1984) Plasma membrane coat and a coated vesicle-associated reticulum of membranes: their structure and possible interrelationship in *Chara corallina*, *J. Cell. Biol.*, **98**, 1537–1545.

Philippar, K., Fuchs, I., Lüthen, H., *et al.* (1999) Auxin-induced $K^+$ channel expression represents an essential step in coleoptile growth and gravitropism, *Proc. Natl. Acad. Sci. U.S.A.*, **96**, 12186–12191.

Roelfsema, M.R., Steinmeyer, R., Staal, M. & Hedrich, R. (2001) Single guard cell recordings in intact plants: light-induced hyperpolarization of the plasma membrane, *Plant J.*, **26**, 1–13.

Savitsky, A. & Golay, M.J.E. (1964) Smoothing and differentiation of data by simplified least-squares procedures, *Anal. Chem.*, **36**, 1627–1639.

Schroeder, J.I. & Keller, B.U. (1992) Two types of anion channel currents in guard cells with distinct voltage regulation, *Proc. Natl. Acad. Sci. U.S.A.*, **89**, 5025–5029.

Thiel, G., MacRobbie, E.A. & Blatt, M.R. (1992) Membrane transport in stomatal guard cells: the importance of voltage control, *J. Membr. Biol.*, **126**, 1–18.

Venkataramanan, L. & Sigworth, F.J. (2002) Applying hidden Markov models to the analysis of single ion channel activity, *Biophys. J.*, **82**, 1930–1942.

Vergani, P. & Blatt, M.R. (1999) Mutations in the ast two pore $K^+$ channel YKC1 identify functional differences between the pore domains, *FEBS Lett.*, **458**, 285–291.

Vergani, P., Hamilton, D., Jarvis, S. & Blatt, M. (1998) Mutations in the pure regions of the yeast $K^+$ channel YKC1 affect gating by extracellular $K^+$. *EMBO J.*, **17**, 7190–7198.

Vogelzang, S.A. & Blom-Zandstra, M. (1998) Preparation of patchable plant cell protoplasts and a procedure for the improvement of Gigaseal formation, in *Signal Transduction: Single Cell Techniques* (eds B. Van Duijn & A. Wilting), Springer, New York, pp. 99–108.

# 3 Structure, function and regulation of primary $H^+$ and $Ca^{2+}$ pumps

Rosa L. López-Marqués, Morten Schiøtt, Mia Kyed Jakobsen and Michael G. Palmgren

## 3.1 Pumps in plants

*Pumps* are proteins fuelled by cellular metabolites that transport solutes across biological membranes against their electrochemical gradients. Among the solutes handled by pumps, the active transport of protons ($H^+$) plays a pivotal role in the interconversion of biological energy forms such as chemical, osmotic and electrical energy. Thus, transport of $H^+$ is a fundamental cellular process. Most membranes in a living plant cell employ pumps to develop and maintain an unequal distribution of $H^+$ across the lipid bilayer. Although different families of $H^+$ pumps all transport the same ion, they have very different structures and mechanisms. Whereas $H^+$ pumps are the major focus of this review, other pumps in the plant, transporting cations such as $Ca^{2+}$, will also be discussed so as to provide a general overview of pumps in a plant cell.

## 3.2 Proton pumps in plant cells

Proton gradients and membrane potentials are maintained over many biological membranes in a typical plant. These are (i) the plasma membrane, (ii) the vacuolar membrane, (iii) the inner mitochondrial membrane and (iv) the thylakoid membrane. Other membranes, such as those of secretory vesicles, probably also establish proton gradients. Proton pumps in the inner mitochondrial membrane and the thylakoid membrane (ATP synthases) harvest already established $H^+$ gradients and use the energy in the electrochemical gradient for synthesis of ATP. However, in this review, we will discuss only proton pumps operating in the other direction. Such pumps establish $H^+$ gradients at the expense of ATP or other high-energy chemical compounds.

In the following three $H^+$ pumps will be discussed in some detail: (i) plasma membrane $H^+$-ATPases, which extrude $H^+$ to the outside of the cell, generating a proton electrochemical gradient, (ii) vacuolar $H^+$-ATPases and (iii) vacuolar $H^+$-pumping pyrophosphatase ($H^+$-PPases, which acidify endomembrane compartments, and the vacuole. These three pumps show no homology to each other and must have evolved independently.

### 3.2.1 Plasma membrane $H^+$-ATPase

#### 3.2.1.1 Physiological role

A key function of the plasma membrane $H^+$-ATPase is to generate a proton electrochemical gradient, thereby providing the driving force for the uptake and efflux of ions and metabolites across the plasma membrane (Fig. 3.1) (reviewed in Serrano, 1989; Morsomme & Boutry, 2000; Palmgren, 2001). The plasma membrane $H^+$-ATPase extrudes $H^+$ from the cell to generate a transmembrane proton gradient and a membrane potential. This pH gradient is typically 1.5–2.0 pH units (acid outside) whereas the membrane potential usually lies between −120 and −220 mV (negative on the inside).

Many essential cations, such as $K^+$, are typically taken up by the cell through channel proteins in the plasma membrane (Very & Sentenac, 2003). This transport is solely driven by the membrane potential. Cations that need to be highly concentrated in the cell are transported by $H^+$-coupled cation symporters. Anionic nutrients,

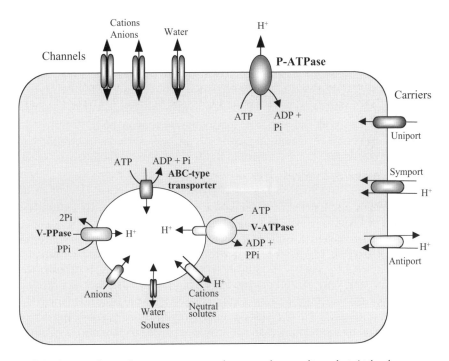

**Fig. 3.1** Primary and secondary transports across plasma membrane and tonoplast. At the plasma membrane, proton gradients generated by a P-ATPase are used by different secondary transporters (carriers and channels) to transport ions and other solutes inside the cell. Two proton-translocating enzymes contribute to acidification of the vacuolar compartment: V-ATPases and V-PPases. Carriers and channels move ionic and non-ionic solutes across the tonoplast, relaxing the proton gradient established by these two pumps. Another type of transporters, ABC-type, transport solutes directly energized by ATP. Aquoporins involved in water uptake in both membranes might be regulated solely by osmotic pressure.

such as nitrate, phosphate and sulfate, are taken into cells against concentration and electrical gradients by $H^+$-coupled anion symporters (Tanner & Caspari, 1996). Anions can leave the cell again through anion channel proteins.

Plasma membrane $H^+$ pumps have been localized by immunolabelling techniques to all living plant cells investigated. However, some cell types have a much higher concentration of pumps than do others. This is revealed in many plant species by strong immunostaining of, for example, the epidermis, root hairs, phloem and guard cells (e.g. Paret-Soler *et al.*, 1990; Villalba *et al.*, 1991; Bouche-Pillon *et al.*, 1994; Fromard *et al.*, 1995, Jahn *et al.*, 1998; Pertl *et al.*, 2001).

Plasma membrane $H^+$ pumps in the root epidermis are thought to function mainly in energizing nutrient uptake from the soil. Pumps in the root endodermis are likely to be required for active loading of solutes into the xylem of the vascular cylinder. Another important hypothetical function in roots is to energize growth of root hairs that are tip-growing systems like pollen tubes (see below).

Photosynthetic assimilates and other organic compounds produced by the plant are translocated into phloem tissues for delivery throughout the plant to sink tissues. These processes require $H^+$-coupled symporters for amino acids and sucrose (Lalonde *et al.*, 1999). Plasma membrane $H^+$ pumps in the phloem companion cells most likely energize phloem loading.

Vessel-associated cells are adjacent to dead xylem vessels rich in plasma membrane $H^+$-ATPase. These cells might serve a function in the xylem very similar to that of phloem companion cells in the phloem. In trees, plasma membrane $H^+$-ATPases in vessel-associated cells are thought to drive uptake of sucrose from the vessels in the spring and reallocate metabolites to bursting buds.

Plasma membrane $H^+$ pumps in guard cells are believed to be involved in controlling the size of the stomatal aperture. Swelling of guard cells following uptake of $K^+$ and water is a prerequisite for opening of the stomatal pore.

In pollen, the male gametophyte of plants, plasma membrane $H^+$ pumps are abundant and are thought to be essential for pollen maturation and for germination and growth of the pollen tube.

According to the acid growth hypothesis (reviewed in Hager, 2003), elongation growth of cotyledons and hypocotyls is the result of auxin-mediated activation of plasma membrane $H^+$ pumps. Acidification of the cell wall by $H^+$ is hypothesized to render the wall soft. This allows for cell expansion as cells swell as a result of ion and water uptake, processes that are also driven by the plasma membrane $H^+$-ATPase. Directional growth of plants during phototropism and gravitropism is most likely regulated by a similar mechanism involving plasma membrane $H^+$ pumps.

### 3.2.1.2 Genetics

There have been very few genetic studies to illuminate the physiological role of specific plant plasma membrane $H^+$-ATPase isoforms. A mutation in one plasma membrane $H^+$-ATPase gene, *AHA4* of *Arabidopsis thaliana*, has been characterized (Vitart *et al.*, 2001). The gene is active in root endodermis and the mutant is sensitive to excess salt in the growth medium. The mutation results from the insertion of a T-DNA in the middle of the *AHA4* gene. However, the mutation is semi-dominant, suggesting that the mRNA or the gene product is dominant negative. In principle, the truncated pump could suppress the activity of full-length AHA4 or possibly other pumps in the root endodermis that play a role in loading of cations into the root vascular tissue.

### 3.2.1.3 Structure and mechanism

The plant plasma membrane H$^+$-ATPase was first studied biochemically in isolated plasma membrane vesicles (for review see Sze, 1985). The plasma membrane constitutes only about 1% of total cell membranes and therefore development of the aqueous two-phase partioning technique (Widell & Larsson, 1981), which yields very pure plasma membranes, was a major breakthrough. The plasma membrane H$^+$-ATPase was first cloned from the yeast *Saccharomyces cerevisiae* in 1986 (Serrano *et al.*, 1986) and from the plants *Arabidopsis* and tobacco in 1989 (Boutry *et al.*, 1989; Harper *et al.*, 1989; Pardo & Serrano, 1989).

The plasma membrane H$^+$ pump is a 'P-type' ATPase, meaning that it forms a *p*hosphorylated intermediate in the reaction cycle (Moller *et al.*, 1996). It is made up of a single polypeptide of 100 kDa (Fig. 3.2) having 10 transmembrane domains and a large hydrophilic loop containing the ATP-binding region situated between transmembrane segments 4 (TM4) and 5 (TM5) (Palmgren, 2001). This loop is facing the cytoplasm and is divided in two distinct domains: (i) the phosphorylation

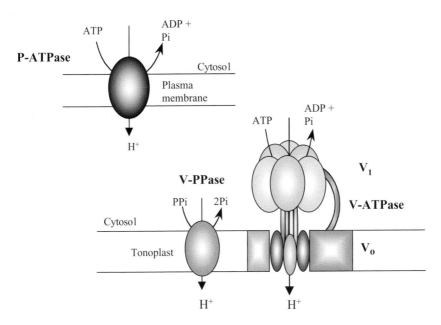

**Fig. 3.2** Subunit composition of the main plant proton pumps. Both plasma membrane P-type ATPase and V-PPase associated to the tonoplast consist of one single polypeptide. In contrast, V-ATPase is a large protein complex (7–12 different subunits) that consists of a peripheral hydrophilic catalytic sector (V$_1$) and a hydrophobic membrane sector (V$_0$). Differential expression and assembly of some of these subunits might be used for regulation under different physiological conditions.

domain (P domain), which contains the conserved aspartyl residue that is phospho-rylated during catalysis, and (ii) the nucleotide-binding domain (N domain), which binds the sugar moiety of the ATP molecule. The N domain is not a continuation of the P domain but rather is inserted into this domain.

In addition to the large cytoplasmic loop, two other hydrophilic domains are exposed to the cytoplasmic side of the membrane. One is the anchor or attenuator domain (A domain), which is formed by about 50 N-terminal residues that associate with a small loop between TM2 and TM3. Another is the C-terminal domain of about 100 residues (R domain), which serve as an autoinhibitor involved in regulation of pump activity (Palmgren, 2001).

Based on thoroughly studied P-type ATPases, such as mammalian $Na^+/K^+$ and $Ca^{2+}$ pumps (Moller et al., 1996; Jorgensen et al., 2003), the proposed reaction mechanism involves a conformational switch between a minimum of two alternate intermediates (termed enzyme 1 and enzyme 2, $E_1$ and $E_2$), although this mechanism has not been demonstrated for $H^+$ pumps. Orthovanadate inhibits P-type ion pumps by competing with phosphate for binding to the $E_2$ intermediate.

Based on mutagenesis experiments with many P-type pumps (Moller et al., 1996; Jorgensen et al., 2003) in combination with structural studies of the $Ca^{2+}$-ATPase from rabbit sarcoplasmic reticulum (Toyoshima et al., 2000; Toyoshima & Nomura, 2002), a transport mechanism has been proposed: the adenosine half of ATP binds to the N domain. Subsequently, the negatively charged phosphate groups of ATP are attracted by a tightly bound $Mg^{2+}$ atom in the P domain. This moves the whole N domain, which is attached to the protein by a very flexible hinge, towards the P domain. ATP now donates its gamma phosphate to a conserved aspartate (Asp-351 in AHA2) in the P domain. Phosphorylation of the P domain triggers a conformational change that involves reorientation of helices in the transmembrane region, where the ion(s) to be transported are bound. The pump thus goes from the $E_1P$ (phosphorylated $E_1$) towards the $E_2P$ conformation, a structural change that is tightly coupled to ion pumping through the membrane. During this conformational change the A domain moves and becomes adjacent to the P domain. The A domain has phosphatase activity and dephosphorylates the P domain. This causes the dephosphorylated $E_2$ to return to the $E_1$ conformation. The pump is now ready for a new catalytic cycle.

In $Na^+/K^+$ and $Ca^{2+}$-ATPases, ion binding occurs in the transmembrane region of the pump (Moller et al., 1996; Jorgensen et al., 2003). In one conformation, probably $E_1$, a half-channel is proposed to lead from the cytoplasmic side of the protein to the middle of the plane of the membrane. In the other conformation, $E_2$, the cytoplasmic half-channel is closed and a new half-channel opens up, leading from the centre of plane the membrane towards the extracellular side. However, in the crystal structures of $Ca^{2+}$-ATPase in the $E_1$ and $E_2$ conformations, respectively, the entrance and exit pathways are not clear.

$H^+$ ions could, in principle, traverse the membrane by jumping between pro-tonable amino acid side chains like in the well-studied bacterial $H^+$ pump bacteri-orhodopsin. However, according to one model, $H^+$ does not bind as free $H^+$, but rather as $H_3O^+$ (protonated water), the hydronium ion (Palmgren, 2001). This ion

## Plate 1

```
                    P                              M2
                     *                                      *
AtGLR1.1   QLSMMLWFGF STIVFAH-RE -KLQKMSSRF LVIVWVFVVL ILTSSYSANL TSTKTISR
AtGLR1.2   QIGVVIWFGF STLVYAH-RE -KLQHNLSRF VVTVWVFAVL ILVTSYTATL TSMMTVQQ
AtGLR1.3   QIGVVLWFGF STLVYAH-RE -KLKHNLSRF VVTVWVFAVL ILTASYTATL TSMMTVQQ
AtGLR1.4   QIGTLLCFGF STLVFAH-RE -RLQHNMSRF VVIVWIFAVL ILTSNYTATL TSVMTVQQ
AtGLR2.1   QLSTIFWFSF SIMVFAP-RE -RVLSFWARV VVIIWYFLVL VLTQSYTASL ASLLTTQH
AtGLR2.2   QASTIFWFAF STMVFAP-RE -RVLSFGARS LVVTWYFVLL VLTQSYTASL ASLLTSQQ
AtGLR2.3   QASTICWFAF STMVFAP-RE -RVFSFWARA LVIAWYFLVL VLTQSYTASL ASLLTSQK
AtGLR2.4   QISTMFWFAF STMVFAP-RE -RVMSFTARV VVITWYFIVL VLTQSYTASL SSLLTTQQ
AtGLR2.5   KISSVFYFSF STLFFAH-RR -PSESFFTRV LVVVWCFVLL ILTQSYTATL TSMLTVQE
AtGLR2.6   KISNVFYFSF STLFFAH-MR -PSESIFTRV LVVVWCFVLL ILTQSYTATL TSMLTVQQ
AtGLR2.7   QIGTSFWFAF STMNFAH-RE -KVVSNLARF VVLVWCFVVL VLIQSYTANL TSFFTVKL
AtGLR2.8   QIGTSFWFSF STMVFAH-RE -KVVSNLARF VVVVWCFVVL VLIQSYTANL TSFLTVQR
AtGLR2.9   QIGTSLWFSF STMVFAH-RE -NVVSNLARF VVVVWCFVVL VLIQSYTANL TSFLTVQS
AtGLR3.1   QIITILWFTF STMFFSH-RE -TTVSTLGRM VLLIWLFVVL IITSSYTASL TSILTVQQ
AtGLR3.2   QIVTILWFSF STMFFSH-RE -NTVSTLGRA VLLIWLFVVL IITSSYTASL TSILTVQQ
AtGLR3.3   QCVTILWFSF STMFFAH-RE -NTVSTLGRL VLIIWLFVVL IINSSYTASL TSILTVQQ
AtGLR3.4   QLITIFWFSF STMFFSH-RE -NTVSSLGRF VLIIWLFVVL IINSSYTASL TSILTIRQ
AtGLR3.5   QIITVFWFSF STMFFSH-RE -NTVSTLGRF VLLVWLFVVL IINSSYTASL TSILTVQQ
AtGLR3.6   QVITTFWFSF STLFFSH-RE -TTTSNLGRI VLIIWLFVVL IINSSYTASL TSILTVHQ
AtGLR3.7   QLSTMLLFSF STLFKRN-QE -DTISNLARL VMIVWLFLLM VLTASYTANL TSILTVQQ
GluR1      GIFNSLWFSL GAFMQQG-CD ISPRSLSGRI VGGVWWFFTL IIISSYTANL AAFLTVER
GluR6      TLLNSFWFGV GALMQQG-SE LMPKALSTRI VGGIWWFFTL IIISSYTANL AAFLTVER
NMDAR1     TLSSAMWFSW GVLLNSGIGE GAPRSFSARI LGMVWAGFAM IIVASYTANL AAFLVLDR
NMDAR2A    TIGKAIWLLW GLVFNNSVPV QNPKGTTSKI MVSVWAFFAV IFLASYTANL AAFMIQEE
d2         TLYNSMWFVY GSFVQQG-GE VPYTTLATRM MMGAWWLFAL IVISSYTANL AAFLTITR
KBP        TLLNSLWYGV GALTLQG-AE PQPKALSARI IAVIWWVFSI TLLAAYIGSF ASYINSNT
GluR0      GVQNGMWFAL VTLTTVGYGD RSPRTKLGQL VAGVWMLVAL LSFSSITAGL ASAFSTAL
```

**Plate 1** Alignment of glutamate receptor sequences. Pore regions, containing the selectivity filter, of plant (AtGlR), rat (GluR1, NR2A), human (GluR6, NR1), frog (KBP), mouse (d2) and *Synechocystis* (GluR0) were aligned with the selectivity filter sequence boxed (corresponding to the GYGD motif of K-selective channels). The 'Q/R/N' RNA editing site in the pore of some receptors is denoted with an asterisk. Colours denote different chemical characteristics of the amino acids: red, basic; pink, histidine; dark blue, acidic; light blue, hydrophilic; yellow, aliphatic; orange, aromatic; green, proline and glycine; purple, cysteine. Reproduced from Davenport (2002), by permission of Oxford University Press.

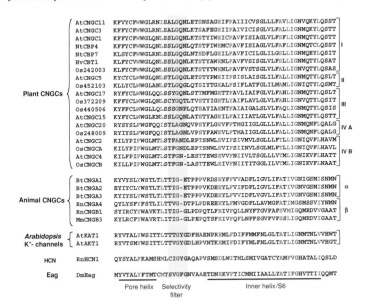

**Plate 2** Sequence alignments of plant CNGCs. Pore and S6 domains of selected plant and animal CNGCs, and plant and animal K channels with a cyclic-nucleotide-binding domain were compared. The selectivity filter in the pore region is highlighted in yellow. Identical amino acids are in red, conserved substitutions in green and semiconserved substitutions in blue. Eag: Ether-a-go-go channel, HCN: hyperpolarisation-activated, cyclic-nucleotide-gated channel. Reproduced from Talke *et al.* (2003), by permission of Elsevier Science.

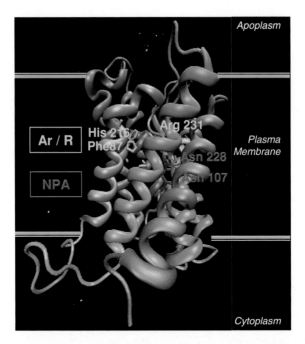

**Plate 3** Structural model of *Arabidopsis* aquaporin PIP2;1. A putative structure was obtained by homology modelling (using the Swiss-Model server at http://www.expasy.org/swissmod/swiss-model.html) based on the X-ray structure of human AQP1 (PDB ID: 1h6i). The six transmembrane domains and the two main pore constrictions, Ar/R and NPA (see text), are represented. The Asparagine residues of the two NPA motifs and the residues forming the Ar/R motif are shown in red and yellow, respectively.

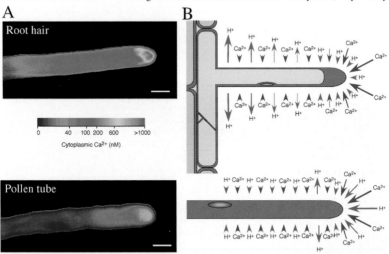

**Plate 4** $Ca^{2+}$ gradients and fluxes in root hairs and pollen tubes. (A) $Ca^{2+}$ distribution in a tip-growing root hair (top) and pollen tube (bottom). Note the pronounced tip-focused gradient in both cases. The root hair and pollen tube were microinjected with the $Ca^{2+}$ reporting fluorescent dye Indo-1 conjugated to a 10 kDa dextran (to ensure cytosolic localization). $Ca^{2+}$ distribution was then monitored by confocal ratio imaging (Bibikova *et al.*, 1997), and color-coded images calculated according to the inset scale. Scale bar, 10 μM. (B) Summary of $Ca^{2+}$ and $H^+$ fluxes around growing root hairs and pollen tubes monitored using the self-referencing (vibrating reed) microelectrode. Length of arrows indicates relative flux magnitude.

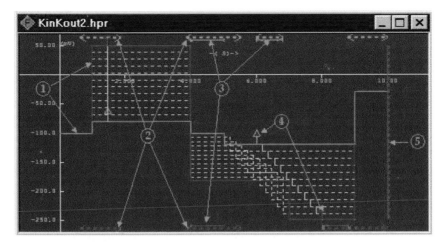

**Plate 5** A protocol editor window, showing the various elements of a protocol (not all of which need be present): (1) The clamp trace for the first (green) and last (red) cycles of the protocol defines the voltage (or current for current clamp mode) values and time course over which the clamp will be driven; the calculated traces for the intermediate cycles are shown as dashed yellow lines. (2) Sampling windows for the first cycle (top of window) and last cycle (bottom of window) define the time periods for sampling and sampling frequencies to be used. (3) Analysis envelopes within sample windows (an optional component) are used primarily for the pre-selection of data within sample windows for run-time and post-run analysis. (4) Triggers/signals (shown for the first and last cycles and expanded to cycles between) permit optional time points at which a specified byte value is sent to the digital output port on the ADC/DAC board for communication and synchronisation with other equipment. (5) The protocol end-marker – this is not actually part of the protocol; rather, it is an editor tool that can be used to 'stretch' or 'shrink' the entire protocol (and all its component elements) along the time axis. Some examples of protocols generated with this tool are shown in Figs. 2.1, 2.2, 2.5 and 2.10.

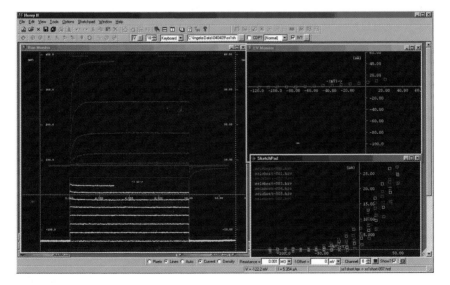

**Plate 6** The 'run-time' display, showing the 'in-progress' clamp monitor (left-hand window: voltage protocol is shown in background grey, recorded voltage in yellow, recorded current in blue), the real-time $IV$ analysis for this data set (upper-right-hand window) and the sketchpad, on which the $IV$ plots of previous voltage clamp data sets are shown (colour coded to the overlaid filename list).

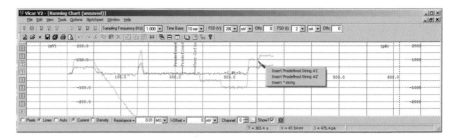

**Plate 7** The Vicar V2 application during recording. Note the vertical toolbar to the left of the picture, whereby the input channels to be monitored can be quickly selected, and the horizontal toolbar under the Tools menu, whereby up to eight predefined comments can be written to the record with a single function key stroke. The pop-up menu shown slightly to the right of centre is activated by clicking the mouse at any point in the record.

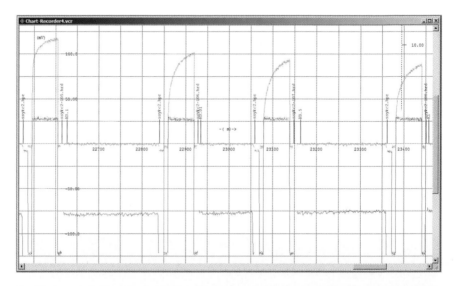

**Plate 8** A section of a chart recorded with the Vicar V2 application, showing how the program can record data collected while running a protocol in Henry II. At the start of each run, the chart displays the name of the protocol file (ssykc2.hpr) and, on completion, the name of the saved data file (e.g. ssykc2-107.hrd for the run to the right of centre). At any stage, the user can insert comments, just as if writing on a chart-paper record (e.g. the 'K0.1', 'K0.01', 'K0.5' remarks, indicating a change in the potassium concentration of the bath solution). Data from two analogue input channels are displayed – the *V* electrode (red) and the *I* electrode (green). Data courtesy of Dr Ingela Johansson, Laboratory of Plant Physiology and Biophysics, IBLS, University of Glasgow, Glasgow, UK.

has approximately the same ionic radius as Ca$^{2+}$ and K$^+$ whereas free H$^+$ is much smaller. It is possible that the water molecule is positioned permanently at the ion-binding site and becomes protonated and deprotonated during turnover of the pump. Alternatively, the water molecule, together with H$^+$ as H$_3$O$^+$, enters and leaves the pump in every pump cycle.

H$^+$ transport through the plasma membrane H$^+$-ATPase involves a conserved aspartate in TM6 (Asp-684 in AHA2) (Buch-Pedersen et al., 2000; Buch-Pedersen & Palmgren, 2003). If H$^+$ is bound as H$_3$O$^+$, this ion is likely to be coordinated at the ion-binding site by four oxygen atoms. Asp-684 could donate one such oxygen atom. Structural predictions based on homology modelling have suggested that three other oxygen atoms in TM4 to be involved in ion coordination (Bukrinsky et al., 2001). These oxygen atoms do not originate from side chains but rather from carbonyl oxygens in the peptide bonds in three hydrophobic residues in TM4.

### 3.2.1.4   Isoforms and expression in the plant

Plasma membrane H$^+$-ATPases in plants are expressed from a multi-gene family. In Arabidopsis, 11 AHA genes are present (Palmgren, 2001). Evolution of isogenes with different promoters allows for transcriptional regulation of the expression of each isoform according to the demand of any given cell during development and environmental challenges. In addition, some isoforms have slightly different biochemical properties compared to others (Palmgren & Christensen, 1994), the biological significance of which is not known.

Cell-specific expression of plasma membrane H$^+$-ATPase genes has been detected in transgenic plants following fusion of the promoter of individual isoforms with the open reading frame of the reporter gene β-glucuronidase (GUS) (Sussman, 1994) and immunocytologically following epitope tagging of H$^+$-ATPases (DeWitt & Sussman, 1995). Some H$^+$-ATPase genes, like AHA3, are preferentially expressed in the companion cells of the phloem (DeWitt et al., 1991; DeWitt & Sussman, 1995) whereas others, like AHA9 and AHA10, are expressed specifically in the anther (Houlne & Boutry, 1994) and the developing seed (Harper et al., 1994), respectively. Some isoforms, such as pma1 of tobacco, are expressed in many cell types, including the root epidermis, stem cortex and guard cells (Michelet et al., 1994). Gene microarrays have demonstrated that gene activity of the various isoforms is regulated differentially by environmental stresses (Maathuis et al., 2003).

### 3.2.1.5   Regulation

Plant hormones, light, phytotoxins and environmental stress influence the electrochemical gradient of protons across the plasma membrane. Thus, many regulatory signals are likely to have plasma membrane H$^+$-ATPase as a prime target (Serrano, 1989).

The activity of the plant plasma membrane H$^+$ pump can be regulated rapidly at the post-translational level by a modification of the C-terminal autoinhibitory domain. The presence of such a domain is supported by several observations, including the finding that the pump is activated in vitro after trypsin removal of the

C-terminal domain (Palmgren *et al.*, 1991). Activation is also the result of removal at the gene level of the 66 C-terminal residues of AHA2 expressed in yeast. The truncated enzyme has an increased affinity for ATP, increased $V_{max}$, and an alkaline shift in pH optimum that suggests an increased affinity for $H^+$ (Regenberg *et al.*, 1995).

The 14-3-3 protein interacts directly with the C-terminal region of the PM $H^+$-ATPase (Jahn *et al.*, 1997; Fullone *et al.*, 1998; Piotrowski *et al.*, 1998). Fusicoccin stabilizes this interaction (Baunsgaard *et al.*, 1998; Wurtele *et al.*, 2003). 14-3-3 binding leads to enzyme activation (Baunsgaard *et al.*, 1998), most likely because binding causes the autoinhibitory portion of the C-terminal domain to become displaced from the rest of the pump molecule. A regulatory role for phosphorylation is to facilitate the binding of the 14-3-3 protein to the $H^+$-ATPase. Thr-947 is a target for protein kinase modification in plants (Olsson *et al.*, 1998) and, when phosphorylated, is part of the 14-3-3 binding site (Fuglsang *et al.*, 1999; Svennelid *et al.*, 1999; Maudoux *et al.*, 2000; Wurtele *et al.*, 2003).

Environmental stress, hormones and light could lead to the activation or inactivation of protein kinases and/or phosphatases, thus altering the phosphorylation state of the $H^+$-ATPase. Depending on the mode of action, this might result in stabilization of either the low-activity state or the high-activity state of the pump. Salt stress might activate the plasma membrane $H^+$-ATPase by a mechanism involving the C-terminal domain (Wu & Seliskar, 1998). In stomatal guard cells, blue light causes phosphorylation of the plasma membrane $H^+$-ATPase, in this way triggering 14-3-3 binding and pump activation (Kinoshita & Shimazaki, 1999).

The yeast plasma membrane $H^+$-ATPase PMA1 is regulated *in vivo* by modulation of the transport coupling ratio, which is defined as the number of protons pumped per ATP hydrolyzed (Venema & Palmgren, 1995). In glucose-starved cells, the pump is partially uncoupled; however, the pump from glucose-activated cells is tightly coupled. Several lines of evidence suggest that the plant plasma membrane $H^+$-ATPase may likewise be regulated by a change in coupling ratio (Baunsgaard *et al.*, 1996).

### 3.2.2   V-ATPases

#### 3.2.2.1   Physiological role
Vacuolar $H^+$-ATPases (or V-ATPases; Pedersen & Carafoli, 1987) have derived their name from the fact that they were originally discovered in the vacuolar membrane. Later they have been identified in many more internal membranes such as provacuoles, the endoplasmic reticulum, the Golgi apparatus, coated vesicles and secretory vesicles (Herman *et al.*, 1994; Matsuoka *et al.*, 1997). These internal membrane systems are all part of the secretory pathway involved in maturation, processing, secretion and intracellular distribution of membrane proteins. Most recently, V-ATPases have been identified in the plasma membrane of many eukaryotic cells (including plants; Robinson *et al.*, 1996). Thus, the term V-ATPase does only refer to a group of proton pumps with common structural and functional properties,

not to their intracellular localization (reviewed in Stevens & Forgac, 1997). Many excellent recent reviews deal with V-ATPases in plants (Sze *et al.*, 1999; Ratajczak, 2000; Dietz *et al.*, 2001; Kluge *et al.*, 2003) and other eukaryotes, mainly the yeast *S. cerevisiae* (Nishi & Forgac, 2002; Kawasaki-Nishi *et al.*, 2003; Nelson, 2003).

The common role of V-ATPases is to acidify intracellular compartments. Thus, many endomembrane compartments maintain a proton electrochemical gradient as a result of the activity of this pump. These gradients do not reach the same values as that across the plasma membrane but are still very significant. Typically, the vacuolar membrane potential reaches +30 mV (positive towards the lumen of the vacuole), and the pH of the vacuolar sap ranges from 3 to 6.

In the vacuole, the main function of the proton electrochemical gradient is to energize secondary active transport (Fig. 3.1). Membrane potentials across the plasma membrane and the vacuolar membrane have opposite values relative to the cytosol (negative for the plasma membrane; positive for the vacuolar membrane). Therefore, the type of transporters used for secondary active transport has to be different at these two membranes. Positively charged ions, such as Ca$^{2+}$, are transported by proton-coupled antiporters from the cytosol into the lumen of the vacuole (Hirschi *et al.*, 1996). Anion influx into the vacuole is facilitated by the electrical gradient (positive inside) and channel proteins can be used for this purpose. In the opposite direction, cation release into the cytosol is favoured via channel proteins, such as ligand-activated Ca$^{2+}$ channels (Sanders *et al.*, 1999). In order for anions to flow from the lumen of the vacuole into the cytosol, they have to overcome the repulsion from the membrane potential, and transport is through proton-coupled anion symporters.

In many plant tissues, the large central vacuole occupies most (>90%) of the space inside cells. The high solute concentration of the vacuole is a prerequisite for water uptake into this compartment. The vacuole serves many essential functions:

1) It generates the turgor pressure that drives cell expansion and is responsible for tissue rigidity.
2) It serves as a compartment for storage of salts and metabolites, such as organic acids, sugars, amino acids and conjugates.
3) It is the waste bag of the cell where macromolecules are broken down for recycling by a battery of proteases and other enzymes being active in an acidic environment.
4) It plays an important role in defence against stresses, which can be biotic (e.g. the vacuole might contain toxins directed against herbivores) or abiotic (e.g. Na$^+$ transport into the vacuole counteracts salinity-induced stress).

All the above functions depend upon the activity of the V-ATPase. Thus, the transport into and out of the vacuole of all its solutes is driven by this pump as explained above. Furthermore, for optimal performance, many of the enzymes in the vacuole depend upon acidic pH generated by the V-ATPase.

Depending upon the membrane system they are present in, V-ATPases have many additional roles. Thus, acidification of vesicles might be important for many activities of the endomembrane system such as membrane trafficking, fusion, protein sorting and secretion (Sze *et al.*, 1999).

### 3.2.2.2    Genetics

In the yeast *S. cerevisiae*, genes encoding V-ATPase subunits are present in single copies only (with one exception: *VPH1* and *STV1* encode homologous proteins). It was expected that abolishment of V-ATPase function in any eukaryote would be lethal. However, disruption of single copy genes in yeast results in a conditional lethal phenotype (Nelson & Nelson, 1990). Yeast cells without a V-ATPase will not survive if pH of the growth medium is kept neutral (above pH 7) but grow quite well at pH 5.5. At this pH, acidic fluid from the external medium might enter the vacuole by endocytosis.

In the plant *Arabidopsis*, the *det3* (*deetiolated* 3) mutant (Schumacher *et al.*, 1999) has reduced V-ATPase activity. This mutant was isolated in a screen for seedlings that do not exhibit rapid hypocotyl expansion during ethiolation in the dark. Map-based cloning revealed that the phenotype is due to a single-point mutation at a putative 3′ splicing site of the *AtVHA-C* gene, encoding subunit C of the V-ATPase. The adult *det3* plants exhibit dwarfism mainly because of deficient cell expansion and exhibit a 60% reduction of V-ATPase activity. This is strong genetic evidence for a role of V-ATPase in cell elongation.

The *det3* mutant furthermore exhibits a deficiency in maintaining $Ca^{2+}$ homeostasis of the cytosol (Allen *et al.*, 2000), possible because the V-ATPase drives uptake of $Ca^{2+}$ into the vacuole via $Ca^{2+}/H^+$ antiporters (Pittman & Hirschi, 2003). In guard cells of the *det3* mutant, external stress such as high $Ca^{2+}$ and oxidation elicit prolonged increases in cytoplasmic $Ca^{2+}$ (Allen *et al.*, 2000). In wild-type guard cells, the $Ca^{2+}$ concentration oscillates in response to external stress and ultimately the stomatal pore closes. In contrast, the concentration of $Ca^{2+}$ in *det3* cells remains constitutively high and stomatal closure is abolished. These data imply a role for V-ATPase in providing the ionic basis for $Ca^{2+}$-dependent signal transduction in plant cells.

### 3.2.2.3    Structure and mechanism

The V-ATPase was first discovered in chromaffin granules isolated from bovine adrenal medulla (for review see Nelson & Harvey, 1999). In plants, biochemical evidence for their presence was the result of work using membrane vesicles of tonoplast origin (for review see Sze, 1985). Later they were purified and the reconstituted enzyme was shown to function as a $H^+$ pump (Parry *et al.*, 1989; Ward & Sze, 1992a,b; Warren *et al.*, 1992).

All V-ATPases share common structural and functional features. In contrast to the single polypeptide that comprises plasma membrane $H^+$ pumps, the V-ATPases are highly complex multi-subunit assemblies (Fig. 3.2). In the model plant *A. thaliana*, 12 subunits are present (Dietz *et al.*, 2001; Sze *et al.*, 2002). In the completed genome of rize, 12 subunits have likewise been identified (Kluge *et al.*, 2003). All subunits are named VHA (vacuolar H$^+$-ATPase) followed by a letter (A to H, and a, c, d and e). Some of the subunits are present in several copies in the complex whereas others are present in single copies. The overall stoichiometry of the V-ATPase in eukaryotes has been proposed to be A$_3$B$_3$CDEFG$_2$ac$_6$d. In the yeast enzyme, and possibly also in some plants, the VHA-c subunit is present in the complex in three variants encoded by separate genes (c, c$'$ and c$''$ present as c$_4$c$_1'$c$_1''$) (Nishi & Forgac, 2002).

The subunits are assembled in two domains or sectors (Nishi & Forgac, 2002; Kawasaki-Nishi *et al.*, 2003; Nelson, 2003). This is easily seen in a low-resolution structure of the plant V-ATPase (Domgall *et al.*, 2002). One domain (V$_1$) is responsible for ATP hydrolysis and the other domain (V$_0$) is responsible for H$^+$ translocation. The V$_1$ sector, facing the cytoplasm, is a peripheral domain of 500–600 kDa consisting of seven subunits, VHA-A to VHA-H. The V$_0$ sector is an integral domain embedded in the membrane. It is of 250–300 kDa and consists of four subunits, VHA-a, VHA-c, VHA-d and VHA-e.

V-ATPases are structurally related to F$_1$F$_0$-ATP synthases in mitochondria and chloroplasts (Nishi & Forgac, 2002; Kawasaki-Nishi *et al.*, 2003; Nelson, 2003). Although operating in the inverse direction and functioning in ATP synthesis, F$_1$F$_0$-ATPases likewise are multi-subunit complexes divided into a peripheral and an integral domain. The structure of F$_1$F$_0$-ATPases is known in considerable detail and their catalytic mechanism is well understood. Based upon structural similarities between V-ATPases and F$_1$F$_0$-ATPases, it has been hypothesized that V-ATPases catalyse ATP fuelled transport of H$^+$ by a rotary mechanism. Thus, hydrolysis of ATP by V$_1$ drives rotation of a ring of subunits in V$_0$. Rotation of this ring is tightly coupled to H$^+$ translocation from the cytosol to the lumen of the vacuole. Recent experiments with the yeast V-ATPase have provided strong evidence for a rotary mechanism (Hirata *et al.*, 2003; Imamura *et al.*, 2003; Yokoyama *et al.*, 2003).

The ATP-binding sites in V-ATPase are located on the A and B subunits, which are each present in three copies per holoenzyme complex. Subunit A contains a catalytic site and corresponds to the β subunit of F$_1$F$_0$-ATPase. Subunit B, which is analogous to the α subunit of F$_1$F$_0$-ATPase, also contains a nucleotide-binding site, but this site is non-catalytic and does not hydrolyze ATP. The A and B subunits alternate and form a spherical complex resembling an orange composed of six pieces.

The other subunits of the V$_1$ complex form (i) the central rotor and (ii) the peripheral stator. The central rotor, composed of subunits D and F, extends up into the central cavity of the A$_3$B$_3$ complex. The central rotor rotates within the A$_3$B$_3$ complex, which is kept in its position by the peripheral stator. Subunit D, which corresponds to the γ subunit of F$_1$F$_0$-ATPase, is rod-like and resembles a hairpin composed of two highly α-helical segments. Subunit F is smaller and outside the A$_3$B$_3$ complex. Its role seems to be bridging subunit D and the V$_0$ sector.

The peripheral stator serves a function much like the hand that holds a wine bottle (the V$_1$ sector) in which a corkscrew (the central rotor) is being inserted. Subunit E binds directly to the outside of subunit B and contributes together with two copies of subunit G to the stalk connecting V$_1$ with V$_0$. In the 22 Å electron microscopic image of the *Kalanchoë* V-ATPase, three peripheral stalks are visible (Domgall *et al.*, 2002), which might represent these subunits. Subunits C and H also contribute to the stator by associating with the stalks and $V_1$ and $V_o$, respectively. The peripheral stator binds to the $V_o$ sector through an interaction with a hydrophilic portion of subunit VHA-a, which is oriented towards the cytoplasm.

The VHA-c subunits of the $V_o$ sector forms a six-membered ring in the plane of the membrane. During $H^+$ pumping, this ring rotates like a carousel driven by the movement of the central rotor of the $V_1$ sector (Hirata *et al.*, 2003; Imamura *et al.*, 2003; Yokoyama *et al.*, 2003). The c subunit is a hydrophobic polypeptide of 16–17 kDa having four transmembrane domains. A variant of c, subunit c″, also present in the *Arabidopsis* genome, has four predicted transmembrane segments. A conserved glutamyl residue of subunit c (Glu-141 in *Arabidopsis* VHA-c), which is present in TM4 (TM3 in the c″ subunit), is essential for proton transport.

The ring of c subunits is adjacent to the large VHA-a subunit of about 90 kDa, which is an integral membrane protein facing both the cytosol and the lumen of the vacuole. The association of VHA-a with the functional plant V-ATPase complex has been debated (Li & Sze, 1999; Ratajczak, 2000) and is still controversial. According to a model (Kawasaki-Nishi *et al.*, 2003), based on an extension of the mechanism of $F_oF_1$-ATPase, the VHA-a subunit is likely to contain to half-channels that allow for the passage of $H^+$. One half-channel extends from the cytosol to a part of subunit a, which is in the middle of the plane of the membrane and in communication with the ring of c subunits. The other half-channel, which also communicates with the c ring, extends from the middle of the membrane in subunit a to the vacuolar side of the membrane. The two half-channels are not connected and therefore do not *per se* form a $H^+$ pathway through the membrane. In addition to subunits a and c, the $V_o$ sector contains two additional subunits (d and e) of unknown function.

The proposed rotary mechanism of V-ATPase involves entrance of an $H^+$ into the $V_o$ sector from the cytosolic half-channel of subunit a. At the middle of the membrane, the $H^+$ is liberated from subunit a and is donated to a c subunit via Glu-141. ATP hydrolysis at one of the A subunits in the $V_1$ sector results in rotation of subunit D and this movement is transmitted to the c ring to which subunit D is attached. The $H^+$ carried by Glu-141 of subunit c is driven out of the vicinity of subunit a and now faces the inner part of the lipid bilayer. Concomitantly, the neighbouring subunit c takes the place adjacent to the cytoplasmic half-channel of subunit a. An additional movement of the carousel is hypothesized to bring the protonated c subunit back to subunit a where it delivers the $H^+$ to the bottom of the other half-channel. The released $H^+$ is now free to diffuse towards the lumen of the vacuole.

### 3.2.2.4   Isoforms and expression in the plant

Several of the genes encoding subunits of V-ATPase are present in multiple copies in the genome (Kluge *et al.*, 2003). Thus, in *Arabidopsis*, VHA-B, VHA-E, VHA-G and VHA-a are present in three copies each encoding a distinct isoform of each of these subunits. VHA-d and VHA-e are present in two copies, whereas VHA-c is present in as many as five copies (among which, one is encoding the slightly different subunit c''). A very similar picture is true for the VHA genes in rize and *Mesembryanthemum crystallinum* (Kluge *et al.*, 2003).

The presence of multiple isoforms allows for differential expression of V-ATPase in different cells at different times of development and in response to external factors. Macroarray experiments with cDNA fragments of all *M. crystallinum* V-ATPase genes suggest that among those subunits encoded by multiple genes only few respond to stress conditions such as high salt (Kluge *et al.*, 2003). This might indicate that these isoforms have specific functions or, alternatively, that they are encoded by genes under control of specific stress-inducible promoters.

### 3.2.2.5 Regulation

V-ATPases are regulated at several levels but only little is known about their regulation in plants. Since V-ATPase is a complex of several subunits, some encoded by multiple genes, and is localized in different membrane systems within the cell, the regulation of V-ATPase during gene expression, protein synthesis, assembly, targeting and post-translational modification is likely to be very complex (Sze *et al.*, 1999).

Regulation of gene expression is especially evident when plants are exposed to salt stress (Barkla & Pantoja, 1996; Lüttge & Ratajczak, 1997). The halophyte *M. crystallinum* transports Na$^+$ into the vacuole by a Na$^+$/H$^+$ antiporter in the vacuolar membrane, which is driven by the activity of the V-ATPase. NaCl induces transcription of the 16-kDa subunit c (Tsiantis *et al.*, 1996).

Regulation of pump activity at the post-translational level might involve dissociation of the soluble V$_1$ sector from the integral V$_o$ sector. Such a mechanism was first identified in yeast in response to glucose depletion (Kane, 1995) and the process apparently involves the cytoplasmic domain of VHA-a (Kawasaki-Nishi *et al.*, 2001). Patch clamping of whole vacuoles reveals that the coupling ratio (i.e. the number of H$^+$ transported per ATP split) of red beet V-ATPase decreases when the pH of the vacuole becomes more acidic relative to the cytoplasm (Davies *et al.*, 1994). These results indicate that the coupling efficiency of the V-ATPase is subject to post-translational regulation. The underlying mechanism is not understood but could, in principle, involve dissociation of the V$_1$ sector from the V$_o$ sector.

V-ATPase activities can also be regulated by the oxidation/reduction of disulfide bonds located at the active site of the VHA-A subunit (Stevens & Forgac, 1997; Nishi & Forgac, 2002). Disulfide bond formation leads to reversible inhibition of pump activity.

### 3.2.3  Vacuolar pyrophosphatase

### 3.2.3.1  Physiological role

Vacuolar pyrophosphatase (V-PPase) translocates H$^+$ across the vacuolar membrane at the expense of inorganic pyrophosphate (PPi). PPi is an abundant by-product of cellular metabolism whose removal allows the shifting of the equilibrium of anabolic

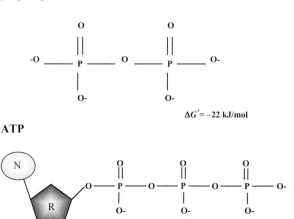

| Pyrophosphate and ATP |
| --- |

**Pyrophosphate (PPi)**

$$\Delta G' = -22 \text{ kJ/mol}$$

**ATP**

$$\Delta G' = -31 \text{ kJ/mol}$$

**Fig. 3.3** The pyrophosphate bridge in PPi and ATP. The structure P–O–P is known as the *pyrophosphate bridge*. This chemical group stores energy that can be readily used for biochemical reactions and is present in both PPi and ATP. PPi hydrolysis could be an important source of metabolic energy for the cells, as this molecule is produced as a by-product in many anabolic reactions (synthesis of polysaccharides, proteins, nucleic acids, lipids).

reactions towards biosynthesis. Although PPi was until recently considered a waste product, an increasing body of evidence indicates that it can play an important role in cellular bioenergetics and it has been suggested to be an alternative 'energy currency', different to ATP, in the bioenergetics of some modern cells (Fig. 3.3) (Lahti *et al.*, 1988). Because ATP is the major energy currency of cells, the activity of V-ATPase is thought to be paramount to that of the V-PPase. However, inorganic phosphate (PPi) is an important energy source for plant cells, and the V-PPase is a ubiquitous and active H$^+$ pump (Rea & Poole, 1993). Still, it is unclear how the two H$^+$ pumps localized to the vacuolar membrane complement one another.

It has been suggested that in mature tissues, where the metabolic rate and consequently PPi levels are low, V-PPase could work in a reversed manner, synthesizing PPi and keeping a constant supply of this metabolite for cytosolic PPi-dependent enzymes, such as pyrophosphate:fructose-6-phosphate 1-phosphotransferase [PFK(PPi), EC 2.7.1.90] or UDP-glucose pyrophosphorylase (Chanson *et al.*, 1985).

Thus, although cytosolic PPi scavenging seems to be the main role for V-PPase (Fig. 3.4), at least two other functions have been suggested:

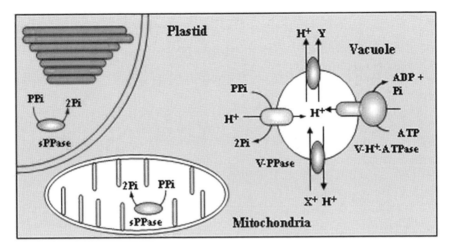

**Fig. 3.4** Physiological relationships between soluble and membrane-bound PPases in photosynthetic cells. PPi-hydrolyzing enzymes (inorganic pyrophosphatases, PPases) can be divided into two big groups. Soluble PPases (sPPases) hydrolyze inorganic pyrophosphate (PPi) releasing heat, thus contributing to the recovery of inorganic phosphate (Pi) pools inside the cell and allowing anabolic reactions to proceed in the direction of biosynthesis. Membrane-bound proton-translocating inorganic pyrophosphatases (H$^+$-PPases or V-PPases, for vacuolar PPases) are enzymes capable of recycling part of the energy contained in the pyrophosphate bond (P–O–P) by coupling the hydrolysis of this molecule to proton pumping across a membrane. In photosynthetic cells, sPPases are restricted to plastids and mitochondria, where PPi is hydrolyzed to recover the Pi pools required for ATP synthesis and to allow anabolic reactions to proceed in the direction of biosynthesis. As no sPPase has been found in the cytosol of these cells, V-PPases associated to the tonoplast are supposed to be the sole enzymes responsible for cytoplasmic PPi hydrolysis coupled to proton translocation, thus contributing to the vacuolar pH gradient (acidic inside).

1) The H$^+$-PPase supplements V-ATPase in creating an electrochemical gradient across the vacuolar membrane.

2) The H$^+$-PPase is involved in PPi synthesis, utilizing the electrochemical gradient across the vacuolar membrane.

The implication of V-PPase in tissue development seems quite clear, as higher levels of V-PPase hydrolytic and pumping activities had been found in young growing tissues of different plants with respect to mature cells (reviewed in Maeshima, 2000). This V-PPase activity decreased during development (in most cases accompanied also by a decrease in protein content), while both V-ATPase protein and activity remained almost constant. Thus, V-PPase was proposed to be responsible for hydrolyzing the high amount of PPi produced by intense cell metabolism in young tissues. This PPi hydrolysis would support the intense proton transport required for vacuolar expansion, so that V-PPase would be the main tonoplast proton pump in the early stages of development. In mature tissues, V-ATPase would assume that role, although both proton pumps remain active.

In the plant *M. crystallinum*, the decrease in V-PPase activity observed after induction of crassulean acid metabolism (CAM), both by ageing or induced by NaCl, was explained according to the lower metabolic rate in the CAM state (Bremberger & Lüttge, 1992); however, in *Kalanchoë daigremontiana*, V-PPase activity seems to be important for energetization of malate transport in the CAM state (Smith *et al.*, 1984).

Smart *et al.* (1998) studied V-PPase expression during cotton fibre development. This process involves three main stages: rapid cell expansion, secondary cell wall deposition and maturation. Both V-ATPase and V-PPase were found to be involved in the acidification of growing vacuoles at the first stage of development, but in this case, V-ATPase seemed to be supporting most of the vacuolar acidification and V-PPase was suggested to be involved in secondary wall deposition. Moreover, Marsh *et al.* (2000, 2001) proposed that V-ATPase was responsible for vacuolar acidification in maturating citrus fruits. In this case, V-PPase would be involved in energy conservation through PPi synthesis during fruit deacidification.

Two main questions are still pending: on one side, the fact that PPi levels remain constant under most stress conditions, in contrast to the dramatic decrease in ATP content (Plaxton, 1999) and the reversibility of both tonoplast proton pumps (Façanha & de Meis, 1998; Hirata *et al.*, 2000), opens the possibility that V-ATPases could be synthesizing ATP under conditions of reduced cellular energy content, by releasing part of the electrochemical gradient generated by V-PPases (thus contributing to the recovery of ATP pools); on the other side, the reduced proton-pumping capacity of V-PPase with respect to V-ATPases (Rea & Poole, 1993) and the need for a constant cytosolic PPi scavenging suggest a regulation of coupling between PPi hydrolysis and $H^+$ pumping. In this sense, V-PPase in acid lime fruits, which hyper-acidify their vacuoles to pH 2.0, were reported to hydrolyze PPi but were unable to pump protons against the high pH gradient generated by V-ATPase (Marsh *et al.*, 2001).

### 3.2.3.2   Structure and mechanism

V-PPases were first discovered biochemically in studies using membrane vesicles (reviewed in Rea & Poole, 1983; Rea *et al.*, 1992b) and was first cloned from *A. thaliana* (Sarafian *et al.*, 1992). V-PPase is characterized as being simple in its structure with a single catalytic subunit (Fig. 3.2). However, in contrast to the plasma membrane $H^+$-ATPase, it does not have large cytosolic hydrophilic domains and does not have a phosphorylated intermediate during catalysis.

All V-PPases characterized to date have been shown to hydrolyze PPi *in vitro* (Maeshima, 2000) and several have been reported to be capable of establishing a $H^+$ gradient of equal or greater magnitude than that generated by the V-$H^+$-ATPase (Rea & Poole, 1993; Terrier *et al.*, 1998). Nevertheless, different exchange experiments using radioactively labelled isotopes have demonstrated the reversibility of V-PPase (Baykov *et al.*, 1994; Façanha & de Meis, 1998), arousing doubts about its possible physiological role. It was not until recently that these enzymes were reported to have a hydrolytic activity *in vivo* using the functional complementation of a conditional yeast mutant for its cytosolic sPPase (Pérez-Castiñeira *et al.*, 2002b).

Two distinct biochemical subclasses of H$^+$-PPases have been characterized to date: (i) K$^+$-stimulated and (ii) K$^+$-insensitive H$^+$-PPases (Pérez-Castiñeira *et al.*, 2001a). For many years, K$^+$-stimulated isoforms were thought to be restricted to plants (Maeshima, 2000) and, more recently, also to parasitic protozoa (McIntosh & Vaidya, 2002), while K$^+$-independent isoforms had been characterized only in prokaryotes (Baltscheffsky & Baltscheffsky, 1992). The cloning and biochemical characterization of a K$^+$-independent V-PPase isoform from *Arabidopsis* (Drozdowicz *et al.*, 2000) and K$^+$-stimulated H$^+$-PPases from the bacteria *Thermotoga maritima* (Pérez-Castiñeira *et al.*, 2001b) and *Carboxydothermus hydrogenoformans* (Belogurov & Lahti, 2002) have shown that K$^+$ dependency is not a specific plant feature. Moreover, sequence comparison analyses allowed the identification of putative K$^+$-independent isoforms in different parasites (McIntosh & Vaidya, 2002; Pérez-Castiñeira *et al.*, 2002a) and a K$^+$-stimulated isoform in the methanogenic archaeon *Methansarcina mazei* (Bäumer *et al.*, 2002).

### 3.2.3.3 Isoforms and expression in the plant

Interestingly, for some plant species it has been possible to clone several cDNAs for V-PPases. Thus, in contrast to the single isoform present in plants like barley or mung bean, two isoforms have been cloned in red beet, at least three in tobacco, two in rice and another two in *Arabidopsis* (AVP1, K$^+$-stimulated; AVP2, K$^+$-insensitive), although the existence of a third isoform has been proposed for this plant (Drozdowicz *et al.*, 2000; Maeshima, 2000). Although most of them seem to be potassium dependent according to their sequence similarity to other known V-PPases, further biochemical characterization is needed in order to clarify the requirements of each isoform for full activity. On the other hand, preliminary studies suggest a different sub-cellular localization or a different expression pattern for different isoforms in the same plant (Lerchl *et al.*, 1995; Mitsuda *et al.*, 2001a,b). The discovery of several isoforms in the same plant arises the possibility that different isoforms with different biochemical characteristics are been used to cope with different physiological conditions.

V-PPases seem to be ubiquitous in plants as PPi-dependent H$^+$-translocation or K$^+$-stimulated PPi hydrolysis has been described in membranes of both monocotyledons and dicotyledons, C3, C4 and CAM plants (Rea & Poole, 1993) and in different tissues, such as leaves, roots, cotyledons, coleoptiles, hypocotyls, seedlings, aleurone tissue, fruits, ovules, pollen and the sieve element (Rea & Poole, 1993; Maeshima, 2000; Marsh *et al.*, 2000; Langhans *et al.*, 2001; Mitsuda *et al.*, 2001b; Wang *et al.*, 2001).

Although some V-PPases have been reported to be located at the Golgi cisternae (Chanson *et al.*, 1985; Mitsuda *et al.*, 2001a) and at the plasma membrane (Maeshima, 2000; Langhans *et al.*, 2001), most V-PPases seem to be associated to the tonoplast, where they coexist with the V-ATPase (Fig. 3.1) (Rea & Poole, 1993).

### 3.2.3.4  Regulation

Not much is known concerning regulation of V-PPase. Further physiological studies should be needed to characterize V-PPase regulation and some regulation properties might be specific of each plant family.

Saline and osmotic stresses have been quite extensively studied owing to their agricultural interest. An increase in V-PPase activity has been reported for salt-treated suspension cultures of *Daucus carota* (Colombo & Cerana, 1993), in salt adapted *Acer pseudoplatanus* cells (Zingarelli *et al.*, 1994) and in sunflower roots (Ballesteros *et al.*, 1996). In halophyte plants, a higher pyrophosphate-dependent $H^+$-pumping activity has also been described in leaves of *Saueda salsa* (Wang *et al.*, 2001) and roots of *Salicornia bigelovii* (Parks *et al.*, 2002) grown at high NaCl concentrations; however, other reports showed the inhibition of V-PPase under salt stress in mung bean roots (Matsumoto & Chung, 1988; Nakamura *et al.*, 1992), wheat roots (Wang *et al.*, 2000), cowpea hypocotyls (Otoch *et al.*, 2001) and *M. crystallinum* leaf cells (Bremberger & Lüttge, 1992). No effect of salt stress upon V-PPase activity was found in the halophyte *S. maritima* (Leach *et al.*, 1990) and cultured cells of *Nicotiana tabacum* (Reuveni *et al.*, 1990).

Some authors have suggested the possibility that these discrepancies in V-PPase response to salt treatments might be due to differences in the experimental conditions used (Ballesteros *et al.*, 1996) or that these responses are dependent on the plant species and, therefore, cannot be generalized (Wang *et al.*, 2001). In most cases, no effect on pyrophosphatase activity was observed after osmotic treatment with poly(ethylene glycol) or sorbitol, suggesting that regulation of V-PPase activity by salt is due to ionic stress *per se*.

Heterologous over-expression of AVP1 in a yeast mutant deficient in $Na^+$ extrusion led to the obtention of salt-tolerant colonies, which were shown to sequestrate $Na^+$ inside their vacuole (Gaxiola *et al.*, 1999). Over-expression of the same enzyme in *Arabidopsis* also conferred salt tolerance to the plants. The higher vacuolar solute content allowed plants to increase water accumulation, making them also drought-tolerant (Gaxiola *et al.*, 2001). Nevertheless, further functional characterization will be required in order to have a clear model for the role and regulation of V-PPase under salt stress. Comparison of glycophyte and halophyte plants responses might be of interest.

The regulation of V-PPase by the presence of other ions, such as $Cl^-$ or $K^+$, has also been studied in mung bean roots (Kasai *et al.*, 1993a). A marked decrease in V-PPase activity was found in membranes isolated from potassium-treated plants, while chloride-treated ones showed a significant increase in activity. $K^+$ had been proposed to cause cytoplasmic alkalinization in root hairs of *Limnobium stoloniferum*, while $Cl^-$ induced cytoplasmic acidification (Ullrich & Novacky, 1990). Thus, these authors propose that V-PPase response is not directly related to the presence of the ions, but to changes in cytoplasmic pH. Regulation of V-PPase through cytoplasmic pH had also been proposed for mung bean seedlings subjected to aluminium stress. Up-regulation of V-PPase activity was related to cytosol acidification caused by the presence of aluminium ions inside the cell (Kasai *et al.*, 1992).

A similar mechanism seems to be involved in V-PPase activity up-regulation under conditions of chilling and anoxia. Both types of stress generate an energy deficiency, caused by interruption of oxidative phosphorylation and mitochondrial disruption, respectively. This causes a drastic reduction in the cellular ATP content that can be somehow alleviated by induction of a fermentative metabolism. As an additional adaptation response to energy stress, part of the ATP-consuming reactions can be replaced by reactions driven by PPi (Carystinos et al., 1995). In contrast to the dramatic loss of ATP during anoxia or chilling, PPi levels seem to be unaffected by changes in the respiration rate or the metabolic state of the cell (Plaxton, 1999), further supporting this idea. Accordingly, an enhancement of V-PPase activity has been reported for mung bean hypocotyls under chilling (Darley et al., 1995) and in rice seedlings under anoxia or chilling (Carystinos et al., 1995).

V-PPase regulation under nutrients deficiency has also been studied. Palma et al. (2000) described the stimulation of V-PPase activity in suspension cell cultures of Brassica napus under phosphate starvation. These conditions are similar to those achieved by anoxia in the sense that cellular ATP pools are significantly reduced, while PPi levels remain unaffected (Plaxton, 1999). Kasai et al. (1998) studied the behaviour of both tonoplast proton pumps under mineral nutrients deficiency (MND) conditions. Enhancement of proton translocation across the tonoplast had been proposed to increase solute uptaking from the cell exterior (González-Reyes et al., 1994). A threefold increase in V-PPase activity was reported with only a slight change in ATPase activity; however, levels of PPi found in this case were significantly lower in MND-grown than in control plants. These authors propose that the high V-PPase activity reduces the PPi levels in the cytosol.

In order to shed some light upon the apparently contradictory results obtained in response to different stress situations, some attempts have been made to identify the factors that regulate V-PPase activity in vivo. The enzyme has been shown to be inhibited by Ca$^{2+}$ (Maeshima, 1991; Rea et al., 1992a) and in vivo treatment with a cytokinin (BA) (Kasai et al., 1993b). An increase in activity was observed when barley aleurone protoplasts were treated with gibberellic acid (Swanson & Jones, 1996); however, both stimulation by in vivo treatment (Kasai et al., 1993b) and inhibition by treatment of protoplasts (Swanson & Jones, 1996) with ABA have been described. In Chenopodium rubrum, V-PPase activity seems to be regulated by a lysolipid (Bille et al., 1992).

## 3.3   Calcium pumps in plant cells

### 3.3.1   Calcium in plant cells

The major part of the Ca$^{2+}$ in a plant is bound to oxalate in the apoplasm and the vacuole, or to pectates in the cell wall (Marschner, 1995). On the contrary, the concentration of free Ca$^{2+}$ in the cytoplasm is very low, in the range of 0.1–0.2 $\mu$M, which is several orders of magnitude lower than in the apoplast. The low cytoplasmic Ca$^{2+}$

concentration may result from the fact that the solubility product of $Ca^{2+}$ with orthophosphate (Pi) is very low. It is thus necessary for the cell to have mechanisms that keep the concentration of free $Ca^{2+}$ in the cytoplasm at a low level in order to avoid precipitation of Pi, which would otherwise interfere with the metabolism of the cell (Sanders et al., 1999). Also, competition with $Mg^{2+}$ for binding sites might be a problem at higher concentrations of $Ca^{2+}$ (Marschner, 1995). The large difference in concentration of $Ca^{2+}$ between the cytoplasm on one hand and the apoplasm and intracellular compartments on the other hand, is probably the reason for the use of $Ca^{2+}$ signals as a link between extracellular stimuli and intracellular responses in many different physiological pathways. Changes in the cytoplasmic concentration of $Ca^{2+}$ has thus been reported to occur in response to several different hormonal and environmental signals such as abscisic acid, gibberellic acid, auxin, NaCl, anoxia, touch, wind, gravity, temperature, drought, osmotic stress, red light, ozone, oxidative stress, aluminium, pathogen elicitors and NOD factors (reviewed by Sanders et al., 1999; Reddy, 2001).

### 3.3.2    $Ca^{2+}$-ATPases ($P_2$ ATPases)

#### 3.3.2.1    Physiological role
$Ca^{2+}$-ATPases are important for maintaining the cytoplasmic $Ca^{2+}$ concentration at a low level. The cytoplasmic concentration of $Ca^{2+}$ is maintained by the activity of $Ca^{2+}/H^+$ antiporters and $Ca^{2+}$-ATPases, which transport $Ca^{2+}$ into intracellular compartments or out of the cell to the apoplasm (Reddy, 2001; Sanders et al., 2002). $Ca^{2+}/H^+$ antiporters have a high capacity but low affinity for $Ca^{2+}$ and are primarily found in the vacuolar membrane, although in some plants they also seem to be located in the plasma membrane (Bush, 1995; Vicente et al., 1995; Kasai & Muto, 1990), while $Ca^{2+}$-ATPases have a low capacity but high affinity for $Ca^{2+}$ and are found in the plasma membrane and in several different endomembranes (Fig. 3.5) (reviewed by Evans & Williams, 1998). It is thus speculated that the physiological role of $Ca^{2+}/H^+$ antiporters is to remove large amounts of $Ca^{2+}$ from the cytosol after a $Ca^{2+}$ signal, while $Ca^{2+}$-ATPases maintain the very low resting stage level of $Ca^{2+}$ (Hirschi, 2001; White & Broadley, 2003). Whether $Ca^{2+}/H^+$ antiporters and/or $Ca^{2+}$-ATPases also have a function in shaping the dynamic form of the $Ca^{2+}$ signals is still debated (Sanders et al., 2002).

#### 3.3.2.2    Genetics
Knockout mutants have been isolated for all Arabidopsis $Ca^{2+}$-ATPases, but only for very few of these knockouts an altered phenotype has been found (Schiøtt et al., 2004).

Knockout mutants of the $Ca^{2+}$ pump ECA1 showed $Ca^{2+}$-deficiency symptoms like small plant size, short roots, small yellowish leaves and lack of bolts, when grown on low levels of $Ca^{2+}$ (0.2–0.4 mM) (Wu et al., 2002). When grown on high levels of $Mn^{2+}$ (0.5 mM), the eca1 mutant showed stunted growth, chlorotic and deformed leaves, decreased content of chlorophyll, inhibited growth and elongation

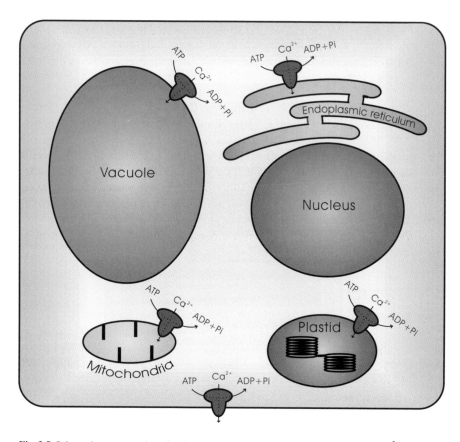

**Fig. 3.5** Schematic representation of a plant cell showing the sub-cellular localization of Ca²⁺-ATPases.

of root hairs. The phenotypes associated with growth on low concentrations of $Ca^{2+}$ may suggest a function for ECA1 in keeping a high concentration of $Ca^{2+}$ in the endoplasmic reticulum necessary for proper function of the secretory pathway. An additional function of ECA1 may be to scavenge excess $Mn^{2+}$ into the endoplasmic reticulum, thus keeping a low cytosolic level. When the gene is disrupted, and the total amount of $Mn^{2+}$ in the plant simultaneously is high owing to high exterior concentrations, the cytosolic level of $Mn^{2+}$ may effectively occlude other divalent cations from their normal binding sites, thus inhibiting transport of these into cellular compartments.

ACA9 is located in the pollen tube plasma membrane, and knockout mutants of ACA9 are found to have reduced growth of pollen tubes (Schiøtt *et al.*, 2004). Mutant pollen tubes have slower growth rate than do wild-type pollen tubes, and the mutant pollen tubes fail to reach the ovules in the lower 20% of the pistil. Furthermore, about 50% of the pollen tubes that reach an ovule fail to rupture and release the sperm cells. Overall, the seed set of homozygous *aca9* plants is reduced to 10–20%

of wild-type plants. $Ca^{2+}$ is known to be important for pollen tube growth (reviewed by Franklin-Tong, 1999; Holdaway-Clarke & Hepler, 2003) and ACA9 might play an important role in controlling cytoplasmic $Ca^{2+}$ during this process.

### 3.3.2.3  Structure and mechanism

$Ca^{2+}$-ATPases are divided into two groups, type IIA and type IIB (Fig. 3.6) (Geisler et al., 2000a; Axelsen & Palmgren, 2001). Both types consist of a single polypeptide of 998–1086 amino acids. Type IIA $Ca^{2+}$-ATPases show similarity to animal $Ca^{2+}$ pumps found in the sarcoplasmic or endoplasmic reticulum, while type IIB $Ca^{2+}$-ATPases show similarity to animal CaM-stimulated $Ca^{2+}$-ATPases found in the plasma membrane (Geisler et al., 2000a). Animal homologues of the type IIB $Ca^{2+}$-ATPases have a C-terminal calmodulin (CaM) binding domain, while the plant type IIB $Ca^{2+}$-ATPases have an N-terminal CaM binding domain.

As $Ca^{2+}$-ATPases belong to the family of P-type ATPases, their reaction mechanism follows the general scheme outlined in the section about the plasma membrane $H^+$-ATPase in this chapter.

### 3.3.2.4  Isoforms and expression in the plant

By homology search in the Arabidopsis Genome Database, four type IIA and ten type IIB $Ca^{2+}$-ATPases have been identified (Axelsen & Palmgren, 2001). ECA1 is located in the endoplasmic reticulum (Liang et al., 1997), ACA1 is located in the plastid inner envelope (Huang et al., 1993), ACA2 is located in the endoplasmic reticulum and perhaps also in the nuclear envelope (Harper et al., 1998), ACA4 is located in endomembranes (Geisler et al., 2000b), while ACA8 and ACA9 are located in the plasma membrane (Bonza et al., 2000; Schiøtt et al., 2004). Plant type IIB $Ca^{2+}$-ATPases are thus not restricted to the plasma membrane as are their

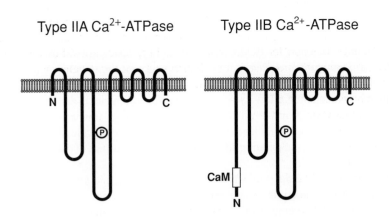

**Fig. 3.6** Schematic model of plant type IIA and type IIB $Ca^{2+}$-ATPases. An encircled P marks the position of the phosphorylated aspartate residue which is formed during the reaction cycle. CaM: calmodulin-binding domain; C: C-terminus; N: N-terminus; P: phosphate.

animal homologues. The remaining *Arabidopsis* Ca$^{2+}$-ATPases have not yet been localized.

The physiological function of Ca$^{2+}$-ATPases can also be deduced from studies of gene expression. In several different plant species like barley (Garciadeblas *et al.*, 2001), *Arabidopsis* (Geisler *et al.*, 2000b), tobacco (Perezprat *et al.*, 1992), tomato (Wimmers *et al.*, 1992) and soybean (Chung *et al.*, 2000), it has thus been reported that expression of Ca$^{2+}$-ATPases increases when the plant is subjected to salt stress. The concentration of cytosolic Ca$^{2+}$ is seen to rise upon salt stress (Lynch *et al.*, 1989; Knight *et al.*, 1997; Kiegle *et al.*, 2000; DeWald *et al.*, 2001; Moore *et al.*, 2002; Halperin *et al.*, 2003), which probably is part of a stress-signalling pathway that ultimately will adapt the cell to the high salt levels. It thus seems logical that expression of Ca$^{2+}$-ATPases must be increased in order for the cell to get rid of the incoming Ca$^{2+}$. Also, Ca$^{2+}$ normally directed to the vacuole might need to be redirected to other compartments in order to release space in the vacuole for sequestration of Na$^+$ (Blumwald *et al.*, 2000). Finally, Ca$^{2+}$ has a stabilizing effect on membranes and cell walls (Caldwell & Haug, 1981; Legge *et al.*, 1982; Marschner, 1995), and one of the toxic effects of salt seems to be the displacement of Ca$^{2+}$ ions by Na$^+$ ions from cation-binding sites on the plasma membrane and in the cell wall (Lynch *et al.*, 1987). This is illustrated by the fact that the toxic effects of salt can be counteracted by supplementation of exogenous Ca$^{2+}$ to the growth medium (Lahaye & Epstein, 1971; Banuls *et al.*, 1991). The increased expression of Ca$^{2+}$-ATPases upon salt stress might lead to increased transport of Ca$^{2+}$ released from intracellular stores into the apoplast, which could be beneficial for the plant in order to occlude Na$^+$ from cation-binding sites in the cell wall and plasma membrane.

### 3.3.2.5   *Regulation*

Plant type IIB Ca$^{2+}$ pumps have a large intracellular N-terminus harbouring a CaM-binding domain (Harper *et al.*, 1998; Bonza *et al.*, 2000; Geisler *et al.*, 2000a,b). At low cytoplasmic concentrations of Ca$^{2+}$, this domain binds to an intramolecular receptor, which keeps the pump in an inactivated state. At high levels of Ca$^{2+}$, CaM will take up a conformation that increases its affinity for the CaM-binding domain of the Ca$^{2+}$ pump, thus occluding the binding to the intramolecular receptor. This brings the pump into an activated state in which Ca$^{2+}$ ions are transported across the membrane under catalysis of ATP. Gel overlay experiments showed that Ca$^{2+}$-dependent CaM binding to the ACA2 isoform was dependent on the presence of the first 36 amino acids (Harper *et al.*, 1998).

In addition to regulation by CaM, the activity of animal type IIB Ca$^{2+}$-ATPases is also regulated by direct phosphorylation by protein kinase A and/or C. Phosphorylation of the CaM-binding domain prevents binding of CaM, and thus inhibits the CaM-stimulated activity of the pump, although the basal activity is increased (Wang *et al.*, 1991; Enyedi *et al.*, 1997). A similar regulation by phosphorylation of type IIB Ca$^{2+}$-ATPases in plants was identified by Hwang *et al.* (2000), who found that the CaM-stimulated activity of ACA2 was inhibited by phosphorylation of Ser-45 near the CaM-binding site by a Ca$^{2+}$-dependent protein kinase. However, in contrary to the findings for animal type IIB $Ca^{2+}$-ATPases, the basal activity of ACA2 was decreased by the phosphorylation.

Finally, the activity of type IIB $Ca^{2+}$-ATPases is also regulated by acidic phospholipids (Bonza *et al.*, 2001). Addition of phosphatidylinositol 4,5-diphosphate or phosphatidylinositol 4-monophosphate to plasma membranes purified from radish (*Raphanus sativus*) thus increased the plasma membrane $Ca^{2+}$-ATPase activity by approximately 300%, which was about the same as observed in CaM-treated plasma membranes, or in membranes treated with trypsin, which cleaves off the CaM-binding domain from the $Ca^{2+}$-ATPase and makes it constitutively active.

Type IIA $Ca^{2+}$-ATPases do not have an autoinhibitory domain like type IIB $Ca^{2+}$-ATPases. Animal type IIA $Ca^{2+}$-ATPases are instead inhibited by the 52-amino acid membrane protein phospholamban, which acts as an external inhibitory domain. The affinity between the $Ca^{2+}$-ATPase and phospholamban is decreased by high concentrations of $Ca^{2+}$ (James *et al.*, 1989; Asahi *et al.*, 2000), and phosphorylation of phospholamban has also been found to relieve the inhibition of the $Ca^{2+}$-ATPase, although the physical interaction with phospholamban may remain (Chiesi *et al.*, 1991; Asahi *et al.*, 2000). The relief of the inhibition includes an increased sensitivity to $Ca^{2+}$ as well as an increase in maximal velocity of the pump (Kargacin *et al.*, 1998). However, phospholamban has not been identified in plants, and no putative phospholamban binding motifs have been identified. The regulation of these ATPases in plants thus remains unknown (Geisler *et al.*, 2000a).

### 3.4   Other plant cation pumps

In addition to the pumps described above, plants harbour a number of additional cation pumps, which all belongs to the P-type ATPase family. P-type ATPases can be divided into five clades ($P_1$ to $P_5$) according to substrate specificity (Axelsen & Palmgren, 1998). Apart from the $H^+$ and $Ca^{2+}$ pumps ($P_3$ and $P_2$ pumps, respectively), there are heavy metal ATPases ($P_1$ pumps; described in Chapter 12), $P_4$ and $P_5$ ATPases.

The substrate specificities of the $P_4$ and $P_5$ ATPases have not yet been determined, and their physiological functions are poorly understood. The pumps seem to be widespread among eukaryotes but are absent from prokaryotes (Axelsen & Palmgren, 1998). In *Arabidopsis*, the $P_4$ subfamily contains 12 members while the $P_5$ subfamily comprises only one protein (Axelsen & Palmgren, 2001).

Evidence from reverse genetics experiments suggests that the $P_4$ ATPases could be important for cold tolerance in *Arabidopsis* (Gomès *et al.*, 2000). The size of adult *ALA1* antisense plants is severely reduced by growth at low temperature (Gomès *et al.*, 2000). The effect on cold sensitivity may occur through regulation of transmembrane phospholipid asymmetry, since ALA1 and a number of $P_4$ ATPases from other organisms have been implicated in transport of phospholipids from one membrane bilayer to the other (Tang *et al.*, 1996; Ding *et al.*, 2000; Gomès *et al.*, 2000;

Pomorski *et al.*, 2003). However, direct biochemical evidence for phospholipid transport activity of the P$_4$ ATPases has not been produced.

The single P$_5$ ATPase in *Arabidopsis* is required for proper development of the male gametes (Jakobsen *et al.*, 2004). T-DNA insertional mutants of the P$_5$ ATPase produce deformed pollen grains that cannot be released from the open anther locules at anthesis. As a result, mutant plants have a markedly reduced fertility. The P$_5$ ATPase is highly expressed in the tapetum at the time of microspore release from the tetrads that result from meiosis (Jakobsen *et al.*, 2004). Pollen development in *Arabidopsis* is highly dependent on the secretory activities of the tapetum and the absence of the P$_5$ ATPase may influence the secretory pathway within these cells.

## Acknowledgements

We thank Drs J.R. Pérez-Castiñeira and A. Serrano for critical reading of the manuscript and M.C. López Martín for valuable discussions.

## References

Allen, G.J., Chu, S.P., Schumacher, K., *et al.* (2000) Alteration of stimulus-specific guard cell calcium oscillations and stomatal closing in *Arabidopsis det3* mutant, *Science*, **289**, 2338–2342.

Asahi, M., McKenna, E., Kurzydlowski, K., Tada, M. & MacLennan, D.H. (2000) Physical interactions between phospholamban and sarco(endo)plasmic reticulum Ca$^{2+}$-ATPases are dissociated by elevated Ca$^{2+}$, but not by phospholamban phosphorylation, vanadate, or thapsigargin, and are enhanced by ATP, *J. Biol. Chem.*, **275**, 15034–15038.

Axelsen, K.B. & Palmgren, M.G. (1998) Evolution of substrate specificities in the P-type ATPase superfamily, *J. Mol. Evol.*, **46**, 84–101.

Axelsen, K.B. & Palmgren, M.G. (2001) Inventory of the superfamily of P-type ion pumps in *Arabidopsis*, *Plant Physiol.*, **126**, 696–706.

Ballesteros, E., Donaire, J.P. & Belver, A. (1996) Effects of salt stress on H$^+$-ATPase and H$^+$-PPase activities of tonoplast-enriched vesicles isolated from sunflower roots, *Physiol. Plant.*, **97**, 259–268.

Baltscheffsky, M. & Baltscheffsky, H. (1992) Inorganic pyrophosphate and inorganic pyrophosphatases, in *Molecular Mechanisms in Bioenergetics* (ed. L. Ernster), Elsevier Science Publishers B.V., Amsterdam, pp. 331–348.

Banuls, J., Legaz, F. & Primomillo, E. (1991) Salinity–calcium interactions on growth and ionic concentration of citrus plants, *Plant Soil*, **133**, 39–46.

Barkla, B.J. & Pantoja, O. (1996) Physiology of ion transport across the tonoplast of higher plants, *Annu. Rev. Plant Physiol. Plant Mol. Biol.* **47**, 159–184.

Bäumer, S., Lentes, S., Gottschalk, G. & Deppenmeier, U. (2002) Identification and analysis of proton-translocating pyrophosphatases in the methanogenic archaeon *Methansarcina mazei*, *Archaea*, **1**, 1–7.

Baunsgaard, L., Fuglsang, A.T., Jahn, T., Korthout, H.A.A., de Boer, A.H. & Palmgren, M.G. (1998) The 14-3-3 proteins associate with the plant plasma membrane H$^+$-ATPase to generate a fusicoccin binding complex and a fusicoccin responsive system, *Plant J.*, **13**, 661–671.

Baunsgaard, L., Venema, K., Axelsen, K.B., *et al.* (1996) Modified plant plasma membrane H$^+$-ATPase with improved transport coupling efficiency identified by mutant selection in yeast, *Plant J.*, **10**, 451–458.

Baykov, A.A., Kasho, V.N., Bakuleva, N.P. & Rea, P.A. (1994) Oxygen exchange reactions catalysed by vacuolar $H^+$-translocating pyrophosphatase. Evidence for reversible formation of enzyme-bound pyrophosphate, *FEBS Lett.*, **350**, 323–327.

Belogurov, G.A. & Lahti, R. (2002) A lysine substitute for $K^+$: a 460K mutation eliminates $K^+$-dependence in $H^+$-pyrophosphatase of *Carboxydothermus hydrogenoformans*, *J. Biol. Chem.*, **277**, 49651–49654.

Bille, J., Weiser, T. & Bentrup, F.-W. (1992) The lysolipid sphingosine modulates pyrophosphatase activity in tonoplast vesicles and isolated vacuoles from a heterotrophic cell suspension culture of *Chenopodium rubrum*, *Physiol. Plant.*, **84**, 250–254.

Blumwald, E., Aharon, G.S. & Apse, M.P. (2000) Sodium transport in plant cells, *Biochim. Biophys. Acta*, **1465**, 140–151.

Bonza, M.C., Luoni, L. & De Michelis, M.I. (2001) Stimulation of plant plasma membrane $Ca^{2+}$-ATPase activity by acidic phospholipids, *Physiol. Plant*, **112**, 315–320.

Bonza, M.C., Morandini, P., Luoni, L., Geisler, M., Palmgren, M.G. & De Michelis, M.I. (2000) At-ACA8 encodes a plasma membrane-localized calcium-ATPase of *Arabidopsis* with a calmodulin-binding domain at the N terminus, *Plant Physiol.*, **123**, 1495–1506.

Bouche-Pillon, S., Fleurat-Lessard, P., Fromont, J.-C., Serrano, R. & Bonnemain, J.-L. (1994) Immunolocalization of the plasma membrane $H^+$-ATPase in minor veins of *Vicia faba* in relation to phloem loading, *Plant Physiol.*, **105**, 691–697.

Boutry, M., Michelet, B. & Goffeau, A. (1989) Molecular cloning of a family of plant genes encoding a protein homologous to plasma membrane $H^+$-translocating ATPases, *Biochem. Biophys. Res. Commun.*, **162**, 567–574.

Bremberger, C. & Lüttge, U. (1992) Dynamics of tonoplast proton pumps and other tonoplast proteins of *Mesembryanthenum crystallinum* L. during the induction of Crassulacean Acid Metabolism, *Planta*, **188**, 575–580.

Buch-Pedersen, M.J. & Palmgren, M.G. (2003) Conserved Asp684 in transmembrane segment M6 of the plant plasma membrane P-type proton pump AHA2 is a molecular determinant of proton translocation, *J. Biol. Chem.*, **278**, 17845–17851.

Buch-Pedersen, M.J., Venema, K., Serrano, R. & Palmgren, M.G. (2000) Abolishment of proton pumping and accumulation in the $E_1P$ conformational state of a plant plasma membrane $H^+$-ATPase by substitution of a conserved aspartyl residue in transmembrane segment 6, *J. Biol. Chem.*, **275**, 39167–39173.

Bukrinsky, J.T., Buch-Pedersen, M.J., Larsen, S. & Palmgren, M.G. (2001) A putative proton binding site of plasma membrane $H^+$-ATPase identified through homology modelling, *FEBS Lett.*, **494**, 6–10.

Bush D.S. (1995) Calcium regulation in plant-cells and its role in signaling, *Annu. Rev. Plant Physiol. Plant Mol. Biol.*, **46**, 95–122.

Caldwell, C.R. & Haug, A. (1981) Temperature-dependence of the barley root plasma membrane-bound $Ca^{2+}$-dependent and $Mg^{2+}$-dependent ATPase, *Physiol. Plant.*, **53**, 117–124.

Carystinos, G.D., MacDonald, H.R., Monroy, A.F., Dhindsa, R.S. & Poole, R.P. (1995) Vacuolar $H^+$-translocating pyrophosphatase is induced by anoxia or chilling in seedlings of rice, *Plant Physiol.*, **108**, 641–649.

Chanson, A., Fichmann, J., Spear, D. & Taiz, L. (1985) Pyrophosphate-driven proton transport by microsomal membranes of corn coleoptiles, *Plant Physiol.*, **79**, 159–164.

Chiesi, M., Vorherr, T., Falchetto, R., Waelchli, C. & Carafoli, E. (1991) Phospholamban is related to the autoinhibitory domain of the plasma-membrane $Ca^{2+}$-pumping ATPase, *Biochemistry*, **30**, 7978–7983.

Chung, W.S., Lee, S.H., Kim, J.C., *et al.* (2000) Identification of a calmodulin-regulated soybean $Ca^{2+}$-ATPase (SCA1) that is located in the plasma membrane, *Plant Cell*, **12**, 1393–1407.

Colombo, R. & Cerana, R. (1993) Enhanced activity of tonoplast pyrophosphatases in NaCl-grown cells of *Daucus carota*, *J. Plant Physiol.*, **142**, 226–229.

Darley, C.P., Davies, J.M. & Sanders, D. (1995) Chill-induced changes in the activity and abundance of the vacuolar proton-pumping pyrophosphatase from mung bean hypocotyls, *Plant Physiol.*, **109**, 659–665.

Davies, J.M., Hunt, I. & Sanders, D. (1994) Vacuolar H$^+$-pumping ATPase: variable transport coupling ratio controlled by pH, *Proc. Natl. Acad. Sci. U.S.A.*, **91**, 8547–8551.

DeWald, D.B., Torabinejad, J., Jones, C.A., *et al.* (2001) Rapid accumulation of phosphatidylinositol 4,5-bisphosphate and inositol 1,4,5-trisphosphate correlates with calcium mobilization in salt-stressed *Arabidopsis*, *Plant Physiol.*, **126**, 759–769.

DeWitt, N.D., Harper, J.F. & Sussman, M.R. (1991) Evidence for a plasma membrane proton pump in phloem cells of higher plants, *Plant J.*, **1**, 121–128.

DeWitt, N.D. & Sussman, M.R. (1995) Immunological localization of an epitope-tagged plasma membrane proton pump (H$^+$-ATPase) in phloem companion cells, *Plant Cell*, **7**, 2053–2067.

Dietz, K.J., Tavakoli, N., Kluge, C., *et al.* (2001) Significance of the V-type ATPase for the adaptation to stressful growth conditions and its regulation on the molecular and biochemical level, *J. Exp. Bot.*, **52**, 1969–1980.

Ding, J.T., Wu, Z., Crider, B.P., *et al.* (2000) Identification and functional expression of four isoforms of ATPase II, the putative aminophospholipid translocase – effect of isoform variation on the ATPase activity and phospholipid specificity, *J. Biol. Chem.*, **275**, 23378–23386.

Domgall, I., Venzke, D., Luttge, U., Ratajczak, R. & Bottcher, B. (2002) Three-dimensional map of a plant V-ATPase based on electron microscopy, *J. Biol. Chem.*, **277**, 13115–13121.

Drozdowicz, Y.M., Kissinger, J.C. & Rea, P.A. (2000) AVP2, a sequence-divergent, K$^+$-insensitive, H$^+$-translocating inorganic pyrophosphatase from *Arabidopsis*, *Plant Physiol.*, **123**, 353–362.

Enyedi, A., Elwess, N.L., Filoteo, A.G., Verma, A.K., Paszty, K. & Penniston, J.T. (1997) Protein kinase C phosphorylates the 'a' forms of plasma membrane Ca$^{2+}$ pump isoforms 2 and 3 and prevents binding of calmodulin, *J. Biol. Chem.*, **272**, 27525–27528.

Evans, D.E. & Williams, L.E. (1998) P-type calcium ATPases in higher plants – biochemical, molecular and functional properties, *Biochim. Biophys. Acta*, **1376**, 1–25.

Façanha, A.R. & de Meis, L. (1998) Reversibility of H$^+$-ATPase and H$^+$-pyrophosphatase in tonoplast vesicles from maize coleoptiles and seeds, *Plant Physiol.*, **116**, 1487–1495.

Franklin-Tong, V.E. (1999) Signaling and the modulation of pollen tube growth, *Plant Cell*, **11**, 727–738.

Fromard, L., Babin, V., Fleurat-Lessard, P., Fromont, J.C., Serrano, R. & Bonnemain, J.L. (1995) Control of vascular sap pH by the vessel-associated cells in woody species (physiological and immunological studies), *Plant Physiol.*, **108**, 913–918.

Fuglsang, A.T., Visconti, S., Drumm, K., *et al.* (1999) Binding of 14-3-3 protein to the plasma membrane H$^+$-ATPase AHA2 involves the three C-terminal residues Tyr$^{946}$-Thr-Val and requires phosphorylation of Thr$^{947}$, *J. Biol. Chem.*, **274**, 36774–36780.

Fullone, M.R., Visconti, S., Marra, M., Fogliano, V. & Aducci, P. (1998) Fusicoccin effect on the in vitro interaction between plant 14-3-3 proteins and plasma membrane H$^+$-ATPase, *J. Biol. Chem.*, **273**, 7698–7702.

Garciadeblas, B., Benito, B. & Rodriguez-Navarro, A. (2001) Plant cells express several stress calcium ATPases but apparently no sodium ATPase, *Plant Soil*, **235**, 181–192.

Gaxiola, R.A., Li, J., Undurraga, S., *et al.* (2001) Drought- and salt-tolerant plants result from overexpression of the AVP1 H$^+$-pump, *Proc. Natl. Acad. Sci. U.S.A.*, **98**, 11444–11449.

Gaxiola, R.A., Rao, R., Sherman, A., Grisafi, P., Alper, S.L. & Fink, G.R. (1999) The *Arabidopsis thaliana* proton transporters, AtNhx1 and AVP1, can function in cation detoxification in yeast, *Proc. Natl. Acad. Sci. U.S.A.*, **96**, 1480–1485.

Geisler, M., Axelsen, K.B., Harper, J.F. & Palmgren, M.G. (2000a) Molecular aspects of higher plant P-type Ca$^{2+}$-ATPases, *Biochim. Biophys. Acta*, **1465**, 52–78.

Geisler, M., Frangne, N., Gomes, E., Martinoia, E. & Palmgren, M.G. (2000b) The *ACA4* gene of *Arabidopsis* encodes a vacuolar membrane calcium pump that improves salt tolerance in yeast, *Plant Physiol.*, **124**, 1814–1827.

Gomès, E., Jakobsen, M.K., Axelsen K.B., Geisler M. & Palmgren, M.G. (2000) Chilling tolerance in *Arabidopsis* involves ALA1, a member of a new family of putative aminophospholipid translocases, *Plant Cell*, **12**, 2441–2453.

González-Reyes, J.A., Hidalgo, A., Caler, J.A., Palos, R. & Navas, P. (1994) Nutrient uptake changes in ascorbate free radical-stimulated onion roots, *Plant Physiol.*, **104**, 271–276.

Hager, A. (2003) Role of the plasma membrane $H^+$-ATPase in auxin-induced elongation growth: historical and new aspects, *J. Plant Res.* **116**, 483–505.

Halperin, S.J., Gilroy, S. & Lynch, J.P. (2003) Sodium chloride reduces growth and cytosolic calcium, but does not affect cytosolic pH, in root hairs of *Arabidopsis thaliana* L., *J. Exp. Bot.*, **54**, 1269–1280.

Harper, J.F., Hong, B., Hwang, I., *et al.* (1998) A novel calmodulin-regulated $Ca^{2+}$-ATPase (ACA2) from *Arabidopsis* with an N-terminal autoinhibitory domain, *J. Biol. Chem.*, **273**, 1099–1106.

Harper, J.F., Manney, L. & Sussman, M.R. (1994) The plasma membrane $H^+$-ATPase gene family in *Arabidopsis*: genomic sequence of *AHA10* which is expressed primarily in developing seeds, *Mol. Gen. Genet.*, **244**, 572–587.

Herman, E.M., Li, X., Su, R.T., Larsen, P., Hsu, H.-T. & Sze, H. (1994) Vacuolar-type $H^+$-ATPases are associated with the endoplasmic reticulum and provacuoles of root tip cells, *Plant Physiol.*, **106**, 1313–1324.

Hirata, T., Iwamoto-Kihara, A., Sun-Wada, G.H., Okajima, T., Wada, Y. & Futai, M. (2003) Subunit rotation of vacuolar-type proton pumping ATPase: relative rotation of the G and C subunits, *J. Biol. Chem.*, **278**, 23714–23719.

Hirata, T., Nakamura, N., Omote, H., Wada, Y. & Futai, M. (2000) Regulation and reversibility of vacuolar $H^+$-ATPase, *J. Biol. Chem.*, **275**, 386–389.

Hirschi K. (2001) Vacuolar $H^+/Ca^{2+}$ transport: who's directing the traffic? *Trends Plant Sci.*, **6**, 100–104.

Hirschi, K.D., Zhen, R.-G., Cunningham, K.W., Rea, P.A. & Fink, G.R. (1996) CAX1, an $H^+/Ca^{2+}$ antiporter from *Arabidopsis*, *Proc. Natl. Acad. Sci. U.S.A.*, **93**, 8782–8786.

Holdaway-Clarke, T.L. & Hepler, P (2003) Control of pollen tube growth: role of ion gradients and fluxes, *New Phytol.*, **159**, 539–563.

Houlne, G. & Boutry, M. (1994) Identification of an *Arabidopsis thaliana* gene encoding a plasma membrane $H^+$-ATPase whose expression is restricted to anther tissues, *Plant J.*, **5**, 311–317.

Huang, L., Berkelman, T., Franklin, A.E. & Hoffman, N.E. (1993) Characterization of a gene encoding a $Ca^{2+}$-ATPase-like protein in the plastid envelope, *Proc. Natl. Acad. Sci. U.S.A.*, **90**, 10066–10070.

Hwang, I., Sze, H. & Harper, J.F. (2000) A calcium-dependent protein kinase can inhibit a calmodulin-stimulated $Ca^{2+}$ pump (ACA2) located in the endoplasmic reticulum of *Arabidopsis*, *Proc. Natl. Acad. Sci. U.S.A.*, **97**, 6224–6229.

Imamura, H., Nakano, M., Noji, H., *et al.* (2003) Evidence for rotation of V1-ATPase, *Proc. Natl. Acad. Sci. U.S.A.*, **100**, 2312–2315.

Jahn, T., Baluska, F., Michalke, W., Harper, J.F. & Volkmann, D. (1998) Plasma membrane $H^+$-ATPase in the root apex: evidence for strong expression in xylem parenchyma and asymmetric localization within cortical and epidermal cells, *Physiol. Plant.*, **104**, 311–316.

Jahn, T., Fuglsang, A.T., Olsson, A., *et al.* (1997) The 14-3-3 protein interacts directly with the C-terminal region of the plant plasma membrane $H^+$-ATPase, *Plant Cell*, **9**, 1805–1814.

Jakobsen, M.K., Moeller, A., Amtmann, A. & Palmgren, M.G. (2004) Male gametogenesis is impaired by disruption of the $P_5$ ATPase in *Arabidopsis thaliana*, in *Abstract P1-26, 13th International Workshop on Plant Membrane Biology*, July 6–10, 2004, Agro Montpellier, France.

James, P., Inui, M., Tada, M., Chiesi, M. & Carafoli, E. (1989) Nature and site of phospholamban regulation of the $Ca^{2+}$ pump of sarcoplasmic reticulum, *Nature*, **342**, 90–92.

Jorgensen, P.L., Hakansson, K.O. & Karlish, S.J. (2003) Structure and mechanism of Na,K-ATPase: functional sites and their interactions, *Annu. Rev. Physiol.*, **65**, 817–849.

Kane, P.M. (1995) Disassembly and reassembly of the yeast vacuolar $H^+$-ATPase *in vivo*, *J. Biol. Chem.*, **270**, 17025–17032.

Kargacin, M.E., Ali, Z. & Kargacin, G.J. (1998) Anti-phospholamban and protein kinase A alter the Ca$^{2+}$ sensitivity and maximum velocity of Ca$^{2+}$ uptake by the cardiac sarcoplasmic reticulum, *Biochem. J.*, **331**, 245–249.

Kasai, M. & Muto, S. (1990) Ca$^{2+}$ pump and Ca$^{2+}$/H$^+$ antiporter in plasma-membrane vesicles isolated by aqueous 2-phase partitioning from corn leaves, *J. Membr. Biol.*, **114**, 133–142.

Kasai, M., Nakamura, T., Kudo, N., Sato, H., Maeshima, M. & Sawada, S. (1998) The activity of the root vacuolar H$^+$-pyrophosphatase in rye plants grown under conditions deficient in mineral nutrients, *Plant Cell. Physiol.*, **39**, 890–894.

Kasai, M., Sasaki, M., Yamamoto, Y. & Matsumoto, H. (1992) Aluminum stress increases K$^+$ efflux and activities of ATP- and PPi-dependent H$^+$ pumps of tonoplast-enriched membrane vesicles from barley roots, *Plant Cell Physiol.*, **33**, 1035–1039.

Kasai, M., Sasaki, M., Yamamoto, Y. & Matsumoto, H. (1993a) In vivo treatments that modulate PPi-dependent H$^+$ transport activity of tonoplast-enriched membrane vesicles from barley roots, *Plant Cell Physiol.*, **34**, 549–555.

Kasai, M., Yamamoto, Y., Maeshima, M. & Matsumoto, H. (1993b) Effects of in vivo treatment with abcisic acid and/or cytokinin on activities of vacuolar H$^+$ pumps of tonoplast-enriched membrane vesicles prepared from barley roots, *Plant Cell Physiol.*, **34**, 1107–1115.

Kawasaki-Nishi, S., Bowers, K., Nishi, T., Forgac, M. & Stevens, T.H. (2001) The amino-terminal domain of the vacuolar proton-translocating ATPase a subunit controls targeting and in vivo dissociation, and the carboxyl-terminal domain affects coupling of proton transport and ATP hydrolysis, *J. Biol. Chem.*, **276**, 47411–47420.

Kawasaki-Nishi, S., Nishi, T. & Forgac, M. (2003) Proton translocation by ATP hydrolysis in V-ATPases, *FEBS Lett.*, **545**, 76–85.

Kiegle, E., Moore, C.A., Haseloff, J., Tester, M.A. & Knight, M.R. (2000) Cell-type-specific calcium responses to drought, salt and cold in the *Arabidopsis* root, *Plant J.*, **23**, 267–278.

Kinoshita, T. & Shimazaki, K. (1999) Blue light activates the plasma membrane H$^+$-ATPase by phosphorylation of the C-terminus in stomatal guard cells, *EMBO J.*, **18**, 5548–5558.

Kluge, C., Lahr, J., Hanitzsch, M., Bolte, S., Golldack, D. & Dietz, K.-J. (2003) New insight into the structure and regulation of the plant vacuolar H$^+$-ATPase, *J. Bioenerg. Biomembr.*, **35**, 377–388.

Knight, H., Trewavas, A.J. & Knight, M.R. (1997) Calcium signalling in *Arabidopsis thaliana* responding to drought and salinity, *Plant J.*, **12**, 1067–1078.

Lahaye, P.A. & Epstein, E. (1971) Calcium and salt toleration by bean plants, *Physiol. Plant.*, **25**, 213–218.

Lahti, R., Pitkäranta, T., Valve, E., Ilta, I., Kukko-Kalse, E. & Heinonen, J. (1988) Cloning and characterization of the gene encoding inorganic pyrophosphatase of *Escherichia coli* K-12, *J. Bacteriol.*, **170**, 5901–5907.

Lalonde, S., Boles, E., Hellmann, H., *et al.* (1999) The dual function of sugar carriers: transport and sugar sensing, *Plant Cell*, **11**, 707–726.

Langhans, M., Ratajczak, R., Lützelschwab, M., *et al.* (2001) Inmunolocalization of plasma membrane H$^+$-ATPase and tonoplast-type pyrophosphatase in the plasma membrane of the sieve element–companion cell complex in the stem of *Ricinus communis* L. *Planta*, **213**, 11–19.

Leach, R.P., Rogers, W.J., Wheeler, K.P., Flowers, T.J. & Yeo, A.R. (1990) Molecular markers for ion compartmentation in cells of higher plants, *J. Exp. Bot.*, **41**, 1079–1087.

Legge, R.L., Thompson, J.E., Baker, J.E. & Lieberman, M. (1982) The effect of calcium on the fluidity and phase properties of microsomal-membranes isolated from post-climacteric golden delicious apples, *Plant Cell Physiol.*, **23**, 161–169.

Lerchl, J., König, S., Zrenner, R. & Sonnewald, U. (1995) Molecular cloning, characterization and expression analysis of isoforms encoding tonoplast-bound proton-translocating inorganic pyrophosphatase in tobacco, *Plant. Mol. Biol.*, **29**, 833–840.

Li, X. & Sze, H. (1999) A 100 kD polypeptide associates with the Vo membrane sector but not with the active oat vacuolar H$^+$-ATPase suggesting a role in assembly, *Plant J.*, **17**, 19–30.

Liang, F., Cunningham, K.W., Harper, J.F. & Sze, H. (1997) ECA1 complements yeast mutants defective in $Ca^{2+}$ pumps and encodes an endoplasmic reticulum-type $Ca^{2+}$-ATPase in *Arabidopsis thaliana*, *Proc. Natl. Acad. Sci. U.S.A.*, **94**, 8579–8584.

Lüttge, U. & Ratajczak, R. (1997) The physiology, biochemistry, and molecular biology of the plant vacuolar ATPase, *Adv. Bot. Res.*, **25**, 253–296.

Lynch, J., Cramer, G.R. & Lauchli, A. (1987) Salinity reduces membrane-associated calcium in corn root protoplasts, *Plant Physiol.*, **83**, 390–394.

Lynch, J., Polito, V.S. & Lauchli, A. (1989) Salinity stress increases cytoplasmic-Ca activity in maize root protoplasts, *Plant Physiol.*, **90**, 1271–1274.

Maathuis, F.J., Filatov, V., Herzyk, P., *et al.* (2003) Transcriptome analysis of root transporters reveals participation of multiple gene families in the response to cation stress, *Plant J.*, **35**, 675–692.

Maeshima, M. (1991) $H^+$ translocating inorganic pyrophosphatase of plant vacuoles. Inhibition by $Ca^{2+}$, stabilization by $Mg^{2+}$ and inmunological comparison with other inorganic pyrophosphatases, *Eur. J. Biochem.*, **196**, 11–17.

Maeshima, M. (2000) Vacuolar $H^+$-pyrophosphatase, *Biochim. Biophys. Acta*, **1465**, 37–51.

Marschner, H. (1995) *Mineral Nutrition of Higher Plants*, 2nd edn, Academic Press, London.

Marsh, K., González, P. & Echeverría, E. (2000) PPi formation by reversal of the tonoplast-bound $H^+$-pyrophosphatase from 'Valencia' orange juice cells, *J. Am. Soc. Hortic. Sci.*, **125**, 420–424.

Marsh, K., González, P. & Echeverría, E. (2001) Partial characterization of $H^+$-translocating inorganic pyrophosphatase from 3 citrus varieties differing in vacuolar pH, *Physiol. Plant.*, **111**, 519–526.

Matsumoto, H. & Chung, G.C. (1988) Increase in proton-transport activity of tonoplast vesicles as an adaptive response of barley roots to NaCl stress, *Plant Cell Physiol.*, **29**, 1133–1140.

Matsuoka, K., Higuchi, T., Maeshima, M. & Nakamura, K. (1997) A vacuolar-type $H^+$-ATPase in a nonvacuolar organelle is required for sorting of soluble vacuolar protein precursors in tobacco cells, *Plant Cell*, **9**, 533–546.

Maudoux, O., Batoko, H., Oecking, C., *et al.* (2000) A plant plasma membrane $H^+$-ATPase expressed in yeast is activated by phosphorylation at its penultimate residue and binding of 14-3-3 regulatory proteins in the absence of fusicoccin, *J. Biol. Chem.*, **275**, 17762–17770.

McIntosh, M.T. & Vaidya, A.B. (2002) Vacuolar type $H^+$ pumping pyrophosphatases of parasitic protozoa, *Int. J. Parasitol.*, **31**, 1343–1353.

Michelet, B., Lukaszewicz, M., Dupriez, V. & Boutry, M. (1994) A plant plasma membrane proton–ATPase gene is regulated by development and environment and shows signs of translational regulation, *Plant Cell*, **6**, 1375–1389.

Mitsuda, N., Enami, K., Nakata, M., Takeyasu, K. & Sato, M.H. (2001a) Novel type of *Arabidopsis thaliana* $H^+$-PPase is localized to the Golgi apparatus, *FEBS Lett.*, **488**, 29–33.

Mitsuda, N., Takeyasu, K. & Sato, M.H. (2001b) Pollen-specific regulation of vacuolar $H^+$-PPase expression by multiple *cis*-acting elements, *Plant Mol. Biol.*, **46**, 185–192.

Moller, J.V., Juul, B. & le Maire, M. (1996) Structural organization, ion transport, and energy transduction of P-type ATPases, *Biochim. Biophys. Acta*, **1286**, 1–51.

Moore, C.A., Bowen, H.C., Scrase-Field, S., Knight, M.R. & White, P.J. (2002) The deposition of suberin lamellae determines the magnitude of cytosolic $Ca^{2+}$ elevations in root endodermal cells subjected to cooling, *Plant J.*, **30**, 457–465.

Morsomme, P. & Boutry, M. (2000) The plant plasma membrane $H^+$-ATPase: structure, function and regulation, *Biochim. Biophys. Acta*, **1465**, 1–16.

Nakamura, Y., Kasamo, K., Shimosato, N., Sakata, M. & Ohta, E. (1992) Stimulation of the extrusion of protons and $H^+$-ATPase activities with the decline in pyrophosphatase activity of the tonoplast in intact mung bean roots under high-NaCl stress and its relation to external levels of $Ca^{2+}$ ions, *Plant Cell Physiol.*, **33**, 139–149.

Nelson, H. & Nelson, N. (1990) Disruption of genes encoding subunits of yeast vacuolar $H^+$-ATPase causes conditional lethality, *Proc. Natl. Acad. Sci. U.S.A.*, **87**, 3503–3507.

Nelson, N. (2003) A journey from mammals to yeast with vacuolar $H^+$-ATPase (V-ATPase), *J. Bioenerg. Biomembr.*, **35**, 281–289.

Nelson, N. & Harvey, W.R. (1999) Vacuolar and plasma membrane proton-adenosinetriphosphatases, *Physiol. Rev.*, **79**, 361–385.

Nishi, T. & Forgac, M. (2002) The vacuolar (H$^+$)-ATPases – nature's most versatile proton pumps, *Nat. Rev. Mol. Cell Biol.*, **3**, 94–103.

Olsson, A., Svennelid, F., Ek, B., Sommarin, M. & Larsson, C. (1998) A phosphothreonine residue at the C-terminal end of the plasma membrane H$^+$-ATPase is protected by fusicoccin-induced 14-3-3 binding, *Plant Physiol.*, **118**, 551–555.

Otoch, M.L.O., Sobreira, A.C.M., de Aragao, M.E.F., Orellano, E.G., Lima, M.G.S. & de Melo, D.F. (2001) Salt modulation of vacuolar H$^+$-ATPase and H$^+$-pyrophosphatase activities in *Vigna unguiculata*, *Plant Physiol.*, **158**, 545–551.

Palma, D.A., Blumwald, E. & Plaxton, W.C. (2000) Upregulation of vacuolar H$^+$-translocating pyrophosphatase by phosphate starvation of *Brassica napus* (rapeseed) suspension cell cultures, *FEBS Lett.*, **486**, 155–158.

Palmgren, M.G. (2001) Plasma membrane H$^+$-ATPases: powerhouses for nutrient uptake, *Annu. Rev. Plant Physiol. Plant Mol. Biol.*, **52**, 817–845.

Palmgren, M.G. & Christensen, G. (1994) Functional comparisons between plant plasma membrane H$^+$-ATPase isoforms expressed in yeast, *J. Biol. Chem.*, **269**, 3027–3033.

Palmgren, M.G., Soummarin, M., Serrano, R. & Larsson, C. (1991) Identification of an autoinhibitory domain in the C terminal region of the plant plasma membrane H$^+$-ATPase, *J. Biol. Chem.*, **266**, 20740–20745.

Pardo, J.M. & Serrano, R. (1989) Structure of a plasma membrane H$^+$-ATPase gene from the plant *Arabidopsis thaliana*, *J. Biol. Chem.*, **264**, 8557–8562.

Paret-Soler, A., Pardo, J.M. & Serrano, R. (1990) Immunolocalization of plasma membrane H$^+$-ATPase, *Plant Physiol.*, **93**, 1654–1658.

Parks, G.E., Dietrich, M.A. & Schumaker, K.S. (2002) Increased vacuolar Na$^+$/H$^+$ exchange activity in *Salicornia bigelovii* Torr. in response to NaCl, *J. Exp. Bot.*, **53**, 1055–1065.

Parry, R.V., Turner, J.C. & Rea, P.A. (1989) High purity preparation of higher plant vacuolar H$^+$-ATPase, *J. Biol. Chem.*, **264**, 20025–20032.

Pedersen, P.L. & Carafoli, E. (1987) Ion motive ATPases, I: Ubiquity, properties and significance to cell function, *Trends Biochem. Sci.*, **12**, 146–150.

Pérez-Castiñeira, J.R., Gómez-García, R., López-Marqués, R.L., Losada, M. & Serrano, A. (2001a) Enzymatic systems of inorganic pyrophosphate bioenergetics in photosynthetic and heterotrophic protists: remnants or metabolic cornerstones? *Int. Microbiol.*, **4**, 135–142.

Pérez-Castiñeira, J.R., López-Marqués, R.L., Losada, M. & Serrano, A. (2001b) A thermostable K$^+$-stimulated vacuolar-type pyrophosphatase from the hyperthermophilic bacterium *Thermotoga maritima*, *FEBS Lett.*, **496**, 6–11.

Pérez-Castiñeira, J.R., López-Marqués, R.L., Losada, M. & Serrano, A. (2002a) Evidence for a wide occurrence of proton-translocating pyrophosphatase genes in parasitic and free-living protozoa, *Biochem. Biophys. Res. Commun.*, **294**, 567–573.

Pérez-Castiñeira, J.R., López-Marqués, R.L., Villalba, J.M., Losada, M. & Serrano, A. (2002b) Functional complementation of yeast cytosolic pyrophosphatase by bacterial and plant H$^+$-translocating pyrophosphatases, *Proc. Natl. Acad. Sci. U.S.A.*, **99**, 15914–15919.

Perezprat, E., Narasimhan, M.L., Binzel, M.L., *et al.* (1992) Induction of a putative Ca$^{2+}$-ATPase messenger-RNA in NaCl-adapted cells, *Plant Physiol.*, **100**, 1471–1478.

Pertl, H., Himly, M., Gehwolf, R., *et al.* (2001) Molecular and physiological characterisation of a 14-3-3 protein from lily pollen grains regulating the activity of the plasma membrane H$^+$ ATPase during pollen grain germination and tube growth, *Planta*, **213**, 132–141.

Piotrowski, M., Morsomme, P., Boutry, M. & Oecking, C. (1998) Complementation of the *Saccharomyces cerevisiae* plasma membrane H$^+$-ATPase by a plant H$^+$-ATPase generates a highly abundant fusicoccin binding site, *J. Biol. Chem.*, **273**, 30018–30023.

Pittman, J.K. & Hirschi, K.D. (2003) Don't shoot the (second) messenger: endomembrane transporters and binding proteins modulate cytosolic Ca$^{2+}$ levels, *Curr. Opin. Plant Biol.*, **6**, 257–262.

Plaxton, W.C. (1999) Metabolic aspects of the phosphate starvation in plants, in *Phosphorus in Plant Biology: Regulatory Roles in Molecular, Cellular, Organismic and Ecosystem Processes* (eds J. Deikman & J. Lynch), American Society of Plant Physiologists, Rockville, MD, pp. 229–241.

Pomorski, T., Lombardi, R., Riezman, H., Devaux, P.F., van Meer, G. & Holthuis, J.C.M. (2003) Drs2p-related P-type ATPases Dnf1p and Dnf2p are required for phospholipid translocation across the yeast plasma membrane and serve a role in endocytosis, *Mol. Biol. Cell*, **14**, 1240–1254.

Ratajczak, R. (2000) Structure, function and regulation of the plant vacuolar $H^+$-translocating ATPase, *Biochim. Biophys. Acta*, **1465**, 17–36.

Rea, P.A., Britten, C.J., Jennings, I.R., *et al.* (1992a) Regulation of vacuolar $H^+$-pyrophosphatase by free calcium, *Plant Physiol.*, **100**, 1706–1715.

Rea, P.A., Kim, Y, Sarafian, V., Poole, R.J., Davies, J.M. & Sanders, D. (1992b) Vacuolar $H^+$-translocating pyrophosphatases: a new category of ion translocase, *Trends Biochem. Sci.*, **17**, 348–353.

Rea, P.A. & Poole, R.J. (1993) Vacuolar $H^+$-translocating pyrophosphatase, *Annu. Rev. Plant Physiol. Plant Mol. Biol.*, **44**, 157–180.

Reddy, A.S. (2001) Calcium: silver bullet in signaling, *Plant Sci.*, **160**, 381–404.

Regenberg, B., Villalba, J.M., Lanfermeijer, F.C. & Palmgren, M.G. (1995) C terminal deletion analysis of plant plasma membrane $H^+$-ATPase: yeast as a model system for solute transport across the plant plasma membrane, *Plant Cell*, **7**, 1655–1666.

Reuveni, M., Bennet, A.B., Bressan, R.A. & Hasegawa, P.M. (1990) Enhanced $H^+$-transport capacity and ATP-hydrolisis activity of the tonoplast $H^+$-ATPase after NaCl adaptation, *Plant Physiol.*, **94**, 524–530.

Robinson, D.G., Haschke, H.-P., Hinz, G., Hoh, B., Maeshima, M. & Marty, F. (1996) Immunological detection of tonoplast polypeptides in the plasma membrane of pea cotyledons, *Planta*, **198**, 95–103.

Sanders, D., Brownlee, C. & Harper, J.F. (1999) Communicating with calcium, *Plant Cell*, **11**, 691–706.

Sanders D., Pelloux J., Brownlee C. & Harper J.F. (2002) Calcium at the crossroads of signaling, *Plant Cell*, **14** (Suppl.), S401–S417.

Sarafian, V., Kim, Y., Poole, R.J. & Rea, P.A. (1992) Molecular cloning and sequence of cDNA encoding the pyrophosphate-energized vacuolar membrane proton pump of *Arabidopsis thaliana*, *Proc.Natl. Acad. Sci. U.S.A.*, **89**, 1775–1779.

Schiøtt, M., Romanowsky, S.M., Baekgaard, L., Jakobsen, M.K., Palmgren, M.G. & Harper, J.F. (2004) A plant plasma membrane $Ca^{2+}$ pump is required for normal pollen tube growth and fertilization. *Proc. Natl. Acad. Sci. U.S.A.*, **101**, 9502–9507.

Schumacher, K., Vafeados, D., McCarthy, M., Sze, H., Wilkins, T. & Chory, J. (1999) The *Arabidopsis det3* mutant reveals a central role for the vacuolar $H^+$-ATPase in plant growth and development, *Genes Dev.*, **13**, 3259–3270.

Serrano, R. (1989) Structure and function of plasma membrane ATPase, *Annu. Rev. Plant Physiol. Plant Mol. Biol.*, **40**, 61–94.

Serrano, R., Kielland-Brandt, M.C. & Fink, G.R. (1986) Yeast plasma membrane ATPase is essential for growth and has homology with $(Na^++K^+)$, $K^+$- and $Ca^{2+}$-ATPases, *Nature*, **319**, 689–693.

Smart, L.B., Vojdani, F., Maeshima, M. & Wilkins, T.A. (1998) Genes involved in osmoregulation during turgor driven cell expansion of developing cotton fibers are differentially regulated, *Plant Physiol.*, **116**, 1539–1549.

Smith, A.C., Uribe, E.G., Ball, E., Heuer, S. & Lüttge, U. (1984) Characterization of the vacuolar ATPase activity of the crassulacean-acid-metabolism plant *Kalanchoë daigremontiana* receptor modulating, *Eur. J. Biochem.*, **141**, 415–420.

Stevens, T.H. & Forgac, M. (1997) Structure, function, and regulation of the vacuolar $H^+$-ATPase, *Annu. Rev. Cell Dev. Biol.*, **13**, 779–808.

Sussman, M.R. (1994) Molecular analysis of proteins in the plasma membrane, *Annu. Rev. Plant Physiol. Plant Mol. Biol.*, **45**, 211–234.

Svennelid, F., Olsson, A., Piotrowski, M., *et al.* (1999) Phosphorylation of Thr-948 at the C terminus of the plasma membrane $H^+$-ATPase creates a binding site for the regulatory 14-3-3 protein, *Plant Cell*, **11**, 2379–2391.

Swanson, S.J. & Jones, R.L. (1996) Gibberelic acid induces vacuolar acidification in barley aleurone, *Plant Cell*, **8**, 2211–2221.

Sze, H. (1985) H$^+$-translocating ATPases: advances using membrane vesicles, *Annu. Rev. Plant Physiol.*, **36**, 175–208.

Sze, H., Li, X. & Palmgren, M.G. (1999) Energization of plant cell membranes by H$^+$-pumping ATPases. Regulation and biosynthesis, *Plant Cell*, **11**, 677–690.

Sze, H., Schumacher, K., Muller, M.L., Padmanaban, S. & Taiz, L. (2002) A simple nomenclature for a complex proton pump: VHA genes encode the vacuolar H$^+$-ATPase, *Trends Plant Sci.*, **7**, 157–161.

Tang, X.J., Halleck, M.S., Schlegel, R.A. & Williamson, P. (1996) A subfamily of P-type ATPases with aminophospholipid transporting activity, *Science*, **272**, 1495–1497.

Tanner, W. & Caspari, T. (1996) Membrane transport carriers, *Annu. Rev. Plant Physiol. Plant Mol. Biol.*, **47**, 595–626.

Terrier, N., Deguilloux, C., Sauvage, F.-X., Martinoia, E. & Romieu, C. (1998) Proton pumps and anion transport in *Vitis vinifera*: the inorganic pyrophosphatase plays a predominant role in the energization of the tonoplast, *Plant Physiol. Biochem.*, **36**, 367–377.

Toyoshima, C., Nakasako, M., Nomura, H. & Ogawa, H. (2000) Crystal structure of the calcium pump of sarcoplasmic reticulum at 2.6 Å resolution, *Nature*, **405**, 647–655.

Toyoshima, C. & Nomura, H. (2002) Structural changes in the calcium pump accompanying the dissociation of calcium, *Nature*, **418**, 605–611.

Tsiantis, M.S., Bartholomew, D.M. & Smith, J.A.C. (1996) Salt regulation of transcript levels for the c subunit of a leaf vacuolar H$^+$-ATPase in the halophyte *Mesembryanthemum crystallinum*, *Plant J.*, **9**, 729–736.

Ullrich, C.I. & Novacky, A.J. (1990) Extra- and intracellular pH and membrane potential changes induced by K$^+$, Cl$^-$, H$_2$PO$_4^-$ and NO$_3^-$ uptake and fusicoccin in root hairs of *Limnobium stoloniferum*, *Plant Physiol.*, **94**, 1561–1567.

Venema, K. & Palmgren, M.G. (1995) Metabolic modulation of transport coupling ratio in yeast plasma membrane H$^+$-ATPase, *J. Biol. Chem.*, **270**, 19659–19667.

Very, A.A. & Sentenac, H. (2003) Molecular mechanisms and regulation of K$^+$ transport in higher plants, *Annu. Rev. Plant Biol.*, **54**, 575–603.

Vicente J.A.F. & Vale M.G.P. (1995) Activities of Ca$^{2+}$ pump and low-affinity Ca$^{2+}$/H$^+$ antiport in plasma-membrane vesicles of corn roots, *J. Exp. Bot.*, **46**, 1551–1559.

Villalba, J.M., Lutzelschwab, M. & Serrano, R. (1991) Immunocytolocalization of plasma membrane H$^+$-ATPase in maize coleoptiles and enclosed leaves, *Planta*, **185**, 458–461.

Vitart, V., Baxter, I., Doerner, P. & Harper, J.F. (2001) Evidence for a role in growth and salt resistance of a plasma membrane H$^+$-ATPase in the root endodermis, *Plant J.*, **27**, 191–201.

Wang, B., Lüttge, U. & Ratajczak, R. (2001) Effects of salt treatment and osmotic stress on V-ATPase and V-PPase in leaves of the halophyte *Suaeda salsa*, *J. Exp. Bot.*, **52**, 2355–2365.

Wang, B.S., Rataiczak, R. & Zhang, J.H. (2000) Activity, amount and subunit composition of vacuolar-type H$^+$-ATPase and H$^+$-PPase in wheat roots under severe NaCl stress, *J. Plant Physiol.*, **157**, 109–116.

Wang, K.K.W., Wright, L.C., Machan, C.L., Allen, B.G., Conigrave, A.D. & Roufogalis, B.D. (1991) Protein-kinase-C phosphorylates the carboxyl terminus of the plasma-membrane Ca$^{2+}$-ATPase from human erythrocytes, *J. Biol. Chem.*, **266**, 9078–9085.

Ward, J.M. & Sze, H. (1992a) Subunit composition and organization of the vacuolar H$^+$-ATPase from oat roots, *Plant Physiol.*, **99**, 170–179.

Ward, J.M. & Sze, H. (1992b) Proton transport activity of the purified vacuolar H$^+$-ATPase from oats, *Plant Physiol.*, **99**, 925–931.

Warren, M., Smith, J.C. & Apps, D.K. (1992) Rapid purification and reconstitution of a plant vacuolar ATPase using Triton X-114: subunit composition and substrate kinetics of the H$^+$-ATPase from the tonoplast of *Kalanchoe daigremontiana*, *Biochim. Biophys. Acta*, **1106**, 117–125.

White, P.J. & Broadley, M.R. (2003) Calcium in plants, *Ann. Bot.*, **92**, 487–511.

Widell, S. & Larsson, C. (1981) Separation of presumptive plasma-membranes from mitochondria by partition in an aqueous polymer 2-phase system, *Physiol. Plant.*, **51**, 368–374.

Wimmers, L.E., Ewing, N.N. & Bennett, A.B. (1992) Higher-plant $Ca^{2+}$-ATPase – primary structure and regulation of messenger-RNA abundance by salt, *Proc. Natl. Acad. Sci. U.S.A.*, **89**, 9205–9209.

Wu, J. & Seliskar, D.M. (1998) Salinity adaptation of plasma membrane $H^+$-ATPase in the salt marsh plant *Spartina patens*: ATP hydrolysis and enzyme kinetics, *J. Exp. Bot.*, **49**, 1005–1013.

Wu, Z.Y., Liang, F., Hong, B.M., *et al.* (2002) An endoplasmic reticulum-bound $Ca^{2+}/Mn^{2+}$ pump, ECA1, supports plant growth and confers tolerance to $Mn^{2+}$ stress, *Plant Physiol.*, **130**, 128–137.

Wurtele, M., Jelich-Ottmann, C., Wittinghofer, A. & Oecking, C. (2003) Structural view of a fungal toxin acting on a 14-3-3 regulatory complex, *EMBO J.*, **22**, 987–994.

Yokoyama, K., Nakano, M., Imamura, H., Yoshida, M. & Tamakoshi, M. (2003) Rotation of the proteolipid ring in the V-ATPase, *J. Biol. Chem.*, **278**, 24255–24258.

Zingarelli, L., Anzani, P. & Lado, P. (1994) Enhanced $K^+$-stimulated pyrophosphatase activity in NaCl-adapted cells of *Acer pseudoplatanus*, *Physiol. Plant.*, **91**, 510–516.

# 4 Ion-coupled transport of inorganic solutes

Malcolm J. Hawkesford and Anthony J. Miller

## 4.1 Introduction

Ion-coupled transport provides the mechanism for uptake of many different types of plant nutrients. By coupling the transport of one ion to the gradient of another ion, large concentration differences can be established across cell membranes. This type of transporter is responsible for the high-affinity uptake of many nutrients from the soil, but these proteins also have important transport functions in all plant tissues. In addition they have major roles in sub-cellular transport across organelle membranes, for example, vacuolar storage of nutrients and compartmentation of toxic ions. This class of transporter is believed to consume a large proportion of the energy budget of cells, especially for nutrients like nitrate, as fluxes of this ion are so large. In order to achieve cellular ion homeostasis, the transporters are strictly regulated, often at the level of expression of the genes. Many gene families have been identified on the basis of sequence homology, but our knowledge of their functions is very sparse and is usually based on the characterization of only one or two family members.

### 4.1.1 Ion gradients and ion-coupled transport mechanisms

Most plant cells sit in an aqueous solution that has concentrations of inorganic solutes that are generally very different from those found inside the cells (see Table 4.1). In some cases, the concentrations inside the cell may be several-fold different from those found outside (for example, potassium), while others like calcium are maintained at much lower concentrations in the cytosol, when compared with the soil (Table 4.1). These concentration differences are maintained by the activity of membrane transporter proteins in both the plasma membrane and the vacuolar membrane, the tonoplast. The direction of transport at the plasma membrane determines whether the concentrations are maintained at higher or lower values than those found in the soil solution. Similarly at the tonoplast, the storage of ions in the vacuole is mediated by inwardly directed flow of nutrient ions and the remobilization is achieved by transport in the reverse direction. The energy for this transport is supplied by the gradients of other ions, chiefly protons. Both the plasma membrane and the tonoplast have pumps that use the energy derived from the hydrolysis of ATP or pyrophosphate to move an ion across the cell membrane. These pumps generate a gradient of these primary ions, mainly protons ($H^+$) that can be coupled to the transport of the inorganic solute. The proton concentration (pH) of the cytosol is typically 2 orders of magnitude different from that measured in the vacuole, the inner compartments of

**Table 4.1** Typical concentrations for some of the main inorganic solutes in soils and in the chief compartments of nutrient replete plant cells[1]

| Inorganic solute | Range in top soil (mM) | Cell (mM) | |
|---|---|---|---|
| | | Cytosol | Vacuole |
| Ammonium ($NH_4^+$) | 2–20 | 0.005–358 | 2–45 |
| Nitrate ($NO_3^-$) | 0.7–2.55 | 2–6 | 0.1–500 |
| Phosphate ($PO_4^{3-}$) | 0.015 | 2 | 18 |
| Sulphate ($SO_4^{2-}$) | 0.6 | 1–10 | 0–100 |
| Calcium ($Ca^{2+}$) | 1.7 | 0.001 | 10–40 |
| Potassium ($K^+$) | 0.5–5.2 | 90 | 10–200 |
| Magnesium ($Mg^{2+}$) | 0.03–0.5 | 0.3–0.7 | |

[1]Nitrate in soil, ranges 1–5 mM (Owen & Jones, 2001); ammonium, averaging 2 mM in some forest soils up to 20 mM in some agricultural soils (Britto & Kronzucker, 2002), and estimates of cytosolic ammonium concentration show a large range of values that largely depend on the method and conditions (Miller et al., 2001). Cytosolic $Mg^{2+}$ measurements are from animal cells (one NMR measurement for plants). Other nutrients in soil are taken from Barber (1984), and NMR for phosphate measurements in cells are taken from Espen et al. (2000).

plastids and mitochondria, and the outside of the cells (Kurkdjian & Guern, 1989). Proton-coupled transport depends on the sum of the electrical and chemical gradients of $H^+$ across cell membranes: the larger the electrochemical gradient, the more the energy that is available to establish gradients of solutes across a membrane.

The mechanism of ion-coupled transport will differ at the plasma membrane compared to the membranes of cellular organelles, such as the tonoplast or chloroplast envelope. At the plasma membrane, the $H^+$concentration is generally higher outside the cell relative to the cytosol and these conditions favour the co-transport of $H^+$ with the inorganic solute into the cell, i.e. symport (see Fig. 4.1). Across the organelle membranes the situation is reversed and these conditions require an exchange of protons with the solute to give accumulation inside the organelle, i.e. antiport (see Fig. 4.1). Although $H^+$ gradients are chiefly used for ion-coupled transport in plants, other ions can drive transport, e.g. $Na^+$ (Jung, 2002), but the importance of this type of co-transport in higher plants for some inorganic solutes, e.g. $K^+$, is controversial (Maathuis et al., 1996; see Chapter 11). Detailed analysis of animal transporters expressed in foreign cells suggests that $H^+$ can substitute for $Na^+$ in driving ion-coupled transport (Hirayama et al., 1994). This may be a more generalized phenomenon such that when suitable ionic gradients exist other cations can substitute for $H^+$ in driving co-transport. For example, in marine environments the high external $Na^+$ concentrations may favour co-transport with this as the driver ion. Accurate measurements of the cellular ion gradients are essential for the determination of likely transport mechanisms at each membrane.

### 4.1.2 Thermodynamics of ion-coupled transport

The establishment of sodium or proton (pH) gradients across membrane provides a potential energy source that can be used to develop gradients of other ions (see also

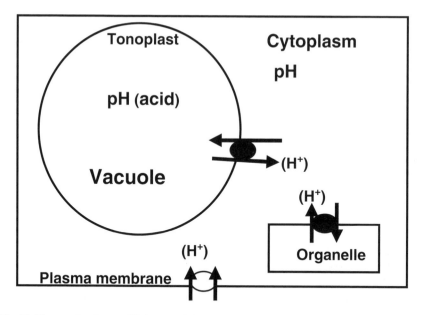

**Fig. 4.1** Diagram showing possible ion-coupled transport mechanisms at intracellular membranes. Open circles and closed circles indicate symport and antiport respectively, while the arrows indicate direction of transport.

Chapter 1). In plants, protons are the main energy source and so only these will be considered here. Primary pumps establish these pH gradients and the free energy change ($\Delta G$) as protons move back into the cell along both chemical and electrical gradients is negative. For protons the relationship can be described mathematically as follows:

$$\Delta G = \Delta\mu_{H^+} = F\Delta\Psi + 2.3RT \log[H^+]_i/[H^+]_o$$

In this equation, $F$ is the Faraday constant [96.5 kJ/(V·mol)], $R$ the gas constant [8.3 J/(mol·K)], $T$ the absolute temperature (K) and $\Delta\Psi$ is the electrical potential difference across the membrane. The term $\Delta\mu_{H^+}$ is defined as the electrochemical difference for protons across the membrane. This is usually expressed in electrical energy units, the units of the membrane potential, volts (more typically mV).

$$\Delta G/F = \Delta\mu_{H^+}/F = \Delta\Psi + (2.3RT/F) \log[H^+]_i/[H^+]_o$$

The log function in the above equation can be simplified to convert the concentration of protons into the pH difference across the membrane ($\Delta pH = pH_i - pH_o$).

$$\Delta\mu_{H^+}/F = \Delta\Psi - (2.3RT/F)\Delta pH$$

This can be further simplified as at 20°C, $2.3RT/F$ equals 56.7 mV, and so the $\Delta \mu_{H^+}/F$ in mV becomes

$$\Delta\mu_{H^+}/F = \Delta\Psi - 56.7\Delta pH$$

This gives a direct measure of the electrochemical driving force or the proton-motive force. The proton-motive force is a measure of the energy that is available to drive ion-coupled transport, and in any given situation it can be used to calculate the equilibrium concentration of a co-transported ion or other molecule that can be achieved by this mechanism. In this way the energetic feasibility of a specific proton co-transport mechanism to achieve the measured concentration gradients can be assessed, provided the pH gradient and electrical potential difference are known.

The membrane potential across a plant cell membrane results from the charge imbalance between one side and the other, with an excess of negative ions inside. Plant cell membranes have low electrical capacitance and so the transfer of only a few thousand protons across the plasma membrane is enough to generate an electrical potential difference of $-60$ mV. The cytoplasm is highly buffered and the number of protons needed to change pH in this compartment is many orders of magnitude greater than that required in the vacuole or outside the cell. Although at the outer face of the plasma membrane, there is also a significant pH buffering capacity from the cell wall. The energy for co-transport is thus divided between the pH gradient and the electrical gradient.

### 4.1.3   Determining the feasibility of co-transport mechanisms

The proton electrochemical gradient across both the tonoplast and the plasma membrane can provide the energy for transport of an anion. As active anion transport at the plasma membrane can occur by symport with protons, measurements of the electrical changes in membrane potential when the anion is supplied outside cells can identify transporter activity. This pH sensitivity and changes in membrane potential associated with the transport of anions like nitrate or sulphate have supported a proton symport model for uptake at the plasma membrane (Lass & Ullrich-Eberius, 1984; McClure et al., 1990). These measurements suggest that the symport must have a stoichiometry of at least 2:1 $H^+/NO_3^-$, as a 1:1 stoichiometry would be electrically neutral and would not cause depolarization of the membrane potential.

The thermodynamic feasibility of a proton symport mechanism over both high- and low-affinity transport ranges can be determined by using measurements of cytosolic anion activities ($A^-$), pH and membrane potential using triple-barreled ion-selective microelectrodes (Miller & Smith, 1996). For $H^+/A^-$ symport at the plasma membrane, the appropriate free energy relationship (see also Chapter 1) for the reaction is

$$\Delta G'/F = 59\{n(pH_o - pH_c) + (p[A^-]_o - p[A^-]_c)\} + (n-1)\Delta\Psi$$

where $n$ is the stoichiometry of protons to $A^-$ for the symport, and $\Delta\Psi$ is the trans-plasma membrane electrical potential difference and the subscripts o and c denote the external solution and cytosol, respectively. The free energy for the symport is expressed numerically in millivolts. The free energy requirement to maintain a cytosolic $A^-$ concentration of 4 mM (for nitrate) can be calculated for different values of $n$. By calculating the free energy at different external $A^-$ and pH values, the ability of different symport mechanisms to maintain cytosolic $A^-$ can be assessed. Such calculations have been used to demonstrate that for nitrate an electrically neutral symport of 1:1 $H^+/NO_3^-$ is not energetically feasible for high-affinity transport (Miller & Smith, 1996). Furthermore, electroneutral transport is energized by only the pH gradient across the plasma membrane, giving the plant cell little flexibility to maintain uptake in response to changes in external nitrate concentration. A 2:1 stoichiometry for nitrate uptake is likely to be energetically feasible for most soil nitrate concentrations (Siddiqi et al., 1990; Miller & Smith, 1996); however, in alkaline soils high-affinity uptake may require a higher ratio or a different mechanism (Miller & Smith, 1996).

Nitrate-elicited changes in membrane potential have been used as an assay for nitrate symport activity. At an external pH of 8 when the resting potential was $-184$ mV, the nitrate-elicited change in the membrane potential of maize root cells disappeared (McClure et al., 1990). This result is consistent with the idea that under these conditions of external pH and membrane potential a 2:1 $H^+/NO_3^-$ symport is no longer mechanistically feasible for uptake. In the leaves of an aquatic plant at an external pH of 8.3 when the membrane potential was $-234$ mV, a nitrate-elicited change in membrane potential could be observed. However, calculations of the thermodynamics under these conditions (Miller & Smith, 1996) suggest that despite the alkaline external pH, there was still sufficient free energy in 2:1 $H^+/NO_3^-$ symport for uptake because of the large value of the membrane potential ($\Delta\Psi$).

The importance of the plasma membrane voltage in supplying energy for nitrate uptake is shown graphically in Fig. 4.2. This figure shows the plasma membrane electrical potential difference that is required for a 2:1 $H^+/NO_3^-$ symport to maintain the cytosolic nitrate concentration at a range of different external pHs and nitrate concentrations. For example, at pH 7.5 and an external nitrate concentration of 0.1 mM, a membrane potential of $-200$ mV is required to maintain a cytosolic nitrate concentration of 4 mM. However, when the external pH is 5.8 or below, there is no requirement for membrane potential as the pH gradient is sufficient for 2:1 $H^+/NO_3^-$ symport to maintain the cytosolic concentration. Furthermore, kinetic models for $H^+/NO_3^-$ symport at the plasma membrane have emphasized the importance of membrane voltage in controlling nitrate transport. This control occurs not only by way of the energy supply but also through the regulation of kinetic parameters such as the affinity of the transporter (Meharg & Blatt, 1995; Zhou et al., 2000b).

The situation for the uptake at the plasma membrane of divalent anions, like sulphate, is more complicated because a 2:1 $H^+/SO_4^{2-}$ stoichiometry is electrically neutral and so thermodynamically equivalent to a 1:1 $H^+/NO_3^-$ symport. Sulphate

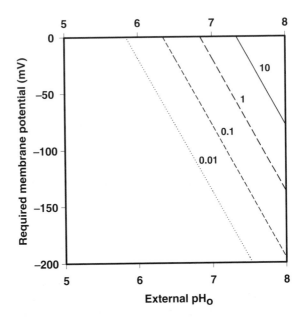

**Fig. 4.2** Graphical representation of the plasma membrane potential difference that is required for a 2:1 $H^+/NO_3^-$ symport to maintain cytosolic nitrate at 4 mM at a range of different external nitrate concentrations (0.01–10 mM) and pH values.

symport is assumed to occur with a 3:1 $H^+/SO_4^{2-}$ stoichiometry because when plant cells are treated with sulphate the membrane potential, like the situation for nitrate, becomes less negative (Lass & Ullrich-Eberius, 1984). A similar situation exists for phosphate where it is also assumed that $HPO_4^{2-}$ transport occurs by symport with three protons because there are electrical changes associated with transport (Dunlop & Gardiner, 1993). However, the thermodynamic feasibility of these mechanisms requires more detailed combined measurements of pH gradients, membrane potentials and cytosolic concentrations of these anions.

Less is known about ion-coupled transport mechanisms at the tonoplast, but one of the best characterized is the $Na^+/H^+$ antiport (see Fig. 4.1), which is important for salt tolerance (see Chapter 11). The large vacuolar accumulations of nitrate, and to a lesser extent sulphate and phosphate, suggest that there are ion-coupled transport mechanisms at the tonoplast. These are also likely to be proton-coupled antiporters, and for nitrate the thermodynamic feasibility has been measured (Miller & Smith, 1992). These antiport systems for anions are electrogenic and should result in the trans-tonoplast potential difference becoming more negative as protons (+) exchange into the cytoplasm for nitrate (−) to the vacuole. A microelectrode inserted into the vacuole of a cell will record these changes and longer term nitrate treatment has been shown to result in more negative membrane potentials (McClure et al., 1990; Glass et al., 1992). However, these electrical changes are generally assumed to occur by increased activity of the plasma membrane proton pump as it is stimulated by the cytosolic acidification resulting from symport activity at the plasma membrane.

### 4.1.4 Functions and relationships to physiology

The membrane bordering every cellular compartment, including organelles, contains ion-coupled transport systems for inorganic solutes. Ion-coupled transport is responsible for nutrient uptake by root cells at the root/soil interface, loading of nutrients into the xylem and subsequent transport around the plant. In later sections of this chapter we will briefly review the role of transporter families for nitrate, ammonium and sulphate ions. This class of transporter has a major role in the stress and tolerance responses of plants, and the balance between nutrition and toxicity is determined by the selectivity of a specific membrane transporter and/or the level of expression of the relevant gene. Specific examples here are antagonism between the micronutrient requirement for some metals and the toxic consequences of excessive uptake (see Chapter 12), or the uptake of $Na^+$ via $K^+$ transport mechanisms during salt stress (see Chapter 11). The activity of this class of transporters is essential for the maintenance of cellular homeostasis and the regulation of the concentrations inside the various compartments including the cytoplasm. Some steps in the assimilation of inorganic solutes occur within specific organelles such as the chloroplast. For example, the assimilation of nitrate requires two ion-coupled transport steps: first, the uptake of $NO_3^-$ across the plasma membrane for reduction to $NO_2^-$ in the cytoplasm by nitrate reductase, and second, $NO_2^-$ is transported into the chloroplast for subsequent further reduction to $NH_4^+$ and synthesis into amino acids. Clearly, the membrane location of these two different types of ion-coupled transport is of fundamental importance for the assimilation of nitrate by plant cells. The coordination of these two transport steps is particularly necessary because $NO_2^-$ is a powerful oxidizing agent and is likely to be toxic if accumulated within the cytosol; therefore, transport into the chloroplast must not be limited. Similarly, the chloroplast is the primary site for reduction of sulphate to sulphide and the subsequent incorporation into cysteine. This gives rise to an absolute requirement for a chloroplast sulphate influx system. Localization of membrane proteins to specific membranes is a key aspect of inorganic solute assimilation and is essential for the function of ion-coupled transporters.

### 4.1.5 Targeting and membrane location

The requirement for transport at different membranes requires targeting of the transporter proteins to specific cellular membranes. The membrane proteins of most organelles (mitochondria, plastids and vacuoles) are encoded in the nucleus and synthesized in the cytoplasm. Molecular machinery at the organelle surfaces recognizes the targeting sequences of specific proteins and mediates their inclusion into the respective membranes (see Chapter 10). Proteins destined for storage in vacuoles are first translocated across the endoplasmic reticulum membrane, packaged into vesicles, transported to the Golgi, where they are sorted into specific vesicles and subsequently carried to the different types of vacuoles. However, the situation for membrane proteins seems to be more complicated. Although plant cells share many features with animal and yeast cells, chloroplasts and distinct lytic and storage vacuoles each have specific targeting sequences (Kessler & Neuhaus, 2003). In the chloroplast, unlike most of the outer-envelope proteins, targeting of proteins to all other plastid compartments (inner-envelope membrane, stroma and thylakoid) is strictly dependent on the presence of a cleavable transit sequence in the precursor N-terminal region (Li & Chen, 1996). Recently, a new protein component was identified for the inner membrane of the chloroplast envelope (Miras *et al.*, 2002). For targeting to the outer envelope of the chloroplast the protein structure is important; for example, a specific seven amino acid region and a transmembrane domain were required for one protein (Lee *et al.*, 2001). At the tonoplast, less is known about membrane protein targeting and until recently the default pathway for all proteins was thought to be the tonoplast (Brandizzi *et al.*, 2002). The specific targeting of a vacuolar membrane protein to the tonoplast has been shown to depend on the promoter region of the gene (Czempinski *et al.*, 2002). The different mechanisms for the sorting of membrane proteins to storage and lytic vacuoles have been reviewed (Jiang & Rogers, 1999).

### 4.1.6    *Transporter expression and nutrient availability*

The synthesis of many transporters is specifically linked to the availability of the particular inorganic solute. The response varies for different types of nutrients; for example, the expression of many nitrate transporter genes is increased (induced) in nitrogen-starved plants by subsequently supplying $NO_3^-$. The reverse situation was found for some ammonium, phosphate and sulphate transporters; for these nutrients the expression of the transporter decreases when the nutrients are supplied after a period of starvation. The reason for this fundamental difference is not clear; however, it has been suggested that this may reflect the relative importance of these different nutrients (Forde & Clarkson, 1999). Nitrogen is the nutrient that limits growth of all plants and is usually available to plants as $NO_3^-$, and consequently plants may have optimized the regulation of their transport systems to acquire this nutrient first. The level of expression of the nitrate transporters involved in acquisition at the root/soil interface is therefore switched on by nitrate *availability*. In contrast, the expression of other nutrient transporters, for example those for phosphate and sulphate, are regulated by the ensuing *demand* for the respective nutrient. With the resolution of families of transporters (see below) for a range of different nutrients, it is becoming clearer that there are exceptions to this generalized rule for induction and repression. The original picture was biased by the fact that the first nutrient transporter genes were often cloned by strategies exploiting the changes in expression that were associated with changes in supply of a specific nutrient. With a more complete picture for the inorganic solute transporters (see sections below on nitrate, ammonium and sulphate), it is becoming clear that this view of induction and repression is too simplistic and there are examples of both types of expression among the families of inorganic solute transporters. Examination of promoter sequences for the different genes within a family will reveal the specific elements responsible for the nutrient regulation.

## 4.2 Types of ion-coupled transporter

The sequencing of entire genomes, such as that of *Arabidopsis*, enables the identification of entire gene families for membrane transporters. Analysis of the *Arabidopsis* sequence predicts that 4589 of 25 470 open reading frames contain two or more membrane-spanning domains, of which 3208 cluster into 628 distinct families (Ward, 2001). These families of proteins, defined by sequence homology, include known transporter groups and many are yet to be defined. Most of the identified transporters contain multiple predicted transmembrane domains, typically 6–12. Many appear to function as monomers, or at least as homo-oligomers, as evidenced by the functional expression in heterologous expression systems. In the case of the high-affinity nitrate transport systems, regulatory subunits are essential for function (see below). Representative families of inorganic ion transporters are listed in Table 4.2. The large number of genes found in some families which have arisen from ancient gene duplication events may give rise to some redundancy; however, it is clear that specialized functions in terms of whole plant nutritional physiology have evolved. These transporters may be responsible for initial uptake of ions, for transport around the plant or for sub-cellular transport, for example across the tonoplast. Phylogenetic relationships within the cation transporter gene families have been defined (Mäser *et al.*, 2001) and family details can be found at the PlantsT Web site (http://plantst.sdsc.edu/). The members of an individual family may transport only single species (e.g. the sulphate transporter family), may have individuals with different transporter subtypes (e.g. the IRT/ZIP family in which branches are primarily responsible for iron or zinc ion transport) or may have relatively broad specificity, as seems to be the case with the NRAMP transporters. NRAMP3, for example, is a vacuolar metal transporter able to transport iron, manganese and zinc ions (Thomine *et al.*, 2003). Further novel transport roles are only just becoming evident as shown in an analysis of the expression profiles of 53 genes encoding known or potential metal transporters, which indicated unexpected transcriptional responses in response to copper, zinc and iron deficiency (Wintz *et al.*, 2003). Individual genes within the families may be expressed in different tissues or in a cell-specific manner (e.g. sulphate transporters, see below; and phosphate transporters, Leggewie *et al.*, 1997, and Gordon-Weeks *et al.*, 2003), typically with individual isoforms expressed selectively in roots, leaves or reproductive organs. As stated above, many of the transporters, particularly those responsible for uptake at the root/soil interface are differentially expressed in response to nutrient availability (most of the uptake

**Table 4.2** Representative putative families of inorganic solute transporters in *Arabidopsis thaliana*

| Mechanism | Inorganic solute transported | Family details Name | Number of genes | Characterized examples | Selected references |
|---|---|---|---|---|---|
| Symport | Ammonium | AMT1 | 5 | | von Wirén et al., 2000a |
| | | AMT2 | 1 | AMT2 | Sohlenkamp et al., 2000 |
| | Chloride | CCC | 1 | CCC1 (tobacco) | Harling et al., 1997 |
| | Copper | COPT | 5 | | Sancenon et al., 2003 |
| | Iron (Fe II), zinc, etc. | IRT/ZIP | 15 | IRT1 | Hall & Williams, 2003 |
| | Magnesium | MGT | 10 | | Li et al., 2001 |
| | Metals (Mn, Fe, Zn, Cd) | NRAMP | 6 | | Mäser et al., 2001 |
| | Nitrate | NRT1 | 51 | NRT1.1 (CHL1) | Guo et al., 2003 |
| | | NRT2 | 7 | NRT2.1 | Orsel et al., 2002 |
| | Phosphate | Pht1 | 9 | | Mudge et al., 2002 |
| | | Pht2 (possibly Na-coupled) | 1 | | Rausch & Bucher, 2002 |
| | Potassium | KUP/HAK/ KT(POT) | 13 | | Mäser et al., 2001 |
| | Sulphate | Sultr | 14 | | Hawkesford, 2003 |
| | Zinc | CDF | 5 | ZAT1 | Hall & Williams, 2003 |
| Antiport | Boron | AE | 7 | BOR1 | Takano et al., 2002 |
| | Cations (CHX family) | CHX, NDH, etc. (undefined) | >40 | | Mäser et al., 2001 |
| | | CAX (calcium) | 12 | | Mäser et al., 2001 |
| | | KEA (potassium) | 6 | | Mäser et al., 2001 |
| | | MHX1 (magnesium, zinc) | 1 | $Na^+/Ca^+$ antiport | Shaul et al., 1999 |
| | | NHX (sodium) | 7 | NHX1 | Xia et al., 2002 |
| | | SOS (sodium) | 1 | SOS1 | Zhu, 2000 |
| Other | Boron | BOR1 | 1 | | Takano et al., 2002 |

symporters: nitrate, e.g. Lauter *et al.*, 1996, Zhuo *et al.*, 1999, Lin *et al.*, 2000, and Ono *et al.*, 2000; phosphate, Smith *et al.*, 1997a, and Muchhal & Raghothama, 1999; sulphate, Smith *et al.*, 1997b). In some cases there are functional differences resulting in closely related transporters possessing different affinities for substrates, or different selectivities. In rare cases it is proposed that different members of an individual family exist on different sub-cellular membranes (see discussion of the sulphate transporter below). Only small differences in sequence are required to substantially alter properties of the transporters, as evidenced by a study involving selective site-directed mutagenesis of IRT1, which altered selectivity between iron, manganese and zinc (Rogers *et al.*, 2000). In short, an overall picture is of functional specialization of these individual genes rather than redundancy.

## 4.3   Nitrate

Nitrogen is the most important nutrient limiting plant growth and yield and so farmers spend more on applying this fertilizer than any other nutrient (http://www.fma.org.uk/). Nitrogen (N) in the soil is chiefly available to crop plants as nitrate and even if it is applied as fertilizer in another form such as urea or ammonium it is readily converted to nitrate in most agricultural soils. These facts have made the study of nitrate transport a focus for research and this effort has resulted in there being more information available on this type of ion-coupled transporter in comparison with most other nutrients.

### 4.3.1   *Physiology of nitrate transport mechanisms*

Nitrate is actively transported across the plasma membranes of plant cells, but net uptake is the balance between active influx and passive efflux. Little is known about the efflux mechanism and it could be mediated by ion-coupled transport or an anion channel as the electrochemical gradient for nitrate is directed out of the cell (Miller & Smith, 1996). Influx requires energy input from the cell over almost the whole range of concentrations encountered in the soil (Zhen *et al.*, 1991; Glass *et al.*, 1992; Miller & Smith, 1996). It is generally accepted that the uptake of $NO_3^-$ is coupled with the movement of two protons down an electrochemical potential gradient (see Section 4.1.3), and is therefore dependent on ATP supply to the $H^+$-ATPase that maintains the $H^+$ gradient across the plasma membrane (McClure *et al.*, 1990; Meharg & Blatt, 1995; Miller & Smith, 1996). For $NO_3^-$ storage in the vacuole, transport at the tonoplast membrane requires a different mechanism because the pH gradient is the reverse of that found across the plasma membrane and an antiport (see Fig. 4.1) with $H^+$ has been suggested (Miller & Smith, 1992). The genetic identity of the tonoplast transport system has not yet been solved but it is likely to be a gene that has nitrate inducible expression. Transport of $NO_3^-$ into the xylem, like plasma membrane efflux, is electrochemically downhill and so can be mediated by channels and these have been physiologically characterized (Kohler & Raschke, 2000; Kohler *et al.*, 2002).

Physiological studies of plant roots have shown the presence of both high- and low-affinity $NO_3^-$-uptake systems operating at different external $NO_3^-$ concentrations (Aslam et al., 1992; Glass & Siddiqi, 1995). There are believed to be two high-affinity transport systems (HATS) taking up $NO_3^-$ at low concentration (generally below 0.5 mM and with low transport capacity) and one low-affinity transport system (LATS) that transports $NO_3^-$ at high concentrations (generally above 0.5 mM and with high transport capacity) (Glass & Siddiqi, 1995).

### 4.3.2　Nitrate transporter gene families

Numerous $NO_3^-$ transporters have been cloned from a variety of species, and two distinct gene families, NRT1 and NRT2, have been identified (Crawford & Glass, 1998; Forde & Clarkson, 1999; Forde, 2000; Williams & Miller, 2001). The Arabidopsis genome contains 52 NRT1 and 7 NRT2 family members (see Table 4.2); but the two groups do not match with the observed physiology, as for example in Arabidopsis where the low-affinity $NO_3^-$ transporter, AtNRT1.1, also functions in the high-affinity range (Liu et al., 1999). These changes in the kinetics of transport are switched by phosphorylation of the protein (Liu & Tsay, 2003). A further complication for the NRT1 family is that they belong to a much larger family of peptide transporters, the POT, or proton-dependent oligopeptide-transport family, which is also known as the PTR or peptide-transport family (Paulsen & Skurray, 1994). Mammalian members of this family can transport peptides of varying sizes and one plant NRT1 protein can transport nitrate, peptides and some basic amino acids (Zhou et al., 1998). The NRT2 family belongs to a larger group of transporters, the nitrate–nitrite permeases (NNP), and this name reflects their ability to transport both these substrates (Forde, 2000). Some of the NRT2 family require a second gene product for functional activity, but it is not known whether there is an interaction between the gene products (Galván et al., 1996; Zhou et al., 2000a).

### 4.3.3　Regulation of expression

In Arabidopsis the pattern of tissue expression for much of the NRT2 family has been mapped (Orsel et al., 2002). Some members of both NRT1 and NRT2 gene families are $NO_3^-$ inducible, are expressed in the root epidermis and in root hairs, and may therefore have a role in the uptake of $NO_3^-$ from the soil (e.g. Lauter et al., 1996; Zhuo et al., 1999; Lin et al., 2000; Ono et al., 2000). Other family members are constitutively expressed. For example, in Arabidopsis, AtNRT1.2 is constitutively expressed in the roots, particularly in root hairs and the epidermis (Huang et al., 1999). The expression of both families can be regulated by feedback from N metabolites in many plant species (Touraine et al., 2001). Various amino acids have been tested for their ability to alter the expression and activity of $NO_3^-$ transporters through feedback regulation. Feeding amino acids to roots decreases the expression of $NO_3^-$ transporters (Vidmar et al., 2000b; Nazoa et al., 2003). However, identifying which amino acids are responsible for the feedback response is difficult, because they are assimilated and converted into different amino acids within the cell. By using chemical inhibitors to block the conversion of amino acids into other forms, glutamine has been identified as an important regulator (Vidmar *et al.*, 2000b).

### 4.3.4   Function in the root

The roles of both *NRT1* and *NRT2* gene families in the uptake of $NO_3^-$ from the soil have been demonstrated using mutant approaches. For example, a mutant of *Arabidopsis* deficient in the expression of an *NRT1* gene led to the identification of the first member of this family (Tsay *et al.*, 1993). Even stronger evidence is available for the *NRT2* family, where double-mutant knockouts of *NRT2* genes in *Arabidopsis* have demonstrated a clear role for these genes in the uptake of $NO_3^-$ from the soil (Filleur *et al.*, 2001). These mutants are deficient in both AtNRT2.1 and AtNRT2.2, and they have lost almost all the $NO_3^-$-inducible HATS, while LATS activity was not altered. Split-root experiments showed that the double mutant lost the ability to up-regulate uptake in one part of the root to compensate for nitrogen starvation in another part of the root (Cerezo *et al.*, 2001). In addition, the supply of $NH_4^+$ to the $NO_3^-$-containing nutrient solution usually inhibits $NO_3^-$ uptake in the wild type, but this does not occur in the mutant (Cerezo *et al.*, 2001). These elegant experiments illustrate the powerful use of gene 'knockout' technology to identify the role of specific transporter genes in N uptake by roots. These results are also important for confirming the function of these genes as $NO_3^-$ transporters, because almost all of the *in planta* expression studies have assumed function on the basis of sequence homology. Sequence similarities may be misleading, especially when a single protein can transport more than one type of ion or molecule, as in the case for both *NRT1* and *NRT2* transporter families. For example, some members of the *NRT1* family can transport amino acids and peptides, and both families can transport $NO_2^-$ when the proteins have been expressed in foreign cells (Zhou *et al.*, 1998; Miller & Zhou, 2000).

### 4.3.5   Function in the leaf

The first nitrate transporter to be identified *NRT1.1* is now known to be highly expressed in the guard cells of stomata where it has a specific role in the uptake of nitrate for stomatal function (Guo *et al.*, 2003). Gene knockout mutants lacking NRT1.1 have decreased stomatal opening and transpiration rates in the light or when deprived of $CO_2$ in the dark, and these effects resulted in enhanced drought tolerance (Guo *et al.*, 2003). Nitrate transporter expression is diurnally regulated, undergoing marked changes in transcript levels and corresponding $NO_3^-$ influx during day/night cycles with a peak towards the end of the light period (e.g. Ono *et al.*, 2000). Sucrose supply in the dark rapidly increases the transcript levels (Lejay *et al.*, 1999), and the diurnal increases in expression of root $NO_3^-$, $NH_4^+$ and $SO_4^{2-}$ transporters seem to be linked to the changes in sucrose supply to the root which results from photosynthesis during the day (Lejay *et al.*, 2003). These observations indicate the close coordination that exists between $NO_3^-$ and sulphate uptake and C metabolism. $Na^+$-coupled $NO_3^-$ symport has been demonstrated in cyanobacteria, e.g. *Anacystis nidulans* (Rodriguez *et al.*, 1994), and $Na^+$-dependent transport in the leaves of the marine angiosperm *Zostera marina* (Garcia-Sanchez *et al.*, 2000) suggests this transport mechanism may be more widely distributed in the plant kingdom. In the carnivorous plant *Nepenthes*, the specialized leaf structure that forms a pitcher trap has been shown to express an NRT1-type transporter in phloem cells and it was suggested that the protein may have a role in peptide transport as nitrate is not usually transported in this vascular tissue (Schulze *et al.*, 1999). This expression pattern in the pitcher plant supports the suggestion that the NRT1 family may transport a range of different N-containing solutes in plants (Zhou *et al.*, 1998).

## 4.4  Sulphate

The sulphur demand of the plant is met by the uptake of the sulphate oxyanion from the soil solution. This is achieved in a high affinity and saturable process with probable non-saturable components at higher concentrations, first described by Leggett and Epstein (1956). Accumulation is driven up a concentration gradient by coupling with the proton gradient and a ratio of $3H^+$/sulphate ion is assumed (Lass & Ullrich-Eberius, 1984). In common with other inorganic nutrients, the sulphate ion is distributed around the plant to meet biosynthetic requirements. Initial transport processes facilitate radial transfer within the root and unloading (efflux) into the xylem. Subsequently in the shoot, further influx/efflux steps are required for xylem unloading, cell-to-cell transfer, phloem loading/unloading and finally transport into the chloroplast, as the site of reductive assimilation. Additionally, an important contribution to cytoplasmic homeostasis is achieved by short-term storage in the vacuole, requiring influx/efflux systems across the tonoplast. Despite the contrasting energetic circumstances of these different membranes, it is postulated that members of a single gene family may be responsible for many of these transport steps. This specialization of function is outlined below (see also Hawkesford, 2003). Differential activity of these various transport systems will determine sulphur use efficiency within the plant (Hawkesford, 2000).

### 4.4.1  The sulphate transporter gene family

Since the first cloning of three sulphate transporter genes from the tropical legume, *Stylosanthes hamata* (Smith *et al.*, 1995a), sulphate transporter genes have been cloned from a wide range of organisms including barley (Smith *et al.*, 1997b; Vidmar *et al.*, 1999), *Sporobolus* (Ng *et al.*, 1996) and tomato (Howarth *et al.*, 2003). In *Arabidopsis* there are 14 members of this family with related gene sequences (see Takahashi *et al.*, 1996, 1997, 2000; Vidmar *et al.*, 2000a; Shibagaki *et al.*, 2002; Yoshimoto *et al.*, 2002, 2003). At present not all these gene products have verifiable function as sulphate transporters; however, no alternative substrates have been described to date. The genes encode strongly hydrophobic membrane proteins with up to 12 predicted transmembrane helices. In most sequences there are long N and C terminal regions and no other large extra membrane loops. In the C-terminal region, a STAS (sulphate transporters and antisigma factor antagonist) domain, potentially involved in post-translational regulation or binding to cytoskeletal elements, has been identified (Aravind & Koonin, 2000). In two sequences the N and C domains are substantially truncated.

A phylogenetic tree of the *Arabidopsis* protein sequences based on sequence similarity, following alignment, is presented in Fig. 4.3. On the basis of sequence alone, the sulphate transporters fall into five clusters. The sulphate transporters within the clusters have characteristics that lead to the designation of the subtypes, referred to as Groups 1–5. All of the other putative sulphate transporter genes sequenced to date, irrespective of species, fall into these five groups and may be assigned as homologues of one of the specific *Arabidopsis* types (see Hawkesford, 2003). The

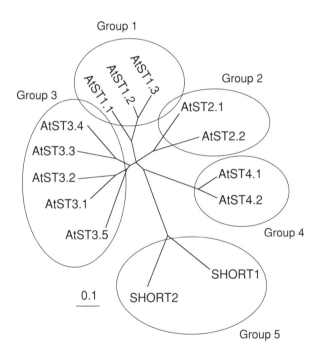

**Fig. 4.3** Phylogenetic representation of the plant sulphate transporter amino acid sequences showing subdivision into five groups. Accession numbers: *Arabidopsis*: AtSultr1.1, AB018695; AtSultr1.2, AB042322; AtSultr1.3, AB049624; AtSultr2.1, AB003591; AtSultr2.2, D85416; AtSultr3.1, D89631; AtSultr3.2, AB004060; AtSultr3.3, AB023423; AtSultr3.4, B054645; AtSultr3.5, AB061739; AtSultr4.1, AB008782; AtSultr4.2, AB052775; AtSultr5.1 (SHORT1), NP_178147; AtSultr5.2 (SHORT2), NP_180139; Alignments were performed using CLUSTAL W program (Thompson *et al.*, 1994), version 1.7, and the tree was drawn using the Treeview32 program (Page, 1996).

only exception is a differentiation within the subgroups between dicotyledonous and monocotyledonous plants; however, the group subdivisions still occur and a similar number of genes exist for all species examined to date.

*Arabidopsis* possesses three Group 1 sulphate transporters (see Takahashi *et al.*, 2000; Vidmar *et al.*, 2000a; Shibagaki *et al.*, 2002; Yoshimoto *et al.*, 2002). The sulphate transporters in Group 1 are characterized by a high affinity for sulphate (typically 1–10 μM). On the basis of expression and localization studies, AtST1.1 and AtST1.2 appear to be responsible for initial uptake in the root; however, expression is found in other tissues. AtST1.1 showed the greatest inducibility by sulphur starvation, indicating a specific role under nutrient-stressed conditions (Yoshimoto *et al.*, 2002). AtST1.3 appears to be localized in the sieve element-companion cell complexes of phloem of both the roots and cotyledons (Yoshimoto *et al.*, 2003). All sulphate transporters in this group show a classical de-repression of expression under S-limiting conditions (Clarkson *et al.*, 1983; see below).

Group 2 comprises two *Arabidopsis* genes that have been localized in vascular tissues (Takahashi *et al.*, 1997, 2000). AtST2.1 was localized in the xylem parenchyma cells of roots and leaves, the root pericycle and leaf phloem. AtST2.2 was localized in root phloem and leaf vascular bundle sheath cells. In contrast to the Group 1 sulphate transporters, all Group 2 transporters examined to date show a lower (>0.1 mM) affinity for sulphate.

There is little information on the Group 3 sequences. It has previously been referred to as the 'leaf group', based on initial localization of AtST3.1–3.3 (Takahashi *et al.*, 1999b, 2000) and the *Sporobolus stapfianus* sulphate transporter that was isolated from shoot tissues (Ng *et al.*, 1996). A Group 4 sulphate transporter was reported to be plastid localized (Takahashi *et al.*, 1999a), and by implication it was suggested that this transporter was responsible for the essential step of import of sulphate into the plastid prior to reduction. A potential chloroplast targeting sequence at the N-terminus is predicted and would support this localization (Takahashi *et al.*, 1999a; Godwin *et al.*, 2003). However, recent data utilizing a range of reporter constructs suggests that this transporter may be localized in the tonoplast (Takahashi *et al.*, 2003). The two *Arabidopsis* sequences tentatively assigned to Group 5 are rather dissimilar to one another and are the least homologous to the rest of the gene family. They are truncated proteins and lack the usual long N- and C-terminal domains that are thought to extend into the cytoplasm.

### 4.4.2 Functional characterization

The cloning strategy adopted to isolate the first plant sulphate transporter cDNAs involved functional expression of a plant cDNA library in a sulphate transport deficient mutant of yeast (Smith *et al.*, 1995b). This type of mutant (see also Cherest *et al.*, 1997) has been useful for the characterization of the transport properties of the cloned plant sulphate transporter gene products in isolation from the complex tissues of the intact root. This approach has enabled the determination of affinities for sulphate and the resolution of the high and low affinity types (see above). In addition, a pH dependency has been demonstrated (Smith *et al.*, 1995a), with highest activity at lower external pH, supporting the idea of a $H^+$-symport mechanism driving uptake. There are no reports of successful functional expression in other heterologous systems.

The roles of any of the putative sulphate transporters in Groups 3–5 remain problematic, as none have been expressed and characterized successfully in yeast or other heterologous systems. It may be speculated that they are responsible for transport across organelle membranes, or alternatively the transport of other related ions. This clearly underlines the requirement for functional characterization of any newly identified putative transporter.

### 4.4.3 Regulation

Many of the sulphate transporters are transcriptionally regulated in response to sulphur nutrition. Sulphate uptake capacity has been documented to be 'de-repressed' during sulphur starvation in algae (Passera & Ferrari, 1975), in intact plants (e.g. Lee, 1982; Clarkson *et al.*, 1983), in cell cultures (Smith, 1975) and in isolated vesicles (Hawkesford *et al.*, 1993). When plants are sulphur deficient, either as a result of interrupted supply or as a result of increased demand (see Lappartient & Touraine, 1996), the capacity for sulphate uptake increases over a period of several days, usually measured as $^{35}S$ tracer influx. Increased sulphate transporter mRNA (Smith *et al.*, 1995a, 1997b) and increased sulphate transporter protein in the plasma membrane (Hawkesford *et al.*, 2003) are observed concomitantly. In parallel, the internal content of sulphate and of reduced-sulphur compounds such as cysteine and glutathione is decreased (Smith *et al.*, 1997b). Upon re-supply of sulphate, the transporter activity is 'repressed' within hours, with parallel decreased mRNA and protein levels, clearly indicating rapid turnover of both of these molecules. Tissue concentrations of sulphate, cysteine and glutathione all increase. On the basis of such results, a simple negative feedback model of regulation may be proposed. However, supply of the cysteine precursor, *O*-acetylserine, to roots of plants adequately supplied with sulphur, also serves to de-repress the transport of sulphate, at the level of gene expression, and there is a resultant increase in sulphate uptake and synthesis of cysteine and glutathione. Under these circumstances, *O*-acetylserine acts as an overriding inducer of gene expression. The two control loops act antagonistically to modulate sulphate uptake and maximize sulphate uptake with fluctuating supply and cellular demand (see Hawkesford *et al.*, 2003).

## 4.5 Ammonium

Ammonium is another form of N that is available to plants in the soil, but in soil it is usually found at lower concentrations than nitrate (Owen & Jones, 2001), and as it is a cation there are very different energy gradients for transport when compared with the anions, nitrate and sulphate.

### 4.5.1  $NH_4^+$ uptake gene family

Many plant $NH_4^+$-transporter (AMT) genes have been identified and their function has been confirmed by their ability to complement a yeast mutant deficient in normal $NH_4^+$ uptake (von Wirén et al., 2000a). In Arabidopsis there are 6 AMT genes (see Table 4.2), while rice has 10. More detailed sequence comparisons have identified two distinct types: AMT1 and AMT2 (Sohlenkamp et al., 2000). When expressed in yeast cells the AMT1 genes have very high affinities for $NH_4^+$, with $K_m$ values typically in the low micromolar range (Sohlenkamp et al., 2000; von Wirén et al., 2000a). The transport mechanism was generally assumed to be by symport with $H^+$, but recently more detailed functional analysis of the AMT1 proteins using heterologous expression in Xenopus oocytes suggests that they have a channel-like structure that can be composed of several different multiples of AMT1 protein units (Ludewig et al., 2002, 2003). The functional activity of the whole protein complex may be modified by altering the AMT1 component units, allowing entry of $NH_4^+$ by a uniport mechanism and thus may not be correctly defined as ion-coupled transporters.

### 4.5.2  Function in the root

Like the $NO_3^-$ transporters, some AMT1-type genes are expressed in root hairs, suggesting that they have a role in uptake of $NH_4^+$ from the soil (Lauter et al., 1996; Ludewig et al., 2002). However, in contrast to the situation for $NO_3^-$, the expression of some AMTs is repressed by the presence of $NH_4^+$, with the mRNA increasing when less $NH_4^+$ is available. Three Arabidopsis AMT1 genes show diurnal changes in expression in roots (Gazzarrini et al., 1999), and the changes in expression during the light period probably result from increases in sucrose availability as a result of photosynthesis during the day (Lejay et al., 2003). The transcript levels of AtAMT1.1, a transporter with an affinity in the nanomolar range, steeply increased with ammonium uptake in roots when nitrogen nutrition became limiting, whereas those of AtAMT1.3 increased only slightly, while AtAMT1.2 was expressed more constitutively (Gazzarrini et al., 1999). More detailed information has been published about the AMT1- than about AMT2-type transporters, and a correlation between transcript level and $NH_4^+$ influx has been observed (Kumar et al., 2003), but the role of neither group in uptake from the soil has been clearly established. For example, Arabidopsis plants deficient in one of the root-expressed AMT1 genes did not show any significant decrease in the uptake of $NH_4^+$ when compared with wild-type plants, and it was suggested that redundancy within the AMT family may compensate for the loss of this transporter (Kaiser et al., 2002). Similarly, inhibiting the mRNA transcript level of the single AMT2 in Arabidopsis failed to significantly alter growth of the plant, although the actual uptake of $NH_4^+$ was not measured (Sohlenkamp et al., 2002). One of the AMT2 transporters is constitutively expressed in the plasma membrane of most tissues including the nodules of a $N_2$-fixing species, suggesting that it may have a general role in the recovery of $NH_4^+$ that is lost by efflux from all tissues, not only the nodule (Simon-Rosin et al., 2003). Some AMTs are constitutively expressed (Suenaga et al., 2003), but for most the expression depends on the availability of $NH_4^+$ (von Wirén et al., 2000b). The expression of one tomato AMT1 gene was induced by the presence of $N_2$-fixing bacteria in the rhizosphere (Becker et al., 2002). In species like paddy rice that chiefly use $NH_4^+$ as the main N source, more of the AMT1 genes show $NH_4^+$-induced expression when compared with Arabidopsis and tomato that chiefly use $NO_3^-$ as a N source (Sonoda et al., 2003).

As for $NO_3^-$, $NH_4^+$ transport in root cells can also be demonstrated by electrophysiology (Wang et al., 1994). Electrophysiology has been used to determine the $NH_4^+$-transporter kinetics, indicating that $NH_4^+$ entry into cells might be mediated by co-transport with protons with Michaelis–Menten-type uptake kinetics (Ayling, 1993; Wang et al., 1994). However, the energy requirements for uptake of a cation (e.g. $NH_4^+$) compared to an anion (e.g. $NO_3^-$) are different. The uptake of $NH_4^+$, like the uptake of $K^+$, could be through a channel, and chiefly driven by the negative membrane potential of the plant cell. Several examples of $K^+$ channels expressed in the root epidermis have been identified (e.g. Hartje et al., 2000) and gene knockout studies should identify if these have a role in $NH_4^+$ uptake. There is evidence from patch clamp studies that $NH_4^+$ ions can enter cells through $K^+$channels (White, 1996), but it can also be a channel blocker (Spalding et al., 1999). It is not clear if the channel route is an important mechanism for the entry of $NH_4^+$ into root cells (see Chapter 6). This topic is worth investigation using plants that have disrupted plasma-membrane $K^+$-channel activity, especially given the lack of direct evidence for the role of AMTs in $NH_4^+$ uptake by root cells.

### 4.5.3 Function in the leaf

The pattern of ammonium transport in leaves changes during development; for example, during growth, $NH_4^+$ influx into cells is required but at senescence, when protein breakdown occurs, N is remobilized from leaves and there is export of $NH_4^+$ from the leaf. Gaseous losses of $NH_3$ from leaves are considered to be of economic importance in agriculture and measurements of the leaf apoplastic concentrations of $NH_4^+$ have been the focus of attention (Nielsen & Schjoerring, 1998). The uptake of $NH_4^+$ into detached leaves of Brassica napus increased in parallel with greater expression of BnAMT1.2 (Pearson et al., 2002), a close relative of AtAMT1.3 showing 97% similarity. Withdrawal of $NH_4^+$ supply had little effect on expression and the addition of amino acids decreased the amounts of mRNA; however, at high $NH_4^+$ concentrations the transcript increased (Pearson et al., 2002). The results, together with proof of function in yeast cells, implicate the protein's role in leaf cell plasma membrane influx. As described for roots, in Arabidopsis the AMT1 genes show diurnal changes in expression that are dependent on changes in carbon supply from photosynthesis (Gazzarrini et al., 1999; Lejay et al., 2003). Transient expression of an AtAMT2–GFP protein fusion in *Arabidopsis* leaf epidermal cells indicated that the protein was located in the plasma membrane (Sohlenkamp *et al.*, 2002). Ammonium transporters are expressed in the pitchers of carnivorous plants where they are thought to have a role in the uptake from the pitcher fluid (Schulze *et al.*, 1999).

## 4.6   Energetic costs of transport

As described previously (Section 4.1.1), the primary $H^+$ pumps provide the energy for $H^+$-coupled uptake of inorganic solutes but efflux processes can provide a significant drain on a plant cell.

### 4.6.1   Nitrate and sulphate efflux

Although the molecular identity of the protein(s)-mediating efflux of nitrate and sulphate has not been demonstrated, the physiological importance of the processes has been identified. For example, efflux is clearly an important aspect of cell-to-cell transport across tissues and xylem loading (reviewed in Forde & Clarkson, 1999). It has been calculated that plants invest a large amount of their respiratory energy for the uptake of anions particularly nitrate (Poorter *et al.*, 1995). As the efflux processes is energetically wasteful and acts against the ion-coupled influx processes, it may have considerable implications for the growth of plants (Scheurwater *et al.*, 1999). For example, when fast- and slow-growing grass species were compared a large efflux of $NO_3^-$ was measured during the night from the slow, but not the fast, growing species (Scheurwater *et al.*, 1999).

### 4.6.2   Ammonium efflux

The energetic requirement of pumping $NH_4^+$ out of cells has been identified as a possible cause for the toxic effect of the ion on some types of plants, for example barley (Britto *et al.*, 2001; see below). The gene(s) responsible for this $NH_4^+$ efflux process has not yet been identified, but the thermodynamic mechanism for such a process requires an ATPase or an antiport exchanging $H^+$ and $NH_4^+$. However, the equilibrium between $NH_4^+$ and $NH_3$ inside the cell may also favour efflux of uncharged $NH_3$ through channels, such as aquaporins (Howitt & Udvardi, 2000). Central to our understanding of these processes are accurate measurements of cytosolic concentrations of $NH_4^+$ and $NH_3$; such information then gives the prevailing transmembrane gradients and likely transport mechanisms can be determined. A large range of values are published for these gradients in plant cells and results seem to depend on the method used (reviewed in Miller *et al.*, 2001). It is not clear why $K^+$ entry and cytosolic concentration should be regulated while those of $NH_4^+$ are poorly regulated, but as for $Na^+$ entry during salt stress, the plant cannot avoid this problem when exposed to high concentrations of these cations. Therefore, accurate measurements of the soil concentrations of $NH_4^+$ may be important for answering these questions for plants growing in soil. The toxic effects of $NH_4^+$ depend on there being high external concentrations of the cation, perhaps greater than 20 mM (Britto & Kronzucker, 2002). The plant *AMT* gene family functions as high-affinity $NH_4^+$-uptake systems when they are expressed in yeast (von Wirén *et al.*, 2000a). The requirement for an active efflux mechanism at high external $NH_4^+$ concentrations does not easily fit with the constitutive expression of some of these genes, and so more expression analysis is needed to clarify this point.

## 4.7 Conclusions and future research

### 4.7.1 Gene families and functional diversity

Physiological studies of the transport of inorganic solutes are no longer dominated by whole plant influx studies. Substantial progress on our understanding of ion transport processes has been aided by the completion of plant genome sequencing projects. These projects have enabled a full inventory of probable transporters to be determined. Initial surprise at the size of gene families is being replaced by an increasing appreciation of the diverse and specialized roles of each family member. Each transporter is likely to have a specific function in the context of whole plant nutrient physiology, with particular roles in organs, cells and even sub-cellular compartments. The intracellular membrane location of very few transporter proteins has been confirmed. This is not an easy task as membrane localization often involves attaching a protein tag to the transporter and this risks preventing the normal targeting of the transporter.

Gene families based on sequence homology are a useful guideline to likely function, but heterologous expression in foreign cells (e.g. yeast or *Xenopus* oocytes) is a better method to identify function of the protein. However, each of these methods has certain limitations, for example the membrane lipid composition may be different or the ionic environment is changed such that the transporter protein properties are altered in this foreign cell. No heterologous expression system is perfect and the safest results are probably acquired using data obtained by comparing at least two different cell expression systems. This approach has not been taken by many authors, but it has been used for some of the *AtAMT1* genes that have been functionally characterized in both yeast and oocyte expression systems (Ludewig *et al.*, 2002). Even this approach has some limitations because in oocytes the membrane voltage is controlled and measured, while in yeast cells this important parameter floats, depending on the health and energy status of the cell. The importance of this parameter in yeast cells was recently demonstrated for a $H^+$-coupled sucrose symporter expressed in yeast. When the yeast cells were fed with a carbon source for increased respiration, the cell membrane potential became more negative and only then could sucrose uptake be measured (Barth *et al.*, 2003). This result shows the importance of membrane voltage as a factor for regulating ion-coupled transporter activity in any cell. Oocyte expression studies have also shown how the membrane voltage can also change the kinetic properties of a transporter protein, for example the affinity of the protein for the inorganic solute (Zhou *et al.*, 1998; Miller & Zhou, 2000). Voltage clamp experiments on plant tissues like root hairs can also be used to demonstrate the importance of membrane voltage *in vivo* (Meharg & Blatt, 1995), but in these experiments the transport properties are likely to represent the sum of several different genes expressed in the cell and this type of experiment can only be performed on a few specialized cell types. Proof of function is likely to be provided by a combination of both heterologous expression systems (yeast and/or oocytes) and gene knockout mutant plants.

The molecular identification of all these transporter genes can provide an unjustified sense of security in their function. Using sequence homology to identify function is very risky as we know that small changes in amino acid sequence, even single residues, can alter function of a particular transport protein (e.g. Rogers *et al.*, 2000). Similarly, categorizing whole families based on a few characterized individuals is likely to be an oversimplification. For example, the recent finding that the post-translational modification of proteins can change their kinetic properties between high and low affinity (Liu & Tsay, 2003) suggests that linking a gene to a specific physiological function can be very difficult. For example, dependent on the phosphorylation state, a single protein may be responsible for both high- and low-affinity transport by a single cell. Furthermore, functional analysis has sometimes identified a diversity of substrates that can be transported by a single protein; for example, a whole range of N-containing substrates can be transported by the *NRT1* family (Zhou *et al.*, 1998). The usefulness of assigning and naming gene families on the basis of their substrate affinity must now be questionable. Transporter function may even be different depending on which membrane and/or in which cell type it is expressed. To date, a limitation of much expression data, including northerns and microarray analysis, is both sensitivity and the lack of resolution at the cellular level. Transcriptional studies at the single-cell level are a way forward.

### 4.7.2 Homeostasis of cell nutrients and nutrient sensors

Ion-coupled transporters have an important role in maintaining the ion environment within cells. The cytosolic ionic environment requires regulation to maintain cell concentrations at non-toxic levels to preserve the function of metabolic processes. For pH and calcium, the homeostasis of cytosol is well established but for other nutrient ions, such as $K^+$ and $NO_3^-$, the situation is more controversial (e.g. Britto & Kronzucker, 2003). The induction or repression of transporter-gene expression requires that there is some nutrient-sensing system(s) within the cell, perhaps in the nucleus or at the cell surface. Membrane proteins have been identified as possible sensors of soil N availability (Redinbaugh & Campbell, 1991). Ion-coupled transporters can provide a sensor at the root/soil interface that may be involved in sensing flux through the transporter protein and/or availability of particular forms of

N at the cell surface. These sensors may have a role in regulating cellular N pools and/or detecting available pools of N both inside cells and in the soil around the root. Homeostasis of cytosolic $NO_3^-$ pools requires some sensors to regulate these concentrations in this compartment (Miller & Smith, 1996; Van der Leij et al., 1998). There are fungal examples of 'transporter' proteins that seem to have this role for $NH_4^+$ and sugar sensing (Lorenz & Heitman, 1998; Ozcan et al., 1998), but the situation in plants is less clear (Barth et al., 2003). The large numbers of particular types of some transporters (e.g. peptide transporters in the PTR family) may be ascribed to gene redundancy, but this may also be because some family members function as sensors. A family of membrane proteins that are related to known glutamate receptors have been identified in plants (Lam et al., 1998), and mutant plants with altered expression of the genes indicate that they have a role in regulating C/N metabolism (Kang & Turano, 2003). There are few studies of sulphate sensing in plants; however, three *Chlamydomonas reinhardtii* mutant lines were identified with aberrant responses to sulphur deficiency, and the genes corresponding to two of these have been isolated: *sac1* has been identified as an integral membrane protein with similarity to sodium dicarboxylate transporters (Yildiz et al., 1996), and *sac3* is a SNF1-like serine threonine kinase (Davies et al., 1999). It appears that these genes may be involved in sensing sulphate status.

### 4.7.3   Conclusions

A wealth of new genomic knowledge has provided new possibilities for the investigation of inorganic nutrient regulation in plants. Gene duplication events and the evolution of specialized functions have created the observed gene families. In some cases, duplications may be so recent that separate function has not evolved, leading to some redundancy, but this seems likely to be the exception rather than the rule. Application of appropriate high-resolution expression analyses and comprehensive physiological studies is the next step forward. The reassuring identification of gene families and the current phylogenic assignment are all based on sequence information, and the sorting of transporters into families on the basis of their function is the future challenge. This system of assignment may yet identify genes with no sequence homology fulfilling the same physiological task in two different species of plant.

### Acknowledgements

M.J. Hawkesford and A.J. Miller are sponsored by grants from the BBSRC, DEFRA (AR0910, AR0911) and by the EU (QLRT-2000-00103 and HPRN-CT-2002-00247). Rothamsted Research receives grant-aided support from the Biotechnology and Biological Sciences Research Council of the United Kingdom.

# References

Aravind, L. & Koonin, E.V. (2000) The STAS domain – a link between anion transporters and antisigma-factor antagonists, *Curr. Biol.*, **10**, R53–R55.

Aslam, M., Travis, R.L. & Huffaker, R.C. (1992) Comparative kinetics and reciprocal inhibition of nitrate and nitrite uptake in roots of uninduced and induced barley (*Hordeum vulgare* L.) seedlings, *Plant Physiol.*, **99**, 1124–1133.

Ayling, S.M. (1993) The effect of ammonium ions on membrane potential and anion flux in roots of barley and tomato, *Plant Cell Environ.*, **16**, 297–303.

Barber, S.A. (1984) *Soil Nutrient Bioavailability: A Mechanistic Approach*, John Wiley & Sons, New York, p. 398.

Barth, I., Meyer, S. & Sauer, N. (2003) PmSUC3: characterization of a SUC2/SUC3-type sucrose transporter from *Plantago major*, *Plant Cell*, **15**, 1375–1385.

Becker, D., Stanke, R., Fendrik, I., *et al.* (2002) Expression of the $NH_4^+$-transporter gene LeAMT1.2 is induced in tomato roots upon association with $N_2$-fixing bacteria, *Planta*, **215**, 424–429.

Brandizzi, F., Frangne, N., Marc-Martin, S., Hawes, C., Neuhaus, J.M. & Paris, N. (2002) The destination for single-pass membrane proteins is influenced markedly by the length of the hydrophobic domain, *Plant Cell*, **14**, 1077–1092.

Britto, D.T. & Kronzucker, H.J. (2002) $NH_4^+$ toxicity in higher plants: a critical review, *J. Plant Physiol.*, **159**, 567–584.

Britto, D.T. & Kronzucker, H.J. (2003) The case for cytosolic $NO_3^-$ heterostasis: a critique of a recently proposed model, *Plant Cell Environ.*, **26**, 183–188.

Britto, D.T., Siddiqi, M.Y., Glass, A.D.M. & Kronzucker, H.J. (2001) Futile transmembrane $NH_4^+$ cycling: a cellular hypothesis to explain ammonium toxicity in plants, *Proc. Natl. Acad. Sci. U.S.A.*, **98**, 4255–4258.

Cerezo, M., Tillard, P., Filleur, S., Munos, S., Daniel-Vedele, F. & Gojon, A. (2001) Major alterations of the regulation of root $NO_3^-$ uptake are associated with the mutation of *Nrt2.1* and *Nrt2.2* genes in *Arabidopsis*, *Plant Physiol.*, **127**, 262–271.

Cherest, H., Davidian, J.-C., Thomas, D., Benes, V., Ansorge, W. & Surdin-Kerjan, Y. (1997) Molecular characterization of two high affinity sulfate transporters in *Saccharomyces cerevisiae*, *Genetics*, **145**, 627–635.

Clarkson, D.T., Smith, F.W. & Vandenberg, P.J. (1983) Regulation of sulfate transport in a tropical legume, *Macroptilium atropurpureum* cv Sirato, *J. Exp. Bot.*, **34**, 1463–1483.

Crawford, N.M. & Glass, A.D.M. (1998) Molecular and physiological aspects of nitrate uptake in plants, *Trends Plant Sci.*, **3**, 389–395.

Czempinski, K., Frachisse, J.M., Maurel, C., Barbier-Brygoo, H. & Müller-Röber, B. (2002) Vacuolar membrane localization of the *Arabidopsis* 'two-pore' $K^+$ channel KCO1, *Plant J.*, **29**, 809–820.

Davies, J.P., Yildiz, F.H. & Grossman, A.R. (1999) Sac3, an snf1-like serine/threonine kinase that positively and negatively regulates the responses of *Chlamydomonas* to sulfur limitation, *Plant Cell*, **11**, 1179–1190.

Dunlop, J. & Gardiner, S. (1993) Phosphate uptake, proton extrusion and membrane electropotentials of phosphorus-deficient *Trifolium repens* L., *J. Exp. Bot.*, **44**, 1801–1808.

Espen, L., Dell'Orto, M., De Nisi, P. & Zocchi, G. (2000) Metabolic responses in cucumber (*Cucumis sativus* L.) roots under Fe-deficiency: a P-31-nuclear magnetic resonance *in-vivo* study, *Planta*, **210**, 985–992.

Filleur, S., Dorbe, M., Cerezo, M., *et al.* (2001) An *Arabidopsis* T-DNA mutant affected in *Nrt2* genes is impaired in nitrate uptake, *FEBS Lett.*, **489**, 220–224.

Forde, B.G. (2000) Nitrate transporters in plants: structure, function and regulation, *Biochim. Biophys. Acta*, **1465**, 219–235.

Forde, B.G. & Clarkson, D.T. (1999) Nitrate and ammonium nutrition of plants: physiological and molecular perspectives, *Adv. Bot. Res.*, **30**, 1–90.

Galván, A., Quesada, A. & Fernández, E. (1996) Nitrate and nitrite are transported by different specific transport systems and by a bispecific transporter in *Chlamydomonas reinhardtii*, *J. Biol. Chem.*, **271**, 2088–2092.

Garcia-Sanchez, M.J., Jaime, M.P., Ramos, A., Sanders, D. & Fernandez, J.A. (2000) Sodium-dependent nitrate transport at the plasma membrane of leaf cells of the marine higher plant *Zostera marina* L., *Plant Physiol.*, **122**, 879–885.

Gazzarrini, S., Lejay, L., Gojon, A., Ninnemann, O., Frommer, W. & von Wirén, N. (1999) Three functional transporters for constitutive, diurnally regulated, and starvation-induced uptake of ammonium into *Arabidopsis* roots, *Plant Cell*, **11**, 937–947.

Glass, A.D.M., Shaff, J. & Kochian, L. (1992) Studies of the uptake of nitrate in barley, IV: Electrophysiology, *Plant Physiol.*, **99**, 456–463.

Glass, A.D.M. & Siddiqi, M.Y. (1995) Nitrogen absorption by plant roots, in *Nitrogen Nutrition in Higher Plants* (eds H. Srivastava & R. Singh), Associated Publishing Co., New Delhi, India, pp. 21–56.

Godwin, R.M., Rae, A.L., Carroll, B.J. & Smith, F.W. (2003) Cloning and characterization of two genes encoding sulfate transporters from rice (*Oryza sativa* L.), *Plant Soil*, **257**, 113–123.

Gordon-Weeks, R., Tong, Y., Davies T.G.E. & Leggewie, G. (2003). Restricted spatial expression of a high-affinity phosphate transporter in potato roots, *J. Cell Sci.*, **116**, 3135–3144.

Guo, F.-Q., Young, J. & Crawford, N.M. (2003) The nitrate transporter AtNRT1.1 (CHL1) functions in stomatal opening and contributes to drought susceptibility in *Arabidopsis*, *Plant Cell*, **15**, 107–117.

Hall, J.L. & Williams, L.E. (2003) Transition metal transporters in plants, *J. Exp. Bot.*, **54**, 2601–2613.

Harling, H., Czaja, I., Schell, J. & Warden, R. (1997) A plant cation-chloride co-transporter promoting auxin-independent tobacco protoplast division, *EMBO J.*, **16**, 5855–5866.

Hartje, S., Zimmerman, S., Klonus, D. & Müller-Röber, B. (2000) Functional characterization of LKT1, a $K^+$ uptake channel from tomato root hairs, and comparison with the closely related potato inwardly rectifying $K^+$ channel after expression in *Xenopus* oocytes, *Planta*, **210**, 723–731.

Hawkesford, M.J. (2000) Plant responses to sulphur deficiency and the genetic manipulation of sulphate transporters to improve S-utilisation efficiency, *J. Exp. Bot.*, **51**, 131–138.

Hawkesford, M.J. (2003) Transporter gene families in plants: the sulphate transporter gene family – redundancy or specialization? *Physiol. Plant.*, **117**, 155–165.

Hawkesford, M.J., Buchner, P., Hopkins, L. & Howarth, J. (2003) The plant sulfate transporter family: specialized function and integration with whole plant nutrition, in *5th Workshop on Sulfur Transport and Assimilation: Regulation, Interaction, Signalling* (eds J.-C. Davidian, D. Grill, L.J. De Kok, I. Stulen, M.J. Hawkesford, E. Schnug & H. Rennenberg), Backhuys, Leiden, The Netherlands, pp. 1–10.

Hawkesford, M.J., Davidian, J.-C. & Grignon, C. (1993) Sulphate/$H^+$ co-transport in plasma membrane vesicles isolated from *Brassica napus*: increased transport in membranes isolated from sulphur-starved plants, *Planta*, **190**, 297–304.

Hirayama, B.A., Loo, D.D.F. & Wright, E.M. (1994) Protons drive sugar transport through the $Na^+$/glucose cotransporter (SGLT1), *J. Biol. Chem.*, **269**, 21407–21410.

Howarth, J.R., Fourcroy, P., Davidian, J.-C., Smith, F.W. & Hawkesford, M.J. (2003) Cloning of two contrasting sulfate transporters from tomato induced by low sulfate and infection by the vascular pathogen *Verticillium dahliae*, *Planta*, **218**, 58–64.

Howitt, S. & Udvardi, M.K. (2000) Structure, function and regulation of ammonium transporters in plants, *Biochim. Biophys. Acta*, **1465**, 152–170.

Huang, N.C., Liu, K.H., Lo, H.J. & Tsay, Y.F. (1999) Cloning and functional characterization of an *Arabidopsis* nitrate transporter gene that encodes a constitutive component of low affinity uptake, *Plant Cell*, **11**, 1381–1392.

Jiang, L.W. & Rogers, J.C. (1999) Sorting of membrane proteins to vacuoles in plant cells, *Plant Sci.*, **146**, 55–67.

Jung, H. (2002) The sodium/substrate symporter family: structural and functional features, *FEBS Lett.*, **529**, 73–77.

Kaiser, B.N., Rawat, S.R., Siddiqi, M.Y., Masle, J. & Glass, A.D.M. (2002) Functional analysis of an *Arabidopsis* T-DNA 'knock-out' of the high affinity transporter AtAMl.1, *Plant Physiol.*, **130**, 1263–1275.

Kang, J. & Turano, F.J. (2003) The putative glutamate receptor 1.1 (AtGLR1.1) functions as a regulator of carbon and nitrogen metabolism in *Arabidopsis thaliana*, *Proc. Natl. Acad. Sci. U.S.A.*, **100**, 6872–6877.

Kessler, F. & Neuhaus, J.M. (2003) Sorting activities in plant cells, *Chimia*, **57**, 634–638.

Kohler, B. & Raschke, K. (2000) The delivery of salts to the xylem. Three types of anion conductance in the plasmamemma of the xylem parenchyma of roots of barley, *Plant Physiol.*, **122**, 243–254.

Kohler, B., Wegner, L.H., Osipov, V. & Raschke, K. (2002) Loading of nitrate into the xylem: apoplastic nitrate controls the voltage dependence of X-QUAC, the main anion conductance in xylem-parenchyma cells of barley roots, *Plant J.*, **30**, 133–142.

Kumar, A., Silim, S.N., Okamoto, M., Siddiqi, M.Y. & Glass, A.D.M. (2003) Differential expression of three members of the *AMT1* gene family encoding putative high-affinity $NH_4^+$ transporters in roots of *Oryza sativa* subspecies indica, *Plant Cell Environ.*, **26**, 907–914.

Kurkdjian, A. & Guern, J. (1989) Intracellular pH: measurement and importance in cell activity, *Annu. Rev. Plant Physiol. Plant Mol. Biol.*, **40**, 271–303.

Lam, H.-M., Chiu, J., Hsieh, M.-H., *et al.* (1998) Glutamate-receptor genes in plants, *Nature*, **396**, 125–126.

Lappartient, A.G. & Touraine, B. (1996) Demand-driven control of root ATP sulphurylase activity and sulfate uptake in intact canola, *Plant Physiol.*, **111**, 147–157.

Lass, B. & Ullrich-Eberius, C.I. (1984) Evidence for proton/sulfate cotransport and its kinetics in *Lemna gibba* G1, *Planta*, **161**, 53–60.

Lauter, F.R., Ninnemann, O., Bucher, M., Riemeier, J.W. & Frommer, W.B. (1996) Preferential expression of an ammonium transporter and of two putative nitrate transporters in root hairs of tomato, *Proc. Natl. Acad. Sci. U.S.A.*, **93**, 8139–8144.

Lee, R.B. (1982) Selectivity and kinetics of ion uptake by barley plants following nutrient deficiency, *Ann. Bot.*, **50**, 529–449.

Lee, Y.J., Kim, D.H., Kim, Y.W. & Hwang, I. (2001) Identification of a signal that distinguishes between the chloroplast outer envelope membrane and the endomembrane system *in vivo*, *Plant Cell*, **13**, 2175–2190.

Leggett, J.E. & Epstein, E. (1956) Kinetics of sulfate absorption by barley roots, *Plant Physiol.*, **31**, 222–226.

Leggewie, G., Willmitzer, L. & Riesmeier, J.W. (1997) Two cDNAs from potato are able to complement a phosphate-deficient yeast mutant: identification of phosphate transporters from higher plants, *Plant Cell*, **9**, 381–392.

Lejay, L., Gansel, X., Cerezo, M., *et al.* (2003) Regulation of root ion transporters by photosynthesis: functional importance and relation with hexokinase, *Plant Cell*, **15**, 2218–2232.

Lejay, L., Tillard, P., Domingo Olive, F., *et al.* (1999) Molecular and functional regulation of two $NO_3^-$ uptake systems by N- and C-status of *Arabidopsis* plants, *Plant J.*, **18**, 509–519.

Li, H.M. & Chen L.J. (1996) Protein targeting and integration signal for the chloroplastic outer envelope membrane, *Plant Cell*, **8**, 2117–2126.

Li, L.G., Tutone, A.F., Drummond, R.S.M., Gardner, R.C. & Luan, S. (2001) A novel family of magnesium transport genes in *Arabidopsis*, *Plant Cell*, **13**, 2761–2775.

Lin, C.M., Koh, S., Stacey, G., Yu, S.M., Lin, T.Y. & Tsay, Y.F. (2000) Cloning and functional characterization of a constitutively expressed nitrate transporter gene, *OsNRT1* from rice, *Plant Physiol.*, **122**, 379–388.

Liu, K.-H., Huang, C.-H. & Tsay, Y.-F. (1999) CHL1 is a dual-affinity nitrate transporter of *Arabidopsis* involved in multiple phases of nitrate uptake, *Plant Cell*, **11**, 865–874.

Liu, K.-H. & Tsay, Y.-F. (2003) Switching between the two action modes of the dual-affinity nitrate transporter CHL1 by phosphorylation, *EMBO J.*, **22**, 1005–1013.

Lorenz, M. & Heitman, J. (1998) The MEP2 ammonium permease regulates pseudohyphal differentiation in *Saccharomyces cervisiae*, *EMBO J.*, **17**, 1236–1247.

Ludewig, U., von Wirén, N. & Frommer, W.B. (2002) Uniport of $NH_4^+$ by the root hair plasma membrane ammonium transporter *LeAMT1.1*, *J. Biol. Chem.*, **277**, 13548–13555.

Ludewig, U., Wilken, S., Wu, B., *et al.* (2003) Homo- and hetero-oligomerization of ammonium transporter-1 $NH_4^+$ uniporters, *J. Biol. Chem.*, **278**, 45603–45610.

Maathuis, F.J.M, Verlin, D., Smith, F.A., Sanders, D., Fernandez, J.A. & Walker, N.A. (1996) The physiological relevance of $Na^+$-coupled $K^+$ transport, *Plant Physiol.*, **112**, 1609–1616.

Mäser, P., Thomine, S., Schroeder, *et al.* (2001) Phylogenetic relationships within the cation transporter families of *Arabidopsis*, *Plant Physiol.*, **126**, 1646–1667.

McClure, P.R., Kochian, L.V., Spanswick, R.M. & Shaff, J.E. (1990) Evidence for cotransport of nitrate and protons in maize roots, I: Effects of nitrate on the membrane potential, *Plant Physiol.*, **93**, 281–289.

Meharg, A.A. & Blatt, M.R. (1995) $NO_3^-$ transport across the plasma membrane of *Arabidopsis thaliana* root hairs: kinetic control by pH and membrane voltage, *J. Membr. Biol.*, **145**, 49–66.

Miller, A.J., Cookson, S.J., Smith, S.J. & Wells, D.M. (2001) The use of microelectrodes to investigate compartmentation and the transport of metabolized inorganic ions in plants, *J. Exp. Bot.*, **52**, 541–549.

Miller, A.J. & Smith, S.J. (1992) The mechanism of nitrate transport across the tonoplast of barley root cells, *Planta*, **187**, 554–557.

Miller, A.J. & Smith, S.J. (1996) Nitrate transport and compartmentation, *J. Exp. Bot.*, **47**, 843–854.

Miller, A.J. & Zhou, J.-J. (2000) *Xenopus* oocytes as an expression system for plant transporters, *Biochim. Biophys. Acta*, **1465**, 343–358.

Miras, S., Salvi, D., Ferro, M., *et al.* (2002) Non-canonical transit peptide for import into the chloroplast, *J. Biol. Chem.*, **277**, 47770–47778.

Muchhal, U.S. & Raghothama, K.G. (1999) Transcriptional regulation of plant phosphate transporters, *Proc. Natl. Acad. Sci. U.S.A.*, **96**, 5868–5872.

Mudge, S.R., Rae, A.L., Diatloff, E. & Smith F.W. (2002) Expression analysis suggests novel roles for members of the Pht1 family of phosphate transporters in *Arabidopsis*, *Plant J.*, **31**, 341–353.

Nazoa, P., Vidmar, J.J., Tranbarger, T.J., *et al.* (2003) Regulation of the nitrate transporter gene *ATNRT2.1* in *Arabidopsis thaliana*: responses to nitrate, amino acids and developmental stage, *Plant Mol. Biol.*, **52**, 689–703.

Ng, A.Y.-N., Blomstedt, C.K., Gianello, R., Hamill, J.D., Neale, A.D. & Gaff, D.F. (1996) Isolation and characterization of a lowly expressed cDNA from the resurrection grass *Sporobolus stapfianus* with homology to eukaryotic sulfate transporter proteins (Accession No. X96761), *Plant Physiol.*, **111**, 651.

Nielsen, K.H. & Schjoerring, J.K. (1998) Regulation of apoplastic $NH_4^+$ concentration in leaves of oilseed rape, *Plant Physiol.*, **118**, 1361–1368.

Ono, F., Frommer, W.B. & von Wirén, N. (2000) Coordinated diurnal regulation of low- and high-affinity nitrate transporters in tomato, *Plant Biol.*, **2**, 17–23.

Orsel, M., Krapp, A. & Daniel-Vedele, F. (2002) Analysis of the NRT2 nitrate transporter family in *Arabidopsis*. Structure and gene expression, *Plant Physiol.*, **129**, 886–896.

Owen, A.G. & Jones, D.L. (2001) Competition for amino acids between wheat roots and rhizosphere microorganisms and the role of amino acids in plant N acquisition, *Soil Biol. Biochem.*, **33**, 651–657.

Ozcan, S., Dover, J. & Johnston, M. (1998) Glucose sensing and signalling by two glucose receptors in the yeast *Saccharomyces cerevisiae*, *EMBO J.*, **17**, 2566–2573.

Page, R.D.M. (1996) TREEVIEW: an application to display phylogenetic trees on personal computers, *Comput. Appl. Biosci.*, **12**, 357–358.

Passera, C. & Ferrari, G. (1975) Sulphate uptake in two mutants of *Chlorella vulgaris* with high and low sulphur amino acid content, *Physiol. Plant.*, **35**, 318–321.

Paulsen, I. & Skurray, R. (1994) The POT family of transport proteins, *Trends Biol. Sci.*, **19**, 404.

Pearson, J.N., Finnemann, J. & Schjoerring, JK. (2002) Regulation of the high-affinity ammonium transporter (BnAMT1.2) in leaves of *Brassica napus* by nitrogen status, *Plant Mol. Biol.*, **49**, 483–490.

Poorter, H., Vandevijver C., Boot, R.G.A & Lambers, H. (1995) Growth and carbon economy of a fast-growing and a slow-growing grass species as dependent on nitrate supply, *Plant Soil*, **171**, 217–227.

Rausch, C. & Bucher, M. (2002) Molecular mechanisms of phosphate transport in plants, *Planta*, **216**, 23–37.

Redinbaugh, M.G. & Campbell, W.H. (1991) Higher plant responses to environmental nitrate, *Physiol. Plant.*, **82**, 640–650.

Rodriguez, R., Guerrero, M. & Lara, C. (1994) Mechanism of sodium/nitrate symport in *Anacystis nidulans*, *Biochim. Biophys. Acta*, **1187**, 250–254.

Rogers, E.E., Eide, D.J. & Guerinot, M.L. (2000) Altered selectivity in an *Arabidopsis* metal transporter, *Proc. Natl. Acad. Sci. U.S.A.*, **97**, 12356–12360.

Sancenón, V., Puig, S., Mira, H., Thiele D.J. & Peòarrubia, L. (2003) Identification of a copper transporter family in *Arabidopsis thaliana*, *Plant Mol. Biol.*, **51**, 577–587.

Scheurwater, I., Clarkson, D.T., Purves, J.V., *et al.* (1999) Relatively large nitrate efflux can account for the high specific respiratory costs for nitrate transport in slow-growing grass species, *Plant Soil*, **215**, 123–134.

Schulze, W., Frommer, W.B. & Ward, J.M. (1999) Transporters for ammonium, amino acids and peptides are expressed in pitchers of the carnivorous plant *Nepenthes*, *Plant J.*, **17**, 637–646.

Shaul, O., Hilgemann, D.W., de-Almeida-Engler, J., Van Montagu, M., Inze, D. & Galili, G. (1999) Cloning and characterization of a novel $Mg^{2+}/H^+$ exchanger, *EMBO J.*, **18**, 3973–3980.

Shibagaki, N., Rose, A., McDermott, J.P., *et al.* (2002) Selenate-resistant mutants of *Arabidopsis thaliana* identify *Sultr1.2*, a sulfate transporter required for efficient transport of sulfate into roots, *Plant J.*, **29**, 475–486.

Siddiqi, M.Y., Glass, A.D.M., Ruth, T. & Rufty, T.J. (1990) Studies of the uptake of nitrate in barley, 2: Energetics, *Plant Physiol.*, **93**, 1585–1589.

Simon-Rosin, U., Wood, C. & Udvardi, M.K. (2003) Molecular and cellular characterization of LjAMT2.1, an ammonium transporter from the model legume *Lotus japonicus*, *Plant Mol. Biol.*, **51**, 99–108.

Smith, F.W., Ealing, P.M., Dong, B. & Delhaize, E. (1997a) The cloning of two *Arabidopsis* genes belonging to a phosphate transporter family, *Plant J.*, **11**, 83–92.

Smith, F.W., Ealing, P.M., Hawkesford, M.J. & Clarkson, D.T. (1995a) Plant members of a family of sulfate transporters reveal functional subtypes, *Proc. Natl. Acad. Sci. U.S.A.*, **92**, 9373–9377.

Smith, F.W., Hawkesford, M.J., Ealing, P.M., *et al.* (1997b) Regulation of expression of a cDNA from barley roots encoding a high affinity sulphate transporter, *Plant J.*, **12**, 875–884.

Smith, F.W., Hawkesford, M.J., Prosser, I.M. & Clarkson, D.T. (1995b) Isolation of a cDNA from *Saccharomyces cerevisiae* that encodes a high affinity sulphate transporter at the plasma membrane, *Mol. Gen. Genet.*, **247**, 709–715.

Smith, I.K. (1975) Sulfate transport in cultured tobacco cells, *Plant Physiol.*, **55**, 303–307.

Sohlenkamp, C., Shelden, M., Howitt, S. & Udvardi, M.K. (2000) Characterization of *Arabidopsis* AtAMT2, a novel ammonium transporter in plants, *FEBS Lett.*, **467**, 273–278.

Sohlenkamp, C., Wood, C.C., Roeb, G.W. & Udvardi, M.K. (2002) Characterization of *Arabidopsis* AtAMT2, a high-affinity ammonium transporter of the plasma membrane, *Plant Physiol.*, **130**, 1788–1796.

Sonoda, Y., Ikeda, A., Saiki, S., von Wirén, N., Yamaya, T. & Yamaguchi, J. (2003) Distinct expression and function of three ammonium transporter genes (OsAMT1.1–1.3) in rice, *Plant Cell Physiol.*, **44**, 726–734.

Spalding, E.P., Hirsch R.E., Lewis, D.R., Qi, Z., Sussman, M.R. & Lewis, B.D. (1999) Potassium uptake supporting plant growth in the absence of AKT1 channel activity. Inhibition by ammonium and stimulation by sodium, *J. Gen. Physiol.*, **113**, 909–918.

Suenaga, A., Moriya, K., Sonoda, Y., *et al.* (2003) Constitutive expression of a novel-type ammonium transporter OsAMT2 in rice plants, *Plant Cell Physiol.*, **44**, 206–211.

Takahashi, H., Asanuma, W. & Saito, K. (1999a) Cloning of an *Arabidopsis* cDNA encoding a chloroplast localizing sulphate transporter isoform, *J. Exp. Bot.*, **50**, 1713–1714.

Takahashi, H., Sasakura, N., Kimura, A., Watanabe, A. & Saito, K. (1999b) Identification of two leaf-specific sulfate transporters in *Arabidopsis thaliana* (Accession No. AB012048 and AB004060) (PGR99-154), *Plant Physiol.*, **121**, 686.

Takahashi, H., Sasakura, N., Noji, M. & Saito, K. (1996) Isolation and characterization of a cDNA encoding a sulfate transporter from *Arabidopsis thaliana*, *FEBS Lett.*, **392**, 95–99.

Takahashi, H., Watanabe-Takahashi, A., Smith, F.W., Blake-Kalff, M., Hawkesford, M.J. & Saito, K. (2000) The roles of three functional sulphate transporters involved in uptake and translocation of sulphate in *Arabidopsis thaliana*, *Plant J.*, **23**, 171–182.

Takahashi, H., Watanabe-Takahashi, A. & Yamaya, T. (2003) T-DNA insertion mutagenesis of sulfate transporters in *Arabidopsis*, in *5th Workshop on Sulfur Transport and Assimilation: Regulation, Interaction, Signalling* (eds J.-C. Davidian, D. Grill, L.J. De Kok, I. Stulen, M.J. Hawkesford, E. Schnug & H. Rennenberg), Backhuys, Leiden, The Netherlands, pp. 339–340.

Takahashi, H., Yamazaki, M., Sasakura, N., *et al.* (1997) Regulation of sulfur assimilation in higher plants: a sulfate transporter induced in sulfate-starved roots plays a central role in *Arabidopsis thaliana*, *Proc. Natl. Acad. Sci. U.S.A.*, **94**, 11102–11107.

Takano, J., Noguchi, K., Yasumori, M., *et al.* (2002) *Arabidopsis* boron transporter for xylem loading, *Nature*, **420**, 337–340.

Thomine, S., Lelièvre, F., Debarbieux, E., Schroeder, J.I. & Barbier-Brygoo, H. (2003) AtNRAMP3, a multispecific vacuolar metal transporter involved in plant responses to iron deficiency, *Plant J.*, **34**, 685–695.

Thompson, J.D., Higgins D.G. & Gibson, T.J. (1994) CLUSTAL W: improving the sensitivity of progressive multiple sequence alignment through sequence weighting, position-specific gap penalties and weight matrix choice, *Nucl. Acids Res.*, **22**, 4673–5680.

Touraine, B., Daniel-Vedele, F. & Forde, B.G. (2001) Nitrate uptake and its regulation, in *Plant Nitrogen* (eds P. Lea & J.-F. Morot-Gaudry), Springer-Verlag, Berlin, pp. 1–36.

Tsay, Y.-F., Schroeder, J.I., Feldmann, K.A. & Crawford, N.M. (1993) The herbicide sensivity gene CHL1 of *Arabidopsis* encodes a nitrate-inducible nitrate transporter, *Cell*, **72**, 705–713.

Van der Leij, M., Smith, S.J. & Miller, A.J. (1998) Remobilisation of vacuolar stored nitrate in barley root cells, *Planta*, **205**, 64–72.

Vidmar, J.J., Schjoerring, J.K., Touraine, B. & Glass, A.D.M. (1999) Regulation of the *hvst1* gene encoding a high-affinity sulfate transporter from *Hordeum vulgare*, *Plant Mol. Biol.*, **40**, 883–892.

Vidmar, J.J., Tagmount, A., Cathala, N., Touraine, B. & Davidian, J.-C. (2000a) Cloning and characterization of a root specific high-affinity sulfate transporter from *Arabidopsis thaliana*, *FEBS Lett.*, **475**, 65–69.

Vidmar, J.J., Zhuo, D., Siddiqi, M.Y., Schjoerring, J., Touraine, B. & Glass, A.D.M. (2000b) Regulation of high-affinity nitrate transporter genes and high-affinity nitrate influx by nitrogen pools in roots of barley, *Plant Physiol.*, **123**, 307–318.

von Wirén, N., Gazzarrini, S., Gojon, A. & Frommer, W.B. (2000a) The molecular physiology of ammonium uptake and retrieval, *Curr. Opin. Plant Biol.*, **3**, 254–261.

von Wirén, N., Lauter, F.R., Ninnemann, O., *et al.* (2000b) Differential regulation of three functional ammonium transporter genes by nitrogen in root hairs and by light in leaves of tomato, *Plant J.*, **21**, 167–175.

Wang, M., Glass, A.D.M., Shaff, J. & Kochian, L. (1994) Ammonium uptake by rice roots, III: Electrophysiology, *Plant Physiol.*, **104**, 899–906.

Ward, J.M. (2001) Identification of novel families of membrane proteins from the model plant *Arabidopsis thaliana*, *Bioinformatics*, **17**, 560–563.

White, P.J. (1996) The permeation of ammonium through a voltage-independent $K^+$ channel in the plasma membrane of rye roots, *J. Membr. Biol.*, **152**, 89–99.

Williams, L.E. & Miller, A.J. (2001) Transporters responsible for the uptake and partitioning of nitrogenous solutes, *Annu. Rev. Plant Physiol. Plant Mol. Biol.*, **52**, 659–688.

Wintz, H., Fox, T., Wu, Y.-Y, *et al.* (2003) Expression profiles of *Arabidopsis thaliana* in mineral deficiencies reveal novel transporters involved in metal homeostasis, *J. Biol. Chem.*, **278**, 47644–47653.

Xia, T., Apse, M.P., Aharon, G.S. & Blumwald, E. (2002) Identification and characterization of a NaCl-inducible vacuolar $Na^+/H^+$ antiporter in *Beta vulgaris*, *Physiol. Plant.*, **116**, 206–212.

Yildiz, F.H., Davies, J.P. & Grossman, A. (1996) Sulfur availability and the SAC1 gene control adenosine triphosphate sulphurylase gene expression in *Chlamydomonas reinhardtii*, *Plant Physiol.*, **112**, 669–675.

Yoshimoto, N., Inoue, E., Saito, K., Yamaya, T. & Takahashi, H. (2003) Phloem-localizing sulfate transporter, Sultr1.3, mediates re-distribution of sulfur from source to sink organs in *Arabidopsis*, *Plant Physiol.*, **131**, 1511–1517.

Yoshimoto, N., Takahashi, H., Smith, F.W., Yamaya, T. & Saito, K. (2002) Two distinct high-affinity sulfate transporters with different inducibilities mediate uptake of sulfate in *Arabidopsis* roots, *Plant J.*, **29**, 465–473.

Zhen, R.G., Koyro, H.W., Leigh, R.A., Tomos, A.D. & Miller, A.J. (1991) Compartmental nitrate concentrations in barley root-cells measured with nitrate-selective microelectrodes and by single-cell sap sampling, *Planta*, **185**, 356–361.

Zhou, J.J., Fernández, E., Galván, A. & Miller, A.J. (2000a) A high affinity nitrate transport system from *Chlamydomonas* requires two gene products, *FEBS Lett.*, **466**, 225–227.

Zhou, J.J., Theodoulou, F., Muldin, I., Ingemarsson, B. & Miller, A.J. (1998) Cloning and functional characterization of a *Brassica napus* transporter which is able to transport nitrate and histidine, *J. Biol. Chem.*, **273**, 12017–12033.

Zhou, J.J., Trueman L.J., Boorer, K.J., Theodoulou, F.L., Forde, B.G. & Miller, A.J. (2000b) A high-affinity fungal nitrate carrier with two transport mechanisms, *J. Biol. Chem.*, **275**, 39894–39899.

Zhu, J.-K. (2000) Genetic analysis of plant salt tolerance using *Arabidopsis*, *Plant Physiol.*, **124**, 941–948.

Zhuo, D., Okamoto, M., Vidmar, J. & Glass, A.D.M. (1999) Regulation of a putative high-affinity nitrate transporter (Nrt2.1At) in roots of *Arabidopsis thaliana*, *Plant J.*, **17**, 563–568.

# 5 Functional analysis of proton-coupled sucrose transport

Daniel R. Bush

## 5.1 Introduction

It has long been known that plants, as photoautotrophic, multicellular organisms, transport sugars and amino acids from sites of primary assimilation to heterotrophic tissues that must import organic nutrients to support growth and development (Gifford & Evans, 1981; Thorne, 1985). For many years, plant biologists viewed the distribution of organic nutrients as a fairly simple and straightforward process involving a small number of transporters that loaded the phloem with sucrose and amide amino acids, and complementary carriers that moved these metabolites across the plasma membrane of cells importing the compounds. Underlying assumptions were that a relatively small number of substrates move systemically in the plant, and that most metabolites were synthesized in place as needed. While this model is a good approximation for the bulk of assimilate partitioning, we now recognize that transport of metabolites and other small molecules is much more complex at the cellular level. For example, sucrose partitioning in *Arabidopsis* relies on transporters that are part of a closely related gene family with nine members (Kuhn *et al.*, 1999); for amino acids there are many more transporters that are represented by several gene families (Fischer *et al.*, 1998; Ortiz-Lopez *et al.*, 2000). In both cases, these transporters are differentiated from one another by differences in their kinetic properties, expression patterns and responses to developmental and environmental cues. Other transporters that facilitate the movement of multiple classes of sugars, peptides, metabolites, osmoprotectants and nucleic acid bases have also been identified (Schwacke *et al.*, 1999; Buttner & Sauer, 2000; Gillissen *et al.*, 2000; Noiraud *et al.*, 2001; Stacey, 2002). Thus, there is a complex and rich exchange of organic molecules between adjacent cells, tissues and organs that not only satisfies nutritional needs, but also conveys molecules that contribute to physiological responses and genetic programs regulating plant growth and development (Chiou & Bush, 1998; Bush, 1999; Fletcher, 2002; Forde, 2002; Kessler & Baldwin, 2002; Lalonde *et al.*, 2003). These new insights regarding the complexity of transport processes in the physiology of plant growth has placed considerable importance on defining the transport properties, expression patterns and functional contributions of each transporter. This chapter outlines some of the key methods and approaches used to define the kinetics, expression patterns and functions of transporters that move organic molecules across the plasma membrane. To simplify the presentation,

I focus on proton–sucrose symporters and some of the seminal methods that have been used to define the functional contributions of this important family of plant transport proteins.

## 5.2    Defining basic properties of transport

An essential part of any description of transport is the defining of its kinetic and thermodynamic properties. Thus, one of the first tasks is to examine substrate accumulation versus time. Some basic characteristics are to be expected for any transport process and can be formalized with a few simplifying assumptions (Segel, 1993). Facilitated transport will be mediated by a finite number of transport proteins in the membrane and, therefore, is expected to show a rate of uptake that saturates with respect to substrate concentration, by contrast with diffusion across the membrane. However, because both facilitated and diffusional uptake rates will be affected also by the substrate gradient across the membrane, the simplest limiting conditions are those in which substrate concentration is effectively zero – or at least constant and low – on the other (trans) side of the membrane. As a consequence, uptake rates at each substrate concentration are usually determined from measurements over short time intervals after adding the substrate when transport should proceed in a unidirectional fashion. A first test of this assumed 'zero trans' assumption lies in demonstrating a linear uptake over time. A nonlinear relationship may reflect a variety of complications, including unstable driving forces, multiple sites of accumulation with different filling rates, or concentration-dependent efflux pathways.

Provided a linear relationship is observed for accumulation over time, a formal description of transport can be defined from a plot of initial transport rate as a function of substrate concentration. With the simplifying assumptions of unidirectional and facilitated transport, uptake rate is usually expected to show Michaelis–Menten-like kinetics with substrate concentration and yield a hyperbolic curve. Such a saturable dependence on substrate concentration is defined by an apparent $K_m$ that provides some insight into the relative affinity of the transporter for the substrate and by a $V_{max}$ that describes the limiting rate for transport, all other conditions being constant. Additional components can appear in the plot, especially if measurements are taken over a wide range of substrate concentrations. If linear, these components are often subtracted out as consistent with simple diffusion in which the substrate partitions into and out of the lipid bilayer as it moves across the membrane.

Evidence supporting a protein-mediated pathway have included demonstrating sensitivity to chemical agents, such as diethyl pyrocarbonate (DEPC), that modify amino acid residues within proteins (Bush, 1993a). Sensitivity to such broad-range inhibitors in intact cells and tissues is ambiguous, at least in uptake and radiotracer flux measurements, because these agents will also modify other targets, such as the plasma membrane $H^+$-ATPase, and thereby impact on transport, even if the effect is indirect (see Chapter 1). For example, the plasma membrane $H^+$-ATPase is sensitive to chemical inhibition by $N,N$-dicyclohexylcarbodiimide (Briskin &

Poole, 1983) and ATP depletion (Slayman *et al.*, 1973), and its blockade will affect transport through a reduction in the membrane proton electrochemical potential difference, $\Delta\mu_{H^+}$ (Slayman & Slayman, 1974). Differentiating between active and passive transport activity in intact cells and tissues using these compounds can be ambiguous for the same reasons. In general, inhibitors are most useful when applied to purified membrane vesicles, simplified expression systems and in combination with electrophysiological measurements (see Chapter 1). The latter point is well illustrated in experiments examining the inhibition of proton–sucrose symporter activity by *p*-chloromercuribenzenesulfonic acid (PCMBS) and $Hg^{2+}$ (Bush, 1993a). These compounds form covalent bonds with protein sulfhydryl groups and both are potent inhibitors of proton–sucrose symporter mediated sucrose accumulation in purified plasma membrane vesicles. Transport inhibition in this case could not distinguish between a direct modification of the symporter or by dissipation of the proton electrochemical potential difference driving transport, and it was necessary to measure the impact of each inhibitor on the imposed $\Delta\mu_{H^+}$. The data showed that PCMBS had no impact on the imposed proton electrochemical potential, while $Hg^{2+}$ rapidly dissipated the gradient. Subsequent experiments with liposomes confirmed that $Hg^{2+}$ was effective in dissipating the transmembrane proton concentration gradient, presumably by partitioning directly into the lipid bilayer (Bush, 1993a).

## 5.3   Intact tissues

While it is desirable to establish the presence of transport in the intact tissue, these measurements often pose special difficulties, particularly when quantifying transport of organic substrates. Experiments often make use of a $^{14}$C- or $^3$H-radiolabeled form of the substrate as a tracer for substrate. Clearly, it is possible to use chemical analysis of cell extracts to measure accumulation, but only if the substrate is not modified or metabolized once transported into the cell. Accumulation can also be quantified by measuring depletion of the substrate (or radiotracer) from the uptake medium. Radiotracer measurements have significant advantages, because it is possible to separate unidirectional influx and efflux. However, both radiotracer flux analysis and net uptake measurements can be complicated by cellular metabolism of the transported substrate. There are a number of other considerations that factor into measurements with intact cells and tissues: (1) Barriers to substrate diffusion, especially cell wall and cuticle, can affect access to the membrane. Local depletion of substrate at the plasma membrane surface then can give a false impression of substrate saturation. The pH and buffering of cations within the cell wall will also affect their availability. (2) Accumulation of substrate in multiple (sub)compartments within the cell may affect the apparent kinetics of uptake, as mentioned above. (3) In nonhomogeneous multicellular tissues, substrate transport may be associated with a specific cell type, although this cannot be determined at a macroscopic level. Even with suspension culture cells, difficulties arise in assessing the transport uniformity from one experiment to the next. (4) Efflux (and for complex tissue, long-distance transport) pathways may need assessing, as these can also contribute to the overall kinetics for transport.

Early studies examining sucrose transport in isolated leaf discs illustrate many of these points. For example, the cuticle and epidermis are major diffusion barriers that must be removed in transport experiments to ensure access of radiolabeled sucrose to the cells of the leaf (Delrot *et al.*, 1980). Even with these precautions, not all cells in the leaf prove to be active in sucrose transport. Measurements of cellular levels of osmotic pressure or autoradiography of leaf tissue have shown that the phloem accumulates the largest amount of sucrose in leaf discs (Geiger *et al.*, 1973; Delrot *et al.*, 1980; Komor *et al.*, 1980). However, other cells such as stomatal guard cells (Talbott & Zeigler, 1996) also accumulate sucrose. Likewise, determining the impact of sucrose efflux from the cut ends of the phloem can be very difficult because the substrate is recycled back into the transport solution. Inhibitor studies are equally difficult to interpret in such experiments, because of the large number of cell types present and the limitations of attributing a response to a specific cell or transport protein. Nevertheless, leaf disc experiments showed that sucrose is accumulated in the phloem of leaf tissue, consistent with energy-dependent sucrose transport activity (Komor, 1977; Delrot *et al.*, 1980). Other studies demonstrated that accumulation was inhibited by chemical modification (Giaquinta, 1976; Delrot *et al.*, 1980), and yielded estimates of the apparent affinities for uptake (Komor *et al.*, 1980; Delrot & Bonnemain, 1981). However, definitive experiments for protein-coupled sucrose accumulation in the phloem required other strategies.

## 5.4 Membrane vesicles

As an alternative to the limitations associated with intact tissue and organ systems, a biochemical approach offers a number of advantages. Ideally, the transporter might be purified to homogeneity and reconstituted into liposome vesicles for subsequent radiotracer or uptake analysis. One difficulty is that most transporters occur at exceedingly low abundance in the plasma membrane and, therefore, this approach has not been applied successfully to more than a small number of plant membrane transport proteins (see, for example, Chapter 3). As an alternative, many studies have turned to isolating native membrane vesicles, purified to minimize contaminants (e.g. from endomembranes), and to measuring transport properties in this simplified, but powerful, experimental system. Membrane vesicles were essential to advances in describing the transport properties of plasma membrane and tonoplast $H^+$-ATPases (Sze, 1985) and $H^+$-coupled sucrose and amino acid transport (Bush & Langston-Unkefer, 1988; Bush, 1989; Bush, 1993b).

Work with native membrane vesicles has several advantages for characterizing transport activity (Sze, 1985; Blumwald, 1987; Bush, 1992). Techniques for plasma membrane purification (Larsson *et al.*, 1988) mean that transport activity can generally be assigned to a specific membrane *in vivo*. With a simple vesicle system, factors of metabolism and subcompartmentation are avoided, and both influx and efflux activities can be evaluated independently as functions of a single membrane phase. It is possible, also, to manipulate solute compositions both inside and outside the vesicles, and, using well-defined ion concentration differences across the membrane, the experimenter can impose defined ion and/or electrical potential differences (Fig. 5.1). Of course, there are other considerations. The vesicle membrane must seal tightly enough to maintain ion and substrate concentration differences across the membrane during the period of an experiment. Vesicles are well suited to characterizing 'slower,' active transport with turnover rates on the order of $10^3$ molecules per second per transporter and can be monitored over periods of tens of seconds to minutes. However, the methods for monitoring uptake often require handling with standard biochemical steps of filtration or centrifugation to separate the vesicle volume, so uptake measurements cannot deal with transport that goes to completion or fills the vesicle volume over shorter periods of time.

The proton–sucrose symporter is an excellent example in which isolated membrane vesicles were used to demonstrate proton-motive force-driven sucrose transport. Such vesicle transport experiments were used to define many of the key transport properties of the symporter, including kinetic constants (Buckhout, 1989; Bush, 1989; Lemoine & Delrot, 1989), showing that transport is electrogenic (Bush, 1990), and defining structural determinants that are involved in substrate binding (Hecht *et al.*, 1992). Other studies (Bush, 1989; Williams *et al.*, 1992) identified DEPC as a potent inhibitor of proton–sucrose symport, and implicated a histidine residue at the substrate-binding site (Bush, 1993a), a conclusion that was verified by site-directed mutagenesis (Lu & Bush, 1998). These techniques also yielded a stoichiometry for proton–sucrose cotransport of 1:1 (Bush, 1990; Slone & Buckhout, 1991). In this case, a sucrose concentration gradient (high sucrose in the transport solution, zero sucrose inside the vesicle at time zero) was used to drive proton flux into the vesicle while monitoring changes in proton concentration with a sensitive pH electrode system. These experiments were also significant

| Imposed proton concentration difference pH-jump | Imposed membrane potential pK-jump |
|---|---|
| 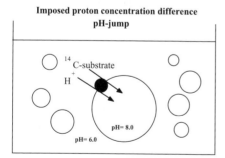 | 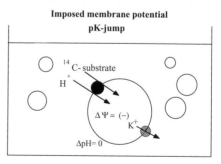 |

**Fig. 5.1** Schematic diagram of an imposed proton-concentration difference (pH-jump) and an imposed membrane electrical potential difference (pK-jump). In each case, concentrated membrane vesicles are diluted into the transport solution to initiate the transport reaction. At desired time intervals, an aliquot of the transport solution is pipetted onto a micropore filter (0.44 μm) to collect the intact vesicles and wash them free of extra-vesicular transport solution. Accumulation of radiolabeled substrate is quantified using liquid scintillation spectroscopy. Black circles represent proton-motive force-driven symporters and the shaded circle represents valinomycin, a potassium ionophore.

because they confirmed the prediction that the symporter can be driven by a favorable thermodynamic gradient of either proton-motive force or sucrose.

It is worth noting that these detailed descriptions of proton–sucrose symport in isolated plasma membrane vesicles benefited from the fact that a single, dominant transporter was present in the purified membranes, especially in material derived from sugar beet. As noted earlier, purified membrane vesicles are derived from several cell types. Thus, not only is there a rich milieu of membrane proteins present for every cell, there is a good chance that different cell types express distinct classes of transporters for a given substrate. Nonetheless, for the sucrose symporter in sugar beet vesicles, kinetic analysis yielded a simple, hyperbolic curve consistent with the activity of a single symporter (Bush, 1989). While additional sucrose transport pathways may have been present, their relative contribution to sucrose transport activity in the isolated vesicles was too low to impact the kinetic curve.

## 5.5  Sucrose sensing

Research with purified plasma membrane vesicles has also proven important to understanding how plants monitor and regulate sucrose transport (Chiou & Bush, 1998). These studies indicated that sucrose accumulation in the leaf was accompanied by a decrease in sucrose transport activity when assayed subsequently in vesicles isolated from the tissue. Sucrose accumulation blocked symporter transcription in the companion cell and, because symporter protein and message is rapidly cycled (turnover half-time, 2 h), its abundance in the plasma membrane dropped proportionally with the accumulation of sucrose (Vaughn et al., 2002). It appears now that transcriptional regulation is controlled by a sucrose-sensing regulatory system mediated, at least in part, by a protein phosphorylation cascade (Ransom-Hodgkins et al., 2003). Thus, symporter abundance in the phloem is directly proportional to transcriptional activity. One interpretation of these data is that the sucrose content of the phloem regulates transporter abundance by direct control through transcription and thus controls phloem-loading capacity (Vaughn et al., 2002). The implication is that sucrose will back up in the leaf phloem when sink demand does not keep pace with photosynthetic activity; increased sink demand, by contrast, should have the opposite effect by lowering sucrose levels in the leaf phloem and, thereby, enhancing symporter transcription rates yielding an increase in symporter abundance.

The nature of a sucrose-sensor in the plant is still unknown. Barker et al. (2000) has suggested that AtSUT2, a member of the sucrose symporter gene family in Arabidopsis, serves to monitor sucrose concentrations in the phloem apoplast. The proposal is based on dissimilarities of protein sequence with other sucrose transporters and its apparent lack of activity in transporting sucrose. In principle, it is an attractive possibility because sucrose transporters include a substrate-binding site with a dissociation constant in the range of 1 M sucrose and, thus, appropriate for sensing the concentration ranges expected inside the phloem. However, subsequent work has shown that AtSUT2 is transport-active, and the unique sequences can be removed with little impact on transport activity (Eckardt, 2003). Nevertheless, given the precedent set by the yeast hexose sensors (Ozcan *et al.*, 1998), the question remains whether a similar function might be found for a plant sucrose transporter.

## 5.6   Heterologous expression systems

One of the key breakthroughs for membrane transport research in the early 1990s was the introduction of functional complementation as a strategy for cloning cDNA's encoding plant membrane transporters. In these experiments, yeast transport mutants are transformed with plant cDNA libraries constructed in yeast expression vectors to place the plant cDNA expression under the control of a constitutive or inducible yeast promoter. The transformed yeast were grown under limiting conditions where the absence of the native transporter inhibited growth because the mutant cell line was not able to take up a required nutrient. Thus, any cell expressing a plant transporter that complemented the normal yeast transport function of the absent carrier rescued yeast growth and could easily be identified (see Fig. 5.2; also Anderson *et al.*, 1992; Riesmeier *et al.*, 1992; Sentenac *et al.*, 1992; Frommer *et al.*, 1993; Hsu *et al.*, 1993).

# Molecular Cloning via Functional Complementation

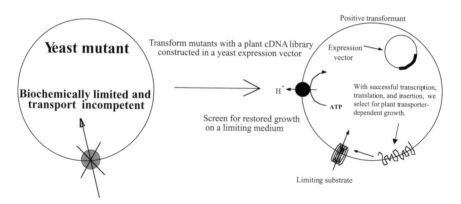

**Fig. 5.2** Schematic outline of functional complementation. Mutant yeast lines that are auxotrophic for a specific metabolite, such as sucrose or an amino acid, and also lack a transporter for that substrate are unable to grow on limiting concentrations of that metabolite because they cannot synthesize it and they are not able to take up sufficient amounts from the medium to allow for growth. That yeast line is transformed with a plant cDNA library constructed in a yeast expression vector and millions of cells are screened for positive transformants that acquire the ability to grow on the limiting medium (i.e. the mutation has been complemented). Positive transformants have acquired the ability to grow because they may be expressing a plant transporter for the metabolite. However, this phenotype may also be attributable to a plant gene encoding a missing synthetic gene for the metabolite, or by a mutation in the yeast genome that results in a new transport function. Each of these possibilities must be eliminated before concluding a plant transporter complemented the mutant phenotype.

While functional complementation is a powerful strategy for cloning plant transporters (Frommer & Ninnemann, 1995), it also gives transport biologists a robust method for determining important transport properties of these newly identified transporters. For example, in addition to describing basic transport characteristics of plant sucrose and amino acid transporters (Riesmeier et al., 1992; Frommer et al., 1993; Hsu et al., 1993; Fischer et al., 1995; Chen & Bush, 1997; Chen et al., 2001), yeast expression has been coupled to site-directed and random mutagenesis to identify critical amino acids and protein domains associated with the active sites of these integral membrane proteins (Lu & Bush, 1998; Ortiz-Lopez et al., 2000). In these experiments, it is possible to alter a single amino acid and ask what impact that change has on transport activity. For example, Lu and Bush (1998) used site-specific mutagenesis to show dramatic changes in transport kinetics when histidine at position 65 was substituted with other amino acids at this position of the symporter. These experiments also confirmed that His-65 is the principle target site conferring DEPC sensitivity in transport. These data strongly suggested that the amino acid residue was associated with sucrose-binding. In a complementary approach, the symporter transcript was randomly mutagenized in a yeast expression vector and then transformed into yeast cells that were screened for an altered transport property. Lu (1999) identified several amino acid residues associated with the sucrose symporter active site by identifying mutations that altered substrate specificity.

Heterologous expression in yeast has also been used to explore the possibility that the sucrose symporters function as homotypic oligomers rather than as monomers. Reinders et al. (2002) used a split-ubiquitin system in yeast in which protein interactions between two integral membrane proteins are detected by activation of a lacZ reporter gene. This assay relies on protein interactions, bringing together two halves of ubiquitin, which then promotes proteosome cleavage of a synthetic transcription factor. Reinders et al. (2000) found LeSUT1, LeSUT2 and LeSUT4 sucrose symporters of tomato to form homo-oligomers as well as heteromeric oligomers in yeast. These findings are intriguing since each of these symporters exhibits unique substrate affinities. Heteromeric interaction between the three suggests that a range of transport properties, depending on their stoichiometric relationship, could be observed when they are coexpressed in the same cell. The authors propose this idea as a mechanism for regulating phloem loading in the sieve elements of tomato.

Xenopus oocytes represent another heterologous expression system that is widely used to investigate plant transporters (Boorer et al., 1992). Commonly, oocytes are injected with a plant transcript that codes for a specific transporter and, once translated, the transport activity is quantified by using labeled substrate or, for charge-carrying transporters, by measuring membrane current with microelectrodes inserted into the cell (see Chapter 1). The electrophysiological approach is especially powerful because it enables direct control of membrane voltage as a kinetic parameter, which, otherwise, goes unaccounted for. Thus, investigations of proton–sucrose symporters expressed in oocytes not only confirmed earlier determinations of electrogenicity, stoichiometry and substrate specificity measured in purified membrane vesicles, but also provided kinetic insights into specific steps of the reaction cycle (Boorer et al., 1996; Zhou et al., 1997).

While heterologous expression is a powerful tool, it can also be limited by a variety of complications. Both yeast and oocytes are living cells with their own suite of conductance pathways that can interfere with or modify the activity of the foreign protein. Likewise, the presence or absence of posttranslational modifications may alter measured activities, and the lipid composition of the membrane may impact the function of the transporter. A thorough discussion of the power and pitfalls of heterologous expression can be found in Dreyer et al. (1999).

## 5.7   Sucrose transport in plant growth and development

Although the kinetic and thermodynamic characteristics are important determinants for a transport process, transport properties alone cannot define the contribution of a transporter to plant growth and development. It is also necessary to determine the cell-specific location of each transporter, as well as its pattern of expression during the course of normal development and in response to environmental challenges. Even with these details, a comprehensive description of transport properties, expression pattern and regulation, definitive insights into the biological function of a given transporter is not always straightforward. In many instances, additional experiments with gene knockouts or antisense expression are needed to gain complete understanding of the biological function of a specific sugar, amino acid, or metabolite transporter.

### 5.7.1   Patterns of gene expression

As noted earlier, there are nine members of a closely related gene family that encode proton–sucrose symporters. In *Arabidopsis*, *AtSUC2* is expressed in the companion cell of leaf phloem where it has been immunolocalized (Stadler & Sauer, 1996). Because of its location, this transporter is thought to be a major contributor to phloem loading. Antisense and insertional knockout of the gene yield stunted plants in which assimilate partitioning appears to be significantly inhibited, supporting a role in phloem loading (Riesmeier et al., 1994; Gottwald et al., 2000). In contrast, *AtSUC3* is also expressed in the leaf companion cell, but it is localized to the sieve element (Meyer et al., 2004). Moreover, expression is highest in several sink tissues and expression is strongly induced after wounding. These observations suggest that AtSUC3 plays an important role in sink metabolism, and has an undefined function in plant cell responses to stress. *AtSUC1*, on the other hand, is expressed exclusively in floral tissues where it is thought to play a role in anther dehiscence and pollen tube growth (Stadler et al., 1999). Detailed descriptions of the expression patterns for the other six members of the gene family have yet to be published.

Gene expression patterns are determined using a variety of techniques, each of which has its own attributes and limitations. RNA gel blot hybridization is most frequently used to gain an initial idea of what the expression pattern is for a given gene among different organs. This approach is limited by sensitivity and its inability to resolve expression to a specific cell or tissue type. RT-PCR is substantially more sensitive than standard RNA gel blots (Freeman *et al.*, 1999) but, without added methods of single-cell sampling, it also cannot yield information about the identity of the cells expressing the gene of interest. Promoter::reporter gene constructs represent another commonly used method for analyzing gene expression patterns. In this case, the promoter of the gene of interest (typically 1000–3000 bp of the promoter region) is fused to a reporter gene whose expression can be monitored using histochemical staining (de Ruijter *et al.*, 2003). The promoter::reporter constructs are transferred into the study plant and promoter-driven expression of the reporter is analyzed in stable transformants. Commonly used reporters include β-glucuronidase (GUS; Jefferson *et al.*, 1987), green fluorescent protein (GFP; Sheen *et al.*, 1995) and luciferase (Welsh & Kay, 1997). GUS expression is examined in nonliving tissue, while GFP and luciferase can be visualized in living plants. While this approach can be used to resolve expression at the level of single cells, it is limited by a variety of factors. For example, *cis*-elements that control cell specificity or expression levels that reside in untranslated regions of the gene or in introns are not present in these constructs and so expression of the reporter gene may not reflect the true expression pattern of the native gene. Patterns of reporter gene expression can also be misleading if the mRNA of the native gene turns over rapidly as compared to the reporter message, or if the mRNA or expressed protein of the gene under investigation traffic to nearby cells. *AtSUC3*, for example, is expressed in phloem companion cells but the expressed protein is found only in adjacent sieve elements (Meyer *et al.*, 2004).

Perhaps the best method for high-resolution analysis of gene expression is *in situ* hybridization (Meyerowitz, 1987). In this approach, tissue is chemically fixed, embedded in wax, and sectioned for subsequent analysis of mRNA abundance that can be resolved at the level of individual cells. This method should yield unambiguous information about the spatial and temporal patterns of expression. However, the technique is limited because it is necessary to sample all tissues under all conditions to avoid missing a critical period of expression, a rather laborious undertaking. Thus, *in situ* hybridizations are usually preceded by RNA gel blot and/or promoter::reporter analysis that focuses the investigator's attention on a specific tissue at a well-defined time point. Even under these conditions, however, *in situ* hybridizations can be compromised if the target gene is a member of a closely related gene family. In those cases, cross-hybridization between family members can severely limit one's ability to determine which gene's expression is being monitored. For example, *AtSUT6* and *AtSUT7* are so closely related (96% identity) that it has not been possible to design gene-specific probes for *in situ* hybridizations in our hands.

With the above points in mind, it is clear that analyzing the expression pattern of any given gene requires a combination of methods to accurately determine the spatial and temporal pattern of expression. Consider *AtSUC3* as an example. Expression of this gene was initially reported to be localized to large cells adjacent to the vascular system in *Arabidopsis* leaves, based on promoter::GUS expression and immunolocalization using AtSUC3-specific antiserum (Meyer *et al.*, 2000). To resolve the exact location of AtSUC3 transcripts and protein, Meyer *et al.* (2004) reexamined the expression of *AtSUC3* using promoter::reporter constructs that expressed different reporters (GUS, GFP, or a membrane-targeted GFP) and a new antiserum directed against the N terminus of AtSUC3 for immunolocalization. Promoter::reporter analysis again identified large cells along the leaf vascular system as the site of gene expression. With a more detailed analysis of the cells involved using multiple reporter genes, the authors concluded that *AtSUC3* is expressed in the phloem companion cell and that these results are consistent with their earlier analysis (Meyer *et al.*, 2000). In contrast, immunolocalization of the AtSUC3 protein showed that it is localized to the sieve elements of the phloem and this localization was confirmed by demonstrating colocalization with a sieve element specific monoclonal RS6 (Meyer *et al.*, 2004). These results are significant, among other reasons, because they illustrate the potential danger of assuming that transcript location will coincide with the final location of the encoded protein.

In addition to resolving *AtSUC3* expression in the leaf vascular tissue more clearly using multiple methods for detecting *AtSUC3* gene expression and protein localization, Meyer *et al.* (2004) discovered that *AtSUC3* is also expressed at high levels in many sink tissues and cells, such as root tips, trichomes, guard cells and stipules. Moreover, they also showed that this symporter gene is strongly induced after tissue wounding, thus implicating sucrose transport activity in plant cell responses to physical stress. These observations raise new and unexpected questions about the functional role of this transporter in sink cells and wounding responses that would not have been posed in the absence of a thorough analysis of its spatial and temporal pattern of gene expression.

### 5.7.2  Antisense expression and gene knockouts in transgenic plants

Molecular genetic methods that alter the expression of a target gene are powerful tools for determining the biological function of the encoded protein in the context of the multicellular plant. The ability to decrease or eliminate the expression of a single gene has revolutionized our understanding of many physiological processes, and the functional contributions of several membrane transport systems have benefited from this approach. For example, initial results describing *AtSUC2* and its potato ortholog *StSUT1* showed that these genes code for a proton–sucrose symporter that is localized in the phloem of leaf vascular system (Riesmeier *et al.*, 1993; Truernit & Sauer, 1995). These data supported the notion that this sucrose symporter plays a vital role in phloem loading. To test that hypothesis, Riesmeier *et al.* (1994) used antisense repression to decrease the abundance of *StSUT1* transcripts. Potato plants were transformed with a *StSUT1* antisense gene under the control of the CaMV 35S promoter. Transformants containing reduced levels of *StSUT1* message displayed a variety of physiological changes that were consistent with a decrease in sucrose export. The leaves of these plants contained 20-fold higher concentrations of soluble sugars and a 5-fold increase in starch. Plant growth was stunted, and the root systems and tuber yield were reduced by a great extent. Taken together, these phenotypes are consistent with those expected for a substantial block in phloem loading. Riesmeier *et al.* concluded that the StSUT1 symporter is the primary transporter responsible for phloem loading and long-distance assimilate transport in potato.

By contrast, genetic evidence for the role of the AtSUC2 symporter in phloem loading was based on a reverse genetic screen. Gottwald *et al.* (2000) screened a population of *Arabidopsis* plants that contain random insertions of T-DNA. Each insertion represents a potential mutagenic event that alters or eliminates the expression of a specific gene if the T-DNA insertion site falls within its promoter or coding region. So, using PCR with one primer designed against the *AtSUC2* gene and the other against the T-DNA, the authors identified three plants out of several thousands that contained T-DNA insertions in the coding region of *AtSUC2*. These null alleles (or knockouts) of *AtSUC2* displayed similar phenotypes to those described for the antisense potato plants. Substantial amounts of starch built up in the leaves and radiolabeled sucrose was not transported to the rest of the plant. Homozygous plants were stunted and chlorotic, and would not fully develop without additions of exogenous sucrose. These data also led to the conclusion that the SUC2/SUT1 symporter is essential to phloem loading.

In conclusion, of the nine putative sucrose symporter genes identified in *Arabidopsis* (Kuhn *et al.*, 1999), the three most well described have been shown to play significantly different roles in plant growth and development based on transport properties, gene expression patterns and insights from antisense and knockout experiments. A similar, comprehensive description of transport properties, expression patterns, regulations and mutants now also is awaited for the remaining members of this gene family.

# References

Anderson, J.A., Huprikar, S.S., Kochian, L.V., Lucas, W.J. & Gaber, R.F. (1992) Functional expression of a probable *Arabidopsis thaliana* potassium channel in *Saccharomyces cerevisiae*, *Proc. Natl. Acad. Sci. U.S.A.*, **89**, 3736–3740.

Barker, L., Kuhn, C., Weise, A., *et al.* (2000) Sut2, a putative sucrose sensor in sieve elements, *Plant Cell*, **12**, 1153–1164.

Blumwald, E. (1987) Tonoplast vesicles as a tool in the study of ion transport at the plant vacuole, *Physiol. Plant.*, **69**, 731–734.

Boorer, K.J., Forde, B.G., Leigh, R.A. & Miller, A.J. (1992) Functional expression of a plant plasma membrane transporter in *Xenopus* oocytes, *FEBS Lett.*, **302**, 166–168.

Boorer, K.J., Loo, D.D.F., Frommer, W.B. & Wright, E.M. (1996) Transport mechanism of the cloned potato H$^+$/sucrose cotransporter stSUT1, *J. Biol. Chem.*, **271**, 25139–25144.

Briskin, D.P. & Poole, R.J. (1983) Characterization of a K$^+$-stimulated adenosone triphosphatase associated with the plasma membrane of red beet, *Plant Physiol.*, **71**, 350–355.

Buckhout, T.J. (1989) Sucrose transport in isolated plasma membrane vesicles from sugar beet, *Planta*, **178**, 393–399.

Bush, D.R. (1989) Proton-coupled sucrose transport in plasmalemma vesicles isolated from sugar beet (*Beta vulgaris* L.) leaves, *Plant Physiol.*, **89**, 1318–1323.

Bush, D.R. (1990) Electrogenicity, pH-dependence, and stoichiometry of the proton-sucrose symport, *Plant Physiol.*, **93**, 1590–1596.

Bush, D.R. (1992) The proton-sucrose symport, *Photosynth. Res.*, **32**, 155–165.

Bush, D.R. (1993a) Inhibitors of the proton-sucrose symport, *Arch. Biochem. Biophys.*, **307**, 355–360.

Bush, D.R. (1993b) Proton-coupled sugar and amino acid transporters in plants, *Annu. Rev. Physiol. Plant. Mol. Biol.*, **44**, 513–542.

Bush, D.R. (1999) Amino acid transport, in *Plant Amino Acids: Biochemistry and Biotechnology* (ed. B.K. Singh), Marcel Dekker, New York, pp. 305–318.

Bush, D.R. & Langston-Unkefer, P.J. (1988) Amino acid transport into membrane vesicles isolated from zucchini: evidence of a proton-amino acid symport in the plasmalemma, *Plant Physiol.*, **88**, 487–490.

Buttner, M. & Sauer, N. (2000) Monosaccharide transporters in plants: structure, function and physiology, *Biochim. Biophys. Acta.*, **1465**, 263–274.

Chen, L.S. & Bush, D.R. (1997) LHT1, a lysine-and histidine-specific amino acid transporter in *Arabidopsis*, *Plant Physiol.*, **115**, 1127–1134.

Chen, L.S., Ortiz-Lopez, A., Jung, A. & Bush, D.R. (2001) ANT1, an aromatic and neutral amino acid transporter in *Arabidopsis*, *Plant Physiol.*, **125**, 1813–1820.

Chiou, T.J. & Bush, D.R. (1998) Sucrose is a signal molecule in assimilate partitioning, *Proc. Natl. Acad. Sci. U.S.A.*, **95**, 4784–4788.

Delrot, S. & Bonnemain, J.L. (1981) Involvement of protons as a substrate for the sucrose carrier during phloem loading in *Vicia faba* leaves, *Plant Physiol.*, **67**, 560–564.

Delrot, S., Despeghel, J. & Bonnemain, J. (1980) Phloem loading in *Vicia faba* leaves: effect of $n$-ethylmaleimide and $p$-chloromercuribenzenesulfonic acid on $H^+$ extrusion, $K^+$ uptake, and sucrose uptake, *Planta*, **149**, 144–148.

de Ruijter, N.C.A., Verhees, J., van Leeuwen, W., & van der Krol, A.R. (2003) Evaluation and comparison of the GUS, LUC and GFP reporter system for gene expression studies in plants, *Plant Biol.*, **5**, 103–115.

Dreyer, I., Horeau, C., Lemaillet, G., *et al.* (1999) Identification and characterization of plant transporters using heterologous expression systems, *J. Exp. Bot.*, **50**, 1073–1087.

Eckardt, N.A. (2003) The function of SUT2/SUC3 sucrose transporters: the debate continues, *Plant Cell*, **15**, 1259–1262.

Fischer, W.N., Andre, B., Rentsch, D., *et al.* (1998) Amino acid transport in plants, *Trends Plant Sci.*, **3**, 188–195.

Fischer, W.N., Kwart, M., Hummel, S. & Frommer, W.B. (1995) Substrate specificity and expression profile of amino acid transporters (AAPs) in *Arabidopsis*, *J. Biol. Chem.*, **270**, 16315–16320.

Fletcher, J.C. (2002) Shoot and floral meristem maintenance in *Arabidopsis*, *Annu. Rev. Plant Biol.*, **53**, 45–66.

Forde, B.G. (2002) Local and long-range signaling pathways regulating plant responses to nitrate, *Annu. Rev. Plant Biol.*, **53**, 203–224.

Freeman, W.M., Walker, S.J., & Vrana, K.E. (1999) Quantitative RT-PCR: pitfalls and potential, *BioTechniques*, **26**, 112–125.

Frommer, W.B., Hummel, S. & Riesmeier, J.W. (1993) Expression cloning in yeast of a cDNA encoding a broad specificity amino acid permease from *Arabidopsis thaliana*, *Proc. Natl. Acad. Sci. U.S.A.*, **90**, 5944–5948.

Frommer, W.B. & Ninnemann, O. (1995) Heterologous expression of genes in bacterial, fungal, animal, and plant cells, *Annu. Rev. Plant Physiol. Plant Mol. Biol.*, **46**, 419–444.

Geiger, D.R., Giaquinta, R.T., Sovonick, S.A. & Fellows, R.J. (1973) Solute distribution in sugar beet leaves in relation to phloem loading and translocation, *Plant Physiol.*, **52**, 585–589.

Giaquinta, R.T. (1976) Evidence for phloem loading from the apoplast, Chemical modification of membrane sulfhydryl groups, *Plant Physiol.*, **57**, 872–875.

Gifford, R.M. & Evans, L.T. (1981) Photosynthesis, carbon partitioning, and yield, *Annu. Rev. Plant Physiol.*, **32**, 485–509.

Gillissen, B., Burkle, L., Andre, B., *et al.* (2000) A new family of high-affinity transporters for adenine, cytosine, and purine derivatives in *Arabidopsis, Plant Cell*, **12**, 291–300.

Gottwald, J.R., Krysan, P.J., Young, J.C., Evert, R.F. & Sussman, M.R. (2000) Genetic evidence for the *in planta* role of phloem-specific plasma membrane sucrose transporters, *Proc. Natl. Acad. Sci. U.S.A.*, **97**, 13979–13984.

Hecht, R., Slone, J.H., Buckhout, T.J., Hitz, W.D. & Vanderwoude, W.J. (1992) Substrate specificity of the $H^+$-sucrose symporter on the plasma membrane of sugar beets (*Beta vulgaris*) – transport of phenylglucopyranosides, *Plant Physiol.*, **99**, 439–444.

Hsu, L.C., Chiou, T.J., Chen, L.S. & Bush, D.R. (1993) Cloning a plant amino acid transporter by functional complementation of a yeast amino acid transport mutant, *Proc. Natl. Acad. Sci. U.S.A.*, **90**, 7441–7445.

Jefferson, R.A., Kavanagh, T.A. & Bevan, M.W. (1987) GUS fusions: β-glucuronidase as a sensitive and versatile gene fusion marker in higher plants, *EMBO J.*, **6**, 3901–3907.

Kessler, A. & Baldwin, I.T. (2002) Plant responses to insect herbivory: the emerging molecular analysis, *Annu. Rev. Plant Biol.*, **53**, 299–328.

Komor, E. (1977) Sucrose uptake by cotyledons of *Ricinus communis, Planta*, **137**, 119–131.

Komor, E., Rotter, M., Waldhauser, J., Martin, E. & Cho, B.H. (1980) Sucrose proton symport for phloem loading in the ricinus seedling, *Ber. Deutsch. Bot. Ges. Bd.*, **93** (Suppl.), 211–219.

Kuhn, C., Barker, L., Burkle, L. & Frommer, W.B. (1999) Update on sucrose transport in higher plants, *J. Exp. Bot.*, **50**, 935–953.

Lalonde, S., Tegeder, M., Throne-Holst, M., Frommer, W.B. & Patrick, J.W. (2003) Phloem loading and unloading of sugars and amino acids, *Plant Cell Environ.*, **26**, 37–56.

Larsson, C., Widell, S. & Sommarin, M. (1988) Inside-out plant plasma membrane vesicles of high purity obtained by aqueous two-phase partitioning, *FEBS Lett.*, **239**, 289–292.

Lemoine, R.E. & Delrot, S. (1989) Pmf-driven sucrose uptake in sugar beet plasma membrane vesicles, *FEBS Lett.*, **249**, 129–133.

Lu, J.M.-Y. (1998) Molecular cloning and structure–function analysis of the plant proton-sucrose symporter, Ph.D. thesis, University of Illinois at Urbana-Champaign.

Lu, J.M.Y. & Bush, D.R. (1998) His-65 in the proton-sucrose symporter is an essential amino acid whose modification with site-directed mutagenesis increases transport activity, *Proc. Natl. Acad. Sci. U.S.A.*, **95**, 9025–9030.

Meyer, S., Lauterbach, C., Niedermeier, M., Barth, I., Sjolund, R.D. & Sauer, S. (2004) Wounding enhances expression of AtSUC3, a sucrose transporter from *Arabidopsis* sieve elements and sink tissues, *Plant Physiol.*, **134**, 684–693.

Meyer, S., Melzer, M., Truernit, E., *et al.* (2000) AtSUC3, a gene encoding a new *Arabidopsis* sucrose transporter, is expressed in cells adjacent to the vascular tissue and in a carpel cell layer, *Plant J.*, **24**, 869–882.

Meyerowitz, E. (1987) In situ hybridization to RNA in plant tissue, *Plant Mol. Biol. Rep.*, **5**, 242–250.

Noiraud, N., Maurousset, L. & Lemoine, R. (2001) Transport of polyols in higher plants, *Plant Physiol. Biochem.*, **39**, 717–728.

Ortiz-Lopez, A., Chang, H.C. & Bush, D.R. (2000) Amino acid transporters in plants, *Biochim. Biophys. Acta*, **1465**, 275–280.

Ozcan, S., Dover, J. & Johnston, M. (1998) Glucose sensing and signaling by two glucose receptors in the yeast *Saccharomyces cerevisiae, EMBO J.*, **17**, 2566–2573.

Ransom-Hodgkins, W.D., Vaughn, M.W. & Bush, D.R. (2003) Protein phosphorylation plays a key role in sucrose-mediated transcriptional regulation of a phloem-specific proton–sucrose symporter, *Planta*, **217**, 483–489.

Reinders, A., Schultze, W., Kuhn, C., *et al.* (2002) Protein–protein interactions between sucrose transporters of different affinities colocalized in the same enucleate sieve elements, *Plant Cell*, **14**, 1567–1577.

Riesmeier, J.W., Hirner, B. & Frommer, W.B. (1993) Potato sucrose transporter expression in minor veins indicates a role in phloem loading, *Plant Cell*, **5**, 1591–1598.

Riesmeier, J.W., Willmitzer, L. & Frommer, W.B. (1992) Isolation and characterization of a sucrose carrier cDNA from spinach by functional expression in yeast, *EMBO J.*, **11**, 4705–4713.

Riesmeier, J.W., Willmitzer, L. & Frommer, W.B. (1994) Evidence for an essential role of the sucrose transporter in phloem loading and assimilate partitioning, *EMBO J.*, **13**, 1–7.

Schwacke, R., Grallath, S., Breitkreuz, K.E., *et al.* (1999) LeProT1, a transporter for proline, glycine betaine, and gamma-amino butyric acid in tomato pollen, *Plant Cell*, **11**, 377–391.

Segel, H. (1993) *Enzyme Kinetics*, Wiley Interscience, New York, pp. 1–957.

Sentenac, H., Bonneaud, N., Minet, M., *et al.* (1992) Cloning and expression in yeast of a plant potassium ion transport system, *Science*, **256**, 663–665.

Sheen, J., Hwang, S.B., Niwa, Y., Kobayashi, H. & Galbraith, D.W. (1995) Green-fluorescent protein as a new vital marker in plant cells, *Plant J.*, **8**, 777–784.

Slayman, C.L., Long, W.S. & Lu, C.Y.-H. (1973) The relationship between ATP and an electrogenic pump at the plasma membrane of *Neurspora crassa*, *J. Membr. Biol.*, **14**, 303–338.

Slayman, C.L. & Slayman, C.W. (1974) Depolarization of the plasma membrane of *Neurospora* during active transport of glucose: evidence for a proton-dependent cotransport system, *Proc. Natl. Acad. Sci. U.S.A.*, **71**, 1935–1939.

Slone, J.H. & Buckhout, T.H. (1991) Sucrose-dependent $H^+$ transport in plasma-membrane vesicles isolated from sugarbeet leaves (*Beta vulgaris* L.) – evidence in support of the $H^+$-symport model for sucrose transport, *Planta*, **183**, 584–589.

Stacey, G., Koh, S., Granger, C. & Becker, J.M. (2002) Peptide transport in plants, *Trends Plant Sci.*, **7**, 257–263.

Stadler, R. & Sauer, N. (1996) The *Arabidopsis thaliana* AtSUC2 gene is specifically expressed in companion cells, *Bot. Acta*, **109**, 299–306.

Stadler, R., Truernit, E., Gahrtz, M. & Sauer, N. (1999) The AtSUC1 sucrose carrier may represent the osmotic driving force for anther dehiscence and pollen tube growth in *Arabidopsis*, *Plant J.*, **19**, 269–278.

Sze, H. (1985) $H^+$-translocating ATPases: advances using membrane vesicles, *Annu. Rev. Plant Physiol.*, **36**, 175–208.

Talbott, L.D. & Zeiger, E. (1996) Central roles for potassium and sucrose in guard-cell osmoregulation, *Plant Physiol.*, **111**, 1051–1057.

Thorne, J.H. (1985) Phloem unloading of c and n assimilates in developing seeds, *Annu. Rev. Plant. Physiol.*, **36**, 317–343.

Truernit, E. & Sauer, N. (1995) The promoter of the *Arabidopsis thaliana* SUC2 sucrose-$H^+$ symporter gene directs expression of beta-glucuronidase to the phloem – evidence for phloem loading and unloading by SUC2, *Planta*, **196**, 564–570.

Vaughn, M.W., Harrington, G.N. & Bush, D.R. (2002) Sucrose-mediated transcriptional regulation of sucrose symporter activity in the phloem, *Proc. Natl. Acad. Sci. U.S.A.*, **99**, 10876–10880.

Welsh, S. & Kay, S.A. (1997) Reporter gene expression for monitoring gene transfer, *Curr. Opin. Biotech.*, **8**, 617–622.

Williams, L.E., Nelson, S.J. & Hall, J.L. (1992) Characterization of solute/proton cotransport in plasma membrane vesicles from *ricinus* cotyledons, and comparison with other tissues, *Planta*, **186**, 541–550.

Zhou, J.J., Theodoulou, F., Sauer, N., Sanders, D. & Miller, A.J. (1997) A kinetic model with ordered cytoplasmic dissociation for SUC1, an *Arabidopsis* $H^+$/sucrose cotransporter expressed in *Xenopus* oocytes, *J. Membr. Biol.*, **159**, 113–125.

# 6 Voltage-gated ion channels

Ingo Dreyer, Bernd Müller-Röber and Barbara Köhler

## 6.1 Introduction

Voltage-regulated ion channels represent one (of several) type(s) of plant transport proteins. The similarity of voltage-gated ion channels found in prokaryotes and eukaryotes suggests that the basic structures responsible for voltage-regulated transmembrane ion transport have evolved more than 2 billion years ago. These similarities facilitated the investigation of plant ion channels. Indeed, in several cases, the knowledge obtained for channels from the different kingdoms served as a good basis for the understanding of the biological role of a plant ion channel. For example, plant potassium channels show the same selectivity filter, i.e. the structural entity that determines the selectivity, as animal $K^+$ channels (see below), and also in their overall topology they are similar. Considered in more detail, however, a number of plant-specific features show divergent patterns in structure and function. Before this chapter portrays voltage-gated ion channels identified in plants (see Section 6.3), some basic structural and mechanistic aspects of voltage-gated ion channels are presented.

A voltage-gated ion channel can be defined as a membrane-spanning protein (i) that facilitates the passive passage of a certain species of ions, and (ii) the activity of which is controlled by the transmembrane voltage. In the following, aspects of this definition will be illustrated exemplarily on the basis of potassium channels since for this channel type most knowledge is available. Figure 6.1 shows schematically the topology of a plant $K^+$ channel of the Shaker-type (Uozumi et al., 1998). This channel is a tetramer composed of four subunits. Each subunit consists of six transmembrane-spanning domains (S1–S6) and a loop between the fifth and sixth domain (Fig. 6.1A). The loops of the four subunits together form a central rigid pore (P), which allows the passage of $K^+$ ions across the membrane (Fig. 6.1B). Usually the pore shows a high selectivity supporting the permeation of a single-ion species (here $K^+$ ions) along its electrochemical gradient while excluding others from transmembrane passage even when they are chemically related (e.g. $Na^+$ ions). The selectivity of the channel is determined by the structure of the pore and the chemical properties of the amino acid side chains facing it (Fig. 6.1C; Doyle et al., 1998). Changes in the P-loop of the plant $K^+$ channel KAT1 by site-directed mutagenesis, for example, caused strong alterations of the selectivity of this channel (Uozumi et al., 1995; Becker et al., 1996; Nakamura et al., 1997; Dreyer et al., 1998).

**Fig. 6.1** Structure of a voltage-gated potassium channel. Potassium channels are multimeric proteins comprising four α-subunits. (A) Each subunit is built of six transmembrane domains (S1–S6) and an amphiphilic linker (P) between S5 and S6. (B) The four subunits assemble in a way that the P domains create a pore in the membrane. Here displayed in a top view. (C) Parts of the four P domains line the ion permeation pathway and determine the selectivity of the channel. Here only the S5–P–S6 segments of two subunits are displayed.

Thus, in a first simple sketch, an ion channel can be considered as a selective hole in a membrane. This hole *per se* would not be of essential benefit for a cell since it only facilitates the passage of a specific ion species along its electrochemical gradient but does not restrain this flow. To control the flux through an ion channel, mechanisms evolved that regulate the activity of the channel, i.e. those that close and open it. In most of these cases, energy from mechanical tension, the binding of a ligand, or the transmembrane voltage is used to cause conformational changes of the channel protein, which in turn result in blocking or opening of the selective permeation pathway.

## 6.2 Voltage gating from a mechanistic point of view

Concepts describing voltage-gated membrane conductivity were already developed before the existence of ion channels had been proven. The first model, describing selective voltage-dependent changes of the membrane permeability for different ion species, postulated the regulation of ion flux by so-called gating particles (Hodgkin & Huxley, 1952; honoured with the Nobel Prize in 1963). During the last half century, the original gating concept has been modified and refined. In particular, the cloning of channel genes in the course of the molecular revolution, e.g. the first cloning of a gene coding for a voltage-dependent ion channel (Noda *et al.*, 1984), gave an identity to structural elements of channel proteins involved in the gating process. The huge efforts in this field and the increasing knowledge were crowned recently by the publication of the first X-ray structure of a voltage-dependent $K^+$ channel (Jiang *et al.*, 2003). R. MacKinnon and P. Agre were honoured with the Nobel Prize in 2003 for their new molecular insights into channel structures enabled by employing crystallographic techniques.

The activity of ion channels can be modulated by the membrane voltage in two ways: indirectly/passively or directly/actively. In the first case, particles (e.g. charged molecules like $Mg^{2+}$) enter the pathway and obstruct the pore. This is identical to a voltage-dependent block of the permeation pathway. In the second case (voltage gating), the channel protein itself senses the transmembrane electrical gradient and undergoes conformational changes. In the case of voltage-dependent $K^+$ channels, for example, one part of the channel (the S4 segment containing a high density of charged amino acid residues) is affected by the transmembrane voltage and is displaced some distance across the electrical gradient. This movement of the voltage sensor in turn provokes structural rearrangements of the entire protein, resulting in channel opening or closing of the permeation pathway (Fig. 6.2).

The consequences of the conformational changes of an ion channel protein for its permeation properties were monitored in detail for the first time with the patch-clamp technique (Neher & Sakmann, 1976; honoured with the Nobel Price in 1991). As illustrated in Fig. 6.3A, an ion channel opens (O) and closes (C) in a stochastic manner. The process of opening and closing proceeds on a timescale that is beyond the experimental resolution of the patch-clamp technique. If we consider a single type of ion channel, each opening of the ion channel results in a current flux of the same amplitude. Under constant environmental conditions, the current flowing through the open channel is constant. The size of the single-channel current amplitude depends on the channel type and on two parameters: the electrical gradient, which influences the movement of the charged ions, and the composition of the extracellular and intracellular solutions, reflecting mainly the chemical gradient of the permeating ion. The voltage dependence of the single-channel amplitude is more or less linear when physiological voltages are applied. Its slope defines the single-channel conductance, which depends strongly on the structure of the permeation pathway, and which is therefore an intrinsic property of the ion channel type.

A living cell does not express a single-ion channel of one type, but instead, an ensemble of channels. In guard cells, for example, 600–2500 potassium uptake

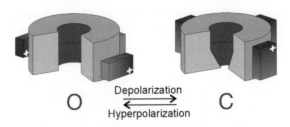

$$O \xrightleftharpoons[\text{Hyperpolarization}]{\text{Depolarization}} C$$

**Fig. 6.2** A principle mechanism of voltage gating. Special parts of the channel protein that are characterised by a high density of charged amino acids (indicated as '+') are able to sense the transmembrane voltage. Upon changes in the electrical gradient, these parts move a bit and force the ion channel to undergo conformational changes, which in turn open or close the ion permeation pathway. Here the gating of inward-rectifying plant potassium channels is represented, postulated on the basis of the X-ray structure of a bacterial voltage-gated potassium channel (Jiang *et al.*, 2003).

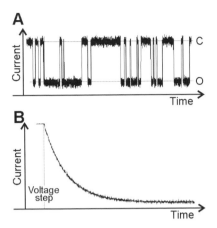

**Fig. 6.3** Stochastic opening of single channels and kinetics of a channel ensemble. (A) A single channel opens (current level O) and closes (current level C) in a stochastic manner. The current amplitude is a constant that depends on the channel type, the ionic composition at both sides of the membrane and the membrane voltage. (B) An ensemble of several channels activate with a characteristic kinetics upon changes of the membrane voltage.

channels were detected (Dietrich *et al.*, 1998). The coordinated action of such a channel ensemble differs strongly from the behaviour of a single channel (Fig. 6.3B). Kinetic components are easily visible upon changes of the membrane voltage.

To comprehend the biological function of ion channels, it is useful to illustrate their behaviour with the help of biophysical/mathematical models. At a membrane voltage $V$ and at time point $t$, the current flowing through an ensemble of channels, $I(V, t)$, can be described mathematically by the equation

$$I(V, t) = N\, i(V) p_O(V, t) \qquad (6.1)$$

where $i(V)$ is the current flowing through a single channel and $p_O(V, t)$ the time- and voltage-dependent open probability, which is defined by the quotient $N_O/N$, i.e. the number of open channels ($N_O$) divided by the total number of channels ($N$). Equation 6.1 is fundamental to describe the dynamic behaviour of ion channels and to analyse their elementary properties. In the following we will provide some more mathematical background in order to fill this equation with life. To reduce the degree of complexity, a simplified channel will be considered that exists in only two conformations (Fig. 6.4), with the pore either open (O) or closed (C). Like an enzyme, the channel in the two configurations can be represented by two distinct levels of free energy: $G_O$ and $G_C$. For a conformational change (O → C, or C → O), the channel has to overcome an energy peak, represented by the energy level $G_P$. Thus, the activation of the channel (C → O) is impeded by the energy barrier $\Delta G_{PC} = G_P - G_C$, and the deactivation (O → C) by the barrier $\Delta G_{PO} = G_P - G_O$.

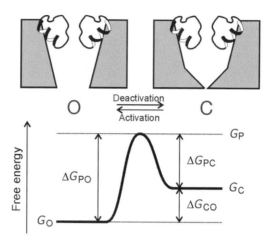

**Fig. 6.4** An idealistic two-state model to illustrate ion channel gating. A simplified ion channel can exist in two states: open and closed. The two states correspond to two distinct conformations that differ in their free intrinsic energy ($G_O$ and $G_C$). For a conformational change (activation or deactivation) the energy barrier ($G_P$) has to be overcome.

### 6.2.1 Static – steady-state equilibrium

The energy levels $G_O$, $G_C$ and $G_P$ depend on the surrounding conditions, e.g. the transmembrane voltage in the case of voltage-gated ion channels. Therefore, a frequently used experimental approach is to keep as many parameters as possible constant (e.g. clamp the voltage to a constant level) and allow the channels to relax into a steady-state equilibrium. Under this condition the probability of finding the channel in the open ($p_{O\infty}$) or in the closed ($p_{C\infty}$) configuration follows the Boltzmann statistics: $p_{C\infty}/p_{O\infty} = \exp(-(G_C - G_O)/(kT))$. Here, $k$ denotes the Boltzmann constant and $T$ the absolute temperature. Taking into account that $p_{O\infty} + p_{C\infty} = 1$, i.e. the probability to find the channel either in the open or in the closed state is 1, the open probability can be directly correlated with the energy difference between the open and the closed state ($\Delta G_{CO} = G_C - G_O$):

$$p_{O\infty} = \frac{1}{1 + e^{-\Delta G_{CO}/(kT)}} \tag{6.2}$$

As indicated above, in voltage-gated ion channels the free energy of the different configurations depends on the transmembrane voltage $V$. Detailed studies on the animal Shaker channel have shown that in the physiological voltage range this dependency can be expressed by a linear function (Perozo *et al.*, 1993): $G(V) = G(0) + V \cdot q$. (Mathematically, this is the first-order Taylor approximation of the function $G(V)$ (Stevens, 1978): $G(V) = G(0) + V \, d/dV[G(V)]|_{V=0} + 0(V^2)$. Higher order terms do not significantly contribute to the free energy.) This implies a voltage dependence of the open probability of the channel:

$$p_{O\infty}(V) = \frac{1}{1 + e^{-(QV+\Delta G_{CO}(0))/(kT)}} = \frac{1}{1 + e^{-Q(V-V_{1/2})/(kT)}} \qquad (6.3)$$

with $\Delta G_{CO}(0) = G_C(0) - G_O(0)$ and $V_{1/2} = -\Delta G_{CO}(0)/Q$, the half-maximal activation voltage ($p_{O\infty}(V_{1/2}) = 0.5$). The parameter $Q$ ($= q_C - q_O$) is often called the (apparent) gating charge and is correlated to the (fractional) charge movement of the channel voltage sensor. The so-called Boltzmann equation (Equation 6.3) describes (in the steady state) the mean behaviour of a single channel observed over a long period of time, or the mean behaviour of a channel ensemble in equilibrium at one time point. As shown in Fig. 6.5, the Boltzmann equation describes hyperpolarisation-activated channels when $Q < 0$ and depolarisation-activated channels when $Q > 0$. The parameter $\Delta G_{CO}(0)$ ($\Delta G_a$, $\Delta G_b$, $\Delta G_c$ and $\Delta G_d$ in Fig. 6.5) is a measure for the activity range of the ion channel. Several physiologically relevant channel-modulatory mechanisms influence this parameter and therefore change the activity range of the channel. We may exemplarily consider the binding of a signalling molecule (ligand) to the ion channel. The ligand binding alters the stability of the different channel conformations. Thus the parameters $G_C(0)$, $G_O(0)$ and hence $\Delta G_{CO}(0)$ change. This value can become larger (Fig. 6.5, $\Delta G_b \rightarrow \Delta G_a$), which results in channel stimulation, or it can become smaller (even negative), which causes an inhibition (Fig. 6.5, $\Delta G_b \rightarrow \Delta G_c$). Therefore, analyses using the Boltzmann equation allow a first biophysical categorisation of the observed channel and its modulation, which often is a prerequisite for assigning a biological function to this protein.

### 6.2.2  Kinetic – relaxation into an equilibrium

In the previous paragraph the idealised channel was considered in thermal equilibrium. Additional characteristic information on the channel can be obtained by

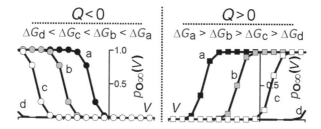

**Fig. 6.5** Illustration of the Boltzmann equation. The Boltzmann equation describes the voltage dependence of the open probability of an ion channel in thermal equilibrium [$p_{O\infty}(V)$]. The graphical representation of the Boltzmann equation varies depending on the parameters $Q$ (gating charge) and $\Delta G$ (difference in the free energy between open and closed state). Eight different $p_{O\infty}(V)$ curves ($Q > 0$, a–d; $Q < 0$, a–d) for eight $Q$-$\Delta G$ parameter sets are displayed.

analysing *transitions* between different equilibriums. Just by changing the environmental circumstances, e.g. in voltage-step protocols, the equilibrium conditions of the channels can be altered almost instantaneously. Upon such a change of the experimental condition, the channels relax more or less slowly into the new equilibrium. Experimentally, this relaxation kinetics is difficult to analyse on a single, isolated channel. Instead, in most cases the mean behaviour of an ensemble of many channels is studied. The experimentally observed relaxation kinetics can be described by an (or a sum of) exponential function(s). The reason why this is the general case will be illustrated again on the basis of the idealised two-state model (Fig. 6.4). A transition of a channel from the closed to the open state (activation) depends on the energy barrier $\Delta G_{PC}$ ($= G_P - G_C$), and a transition from the open to the closed state (deactivation) on the barrier $\Delta G_{PO}$ ($= G_P - G_O$). The corresponding transition rates, $k_a$ and $k_d$, can be expressed according to the Eyring rate theory (Glasstone *et al.*, 1941), which is based on quantum mechanics, as

$$k_a = kT/he^{-\Delta G_{PC}/(kT)} \qquad (6.4)$$
$$k_d = kT/he^{-\Delta G_{PO}/(kT)} \qquad (6.5)$$

With these transition rates, the changes of the open ($dp_O$) and closed probability ($dp_C$) of the channel in the small time interval $dt$ are calculated as

$$\left.\begin{array}{l} dp_O = (-k_d p_O + k_a p_C)\, dt \\ dp_C = (\ \ k_d p_O - k_a p_C)\, dt \end{array}\right\} \qquad (6.6)$$

This system of differential equations determines the dynamic behaviour of the channel. Under certain conditions, e.g. when the voltage-dependent parameters $k_a$ and $k_d$ are kept constant, an analytic solution for the time course of the relaxation of the channel ensemble into a new equilibrium can be deduced:

$$p_O(t) = p_{O\infty} + (p_O(0) - p_{O\infty})\, e^{-\lambda t} \qquad (6.7)$$

Here $p_{O\infty} = 1/(1 + k_d/k_a) = 1/(1 + \exp(-(G_C - G_O)/(kT)))$ is the open probability in thermal equilibrium (compare Equation 6.2), $\lambda = k_a + k_d$ the characteristic (voltage-dependent) time constant of the channel, and $p_O(0)$ the open probability at $t = 0$ (e.g. at the time of the voltage step). Since the measured current mediated by the channel ensemble is proportional to the open probability, the current amplitude is expected to relax in an exponential manner. From the analysis of the decay constant $\lambda$, further insights into the energetic properties of the ion channel protein can be obtained. Figure 6.6 illustrates the consequences of a change in the free energy of the peak, $G_P$. While the steady-state characteristics remain unaffected, the higher the peak energy is, the smaller is the rate of channel activation and deactivation. Such a

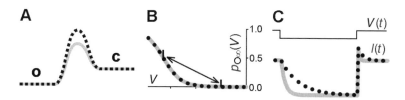

**Fig. 6.6** Changes in the peak energy (A) do not affect open probability (B) but alter channel kinetics (C). For the simplified two-state model, the predictions for a higher free peak energy $G_P$ (*dotted lines*) were compared with those for a lower free peak energy (*grey lines*).

difference in the peak energy can explain the variation observed in the kinetics between the two guard-cell-expressed $K^+$ channels KAT1 and KST1 (Mueller-Roeber *et al.*, 1995; Dietrich *et al.*, 1998). In comparison to KAT1, KST1 relaxes more slowly into equilibrium. Both channels exhibit, however, similar steady-state open probabilities. At this point an important aspect should be emphasised: slower relaxation kinetics does not mean that a single channel opens or closes more slowly. A single channel is still gating rapidly in a stochastic manner (Fig. 6.3A). Instead, slower relaxation kinetics means that the ensemble of channels relaxes more slowly into the new equilibrium because the number of transitions (O → C, or C → O) is smaller.

### 6.2.3 *Comparison of the model with the* in vivo *situation*

*In vivo*, an ion channel can exist in more than two configurations, adding to the complexity of the mathematics presented above (see Colquhoun & Hawkes, 1995). However, even if an ion channel can exist in more than two stable conformations, in general only one open state is observed and multiple closed states are manifested indirectly through kinetic relaxations with several time constants $\lambda_i$. In this case the open probability at equilibrium is described by:

$$p_{O\infty} = \frac{1}{1 + \sum_i e^{-\Delta G_{iO}/(kT)}} \qquad (6.8)$$

where $\Delta G_{iO}$ is the difference in the free energy between the $i$th closed state and the open state.

In most cases, however, it is sufficient to describe an ion channel by a two-state model to explain qualitatively the observed effects and to get a good idea about their biological relevance. This is the reason why in a large number of publications the Boltzmann function is used instead of a more precise, but also more complex, function as exemplified by Equation 6.8. An interesting comparison between several alternative models for a plant channel *in vivo* is discussed in Blatt and Gradmann (1997).

## 6.3 Voltage-gated ion channels uncovered in plants and their involvements in physiological processes

In 1976, the patch-clamp technique allowed for the first time to monitor the action of a single transporter molecule in denervated frog muscle fibres (Neher & Sakmann, 1976). Two years later, a similar technique was employed to document single voltage-gated ion channels in the plasma membrane of the alga *Nitellopsis obtusa* (Krawczyk, 1978), and in 1984 the first single-channel recordings from angiosperms were reported (Moran *et al.*, 1984; Schroeder *et al.*, 1984). Combined use of diverse electrophysiological techniques (see Chapter 1) has since increased tremendously our knowledge of plant transport systems and especially of plant ion channels. The following presentation categorises the different types of channels in a more general way in terms of subcellular localisation, selectivity and voltage-regulation. In most of the cases, this classification already allows to illustrate the physiological function of a channel, independent of its exact spatial expression pattern at the tissue level. In this review we will refer to the membrane voltage according to the convention $E = \varphi_{cytosol} - \varphi_{lumen}$ (Bertl *et al.*, 1992). Here, $\varphi_{cytosol}$ and $\varphi_{lumen}$ denote the electrical potentials in the cytosol and the lumen (e.g. vacuole or extracellular space), respectively. This convention unifies currents and flux directions with respect to the cytosol. For example, a negative current of cations denotes an influx of these ions into the cytosol, independent of whether the ions come from the extracellular space (across the plasma membrane) or from endocellular compartments.

### 6.3.1 Plasma membrane potassium channels

Potassium is the most abundant cation in plants and is involved in many physiological processes. Therefore, it is not unexpected that the first single-ion channels recorded from a land plant were voltage-gated potassium channels (Schroeder *et al.*, 1984). Potassium channels have been identified in every living tissue and cell type investigated, including mesophyll cells, root cells, xylem parenchyma cells, pollen, coleoptiles and, notably, guard cells. Voltage-gated potassium channels are involved in membrane voltage control, in osmotically driven movements, in cytosolic volume control, in cation nutrition and in $K^+$ redistribution at the intracellular-, tissue- and the whole plant-level (see Maathuis *et al.*, 1997, and Chapter 11). The categorisation as 'potassium channels' is based on the strong selectivity of these channels for $K^+$ against, e.g. $Na^+$. In the plasma membrane of plant cells, three distinct types of voltage-gated potassium channels were identified, which are active in different voltage ranges: hyperpolarisation-activated, depolarisation-activated and weakly rectifying $K^+$ channels (see Very & Sentenac, 2002, 2003, for recent reviews).

### 6.3.1.1 Hyperpolarisation-activated $K^+$ channels – $K_{in}$ channels

Hyperpolarisation-activated $K^+$ channels ($K_{in}$ channels) open at membrane voltages more negative (cytosol) than around $-100$ mV, whereas at more positive voltages they are closed (Fig. 6.7A; Schroeder *et al.*, 1987; Blatt, 1992). In terms of the Boltzmann

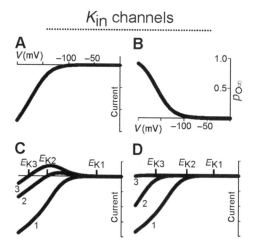

**Fig. 6.7** Characteristics of $K_{in}$ channels. (A) Current–voltage characteristics in the steady state. (B) The corresponding open probability in thermal equilibrium. (C and D) Two types of $K_{in}$ channels have been reported: (C) One type does not sense the $K^+$ concentration, activates also positive of $E_K$, and thus mediates at certain voltages potassium efflux as well (Bruggemann *et al.*, 1999). (D) The other type senses the extracellular potassium concentration and activates only negative of $E_K$, the equilibrium potential for potassium (Maathuis & Sanders, 1995). Curves representing three different extracellular potassium concentrations (1–3) are displayed.

equation (Equation 6.3), this is equivalent to $Q < 0$ (Fig. 6.7B). Because of this activity range, $K_{in}$ channels serve as potassium-uptake channels. They equip plant cells with the ability to accumulate high amounts of potassium in a reasonably short time. The driving force for potassium uptake is provided by the negative membrane voltage established by the activity of the $H^+$-ATPase that pumps $H^+$ ions out of the cell. Well-known examples for hyperpolarisation-activated potassium channels are guard-cell potassium-uptake channels playing a role in stomatal movement (Dietrich *et al.*, 1998), or root $K_{in}$ channels playing a role in potassium uptake from the soil (Maathuis & Sanders, 1995). In general, hyperpolarisation-activated potassium channels show slow activation kinetics upon voltage steps (Equation 6.7; $1/\lambda$ is in the range of hundreds of milliseconds to seconds), and do not inactivate even under sustained hyperpolarisation (Schroeder, 1988). A characteristic fingerprint for this channel type is the activation upon extracellular acidification and the block by extracellular $Cs^+$, $Ba^{2+}$ and tetraethylammonium ($TEA^+$).

The first $K^+$ channel genes from plants were cloned in 1992 by two groups independently. cDNA libraries from *Arabidopsis thaliana* were expressed in mutant strains of the yeast *Saccharomyces cerevisiae*, lacking the endogenous potassium-uptake transport systems and, hence, unable to grow on media containing less than 1 mM $K^+$. Screening on low-$K^+$ media identified two cDNA clones that complemented this deficiency. These clones encoded the two $K_{in}$ channels AKT1 and KAT1, which both belong to the Shaker family of $K^+$ channels (Anderson *et al.*,

1992; Sentenac *et al.*, 1992). We know now that the genome of *Arabidopsis thaliana* contains at least four genes (*KAT1*, *AKT1*, *KAT2*, *SPIK*) coding for potassium-uptake channels (Very & Sentenac, 2002). *KAT1* and *KAT2* are mainly expressed in guard cells (Nakamura *et al.*, 1995; Pilot *et al.*, 2001; Szyroki *et al.*, 2001), *SPIK* in pollen grains (Mouline *et al.*, 2002) and *AKT1* in roots (Lagarde *et al.*, 1996). The function of a fifth putative $K_{in}$ channel encoded by the gene *AKT6* is still to be proven. Additionally, a sixth $K_{in}$-related gene has been identified (*AtKC1*), which shows an expression pattern similar to that of AKT1 (Reintanz *et al.*, 2002). AtKC1 appears not to be functional alone, but rather to modulate the current carried by other $K^+$ channels (Dreyer *et al.*, 1997; Reintanz *et al.*, 2002).

Experiments employing heterologous expression systems and biochemical studies revealed that $K_{in}$ channels are multimeric proteins consisting of four α-subunits (Daram *et al.*, 1997; Dreyer *et al.*, 1997). This result and fundamental pharmacological differences, observed for example between guard-cell potassium-uptake channels *in vivo* and their putative molecular counterparts, led to the model that *in planta* hyperpolarisation-activated potassium channels are (at least in part) heteromultimeric proteins (Dreyer *et al.*, 1997; Bruggemann *et al.*, 1999b; Pilot *et al.*, 2001).

### 6.3.1.2 Depolarisation-activated $K^+$ channels – $K_{out}$ channels

Depolarisation-activated $K^+$ channels ($K_{out}$ channels) open at membrane voltages more positive than the equilibrium potential for potassium, $E_K$. At more negative voltages they are closed (Fig. 6.8A). In terms of the Boltzmann equation (Equation 6.3), this is equivalent to $Q > 0$ (Fig. 6.8B; Blatt, 1988; Hosoi *et al.*, 1988; Schroeder, 1988). To guarantee that the channel is closed at voltages negative of $E_K$, its activation threshold varies with the extracellular potassium concentration (Fig. 6.8; see Section 6.4.5 for details). Because of their activity range, $K_{out}$ channels serve as potassium release channels. In general, depolarisation-activated potassium channels exhibit slow delayed activation kinetics upon voltage-steps (Equation 6.7; $1/\lambda$ is in the range of hundreds of milliseconds to seconds), and do not inactivate even under sustained depolarisation (Schroeder, 1988). A characteristic fingerprint

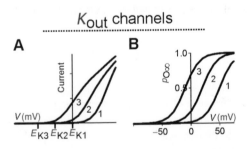

**Fig. 6.8** Characteristics of $K_{out}$ channels. (A) Current–voltage characteristics in the steady state. (B) The corresponding open probability in thermal equilibrium. $K_{out}$ channels sense the extracellular potassium concentration. Curves representing three different extracellular potassium concentrations (1–3) are displayed.

for this channel type is the activation upon extracellular alkalinisation and the block by extracellular $Ba^{2+}$ (Schroeder *et al.*, 1987).

The first gene coding for a plant $K_{out}$ channel was cloned on the basis of sequence homologies to $K_{in}$ channel genes. A DNA-based strategy, employed to identify further proteins belonging to the Shaker $K^+$ channel family, resulted in the identification of SKOR, which turned out to be a potassium release channel (Gaymard *et al.*, 1998). In total, the genome of *Arabidopsis thaliana* contains two genes coding for potassium release channels of the Shaker type: SKOR and GORK (Gaymard *et al.*, 1998; Ache *et al.*, 2000). For these two genes the expression patterns have been determined, and null-allele mutants of *Arabidopsis* have been investigated. SKOR has been proposed to play a major role in $K^+$ release from the stelar parenchyma into the xylem and therefore in the transport of potassium from roots towards shoots (Gaymard *et al.*, 1998). GORK is the potassium release channel of guard cells and plays an important role in stomatal closure (Hosy *et al.*, 2003).

### 6.3.1.3  Weakly rectifying $K^+$ channels – $K_{weak}$ channels

$K_{weak}$ channels belong to the Shaker $K^+$ channel family. These channels appear to be active in the entire physiological voltage range and thus mediate both, potassium inward and potassium outward fluxes. A characteristic fingerprint of weakly rectifying $K^+$ channels is the block by extracellular protons and $Ca^{2+}$. *In vivo*, this channel type has been unambiguously uncovered in protoplasts derived from vascular tissues of maize seedlings (Bauer *et al.*, 2000) and possibly in protoplasts from *Arabidopsis* mesophyll cells (Spalding *et al.*, 1992). Most knowledge, however, was obtained by investigating the cloned molecular counterparts (AKT2, ZmKT2) after expression in heterologous expression systems (Marten *et al.*, 1999; Philippar *et al.*, 1999; Lacombe *et al.*, 2000). As is the case for AKT2 from *Arabidopsis*, $K_{weak}$ channels can act in two different gating modes: while in mode 1, AKT2 activates like a hyperpolarisation-activated $K^+$ channel, in mode 2 it is open in the entire physiological voltage range (Dreyer *et al.*, 2001; Fig. 6.9). It was proposed, that the setting of the gating mode is determined by the phosphorylation status of the protein. Thus, AKT2 can be switched from an inward-rectifying channel to a leak-like $K^+$ channel by phosphorylation events (Dreyer *et al.*, 2001; Cherel *et al.*,

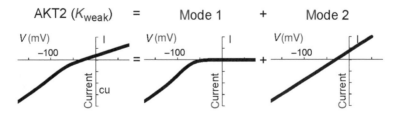

**Fig. 6.9** Currents mediated by $K_{weak}$ channels are composed of two components. The $K_{weak}$ channel AKT2 can exist in two gating modes. In mode 1, AKT2 is comparable to $K_{in}$ channels mediating potassium inward currents. In mode 2, AKT2 mediates leak-like potassium currents.

2002). At the present time, $K_{weak}$ channels are supposed to provide a background conductance that stabilises the membrane voltage, e.g. in phloem loading processes (Ache et al., 2001; Deeken et al., 2002).

### 6.3.2    Vacuolar potassium channels

Vacuoles are storage organelles that typically comprise 90–95% of the cell volume. They play important roles in the osmotic adjustment of plant cells. For example, during unidirectional and reversible turgor-driven processes, like cell elongation or motor cell and guard cell movements, vacuoles accumulate (and release during stomatal closure) the additional osmotically active molecules (e.g. ions). The cytoplasm is mainly only a thoroughfare in these processes. Additionally, vacuoles seem to serve as a buffer for cytosolic potassium. It has been observed that, under conditions of severe $K^+$ deficiency, the cytosolic potassium level in a variety of plant cells is kept constant at the expense of the vacuolar $K^+$ pool (Leigh & Wyn Jones, 1984; Walker et al., 1996). In most plant cells, ATP- and $PP_i$-dependent $H^+$-pumps generate a small cytosol-negative electrical voltage across the vacuolar membrane (Rea & Sanders, 1987). Therefore, under most conditions vacuolar cation uptake requires active transport (for recent reviews on vacuolar transporters see, e.g. Luttge et al., 2000; Martinoia et al., 2000; Maeshima, 2001; Gaxiola et al., 2002), whereas cation release is proposed to occur via ion channels.

#### 6.3.2.1    Slow-activating vacuolar channel

The slow-activating vacuolar (SV) channel was the first tonoplast channel that was characterised (Hedrich & Neher, 1987). The SV channel is ubiquitous in all higher plants tested, and in view of the multitude of factors that regulate this channel, it is likely an essential element in most cells. The SV channel is non-selective for monovalent cations (Kolb et al., 1987; Colombo et al., 1988), and shows significant permeabilities for the divalent cations $Ca^{2+}$, $Ba^{2+}$ and $Mg^{2+}$ (Pantoja et al., 1992b; Ward & Schroeder, 1994; Allen & Sanders, 1995; Gambale et al., 1996). A replacement of cytosolic $Cl^-$ by gluconate led to an increase of the permeability ratio of $Ca^{2+}$ over $Na^+$ (Miedema & Pantoja, 2001). The SV channel is blocked by $TEA^+$, $Ba^{2+}$ and $Zn^{2+}$ (Hedrich & Kurkdjian, 1988). It activates slowly after the application of (cytosol) positive voltages (Fig. 6.5, $Q > 0$) and does not inactivate. However, SV channel activity is strongly modulated. The channels are blocked by luminal $Ca^{2+}$ and cytosolic protons, and are activated by cytosolic $Ca^{2+}$, calmodulin, cytosolic $Mg^{2+}$ and reducing agents, and are bimodally regulated by phosphorylation (Miedema & Pantoja, 2001; and references therein; for details, see Section 6.4). A putative candidate for the SV channel is the two-pore $K^+$ channel KCO1, which has been shown to be targeted to the vacuolar membrane (Czempinski et al., 2002). In a kco1 knockout line of Arabidopsis, the current density of SV currents, measured on vacuoles from mesophyll cells, was reduced by a factor of 3 (Schonknecht et al., 2002).

Despite extensive research over the past years, the physiological function of the SV channels is not clear yet. SV channels are usually closed at resting (low) cytosolic $Ca^{2+}$ levels and physiological tonoplast membrane voltages. Because of the activation of SV channels by increasing cytosolic $Ca^{2+}$, this channel has been proposed to play a role in $Ca^{2+}$-induced vacuolar $Ca^{2+}$-release (Ward & Schroeder, 1994) although this concept has been questioned (Pottosin et al., 1997). Recent investigations on SV channels with gating-modulatory effectors, like $Mg^{2+}$ and reducing agents (Carpaneto et al., 1999, 2001; Pei et al., 1999), however, have renewed the discussion of such a role (Bewell et al., 1999).

### 6.3.2.2 Fast-activating vacuolar channel

Fast-activating vacuolar (FV) channels were observed as a second type of vacuolar cation channels (Hedrich & Neher, 1987) that activated without delay by extreme negative or positive voltages. At around (cytosolic) −40 mV, they mainly reside in the closed state (Allen & Sanders, 1996; Tikhonova et al., 1997). In contrast to SV channels, FV channels are active at cytosolic $Ca^{2+}$ levels <100 nM. Therefore, FV channels are proposed to represent the principle passive pathway for cation uptake and release across the vacuolar membrane under resting cytosolic $Ca^{2+}$ conditions. In the sugar beet tonoplast, several fast-activating channels have been characterised with slightly different single-channel amplitudes (Gambale et al., 1996). Thus, it is tempting to speculate that 'the FV channel' in fact comprises a class of several similar but still distinct channels. FV channels poorly select between monovalent cations (Bruggemann et al., 1999c). They are blocked by cytosolic polyamines (Bruggemann et al., 1998), by cytosolic and luminal $Mg^{2+}$ and by luminal $Ca^{2+}$ (Tikhonova et al., 1997; Pei et al., 1999).

### 6.3.2.3 Vacuolar $K^+$ channels

Vacuolar $K^+$ (VK) channels have been observed in vacuoles from Vicia faba guard cells and Beta vulgaris taproots (Ward & Schroeder, 1994; Allen & Sanders, 1996; Pottosin et al., 2003). They are $K^+$-selective, voltage-independent and active at cytosolic $Ca^{2+}$ levels greater than 100 nM. The conductance of VK channels is pH-dependent, with an optimum at around pH 6.5.

### 6.3.3 Plasma membrane calcium channels

The importance of $Ca^{2+}$ as a second messenger in response to a number of stimuli and in regulating the growth of 'tip-growing' systems like pollen tubes and root hairs is beyond dispute (Sanders et al., 1999). Activation of plasma membrane calcium channels and resulting $Ca^{2+}$ influx contributes to the observed increase in cytosolic $Ca^{2+}$ concentration. Direct evidence for calcium channels was provided in the mid 1990s (Thuleau et al., 1994b; Gelli & Blumwald, 1997). To date, hyperpolarisation- and depolarisation-activated calcium channels from a number of cell types have been characterised electrophysiologically (see below). Since the cytosolic $Ca^{2+}$ concentration is buffered in the nanomolar range, the equilibrium potential of $Ca^{2+}$

is positive of zero, so that both types serve as $Ca^{2+}$ influx channels. If charges are not compensated, activation of calcium channels leads to membrane depolarisation.

Calcium channels conduct several divalent ions including $Mg^{2+}$ and $Ba^{2+}$ and are selective for divalent over monovalent ions. This clearly distinguishes them from non-selective ion channels (see Chapter 7), which can also serve as $Ca^{2+}$ entry pathways. $Ba^{2+}$ often permeates even better than $Ca^{2+}$ and is therefore used in experiments. In contrast, non-selective ion channels are inhibited by $Ba^{2+}$. Another difference is the inhibition of calcium channels by $TEA^+$ (Gelli & Blumwald, 1997; Pineros & Tester, 1997; Hamilton *et al.*, 2000, 2001; Kiegle *et al.*, 2000; Pei *et al.*, 2000; Very & Davies, 2000; White, 2000). The mechanism of permeation for $Ca^{2+}$ channels is thought to be different from $K^+$ channels. Selectivity of animal calcium channels is explained by the existence of two specific binding sites for $Ca^{2+}$ in the pore so that $Na^+$ entry is excluded. Ion permeation occurs by electrostatic repulsion between two $Ca^{2+}$ ions. If there is no $Ca^{2+}$ in the pore, the channel becomes much more permeable to $Na^+$ (Sather & McCleskey, 2003). Indeed, depolarisation-activated $Ca^{2+}$ channels from plants conduct monovalent ions in the absence of $Ca^{2+}$ (Pineros & Tester, 1997; White, 1998, 2000).

The molecular structure of plant voltage-gated $Ca^{2+}$ channels is still unknown. In *Arabidopsis thaliana*, a homologue to the two-pore channel (TPC1) from rat has been cloned, and indirect evidence has been given that this channel is voltage-dependent and opens with depolarisation (Furuichi *et al.*, 2001). However, a clear correlation between the molecular structure and calcium currents through depolarisation- and hyperpolarisation-activated $Ca^{2+}$ channels is lacking. For an overview about calcium channels and candidate genes, the reader is referred to White (2000) and White *et al.* (2002).

### 6.3.3.1 *Hyperpolarisation-activated $Ca^{2+}$ channels*

Hyperpolarisation-activated $Ca^{2+}$ channels (HACCs) open at membrane voltages more negative than −100 mV (Fig. 6.10A, HACC; Fig. 6.5, $Q < 0$, $\Delta G_c$). The

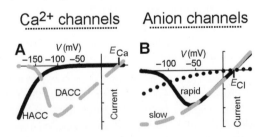

**Fig. 6.10** Current–voltage characteristics of plant ion channels in the steady state. (A) Hyperpolarisation-activated (HACC; *solid line*) and depolarisation-activated $Ca^{2+}$ channels (DACC; *grey dashed line*). (B) Rapid-activating anion channels (QUAC, R-type, GCAC1; *rapid; solid line*), slow-activating anion channels (SLAC, S-type; *slow; grey dashed line*) and inward-rectifying anion channels (*dotted line*).

first HACC was described in tomato (Gelli & Blumwald, 1997). Reports from hyperpolarisation-activated calcium channels in roots, guard cells and mesophyll cells followed (Hamilton et al., 2000, 2001; Kiegle et al., 2000; Pei et al., 2000; Very & Davies, 2000; Stoelzle et al., 2003). The hyperpolarisation-activated $Ca^{2+}$ channels characterised so far are similar with respect to activation threshold, kinetics, selectivity and pharmacology. It is striking that although at negative voltages the open probability is largest, the absolute values remain low. This suggests a strong regulation preventing huge $Ca^{2+}$ influx. Indeed, open probability is suppressed by elevated $Ca^{2+}$ inside (Hamilton et al., 2000, 2001) and is increased dramatically by various stimuli like ABA (abscisic acid), $H_2O_2$, elicitors and blue light (Gelli et al., 1997; Hamilton et al., 2000; Pei et al., 2000; Klusener et al., 2002; Stoelzle et al., 2003). Activation comes along with a shift in the activation threshold towards more positive potentials (see Section 6.4). HACC activation and deactivation kinetics are fast. Mean open times are in the range of milliseconds (Hamilton et al., 2001).

HACCs are involved in $Ca^{2+}$ signalling. The observation in fluorescence-based imaging experiments under voltage control, that a hyperpolarisation-activated $Ca^{2+}$ influx (not a depolarisation-activated one) was important in ABA signalling (Grabov & Blatt, 1998b) and the later identification of such a channel which is activated by ABA (Hamilton et al., 2000; see Section 6.4.7), highlights its role in ABA-mediated cytosolic $Ca^{2+}$ increase. HACC from guard cells is a target for several stimuli like phytohormones and elicitors (Hamilton et al., 2000; Pei et al., 2000; Klusener et al., 2002) and might serve as a focal point, integrating signal transduction pathways and, thus, linking them to membrane voltage (Blatt, 2000b). HACCs also are well suited for nutritive $Ca^{2+}$ uptake. They could maintain a sustained $Ca^{2+}$, $Mg^{2+}$ and $Mn^{2+}$ influx, provided the charge is compensated for. The voltage sensitivity of HACCs and the $H^+$-pump can ensure an effective 'clamp' at hyperpolarisation, preventing a depolarisation of the membrane potential, which in turn would lead to deactivation of HACCs (Miedema et al., 2001). This mechanism is likely in root hairs, where tip growth is associated with an apex-high cytosolic-free calcium gradient generated by a local $Ca^{2+}$ influx (Very & Davies, 2000).

### 6.3.3.2 Depolarisation-activated $Ca^{2+}$ channels

Depolarisation-activated calcium channels (DACCs) have been characterised from suspension cells, mesophyll cells and root cells (Thuleau et al., 1994a,b; Pineros & Tester, 1997; White, 1998, 2000). All of them activate at voltages above $-140$ mV (Fig. 6.10A, DACC; Fig. 6.5, $Q > 0$, $\Delta G_a$) but differ in kinetics with open times ranging from tens and hundreds of milliseconds up to several seconds. Because of their voltage dependence, DACCs are assumed to function in $Ca^{2+}$ uptake during $Ca^{2+}$ signalling, mediating $Ca^{2+}$ spikes and transient depolarisation. The fact that channel activity was detected in few cells only and that measurements were not stable owing to loss of cytosolic factors indicate tight control of DACCs (Thuleau et al., 1994a,b).

### 6.3.4 Vacuolar calcium release channels

In addition to the $Ca^{2+}$-permeable SV channel, several types of $Ca^{2+}$ release channels have been characterised in vacuolar membranes that open over the physiological range of cytosol negative voltages (Fig. 6.5, $Q < 0$; Alexandre *et al.*, 1990; Johannes *et al.*, 1992; Gelli & Blumwald, 1993; Allen & Sanders, 1994). They are permeable for divalent cations ($Ba^{2+}$, $Sr^{2+}$, $Ca^{2+}$, $Mg^{2+}$), and selective for divalents over $K^+$. Physiologically, the gating characteristics and the permeation properties suggest a role of these channels in vacuolar $Ca^{2+}$ release.

### 6.3.5 Calcium channels in the endoplasmatic reticulum

The endoplasmatic reticulum (ER) serves as an intracellular $Ca^{2+}$ pool. Voltage-dependent $Ca^{2+}$ channels, which allow the passage of $Ca^{2+}$ from the lumen of the ER into the cytosol, have been identified by functional reconstitution in planar lipid bilayers from touch-sensitive tendrils (BCC1, Klusener *et al.*, 1995) and from root tips (LCC1, Klusener & Weiler, 1999). $Ca^{2+}$ channels from the ER may play an important role in regulation of the cytosolic-free $Ca^{2+}$ concentration. Their role in $Ca^{2+}$ signalling is supported by the fact that $Gd^{3+}$ inhibits both calcium channel activity and the response of tendrils to touch (Klusener *et al.*, 1995). Furthermore, channel activity is modulated by cytosolic $H_2O_2$ and pH (Klusener *et al.*, 1997).

### 6.3.6 Plasma membrane anion channels

One of the first anion channels identified in higher plants was the quickly activating anion channel from guard cells (Keller *et al.*, 1989). After the establishment of the role of potassium channels for $K^+$ uptake and $K^+$ loss in the opening and closing of stomata, the presence of anion channels had already been postulated, because in processes in which massive salt flux is involved, electroneutrality has to be maintained by an equivalent flow of cations and anions. To date, anion channels are known from various tissues and cell types like guard cells, roots (cortex and stele), hypocotyls and suspension cells (for compilations, see Barbier-Brygoo *et al.*, 2000; Krol & Trebacz, 2000; White & Broadley, 2001). Anion channels can be classified into depolarisation-activated and inward-rectifying anion channels (Fig. 6.10B). Except for the anion channel in wheat roots, which only mediate anion influx (Skerrett & Tyerman, 1994), they serve as anion efflux pathways. Anion channels are highly selective for anions over cations. A larger permeability for $NO_3^-$ than for $Cl^-$ is a common feature of plant anion channels. Some are permeable for malate, citrate and sulphate. Anion channels play a role in volume regulation (as an example stomatal movement was already mentioned), in plant nutrition and in membrane potential control.

Animal voltage-dependent chloride channels (CLCs) show amazing differences to voltage-gated cation channels. CLCs are thought to be homodimers, with each subunit forming a pore, and gating exhibits tight coupling to $Cl^-$ permeation (for a review see, e.g. Pusch, 2004). Plant homologues of CLCs exist, but the ones investigated so far are located in endomembranes (Hechenberger et al., 1996; Barbier-Brygoo et al., 2000). Furthermore, it has been suggested that the slow anion channel from guard cells is an ATP-binding cassette transporter (Leonhardt et al., 1999). However, aside from the existence of some candidate genes, the molecular identity of plant anion channels remains a matter of speculation.

### 6.3.6.1 Depolarisation-activated anion channels

Depolarisation-activated anion channels belong to the best-investigated plant anion channels. Interestingly, the presence of two types of depolarisation-activated anion channels in one cell type is common. The quickly activating anion channel (QUAC; R-type, GCAC1) and the slowly activating anion channel (SLAC; S-type) have been identified in guard cells, xylem parenchyma cells and epidermal cells (Keller et al., 1989; Linder & Raschke, 1992; Schroeder & Keller, 1992; Thomine et al., 1995; Frachisse et al., 2000; Köhler & Raschke, 2000). QUAC (Fig. 6.10B, rapid; Fig. 6.5, $Q > 0$, $\Delta G_b$) is characterised by steep voltage dependence. Activation occurs positive of $-100$ mV and is completed within milliseconds. During sustained depolarisation, QUAC inactivates. A new activation is possible only after QUAC deactivates at negative membrane voltages. In contrast, SLAC (Fig. 6.10B, slow) displays weak voltage dependence. Complete activation and deactivation need seconds. Although there is no proof, it has been proposed that SLAC and QUAC from guard cells represent two switching modes of the same anion channel (Dietrich & Hedrich, 1994; Raschke, 2003; Raschke et al., 2003). SLAC and QUAC are probably dedicated to specific functions: the first channel is able to mediate sustained anion efflux, while the second channel is a good candidate to be involved in fast electrical signalling. Oscillations in the free-running membrane voltage in guard cells were correlated with periodic activations and inactivations of QUAC (Dreyer & Hedrich, 1996; Raschke, 2003; see also Section 6.5).

A dominant theme among depolarisation-activated anion channels is the activation by a rise in cytosolic calcium and the need of nucleotides (Hedrich et al., 1990; Skerrett & Tyerman, 1994; Thomine et al., 1995, 1997; Schulz-Lessdorf et al., 1996; Frachisse et al., 2000). An exception is QUAC from xylem parenchyma cells, which is preferably active at low cytosolic $Ca^{2+}$ concentrations matching its role in xylem loading (Köhler & Raschke, 2000). Raising the cytosolic ATP concentration shifts the activation curve towards less negative potentials (Fig. 6.5, $Q > 0$, $\Delta G_a \rightarrow \Delta G_b$; Colcombet et al., 2001). Other major regulatory mechanisms involve protein phosphorylation and shift of the voltage dependence by auxin and anions (Marten et al., 1991; Hedrich & Marten, 1993; Skerrett & Tyerman, 1994; Zimmermann et al., 1994; Schmidt et al., 1995; Grabov et al., 1997; Dietrich & Hedrich, 1998; Frachisse et al., 2000; Köhler et al., 2002; see also Sections 6.4.6 and 6.4.7).

### 6.3.6.2  Inward-rectifying anion channels

Inward-rectifying anion channels have been described in mesophyll cells, suspension cells and xylem parenchyma cells (Schauf & Wilson, 1987; Terry et al., 1991; Barbara et al., 1994; Elzenga & Van Volkenburgh, 1997; Köhler & Raschke, 2000). This type of channel activates with hyperpolarisation (Fig. 6.10B, dotted line). Voltage dependence can be weak. Activators are high cytosolic $Ca^{2+}$ concentration and light, in which the effect of light could be indirect (Cho & Spalding, 1996; Elzenga & Van Volkenburgh, 1997; Köhler & Raschke, 2000). Inward-rectifying anion channels are considered to provide a path for counter ions, e.g. for $H^+$ during the activity of the $H^+$-ATPase, and to participate in membrane voltage control.

Recently, much attention was paid to a new class of inward-rectifying anion channels in root cells, which are activated by extracellular $Al^{3+}$ possibly indirectly via signalling cascades involving $Ca^{2+}$ or inositol 1,4,5-trisphosphate (Ryan et al., 1997; Kollmeier et al., 2001; Pineros et al., 2002; Zhang et al., 2002). $Al^{3+}$-activated anion channels were proposed to be important in aluminium tolerance, since they mediate $Al^{3+}$-induced release of $Al^{3+}$-chelating ligands (primarily malate and citrate) into the rhizosphere from the root apex.

### 6.3.7  Vacuolar anion channels

In contrast to cation uptake into the vacuole (see introduction to Section 6.3.2), ion channels are well suited to facilitate anion uptake. Chloride and malate are the major charge-balancing anions present in the vacuole.

In the vacuolar membrane of both C3 and CAM (crassulacean acid metabolism) plants, malate-selective channels (VMAL) have been identified (Iwasaki et al., 1992; Pantoja et al., 1992a; Cerana et al., 1995). These channels activate with an instantaneous and a slow, time-dependent component upon (cytosol) negative voltages (Fig. 6.5, $Q < 0$, $\Delta G_a$). VMAL channels are mainly permeable for fumarate$^{2-}$ and malate$^{2-}$, and to a smaller extent for $Cl^-$. VMAL channels are insensitive to changes in cytosolic $Ca^{2+}$ and in the vacuolar proton concentration. However, decreasing cytosolic pH results in strong current reduction very likely caused by a block mechanism (Pantoja & Smith, 2002; Hafke et al., 2003). The distinctive properties of VMAL suggest that this channel is likely to present the principal route for malate$^{2-}$ uptake into the vacuole. Especially in plants performing CAM, the vacuole acts as a temporary repository for large amounts of malic acid synthesised as a result of nocturnal $CO_2$ fixation. In the vacuole, malate$^{2-}$ is chelated or protonated, ensuring that the malate$^{2-}$ gradient across the vacuolar membrane remains directed into the vacuole, which is required for passive transport.

Another $Cl^-$ and malate$^{2-}$ permeable channel (VCL) activated upon (cytosol) negative potentials (Fig. 6.5, $Q < 0$, $\Delta G_a$). Activation was dependent on the phosphorylation status of the channel and was mediated specifically by CDPK, a $Ca^{2+}$-dependent protein kinase (Pei et al., 1996). VCL channels were proposed to provide a pathway for kinase-dependent anion uptake into the vacuole.

## 6.4  Gating modifiers

Research during the past decade revealed that ion channels are part of diverse signalling cascades and/or serve as their final targets (MacRobbie, 1998; Blatt, 2000b; Schroeder *et al.*, 2001). Since most of these signals are transmitted chemically (by direct interaction with other molecules) rather than electrically (by the membrane voltage), the activity of voltage-gated ion channels is, in most cases, additionally modulated by chemical stimuli. The following sections present specific stimuli, and illustrate their modulatory influence on the biophysical gating properties of the target ion channel. In order to focus on the principles of the direct interaction of voltage-gated ion channels with signalling molecules, all possible further upstream signalling steps are mostly left aside. The interested reader is referred to MacRobbie (1998), Blatt (2000b), Schroeder *et al.* (2001) and Assmann (2002) (see also Chapter 8).

### 6.4.1  Phosphorylation

Phosphorylation (adding of $PO_4^{2-}$ to OH groups of, e.g., serines) and dephosphorylation events are all-purpose tools enabling to switch a channel on or off. Mechanistically, this is a rather simple process. Attaching (detaching) of phosphate groups to (from) the channel protein by kinases (phosphatases) changes the stability of the different channel conformations. For phospho-activation in terms of the idealised two-state model, the removal of phosphate groups reduces strongly $\Delta G_{CO}(0)$, in some cases even to negative values (Fig. 6.5, $\Delta G_b \rightarrow \Delta G_d$). Thus, the open configuration is energetically disfavoured. The activity range of the channel is shifted to far more negative ($Q < 0$), or far more positive ($Q > 0$), voltages. The shift may have an extent that, theoretically, very high membrane voltages would be needed to open the channel (Fig. 6.5, $\Delta G_d$). In practice, the channel is closed (switched off) at all times.

Channel control by phosphorylation is a widely observed phenomenon. Several voltage-gated ion channels undergo a rundown upon ATP depletion at the cytosolic face (see, e.g., Schmidt *et al.*, 1995; Moran, 1996; Köhler & Blatt, 2002). Since ATP is the substrate of kinases, ATP depletion disturbs the equilibrium of kinase/phosphatase reactions and results in dephosphorylated channel proteins.

The control of ion channel activity by phosphorylation exposes these transporter proteins as downstream targets in several diverging and converging signalling cascades. However, despite the simple mechanism, the tight control of channel activity by (de)phosphorylation is far more complicated and the understanding of these processes is just in the initial stage. An example is the complex phosphoregulation of guard cell $K_{in}$ channels. On the one hand, they are inhibited by the $Ca^{2+}$-stimulated phosphatase 2B (calcineurin; Luan *et al.*, 1993). On the other hand, the application of okadaic acid and calyculin, inhibitors of protein phosphatases 1 and 2A, strongly reduced $K_{in}$-mediated currents, indicating a negative regulation of $K_{in}$ channels also by phosphorylation events (Li *et al.*, 1994; Thiel & Blatt, 1994).

A similar complexity to (de)phosphorylation control also applies to the SV channel (Allen & Sanders, 1995; Bethke & Jones, 1997).

Phosphorylation is not only used to switch a channel on or off. The different gating modes of weakly rectifying potassium channels, $K_{weak}$, are hypothesised to be determined by the phosphorylation status of the channel protein (Fig. 6.9; Dreyer et al., 2001; Cherel et al., 2002). Additionally, it is discussed that the slow and the rapid guard cell anion channels represent two modes of the same channel, which are distinguished by their phosphorylation status (Dietrich & Hedrich, 1994).

Despite the strong evidence for channel regulation by phosphorylation, the knowledge on enzymes involved in these processes is still poor. To date, a direct interaction with voltage-gated channels was proven for the kinase AAPK and a CDPK (Li & Assmann, 1996; Pei et al., 1996; Li et al., 1998, 2000) and the phosphatase AtPP2CA (Cherel et al., 2002).

### 6.4.2   Nitrosylation and other redox reactions

Analogous to phosphorylation, it has been proposed that channel activity can be controlled via S-nitrosylation of cysteine residues by nitric oxide (NO) or via other redox reactions mediated by, e.g., glutathione or $H_2O_2$. In the case of animal channels, it has been demonstrated that redox compounds regulate important ion channel properties such as channel gating, inactivation and permeation (see references in Carpaneto et al., 1999, and Garcia-Mata et al., 2003). Redox agents are known to affect the ER $Ca^{2+}$ channel BCC1 (Klusener et al., 1997), the SV channel (Carpaneto et al., 1999) and the plasma membrane $K_{in}$ and $K_{out}$ channels in guard cells (Köhler et al., 2003). All are inactivated by oxidising agents. The physiological role of redox regulation is still speculative, but it may have fundamental physiological implications that wait to be uncovered.

Most effects of reactive oxygen species (ROS) on plant ion channels are indirect. In guard cells, for example, it was observed that NO increases the cytosolic $Ca^{2+}$ level via a set of signalling cascades, which in turn has a direct regulatory effect on potassium and chloride channels (Garcia-Mata et al., 2003; for the effect of $Ca^{2+}$, see Section 6.4.3). Additionally, recent reports identified a hyperpolarisation-activated plasma membrane $Ca^{2+}$ channel, which is activated by the oxidising agent $H_2O_2$ indirectly via a phosphatase 2C (Pei et al., 2000; Murata et al., 2001; Köhler et al., 2003).

### 6.4.3   Calcium ions

Calcium is a multifunctional signal transducer in plant cells. Usually, plant cells, like many other cells, maintain their cytosolic-free $Ca^{2+}$ concentration between $10^{-7}$ and $10^{-8}$ M by actively extruding $Ca^{2+}$ from the cytosol towards the vacuole or out of the cell and sequestering it within the ER and mitochondria. Diverse external stimuli (e.g., application of abscisic acid) mobilise $Ca^{2+}$ from different subcellular pools and induce an array of responses including dark-induced inhibition of sucrose synthesis, bending responses to gravity and blue light, turgor regulation and stomatal closure (for review see, e.g., Trewavas & Malho, 1998; Sanders *et al.*, 1999; Blatt, 2000a; Sanders *et al.*, 2002). Some of the responses are mediated by ion channels. In guard cells, for example, increased cytosolic $Ca^{2+}$ levels regulate voltage-gated ion channels in different ways, which results in a depolarisation of the plasma membrane. Transporters that serve for ion accumulation close and those being responsible for ion release open. These are prerequisites for stomatal closure.

$K_{in}$ channels are downregulated upon a rise in cytosolic-free $Ca^{2+}$ (Fig. 6.5, $Q < 0$, $\Delta G_b \rightarrow \Delta G_c \rightarrow \Delta G_d$; Schroeder & Hagiwara, 1989; Grabov & Blatt, 1999). Whether this is a direct effect of $Ca^{2+}$-binding to the channel protein, whether downregulation is mediated via $Ca^{2+}$-dependent kinases/phosphatases or whether other modulatory effectors sensitise $K_{in}$ channels to cytosolic calcium remains to be investigated with cloned $K_{in}$ channels after heterologous expression. It should be emphasised that under certain conditions, $K_{in}$ channels appear to be insensitive to elevated cytosolic $Ca^{2+}$ concentrations (Luan *et al.*, 1993; Armstrong *et al.*, 1995; Wang *et al.*, 1998). $K_{out}$ channels in mesophyll cells also appear to be downregulated by increasing cytosolic $Ca^{2+}$ concentrations (Li & Assmann, 1993). However, in general, guard cell $K_{out}$ channels are known to be $Ca^{2+}$-insensitive (Hosoi *et al.*, 1988; Schroeder & Hagiwara, 1989; Blatt *et al.*, 1990; Blatt & Armstrong, 1993; Lemtiri-Chlieh & MacRobbie, 1994; Grabov & Blatt 1999). By contrast, increased cytosolic $Ca^{2+}$ levels activate both slowly and rapidly activating guard cell anion currents (SLAC, S-type, Schroeder & Hagiwara, 1989; GACA1, QUAC, R-type, Hedrich *et al.*, 1990), apparently by increasing the number of actively gated channels, $N$ (see Equation 6.1).

Control of hyperpolarisation-activated $Ca^{2+}$ channels by cytosolic $Ca^{2+}$ apparently is connected to their physiological function. While the open probability of the HACC from guard cells is suppressed by micromolar cytosolic $Ca^{2+}$, the HACC from root hairs is stimulated. Stimulation results from a shift of the activation threshold towards more positive membrane potentials ($Q < 0$, $\Delta G_c \rightarrow \Delta G_b$; Hamilton *et al.*, 2000; Very & Davies, 2000). Thus, *in vivo*, elevated cytosolic $Ca^{2+}$ may form a negative feedback on $Ca^{2+}$ influx in guard cells in accordance with its role in signalling and a positive feedback in root hairs for continued apical $Ca^{2+}$ influx during tip growth. Interestingly, extracellular divalent cations activate the HACC from guard cells by shifting the voltage dependence towards more positive voltages (Hamilton *et al.*, 2000, 2001).

Cytosolic $Ca^{2+}$ also influences voltage-gated ion channels in endomembranes. The SV channel is activated by cytosolic $Ca^{2+}$. Increasing cytosolic $Ca^{2+}$ from 0.1 to 10 μM shifts the activation threshold of the channel to less positive voltages (Hedrich & Neher, 1987; Fig. 6.5, $Q > 0$, $\Delta G_d \rightarrow \Delta G_c$). Thus, SV channels mostly appear to be closed at physiological tonoplast voltages. Recent studies, however, have indicated that cytosolic $Mg^{2+}$ and reducing agents also serve as gating-modulatory effectors, and thus may sensitise the SV channel to physiological cytosolic $Ca^{2+}$ concentrations. Cytosolic $Mg^{2+}$ stimulates SV channels (Fig. 6.5, $Q > 0$, $\Delta G_d \rightarrow \Delta G_c$; Pei *et al.*, 1999; Carpaneto *et al.*, 2001).

Voltage-gated vacuolar calcium release channels separate into two classes with respect to their susceptibility towards cytosolic calcium. While one class is activated by increasing vacuolar $Ca^{2+}$ (Fig. 6.5, $Q < 0$, $\Delta G_d \rightarrow \Delta G_c \rightarrow \Delta G_b \rightarrow \Delta G_a$; Johannes & Sanders, 1995), however insensitive towards cytosolic $Ca^{2+}$ (Johannes et al., 1992; Allen & Sanders, 1994), the other inactivates when $Ca^{2+}$ reaches 1 $\mu M$ (Gelli & Blumwald, 1993). Also the ER $Ca^{2+}$ channel BCC1 is modulated by $Ca^{2+}$ ions. The channel's open probability is governed largely by the chemical gradient of $Ca^{2+}$. Applying a 100-fold $Ca^{2+}$ gradient, which reflects more the physiological situation, shifted the activation potential towards more negative membrane potentials (Fig. 6.5, $Q > 0$, $\Delta G_c \rightarrow \Delta G_b$; Klusener et al., 1995). The different regulations of endomembrane $Ca^{2+}$ channels may point to a convergence of diverse signalling cascades and a fine-tuning at the level of channel-mediated $Ca^{2+}$-release from subcellular pools.

### 6.4.4 Protons

A physiologically important regulatory trigger for ion channel activity is the proton concentration on one or both sides of the respective membrane. Ion channels can be very sensitive towards changes in the pH. The protonation of a channel protein at susceptible amino acid residues (e.g. histidines) alters its surface charge, which in turn may affect the permeation properties or the stability of certain channel conformations thereby affecting gating.

#### 6.4.4.1 Cytosolic pH changes

The cytosolic pH in plant cells is strongly buffered at values around pH 7.0. Guard cells, for example, have a buffer capacity that corresponds to an apparent buffer concentration of 275 mM with $pK_a = 6.9$ (Grabov & Blatt 1997, 1998a). Nevertheless, external stimuli, e.g. application of hormones like ABA or auxins, are capable of provoking changes in the cytosolic pH in the range of 0.2–0.4 pH units. These apparently moderate changes have dramatic regulatory effects on diverse voltage-gated ion channels. An increase in cytosolic pH activates $K_{out}$ channels and inactivates $K_{in}$ channels (Blatt, 1992; Blatt & Armstrong, 1993; Miedema & Assmann, 1996). Similarly, a decrease in cytosolic pH activates $K_{in}$ channels (Fig. 6.5, $Q < 0$, $\Delta G_c \rightarrow \Delta G_b$; Blatt, 1992; Blatt & Armstrong, 1993; Hoshi, 1995) and inactivates $K_{out}$ channels, SV channels (Fig. 6.5, $Q > 0$, $\Delta G_c \rightarrow \Delta G_d$; Schulz-Lessdorf & Hedrich, 1995) as well as vacuolar voltage-gated calcium release channels (Fig. 6.5, $Q < 0$, $\Delta G_b \rightarrow \Delta G_c \rightarrow \Delta G_d$; Allen & Sanders, 1994). Changes in cytosolic pH appeared not to affect voltage gating of $K_{out}$ channels, but rather the number $N$ (Equation 6.1) of actively gated channels. Whether this effect is the result of a 'switch on/off' shift (Fig. 6.5, $Q > 0$, $\Delta G_d \rightarrow \Delta G_c$; cf. Section 6.4.1) remains to be investigated. Physiologically, the different responses to cytosolic pH adjust the guard cell cation channels for cation uptake when the pH drops and for cation release when the pH increases.

The guard cell anion channel GCAC1 is also regulated by cytosolic pH, namely it is upregulated by a more acidic cytoplasmic pH. More importantly, extracellular acidification slows down the inactivation process of GCAC1 and this modulation depends on the transmembrane pH gradient (Schulz-Lessdorf et al., 1996). Thus, in the presence of a small transmembrane gradient (e.g. in closed stomata when $H^+$-ATPase activity is low), the inactivation process of GCAC1 would be accelerated compared to the presence of a pronounced pH gradient observed in open stomata.

### 6.4.4.2 Extracellular/luminal pH changes

In contrast to the moderate changes in cytosolic pH, the apoplastic pH can vary from pH 7.2 to pH 5.1 (Edwards et al., 1988). The inactivation of guard cell $K_{out}$ channels by lowering pH arises in part from a shift of the activation threshold (Fig. 6.5, $Q > 0$, $\Delta G_b \rightarrow \Delta G_c$; Ilan et al., 1994), but mainly from a decrease in $N$, the number of actively gated channels (Blatt, 1992). Conversely, guard cell $K_{in}$ channels are activated on lowering extracellular pH over this same range and the effect results in a shift of the activation threshold of these channels towards more positive voltages (Fig. 6.5, $Q < 0$, $\Delta G_b \rightarrow \Delta G_a$; Blatt, 1992; Ilan et al., 1994; Hedrich et al., 1995; Mueller-Roeber et al., 1995). In biophysical terms this means that a protonation of specific amino acid residues of the outer part of the channel protein changes the stability of the different channel conformations to favour the open channel, i.e. $\Delta G_{CO}(0)$ increases. In the guard cell potassium channel KST1 from potato, two histidine residues were identified to be involved in the pH modulation (Hoth et al., 1997). Physiologically, the acid activation of $K_{in}$ channels guarantees a less negative membrane potential during long-term proton pump-driven potassium accumulation (Blatt, 1992), and therefore a more efficient use of ATP-hydrolysis energy.

For channels in the vacuolar membrane, the equivalent process to an apoplastic acidification is the decrease of the pH in the vacuolar lumen (physiological range: $3.8 \leq$ vacuolar pH $\leq 6$; Felle, 1988). The activity of the SV channel was reduced at low vacuolar pH values (Fig. 6.5, $Q > 0$, $\Delta G_c \rightarrow \Delta G_d$; Schulz-Lessdorf & Hedrich, 1995). The importance of this regulation can be understood when considering the physiology of vacuolar solute uptake. Solute transport into the vacuole is an active process. Several vacuolar solute transporters are antiporters and use the tonoplastic proton gradient as energy source (Martinoia et al., 2000; Maeshima, 2001). An open cation release channel (SV) would serve as a counteractive shunt pathway in vacuolar solute accumulation.

### 6.4.5 Potassium ions

So far potassium has been presented as a transported substrate. However, in certain cases this permeating ion modulates the activity of its transporting channel by itself. The effect is illustrated here exemplarily on depolarisation-activated $K^+$ channels. The activation threshold of these channels varies with the extracellular potassium concentration (Fig. 6.8B). Thus, depolarisation-activated $K^+$ channels are closed at voltages negative of $E_K$ (Fig. 6.8A), at which the driving force for potassium is directed inwardly (Blatt, 1988, 1991; Schroeder, 1988). Biophysically, this means that potassium ions interact with the channel protein and favour energetically a closed conformation (Blatt & Gradmann, 1997).

Physiologically, the dependency of the activation threshold on the extracellular potassium concentration guarantees that depolarisation-activated $K^+$ channels are active over a wide range of physiological $K^+$ concentrations without providing a pathway for potassium leakage into the cell. Similarly, the $K^+$-dependency of an unusual hyperpolarisation-activated $K^+$ channel of *Arabidopsis* protects root cells from potassium leakage out of the cell when the extracellular $K^+$ concentration is very low (Fig. 6.7D; Maathuis & Sanders, 1995). Usually, hyperpolarisation-activated $K^+$ channels also mediate potassium efflux under unfavourable conditions (Fig. 6.7C; Bruggemann *et al.*, 1999a). Besides the modulation of the gating, another effect is observed when reducing the potassium concentration to extremely low levels (e.g. to nominal zero): some plant $K^+$ channels do not mediate currents in the absence of extracellular potassium. Even though the electrochemical gradient for potassium would favour strong outward currents, nil or only tiny $K^+$ efflux is registered. This observation has fuelled the hypothesis that $K_{in}$ channels also can sense the $K^+$ concentration (Maathuis *et al.*, 1997). However, this type of sensing differs from the sensing mechanism described above for $K_{out}$ channels, in which the gating has been modified. In this case the gating appears to be unaffected. It is rather the pore that adopts a different conformation or even collapses, when the extracellular potassium concentration is lowered (Zhou *et al.*, 2001). At which $K^+$ concentration this collapse occurs apparently differs among the different channels.

Vacuolar potassium modulates the activity of FV channels. Lowering luminal potassium reduces the channel activity at (cytosol) negative voltages (Fig. 6.5, $Q < 0$; $\Delta G_a \rightarrow \Delta G_b \rightarrow \Delta G_c$), whereas the other branch of the bimodal gating ($Q > 0$) remains unaffected. Physiologically, this feedback mechanism likely guarantees that at low vacuolar $K^+$ concentrations no potassium leaks from the cytosol into the $K^+$-starved vacuole (Pottosin & Martinez-Estevez, 2003). This process may contribute to stabilise the cytosolic potassium concentration.

### 6.4.6 Anions

Anions like malate, chloride, nitrate and sulphate modulate the activity of plasma membrane anion channels. Slight increases in the extracellular malate$^{2-}$ concentration in the physiological concentration range shift the activation threshold of the guard cell anion channel GCAC1 towards hyperpolarised potentials (Fig. 6.5, $Q > 0$, $\Delta G_b \rightarrow \Delta G_a$; Hedrich & Marten, 1993; Raschke, 2003). Since cytosolic malate$^{2-}$ proved ineffective, the binding site for shifting the gate must be located on the extracellular face of the channel. Therefore, malate$^{2-}$ activates GCAC1 to feed forward anion release and promote stomatal closure. Since alterations in the ambient $CO_2$ level can modify the extracellular malate$^{2-}$ level, it has been suggested that this mechanism serves as a sensor of the apoplastic $CO_2$ concentration and thus of the metabolic status of the photosynthetic tissue to adjust stomatal aperture in relation to the photosynthetic capacity (Hedrich *et al.*, 1994). Interestingly, extracellular malate$^{2-}$ caused a conversion of the slowly activating anion channel from guard cells to the quickly activating anion channel GCAC1 and initiated a depolarisation of the membrane voltage (Raschke, 2003). Similar to the malate effect, an increase in extracellular $Cl^-$ also shifted the voltage dependence of GCAC1 towards negative potentials (Fig. 6.5, $Q > 0$, $\Delta G_b \rightarrow \Delta G_a$) supporting anion efflux (Dietrich & Hedrich, 1998). Chloride reduces the gate-shifting effect of extracellular malate and therefore seems to compete for the same binding site (Hedrich $et\,al.$, 1994; Dietrich & Hedrich, 1998).

The quickly activating anion conductance (X-QUAC) from root xylem parenchyma cells is a potential control point for $NO_3^-$ release into the xylem (Köhler & Raschke, 2000; Köhler $et\,al.$, 2002). In contrast to GCAC1, X-QUAC appeared to be insensitive to malate$^{2-}$. However, extracellular $NO_3^-$ strongly affected its voltage dependence. Extracellular, but not cytosolic, $NO_3^-$ caused a shift of the activation threshold towards more negative voltages (Fig. 6.5, $Q > 0$, $\Delta G_b \rightarrow \Delta G_a$), increasing the transport capacity of the membrane for $NO_3^-$ and all other permeating ions. X-QUAC may possess a binding site for $NO_3^-$ that is exposed to the apoplast.

Another example for anion regulation of an ion channel is the anion channel from hypocotyl cells. Sulphate is not only a substrate of this channel but has a strong regulatory effect on its activity (Frachisse $et\,al.$, 1999). It prevented channel rundown and shifted the activation potential towards negative potentials (Fig. 6.5, $Q > 0$, $\Delta G_b \rightarrow \Delta G_a$). Remarkably, sulphate acts from the cytosolic site. Channel activation by sulphate is speculated to be part of a mechanism of cellular ion homeostasis, namely to avoid the cytosolic accumulation of $SO_4^{2-}$ when it cannot be extensively metabolised.

Additional to these examples for plasma membrane anion channels, a modification of the gating by $Cl^-$ is also known from the anion channel of vacuoles (Plant $et\,al.$, 1994). Rising vacuolar chloride concentrations increased the open probability of the channel and induced increases in the levels of nitrate and phosphate inward currents. Remarkably, malate currents are reduced. Vacuolar chloride concentrations regulate the influx of anions into the vacuole, favouring the uptake of nutrients for storage, but reducing vacuolar malate uptake, leaving it for use in mitochondrial oxidation and cytosolic pH control (Plant $et\,al.$, 1994).

### 6.4.7   Phytohormones

Plant hormones, like auxins, giberillic acids, cytokinines, abscisic acid, brassinosteroids and ethylene, serve an extremely broad spectrum of plant development and abiotic stress responses. Phytohormones are transported throughout the plant body, or from cell to cell, and reactions are evoked in the perceiving cell(s) through an intracellular signal transduction cascade. The specific 'make-up' of the cell's interior that is required to pass on the hormone signal towards a cellular response is highly diverse, allowing cell- and hormone-specific reactions to occur. The action of hormones may involve protein kinases and phosphatases that alter the phosphorylation status, and hence the activity, of target proteins (see Section 6.4.1). Also phospholipases can be involved (see Section 6.4.8). Cellular reactions may involve transient or sustained, and often rhythmic, increases in the cytosolic $Ca^{2+}$

concentration, which is, for example, a frequent response in ABA-mediated signal transduction.

### 6.4.7.1  Auxins

Hormones of the auxin class influence root initiation, cell expansion and division, apical dominance and responses to light and gravity (for reviews see, e.g., Kaldewey, 1976; Leyser, 2002). Closely connected to auxin-mediated stimuli are changes of the electrical properties of the plasma membrane (Barbier-Brygoo et al., 1991). The effect of auxins on channel activity, however, is mostly indirect, e.g. via auxin-induced transients in the cytosolic pH (Blatt & Thiel, 1993, 1994; Thiel et al., 1993) but also by changing the expression level of channel genes (Philippar et al., 1999). In exceptional contrast, plasma membrane chloride channels of the rapid type (GCAC1) appear to be directly modulated by auxins (Fig. 6.5, $Q > 0$, $\Delta G_b \rightarrow \Delta G_a$) when applied from the extracellular side (Marten et al., 1991).

### 6.4.7.2  Abscisic acid

The hormone $(RS)$-2-cis,4-trans-abscisic acid (ABA) is generally recognised to signal conditions of water stress. It is synthesised during periods of drought and accumulates in the leaf tissues, causing alterations in gene expression and evoking the closure of stomata, thereby reducing transpirational water loss. Directly applied to stomatal guard cells, ABA causes increases in the cytosolic $Ca^{2+}$ concentration and pH via a cascade of signalling events (Gehring et al., 1990; McAinsh et al., 1990; Schroeder & Hagiwara, 1990; Irving et al., 1992; Blatt & Armstrong, 1993). Thus, ABA mainly modulates membrane conductance and therefore channel activity indirectly. The rise in cytosolic-free $Ca^{2+}$ inactivates $K_{in}$ channels and activates anion channels, and the parallel rise in cytosolic pH activates outward-rectifying potassium channels, biasing in combination the plasma membrane for solute efflux and stomatal closure (see MacRobbie, 1998; Blatt, 2000b; Schroeder et al., 2001; and references therein). An initial trigger being responsible for a part of the ABA-induced rise in cytosolic $Ca^{2+}$ might be a hyperpolarisation-activated $Ca^{2+}$ channel in the plasma membrane. This channel activated upon direct application of ABA (Fig. 6.5, $Q < 0$, $\Delta G_d \rightarrow \Delta G_b$; Hamilton et al., 2000). The activation process might occur either by direct binding of ABA to the channel or via modulation by an ABA-sensitive protein that is closely attached to the $Ca^{2+}$ channel. The action of protein phosphatase 1 and 2A antagonists and ABA is remarkably similar, including the displacement of the voltage sensitivity. This indicates that phosphorylation of sites on or close to the $Ca^{2+}$ channel facilitate channel opening (Kohler & Blatt, 2002).

### 6.4.8  Lipids and their hydrolysis products

Lipid signalling is widespread in plants (for reviews see, e.g., Mueller-Roeber & Pical, 2002; Weber, 2002; Bogre et al., 2003). Phospholipases produce phosphatidic acid and choline (in the case of phospholipase D), inositol 1,4,5-trisphophate and diacylglycerol (in the case of phospholipase C) or arachidonate and lysophosphatidylcholine (in the case of phospholipase $A_2$), which then have structural and regulatory functions including ion channel control.

The fatty acids linolenic acid (LN) and arachidonic acid (AA) activate guard cell $K_{in}$ channels and inactivate $K_{out}$ channels. Likewise, LN and AA induce stomatal opening and inhibit stomatal closure (Lee *et al.*, 1994). Other fatty acids, like linoleic acid (LA), have failed to induce a similar response. Whether LN and AA directly act on $K_{in}$ and $K_{out}$ channels or via signalling pathways remains to be investigated.

A voltage-dependent $Ca^{2+}$-release channel in the tonoplast was specifically activated by inositol 1,4,5-trisphosphate ($Ins(1,4,5)P_3$; Fig. 6.5, $Q < 0$, $\Delta G_d \rightarrow \Delta G_c \rightarrow \Delta G_b \rightarrow \Delta G_a$; Alexandre *et al.*, 1990), the product of phospholipase C activity. The channel was insensitive to other polyphosphoinositides including $Ins(1,4)P_2$, $Ins(1,3,4)P_3$, $Ins(2,4,5)P_3$ and $Ins(1,3,4,5)P_4$. This high selectivity towards the agonist may reflect a role of the $Ca^{2+}$-release channel in specific signalling processes while not being involved in others. This $Ins(1,4,5)P_3$-induced $Ca^{2+}$ release into the cytoplasm in turn provokes secondary effects on other voltage-gated channels as, e.g., the inactivation of $K_{in}$ channels (Section 6.4.3; Blatt *et al.*, 1990), and therefore initiates stomatal closure (Gilroy *et al.*, 1990). A similar $Ca^{2+}$-mediated effect on hyperpolarisation-activated guard cell potassium channels was evoked by inositol hexakisphosphate ($InsP_6$). Actually, $InsP_6$ was around 100-fold more potent than $Ins(1,4,5)P_3$ in ($Ca^{2+}$-mediated) modulating $K_{in}$ channel activity (Lemtiri-Chlieh *et al.*, 2000, 2003). Since $InsP_6$ levels were increased upon guard cell stimulation with ABA, these results indicate a role of phosphoinositides in the ABA-induced signalling cascades preceding stomatal closure.

An exciting new finding is that sphingolipids, which play important signalling functions in animals and yeast, also regulate important processes in plants (Worrall *et al.*, 2003). More specifically, it has been discovered that the sphingosine metabolite sphingosine-1-phosphate (S1P) is involved in ABA regulation of the guard cell turgor (Ng *et al.*, 2001). Recently, S1P was shown to inhibit $K_{in}$ channels in *Arabidopsis thaliana* guard cell protoplasts. This inhibition was dependent on the presence of the G-protein α-subunit gene *gpa1* (Coursol *et al.*, 2003). S1P was also able to stimulate slow anion currents in wild-type guard cells, but not in the *gpa1* null mutant (Coursol *et al.*, 2003).

### 6.4.9   Proteins and peptides

In addition to kinases and phosphatases, other proteins and peptides have been identified to modulate the gating behaviour of voltage-gated ion channels. Among them are G-proteins, 14-3-3 proteins and calmodulin.

### 6.4.9.1   G-proteins

Heterotrimeric G-proteins are guanine nucleotide-binding proteins composed of three subunits (α, β, γ). They are generally associated with plasma membrane receptors. Receptor activation stimulates the exchange of GTP for GDP and induces the dissociation into the Gα and Gβγ components. These activated components serve as second messengers, which then in turn activate messengers of higher order (Assmann, 2002). Several reports indicate that G-protein-mediated signalling modulates voltage-gated potassium channels indirectly via cytosolic factors (e.g. changes in free $Ca^{2+}$ concentrations; Fairley-Grenot & Assmann, 1991; Li & Assmann, 1993; Kelly $et\,al.$, 1995). Anion channels are affected by G-proteins as well, probably in an indirect way. Guard cells defective in the sole prototypical heterotrimeric G-protein α-subunit gene lack pH-independent ABA activation of anion channels (Wang $et\,al.$, 2001).

There is evidence for a more direct (cytosol-independent) modulation. Single-channel experiments indicate a membrane-delimited stimulation of ion channels by G-proteins. The activation of G-proteins by the non-hydrolysable GTP analogue guanosine $5'$-$O$-(3-thiotriphosphate) (GTP-γ-S), or by the mastoparan analogue mas7, inhibited guard cell $K_{in}$ channels, favouring stomatal closure (Wu & Assmann, 1994; Fig. 6.5, $Q < 0$, $\Delta G_b \rightarrow \Delta G_c$; Armstrong & Blatt, 1995; Kelly $et\,al.$, 1995). In contrast, certain types of $K_{in}$ channels in xylem parenchyma cells are activated by non-hydrolysable derivatives of GTP, enabling $K^+$ uptake (Wegner & de Boer, 1997). Also hyperpolarisation-activated $Ca^{2+}$ channels were activated by GTP-γ-S and by a recombinant G-protein α-subunit. In fact, GTP-γ-S produced a similar response like a fungal elicitor. Since channel activation by the elicitor was abolished by locking G-proteins in their inactivated state, a role of heterotrimeric G-proteins in response to pathogens has been proposed (Gelli $et\,al.$, 1997; Aharon $et\,al.$, 1998).

### 6.4.9.2   14-3-3 Proteins

14-3-3 proteins are phosphoserine-binding proteins that regulate the activities of a variety of targets via direct protein–protein interactions. Recent reports indicate that 14-3-3 proteins also interact with plant voltage-gated ion channels (for reviews see, e.g., Bunney $et\,al.$, 2002; Roberts, 2003). Overexpression of plant 14-3-3 proteins in tobacco doubled the number of active plant $K_{out}$ channels in mesophyll cells (Saalbach $et\,al.$, 1997). Likewise, the application of recombinant 14-3-3 via the patch-pipette increased the measured potassium efflux currents in tomato cell protoplast (Booij $et\,al.$, 1999). In contrast, the application of recombinant barley 14-3-3B to mesophyll vacuoles resulted in a reduction of SV-channel-mediated currents. The effects on $K_{out}$ and SV channels could not be explained by a regulation of the gating, but rather by an increase/reduction of the number of functional channels. It is concluded that 14-3-3 proteins either interacted directly with $K_{out}$ and SV channels or indirectly via an intermediate protein (van den Wijngaard $et\,al.$, 2001). The physiological implication of this interaction is yet unknown.

### 6.4.9.3   Calmodulin

In plants many $Ca^{2+}$-induced processes are amplified by $Ca^{2+}$-binding proteins like cytosolic or membrane-associated calmodulins (CaMs; for a recent review see, e.g., Luan $et\,al.$, 2002). CaM acts in concert with cytoplamic $Ca^{2+}$ and intriguing evidence suggests that CaM is part of the transduction chain for environmental signals.

The application of calmodulin at high concentrations of about 3.5 μM sensitised the SV channel to lower cytosolic $Ca^{2+}$ concentrations (Bethke & Jones, 1994).

Note that lower calmodulin concentrations ranging from 100 nM to 1 μM were without effect (Allen & Sanders, 1995; Schulz-Lessdorf & Hedrich, 1995). Likewise, calmodulin antagonists applied from the cytosolic side inactivated SV channels via a modulation of the gating (e.g., Fig. 6.5, $Q > 0$, $\Delta G_c \rightarrow \Delta G_d$; Weiser et al., 1991; Bethke & Jones, 1994; Schulz-Lessdorf & Hedrich, 1995). It is still unclear whether CaM antagonists interacted directly with the SV channel or via channel-associated CaMs.

## 6.5   Outlook – voltage-gated ion channels in 'Systems Biology'

The past years were dominated by the separate consideration of isolated voltage-gated ion channels. This approach provided valuable detailed information on the different transporters, facilitating in turn to assign certain physiological roles to them. In some cases the concerted action of different voltage-gated ion channels have provided intuitive models for more complex physiological processes, for example, aspects of long-term ion fluxes in guard cells accompanying stomatal movement (MacRobbie, 1998; Blatt, 2000a; Assmann & Wang, 2001; Schroeder et al., 2001). Additionally, the electro-coupling of plant ion transporters in computer models provided some concepts to explain alterations in the time course of the plasma membrane voltage in guard cells on a shorter timescale (Gradmann et al., 1993; Dreyer & Hedrich, 1996; Gradmann & Hoffstadt, 1998). In the following a simple example is presented, in order to illustrate that computer models may help to understand more complex behaviours.

Oscillations in the free-running membrane potential are observed in guard cells (Fig. 6.11A; Thiel et al., 1992; Gradmann et al., 1993; Blatt & Thiel, 1994; Roelfsema & Prins, 1998). These action potential-like membrane potential oscillations can be simulated in computer models by combining the electrical features of the plasma membrane $H^+$-ATPase, the guard cell $K_{in}$ channel, the guard cell $K_{out}$ channel and the inactivating guard cell anion channel GCAC1 (Dreyer & Hedrich, 1996; Fig. 6.11). The computer model now allows a closer look at the different phases of the action potentials, to separate the contributions of the distinct transporters and to understand their concerted action (Fig. 6.11C, a to f). Initially, the membrane potential is more negative than $-150$ mV. The positive current generated by proton-ATPases, pumping $H^+$ out of the cell, is compensated by $K^+$ influx through open $K_{in}$ channels. This situation is rather stabile and represents the general case when cells accumulate relative high amounts of potassium (e.g. during stomata opening). Some disturbance, e.g. $Ca^{2+}$ influx (Grabov & Blatt, 1998b), can destabilise the equilibrium, resulting in a negative net current, which then in turn induces a rapid membrane depolarisation (Fig. 6.11C, a). In the course of this depolarisation, rapid-activating anion channels GCAC1 activate and $K_{in}$ channels deactivate. The depolarisation stops at a voltage at which the proton current of the $H^+$-ATPases compensates for the chloride efflux through GCAC1 (Fig. 6.11C, b, *white circle*). This condition is semi-stabile, since at these voltages $K_{out}$ channels open with a delay

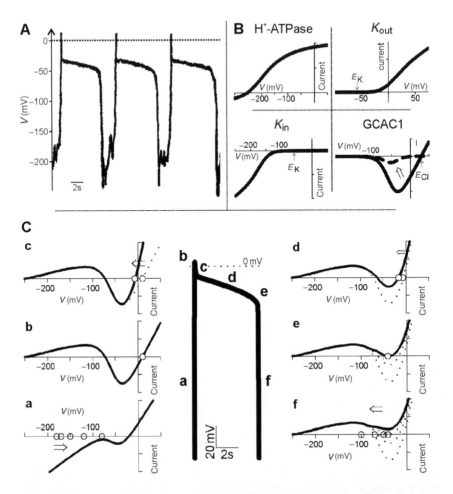

**Fig. 6.11** Action potentials in guard cells. (A) In guard cells, membrane potential oscillations have been observed (data provided by G. Thiel, Darmstadt). (B) In computer simulations, the kinetical features of four guard cell plasma membrane ion transporters were combined: the $H^+$-ATPase (*top left*), the $K_{out}$ channel (*top right*), the $K_{in}$ channel (*bottom left*) and the inactivating rapid anion channel GCAC1 (*bottom right*). (C) The shape of computer-simulated action potentials strongly resembles the measured membrane voltage oscillations (compare C, *middle*, with A). The corresponding current–voltage curves at different phases of an action potential (a to f) are explained in the text.

and provoke a slight negative shift of the membrane potential (Fig. 6.11C, c). Now, a larger $Cl^-$ efflux compensates for both, pumped $H^+$ and $K^+$ efflux. Without alterations in the activity of the ion transporters, this condition would remain stable over several minutes. GCAC1, however, inactivates slowly. In the voltage time course this inactivation is visible as a slow slight negative shift of the membrane potential (Fig. 6.11C, d). The electro-neutral compensation of a reduced $Cl^-$ efflux for $K^+$ efflux drives the membrane potential to voltages at which $K_{out}$ channels also show a lower activity (plateau-phase, Fig. 6.11C, d). The ongoing inactivation of GCAC1 proceeds to an extent that the $Cl^-$ efflux cannot compensate electrically anymore for $H^+$ and $K^+$ efflux (Figs. 6.11C, e and 6.11C, f). In this case, the current that flows across the membrane is positive. The positive net current induces a rapid membrane hyperpolarisation to voltages more negative than $-150$ mV (Fig. 6.11C, f). In the course of this hyperpolarisation, GCAC1 and $K_{out}$ channels deactivate and $K_{in}$ channels activate with a slight delay.

The prediction of the computer model that the inactivating chloride channel plays an important role in membrane potential oscillations was recently supported experimentally (Raschke *et al.*, 2003). Thus, computer models may provide valuable ideas for the understanding of more complex behaviours. At the present time the use of computer models is still an exception. In the future, however, more and detailed computer models are unavoidable. Voltage-gated ion channels, for instance, are entities of 'Ohmic networks', i.e. they are electrically coupled via the transported charges and the voltage-dependent gating. Additionally, the chemical regulation links them to further signalling cascades and even metabolic pathways. Another degree of complexity adds when taking into account that a plant cell is a multicompartment system. The comprehension of this increasing complexity with the increasing knowledge on details is certainly beyond the limits of intuitive qualitative models.

After the reductionistic view in the past, the new beginning epoch in biology research is headlined 'Systems Biology'. Although this terminus is often misinterpreted as the systematic, massive-parallel gathering of data, the aim of systems biology is to understand biological processes as whole systems instead of as isolated parts (Kitano, 2001). This challenging goal can be achieved only by combining experimental, theoretical and modelling techniques. In the past, computational power set limits, and the creation and use of computer models were restricted to researchers with deeper knowledge in programming. Recent developments helped to overcome these restrictions. Now new computational tools are freely accessible to the scientific community, which also allow less experienced computer users to generate really complex models and to perform computational simulations (Tomita, 2001). Electrophysiological simulations can be combined with metabolic and signalling simulations, and even the inclusion of subcompartments does not cause any problem (in contrast to self-made computer-based models). Two modelling environments should be mentioned here explicitly. The first one is the 'E-Cell Project' developed by the Institute for Advanced Biosciences in Japan (http://www.e-cell.org). And the second one is the Virtual Cell developed by the National Resource for Cell Analysis and Modelling (http://www.nrcam.uchc.edu). With these tools, a first detailed computer model for a plant cell comes some large steps closer.

## References

Ache, P., Becker, D., Deeken, R., *et al.* (2001) VFK1, a *Vicia faba* $K^+$ channel involved in phloem unloading, *Plant J.*, **27**, 571–580.

Ache, P., Becker, D., Ivashikina, N., Dietrich, P., Roelfsema, M.R. & Hedrich, R. (2000) GORK, a delayed outward rectifier expressed in guard cells of *Arabidopsis thaliana*, is a K$^+$-selective, K$^+$-sensing ion channel, *FEBS Lett.*, **486**, 93–98.

Aharon, G.S., Gelli, A., Snedden, W.A. & Blumwald, E. (1998) Activation of a plant plasma membrane Ca$^{2+}$ channel by TGalpha1, a heterotrimeric G protein alpha-subunit homologue, *FEBS Lett.*, **424**, 17–21.

Alexandre, J., Lassalles, J.P. & Kado, R.T. (1990) Opening of Ca$^{2+}$ channels in isolated red beet vacuole membrane by inositol 1,4,5-trisphosphate, *Nature*, **343**, 567–570.

Allen, G.J. & Sanders, D. (1994) Two voltage-gated, calcium release channels coreside in the vacuolar membrane of broad bean guard cells, *Plant Cell*, **6**, 685–694.

Allen, G.J. & Sanders, D. (1995) Calcineurin, a type 2B protein phosphatase, modulates the Ca$^{2+}$-permeable slow vacuolar ion channel of stomatal guard cells, *Plant Cell*, **7**, 1473–1483.

Allen, G.J. & Sanders, D. (1996) Control of ionic currents in guard cell vacuoles by cytosolic and luminal calcium, *Plant J.*, **10**, 1055–1069.

Anderson, J.A., Huprikar, S.S., Kochian, L.V., Lucas, W.J. & Gaber, R.F. (1992) Functional expression of a probable *Arabidopsis thaliana* potassium channel in *Saccharomyces cerevisiae*, *Proc. Natl. Acad. Sci. U.S.A.*, **89**, 3736–3740.

Armstrong, F. & Blatt, M.R. (1995) Evidence for K$^+$ channel control in *Vicia* guard cells coupled by G-proteins to a 7TMS receptor mimetic, *Plant J.*, **8**, 187–198.

Armstrong, F., Leung, J., Grabov, A., Brearley, J., Giraudat, J. & Blatt, M.R. (1995) Sensitivity to abscisic acid of guard cell K$^+$ channels is suppressed by *abi1-1*, a mutant *Arabidopsis* gene encoding a putative protein phosphatase, *Proc. Natl. Acad. Sci. U.S.A.*, **92**, 9520–9524.

Assmann, S.M. (2002) Heterotrimeric and unconventional GTP binding proteins in plant cell signaling, *Plant Cell*, **14** (Suppl.), S355–S373.

Assmann, S.M. & Wang, X.Q. (2001) From milliseconds to millions of years: guard cells and environmental responses, *Curr. Opin. Plant Biol.*, **4**, 421–428.

Barbara, J.G., Stoeckel, H. & Takeda, K. (1994) Hyperpolarization-activated inward chloride current in protoplasts from suspension-cultured carrot cells, *Protoplasma*, **180**, 136–144.

Barbier-Brygoo, H., Ephritikhine, G., Klämbt, D., *et al.* (1991) Perception of the auxin signal at the plasma membrane of tobacco mesophyll protoplasts, *Plant J.*, **1**, 83–93.

Barbier-Brygoo, H., Vinauger, M., Colcombet, J., Ephritikhine, G., Frachisse, J.M. & Maurel, C. (2000) Anion channels in higher plants: functional characterization, molecular structure and physiological role, *Biochim. Biophys. Acta*, **1465**, 199–218.

Bauer, C.S., Hoth, S., Haga, K., Philippar, K., Aoki, N. & Hedrich, R. (2000) Differential expression and regulation of K$^+$ channels in the maize coleoptile: molecular and biophysical analysis of cells isolated from cortex and vasculature, *Plant J.*, **24**, 139–145.

Becker, D., Dreyer, I., Hoth, S., *et al.* (1996) Changes in voltage activation, Cs$^+$ sensitivity, and ion permeability in H5 mutants of the plant K$^+$ channel KAT1, *Proc. Natl. Acad. Sci. U.S.A.*, **93**, 8123–8128.

Bertl, A., Blumwald, E., Coronado, R., *et al.* (1992) Electrical measurements on endomembranes, *Science*, **258**, 873–874.

Bethke, P.C. & Jones, R.L. (1994) Ca$^{2+}$-calmodulin modulates ion channel activity in storage protein vacuoles of barley aleurone cells, *Plant Cell*, **6**, 277–285.

Bethke, P.C. & Jones, R.L. (1997) Reversible protein phosphorylation regulates the activity of the slow-vacuolar ion channel, *Plant J.*, **11**, 1227–1235.

Bewell, M.A., Maathuis, F.J., Allen, G.J. & Sanders, D. (1999) Calcium-induced calcium release mediated by a voltage-activated cation channel in vacuolar vesicles from red beet, *FEBS Lett.*, **458**, 41–44.

Blatt, M.R. (1988) Potassium-dependent bipolar gating of potassium channels in guard cells, *J. Membr. Biol.*, **102**, 235–246.

Blatt, M.R. (1991) Ion channel gating in plants: physiological implications and integration for stomatal function, *J. Membr. Biol.*, **124**, 95–112.

Blatt, M.R. (1992) $K^+$ channels of stomatal guard cells. Characteristics of the inward rectifier and its control by pH, *J. Gen. Physiol.*, **99**, 615–644.

Blatt, M.R. (2000a) $Ca^{2+}$ signalling and control of guard-cell volume in stomatal movements, *Curr. Opin. Plant Biol.*, **3**, 196–204.

Blatt, M.R. (2000b) Cellular signaling and volume control in stomatal movements in plants, *Annu. Rev. Cell Dev. Biol.*, **16**, 221–241.

Blatt, M.R. & Armstrong, F. (1993) $K^+$ channels of stomatal guard cells: abscisic-acid-evoked control of the outward rectifier mediated by cytoplasmic pH, *Planta*, **191**, 330–341.

Blatt, M.R. & Gradmann, D. (1997) $K^+$-sensitive gating of the $K^+$ outward rectifier in *Vicia* guard cells, *J. Membr. Biol.*, **158**, 241–256.

Blatt, M.R. & Thiel, G. (1993) Hormonal control of ion channel gating, *Annu. Rev. Plant Physiol. Plant Mol. Biol.*, **44**, 543–567.

Blatt, M.R. & Thiel, G. (1994) $K^+$ channels of stomatal guard cells: bimodal control of the $K^+$ inward-rectifier evoked by auxin, *Plant J.*, **5**, 55–68.

Blatt, M.R., Thiel, G. & Trentham, D.R. (1990) Reversible inactivation of $K^+$ channels of *Vicia* stomatal guard cells following the photolysis of caged inositol 1,4,5-trisphosphate, *Nature*, **346**, 766–769.

Bogre, L., Okresz, L., Henriques, R. & Anthony, R.G. (2003) Growth signalling pathways in *Arabidopsis* and the AGC protein kinases, *Trends Plant Sci.*, **8**, 424–431.

Booij, P.P., Roberts, M.R., Vogelzang, S.A., Kraayenhof, R. & de Boer, A.H. (1999) 14-3-3 proteins double the number of outward-rectifying $K^+$ channels available for activation in tomato cells, *Plant J.*, **20**, 673–683.

Bruggemann, L.I., Dietrich, P., Becker, D., Dreyer, I., Palme, K. & Hedrich, R. (1999a) Channel-mediated high-affinity $K^+$ uptake into guard cells from *Arabidopsis*, *Proc. Natl. Acad. Sci. U.S.A.*, **96**, 3298–3302.

Bruggemann, L.I., Dietrich, P., Dreyer, I. & Hedrich, R. (1999b) Pronounced differences between the native $K^+$ channels and KAT1 and KST1 alpha-subunit homomers of guard cells, *Planta*, **207**, 370–376.

Bruggemann, L.I., Pottosin, I.I. & Schonknecht, G. (1999c) Selectivity of the fast activating vacuolar cation channel, *J. Exp. Bot.*, **50**, 873–876.

Bruggemann, L.I., Pottosin, I.I. & Schonknecht, G. (1998) Cytoplasmic polyamines block the fast-activating vacuolar cation channel, *Plant J.*, **16**, 101–105.

Bunney, T.D., van den Wijngaard, P.W., & de Boer, A.H. (2002) 14-3-3 protein regulation of proton pumps and ion channels, *Plant Mol. Biol.*, **50**, 1041–1051.

Carpaneto, A., Cantu, A.M. & Gambale, F. (1999) Redox agents regulate ion channel activity in vacuoles from higher plant cells, *FEBS Lett.*, **442**, 129–132.

Carpaneto, A., Cantu, A.M. & Gambale, F. (2001) Effects of cytoplasmic $Mg^{2+}$ on slowly activating channels in isolated vacuoles of *Beta vulgaris*, *Planta*, **213**, 457–468.

Cerana, R., Giromini, L. & Colombo, R. (1995) Malate-regulated channels permeable to anions in vacuoles of *Arabidopsis thaliana*, *Aust. J. Plant Physiol.*, **22**, 115–121.

Cherel, I., Michard, E., Platet, N., *et al.* (2002) Physical and functional interaction of the *Arabidopsis* $K^+$ channel AKT2 and phosphatase AtPP2CA, *Plant Cell*, **14**, 1133–1146.

Cho, M.H. & Spalding, E.P. (1996) An anion channel in *Arabidopsis* hypocotyls activated by blue light, *Proc. Natl. Acad. Sci. U.S.A.*, **93**, 8134–8138.

Colcombet, J., Thomine, S., Guern, J., Frachisse, J.M. & Barbier-Brygoo, H. (2001) Nucleotides provide a voltage-sensitive gate for the rapid anion channel of *Arabidopsis* hypocotyl cells, *J. Biol. Chem.*, **276**, 36139–36145.

Colombo, R., Cerana, R., Lado, P. & Peres, A. (1988) Voltage-dependent channels permeable to $K^+$ and $Na^+$ in the membrane of *Acer pseudoplatanus* vacuoles, *J. Membr. Biol.*, **103**, 227–236.

Colquhoun, D. & Hawkes, A.G. (1995) The principles of the stochastic interpretation of ion-channel mechanisms, in *Single-Channel Recording*, 2nd edn (eds B. Sakmann & E. Neher), Plenum Press, New York, pp. 397–482.

Coursol, S., Fan, L.M., Le Stunff, H., Spiegel, S., Gilroy, S. & Assmann, S.M. (2003) Sphingolipid signalling in *Arabidopsis* guard cells involves heterotrimeric G proteins, *Nature*, **423**, 651–654.

Czempinski, K., Frachisse, J.M., Maurel, C., Barbier-Brygoo, H. & Mueller-Roeber, B. (2002) Vacuolar membrane localization of the *Arabidopsis* 'two-pore' $K^+$ channel KCO1, *Plant J.*, **29**, 809–820.

Daram, P., Urbach, S., Gaymard, F., Sentenac, H. & Cherel, I. (1997) Tetramerization of the AKT1 plant potassium channel involves its C-terminal cytoplasmic domain, *EMBO J.*, **16**, 3455–3463.

Deeken, R., Geiger, D., Fromm, J., *et al.* (2002) Loss of the AKT2/3 potassium channel affects sugar loading into the phloem of *Arabidopsis*, *Planta*, **216**, 334–344.

Dietrich, P., Dreyer, I., Wiesner, P. & Hedrich, R. (1998) Cation sensitivity and kinetics of guard-cell potassium channels differ among species, *Planta*, **205**, 277–287.

Dietrich, P. & Hedrich, R. (1994) Interconversion of fast and slow gating modes of GCAC1, a guard-cell anion channel, *Planta*, **195**, 301–304.

Dietrich, P. & Hedrich, R. (1998) Anions permeate and gate GCAC1, a voltage-dependent guard cell anion channel, *Plant J.*, **15**, 479–487.

Doyle, D.A., Morais, C.J., Pfuetzner, R.A., *et al.* (1998) The structure of the potassium channel: molecular basis of $K^+$ conduction and selectivity, *Science*, **280**, 69–77.

Dreyer, I., Antunes, S., Hoshi, T., *et al.* (1997) Plant $K^+$ channel alpha-subunits assemble indiscriminately, *Biophys. J.*, **72**, 2143–2150.

Dreyer, I., Becker, D., Bregante, M., *et al.* (1998) Single mutations strongly alter the $K^+$-selective pore of the $K_{in}$ channel KAT1, *FEBS Lett.*, **430**, 370–376.

Dreyer, I. & Hedrich, R. (1996) Action potentials in guard cells – the molecular basis of oscillations, Botanikertagung, 25.8.–31.8., Düsseldorf, Germany, p. P-9.060.

Dreyer, I., Michard, E., Lacombe, B. & Thibaud, J.B. (2001) A plant Shaker-like $K^+$ channel switches between two distinct gating modes resulting in either inward-rectifying or 'leak' current, *FEBS Lett.*, **505**, 233–239.

Edwards, M.C., Smith, G.N. & Bowling, D.J.F. (1988) Guard cells extrude protons prior to stomatal opening. A study using fluorescence microscopy and pH-micro-electrode, *J. Exp. Bot.*, **39**, 1541–1547.

Elzenga, J.T. & Van Volkenburgh, E. (1997) Kinetics of $Ca^{2+}$- and ATP-dependent, voltage-controlled anion conductance in the plasma membrane of mesophyll cells of *Pisum sativum*, *Planta*, **201**, 415–423.

Fairley-Grenot, K.A. & Assmann, S.M. (1991) Evidence for G-protein regulation of inward $K^+$ channel current in guard cells of *Fava* bean, *Plant Cell*, **3**, 1037–1044.

Felle, H.H. (1988) Short-term pH regulation in plants, *Physiol. Plant.*, **74**, 583–591.

Frachisse, J.M., Colcombet, J., Guern, J. & Barbier-Brygoo, H. (2000) Characterization of a nitrate-permeable channel able to mediate sustained anion efflux in hypocotyl cells from *Arabidopsis thaliana*, *Plant J.*, **21**, 361–371.

Frachisse, J.M., Thomine, S., Colcombet, J., Guern, J. & Barbier-Brygoo, H. (1999) Sulfate is both a substrate and an activator of the voltage-dependent anion channel of *Arabidopsis* hypocotyl cells, *Plant Physiol.*, **121**, 253–262.

Furuichi, T., Cunningham, K.W. & Muto, S. (2001) A putative two pore channel AtTPC1 mediates $Ca^{2+}$ flux in *Arabidopsis* leaf cells, *Plant Cell Physiol.*, **42**, 900–905.

Gambale, F., Bregante, M., Stragapede, F. & Cantu, A.M. (1996) Ionic channels of the sugar beet tonoplast are regulated by a multi-ion single-file permeation mechanism, *J. Membr. Biol.*, **154**, 69–79.

Garcia-Mata, C., Gay, R., Sokolovski, S., Hills, A., Lamattina, L. & Blatt, M.R. (2003) Nitric oxide regulates $K^+$ and $Cl^-$ channels in guard cells through a subset of abscisic acid-evoked signaling pathways, *Proc. Natl. Acad. Sci. U.S.A.*, **100**, 11116–11121.

Gaxiola, R.A., Fink, G.R. & Hirschi, K.D. (2002) Genetic manipulation of vacuolar proton pumps and transporters, *Plant Physiol.*, **129**, 967–973.

Gaymard, F., Pilot, G., Lacombe, B., *et al.* (1998) Identification and disruption of a plant Shaker-like outward channel involved in $K^+$ release into the xylem sap, *Cell*, **94**, 647–655.

Gehring, C.A., Irving, H.R. & Parish, R.W. (1990) Effects of auxin and abscisic acid on cytosolic calcium and pH in plant cells, *Proc. Natl. Acad. Sci. U.S.A.*, **87**, 9645–9649.

Gelli, A. & Blumwald, E. (1993) Calcium retrieval from vacuolar pools (characterization of a vacuolar calcium channel), *Plant Physiol.*, **102**, 1139–1146.

Gelli, A. & Blumwald, E. (1997) Hyperpolarization-activated $Ca^{2+}$-permeable channels in the plasma membrane of tomato cells, *J. Membr. Biol.*, **155**, 35–45.

Gelli, A., Higgins, V.J. & Blumwald, E. (1997) Activation of plant plasma membrane $Ca^{2+}$-permeable channels by race-specific fungal elicitors, *Plant Physiol.*, **113**, 269–279.

Gilroy, S., Read, N.D. & Trewavas, A.J. (1990) Elevation of cytoplasmic calcium by caged calcium or caged inositol triphosphate initiates stomatal closure, *Nature*, **346**, 769–771.

Glasstone, S., Laidler, K.J. & Eyring, H. (1941) *The Theory of Rate Processes*, McGraw-Hill, New York.

Grabov, A. & Blatt, M.R. (1997) Parallel control of the inward-rectifier $K^+$ channel by cytosolic-free $Ca^{2+}$ and pH in *Vicia* guard cells, *Planta*, **201**, 84–95.

Grabov, A. & Blatt, M.R. (1998a) Co-ordination of signalling elements in guard cell ion channel control, *J. Exp. Bot.*, **49**, 351–360.

Grabov, A. & Blatt, M.R. (1998b) Membrane voltage initiates $Ca^{2+}$ waves and potentiates $Ca^{2+}$ increases with abscisic acid in stomatal guard cells, *Proc. Natl. Acad. Sci. U.S.A.*, **95**, 4778–4783.

Grabov, A. & Blatt, M.R. (1999) A steep dependence of inward-rectifying potassium channels on cytosolic free calcium concentration increase evoked by hyperpolarization in guard cells, *Plant Physiol.*, **119**, 277–288.

Grabov, A., Leung, J., Giraudat, J. & Blatt, M.R. (1997) Alteration of anion channel kinetics in wild-type and *abi1-1* transgenic *Nicotiana benthamiana* guard cells by abscisic acid, *Plant J.*, **12**, 203–213.

Gradmann, D., Blatt, M.R. & Thiel, G. (1993) Electrocoupling of ion transporters in plants, *J. Membr. Biol.*, **136**, 327–332.

Gradmann, D. & Hoffstadt, J. (1998) Electrocoupling of ion transporters in plants: interaction with internal ion concentrations, *J. Membr. Biol.*, **166**, 51–59.

Hafke, J.B., Hafke, Y., Smith, J.A., Luttge, U. & Thiel, G. (2003) Vacuolar malate uptake is mediated by an anion-selective inward rectifier, *Plant J.*, **35**, 116–128.

Hamilton, D.W., Hills, A. & Blatt, M.R. (2001) Extracellular $Ba^{2+}$ and voltage interact to gate $Ca^{2+}$ channels at the plasma membrane of stomatal guard cells, *FEBS Lett.*, **491**, 99–103.

Hamilton, D.W., Hills, A., Kohler, B. & Blatt, M.R. (2000) $Ca^{2+}$ channels at the plasma membrane of stomatal guard cells are activated by hyperpolarization and abscisic acid, *Proc. Natl. Acad. Sci. U.S.A.*, **97**, 4967–4972.

Hechenberger, M., Schwappach, B., Fischer, W.N., Frommer, W.B., Jentsch, T.J. & Steinmeyer, K. (1996) A family of putative chloride channels from *Arabidopsis* and functional complementation of a yeast strain with a CLC gene disruption, *J. Biol. Chem.*, **271**, 33632–33638.

Hedrich, R., Busch, H. & Raschke, K. (1990) $Ca^{2+}$ and nucleotide dependent regulation of voltage dependent anion channels in the plasma membrane of guard cells, *EMBO J.*, **9**, 3889–3892.

Hedrich, R. & Kurkdjian, A. (1988) Characterization of an anion-permeable channel from sugar beet vacuoles: effect of inhibitors, *EMBO J.*, **7**, 3661–3666.

Hedrich, R. & Marten, I. (1993) Malate-induced feedback regulation of plasma membrane anion channels could provide a $CO_2$ sensor to guard cells, *EMBO J.*, **12**, 897–901.

Hedrich, R., Marten, I., Lohse, G., *et al.* (1994) Malate-sensitive anion channels enable guard cells to sense changes in the ambient $CO_2$ concentration, *Plant J.*, **6**, 741–748.

Hedrich, R., Moran, O., Conti, F., *et al.* (1995) Inward rectifier potassium channels in plants differ from their animal counterparts in response to voltage and channel modulators, *Eur. Biophys. J.*, **24**, 107–115.

Hedrich, R. & Neher, E. (1987) Cytoplasmic calcium regulates voltage-dependent ion channels in plant vacuoles, *Nature*, **329**, 833–835.

Hodgkin, A.L. & Huxley, A.F. (1952) A quantitative description of membrane current and its application to conduction and excitation in nerve, *J. Physiol.*, **117**, 500–544.

Hoshi, T. (1995) Regulation of voltage dependence of the KAT1 channel by intracellular factors, *J. Gen. Physiol.*, **105**, 309–328.

Hosoi, S., Lino, M. & Shimozaki, K. (1988) Outward-rectifying $K^+$ channels in stomatal guard cell protoplasts, *Plant Cell Physiol.*, **29**, 907–911.

Hosy, E., Vavasseur, A., Mouline, K., *et al.* (2003) The *Arabidopsis* outward $K^+$ channel GORK is involved in regulation of stomatal movements and plant transpiration, *Proc. Natl. Acad. Sci. U.S.A.*, **100**, 5549–5554.

Hoth, S., Dreyer, I., Dietrich, P., Becker, D., Mueller-Roeber, B. & Hedrich, R. (1997) Molecular basis of plant-specific acid activation of $K^+$ uptake channels, *Proc. Natl. Acad. Sci. U.S.A.*, **94**, 4806–4810.

Ilan, N., Schwartz, A. & Moran, N. (1994) External pH effects on the depolarization-activated K channels in guard cell protoplasts of *Vicia faba*, *J. Gen. Physiol.*, **103**, 807–831.

Irving, H.R., Gehring, C.A. & Parish, R.W. (1992) Changes in cytosolic pH and calcium of guard cells precede stomatal movements, *Proc. Natl. Acad. Sci. U.S.A.*, **89**, 1790–1794.

Iwasaki, I., Arata, H., Kijima, H. & Nishimura, M. (1992) Two types of channels involved in the malate ion transport across the tonoplast of a crassulacean acid metabolism plant, *Plant Physiol.*, **98**, 1494–1497.

Jiang, Y., Lee, A., Chen, J., *et al.* (2003) X-ray structure of a voltage-dependent $K^+$ channel, *Nature*, **423**, 33–41.

Johannes, E., Brosnan, J.M. & Sanders, D. (1992) Parallel pathways for intracellular $Ca^{2+}$ release from the vacuole of higher plants, *Plant J.*, **2**, 97–102.

Johannes, E. & Sanders, D. (1995) Lumenal calcium modulates unitary conductance and gating of a plant vacuolar calcium release channel, *J. Membr. Biol.*, **146**, 211–224.

Kaldewey, H. (1976) Considerations of geotropism in plants, *Life Sci. Space Res.*, **14**, 21–36.

Keller, B.U., Hedrich, R. & Raschke, K. (1989) Voltage-dependent anion channels in the plasma-membrane of guard-cells, *Nature*, **341**, 450–453.

Kelly, W.B., Esser, J.E. & Schroeder, J.I. (1995) Effects of cytosolic calcium and limited, possible dual, effects of G protein modulators on guard cell inward potassium channels, *Plant J.*, **8**, 479–489.

Kiegle, E., Gilliham, M., Haseloff, J. & Tester, M. (2000) Hyperpolarisation-activated calcium currents found only in cells from the elongation zone of *Arabidopsis thaliana* roots, *Plant J.*, **21**, 225–229.

Kitano, H. (2001) *Foundations of Systems Biology*, 1st edn, The MIT Press, Cambridge, MA.

Klusener, B., Boheim, G., Liss, H., Engelberth, J. & Weiler, E.W. (1995) Gadolinium-sensitive, voltage-dependent calcium release channels in the endoplasmic reticulum of a higher plant mechanoreceptor organ, *EMBO J.*, **14**, 2708–2714.

Klusener, B., Boheim, G. & Weiler, E.W. (1997) Modulation of the ER $Ca^{2+}$ channel BCC1 from tendrils of *Bryonia dioica* by divalent cations, protons and $H_2O_2$, *FEBS Lett.*, **407**, 230–234.

Klusener, B. & Weiler, E.W. (1999) A calcium-selective channel from root-tip endomembranes of garden cress, *Plant Physiol.*, **119**, 1399–1406.

Klusener, B., Young, J.J., Murata, Y., *et al.* (2002) Convergence of calcium signaling pathways of pathogenic elicitors and abscisic acid in *Arabidopsis* guard cells, *Plant Physiol.*, **130**, 2152–2163.

Köhler, B. & Blatt, M.R. (2002) Protein phosphorylation activates the guard cell $Ca^{2+}$ channel and is a prerequisite for gating by abscisic acid, *Plant J.*, **32**, 185–194.

Köhler, B., Hills, A. & Blatt, M.R. (2003) Control of guard cell ion channels by hydrogen peroxide and abscisic acid indicates their action through alternate signaling pathways, *Plant Physiol.*, **131**, 385–388.

Köhler, B. & Raschke, K. (2000) The delivery of salts to the xylem. Three types of anion conductance in the plasmalemma of the xylem parenchyma of roots of barley, *Plant Physiol.*, **122**, 243–254.

Köhler, B., Wegner, L.H., Osipov, V. & Raschke, K. (2002) Loading of nitrate into the xylem: apoplastic nitrate controls the voltage dependence of X-QUAC, the main anion conductance in xylem-parenchyma cells of barley roots, *Plant J.*, **30**, 133–142.

Kolb, H.A., Kohler, K. & Martinoia, E. (1987) Single potassium in the membrane of isolated mesophyll barley vacuoles, *J. Membr. Biol.*, **95**, 163–169.

Kollmeier, M., Dietrich, P., Bauer, C.S., Horst, W.J. & Hedrich, R. (2001) Aluminum activates a citrate-permeable anion channel in the aluminum-sensitive zone of the maize root apex. A comparison between an aluminum-sensitive and an aluminum-resistant cultivar, *Plant Physiol.*, **126**, 397–410.

Krawczyk, S. (1978) Ionic channel formation in a living cell membrane, *Nature*, **273**, 56–57.

Krol, E. & Trebacz, K. (2000) Ways of ion channel gating in plant cells, *Ann. Bot. (Lond.)*, **86**, 449–469.

Lacombe, B., Pilot, G., Michard, E., Gaymard, F., Sentenac, H. & Thibaud, J.B. (2000) A Shaker-like K$^+$ channel with weak rectification is expressed in both source and sink phloem tissues of *Arabidopsis*, *Plant Cell*, **12**, 837–851.

Lagarde, D., Basset, M., Lepetit, M., *et al.* (1996) Tissue-specific expression of *Arabidopsis* AKT1 gene is consistent with a role in K$^+$ nutrition, *Plant J.*, **9**, 195–203.

Lee, Y., Lee, H.J., Crain, R.C., Lee, A. & Korn, S.J. (1994) Polyunsaturated fatty acids modulate stomatal aperture and two distinct K$^+$ channel currents in guard cells, *Cell. Signal.*, **6**, 181–186.

Leigh, R.A. & Wyn Jones, R.G. (1984) A hypothesis relating critical potassium concentrations for growth to the distribution and functions of this ion in the plant cell, *New Phytol.*, **97**, 1–13.

Lemtiri-Chlieh, F. & MacRobbie, E.A. (1994) Role of calcium in the modulation of *Vicia* guard cell potassium channels by abscisic acid: a patch-clamp study, *J. Membr. Biol.*, **137**, 99–107.

Lemtiri-Chlieh, F., MacRobbie, E.A. & Brearley, C.A. (2000) Inositol hexakisphosphate is a physiological signal regulating the K$^+$-inward rectifying conductance in guard cells, *Proc. Natl. Acad. Sci. U.S.A.*, **97**, 8687–8692.

Lemtiri-Chlieh, F., MacRobbie, E.A., Webb, A.A., *et al.* (2003) Inositol hexakisphosphate mobilizes an endomembrane store of calcium in guard cells, *Proc. Natl. Acad. Sci. U.S.A.*, **100**, 10091–10095.

Leonhardt, N., Vavasseur, A. & Forestier, C. (1999) ATP binding cassette modulators control abscisic acid-regulated slow anion channels in guard cells, *Plant Cell*, **11**, 1141–1152.

Leyser, O. (2002) Molecular genetics of auxin signaling, *Annu. Rev. Plant Biol.*, **53**, 377–398.

Li, J. & Assmann, S.M. (1996) An abscisic acid-activated and calcium-independent protein kinase from guard cells of *Fava* bean, *Plant Cell*, **8**, 2359–2368.

Li, J., Lee, Y.R. & Assmann, S.M. (1998) Guard cells possess a calcium-dependent protein kinase that phosphorylates the KAT1 potassium channel, *Plant Physiol.*, **116**, 785–795.

Li, J., Wang, X.Q., Watson, M.B. & Assmann, S.M. (2000) Regulation of abscisic acid-induced stomatal closure and anion channels by guard cell AAPK kinase, *Science*, **287**, 300–303.

Li, W. & Assmann, S.M. (1993) Characterization of a G-protein-regulated outward K$^+$ current in mesophyll cells of *Vicia faba* L., *Proc. Natl. Acad. Sci. U.S.A.*, **90**, 262–266.

Li, W., Luan, S., Schreiber, S.L. & Assmann, S.M. (1994) Evidence for protein phosphatase 1 and 2A regulation of K$^+$ channels in two types of leaf cells, *Plant Physiol.*, **106**, 963–970.

Linder, B. & Raschke, K. (1992) A slow anion channel in guard-cells, activating at large hyperpolarization, may be principal for stomatal closing, *FEBS Lett.*, **313**, 27–30.

Luan, S., Kudla, J., Rodriguez-Concepcion, M., Yalovsky, S. & Gruissem, W. (2002) Calmodulins and calcineurin B-like proteins: calcium sensors for specific signal response coupling in plants, *Plant Cell*, **14** (Suppl.), S389–S400.

Luan, S., Li, W., Rusnak, F., Assmann, S.M. & Schreiber, S.L. (1993) Immunosuppressants implicate protein phosphatase regulation of K$^+$ channels in guard-cells, *Proc. Natl. Acad. Sci. U.S.A.*, **90**, 2202–2206.

Luttge, U., Pfeifer, T., Fischer-Schliebs, E. & Ratajczak, R. (2000) The role of vacuolar malate-transport capacity in crassulacean acid metabolism and nitrate nutrition. Higher malate-transport capacity in ice plant after crassulacean acid metabolism-induction and in tobacco under nitrate nutrition, *Plant Physiol.*, **124**, 1335–1348.

Maathuis, F.J., Ichida, A.M., Sanders, D. & Schroeder, J.I. (1997) Roles of higher plant K$^+$ channels, *Plant Physiol.*, **114**, 1141–1149.

Maathuis, F.J. & Sanders, D. (1995) Contrasting roles in ion transport of two K$^+$-channel types in root cells of *Arabidopsis thaliana*, *Planta*, **197**, 456–464.

MacRobbie, E.A. (1998) Signal transduction and ion channels in guard cells, *Philos. Trans. R. Soc. Lond. B Biol. Sci.*, **353**, 1475–1488.

Maeshima, M. (2001) Tonoplast transporters: organization and function, *Annu. Rev. Plant Physiol. Plant Mol. Biol.*, **52**, 469–497.

Marten, I., Hoth, S., Deeken, R., *et al.* (1999) AKT3, a phloem-localized $K^+$ channel, is blocked by protons, *Proc. Natl. Acad. Sci. U.S.A.*, **96**, 7581–7586.

Marten, I., Lohse, G. & Hedrich, R. (1991) Plant growth hormones control voltage-dependent activity of anion channels in plasma membrane of guard cells, *Nature*, **353**, 758–762.

Martinoia, E., Massonneau, A. & Frangne, N. (2000) Transport processes of solutes across the vacuolar membrane of higher plants, *Plant Cell Physiol.*, **41**, 1175–1186.

McAinsh, M.R., Brownlee, C. & Hetherington, A.M. (1990) Abscisic acid-induced elevation of guard cell cytosolic $Ca^{2+}$ precedes stomatal closure, *Nature*, **343**, 186–188.

Miedema, H. & Assmann, S.M. (1996) A membrane-delimited effect of internal pH on the $K^+$ outward rectifier of *Vicia faba* guard cells, *J. Membr. Biol.*, **154**, 227–237.

Miedema, H., Bothwell, J.H., Brownlee, C. & Davies, J.M. (2001) Calcium uptake by plant cells – channels and pumps acting in concert, *Trends Plant Sci.*, **6**, 514–519.

Miedema, H. & Pantoja, O. (2001) Anion modulation of the slowly activating vacuolar channel, *J. Membr. Biol.*, **183**, 137–145.

Moran, N. (1996) Membrane-delimited phosphorylation enables the activation of the outward-rectifying K channels in motor cell protoplasts of *Samanea saman*, *Plant Physiol.*, **111**, 1281–1292.

Moran, N., Ehrenstein, G., Iwasa, K., Bare, C. & Mischke, C. (1984) Ion channels in plasmalemma of wheat protoplasts, *Science*, **226**, 835–838.

Mouline, K., Very, A.A., Gaymard, F., *et al.* (2002) Pollen tube development and competitive ability are impaired by disruption of a Shaker K(+) channel in *Arabidopsis*, *Genes Dev.*, **16**, 339–350.

Mueller-Roeber, B., Ellenberg, J., Provart, N., *et al.* (1995) Cloning and electrophysiological analysis of KST1, an inward rectifying $K^+$ channel expressed in potato guard cells, *EMBO J.*, **14**, 2409–2416.

Mueller-Roeber, B. & Pical, C. (2002) Inositol phospholipid metabolism in *Arabidopsis*. Characterized and putative isoforms of inositol phospholipid kinase and phosphoinositide-specific phospholipase C, *Plant Physiol.*, **130**, 22–46.

Murata, Y., Pei, Z.M., Mori, I.C. & Schroeder, J.I. (2001) Abscisic acid activation of plasma membrane $Ca^{2+}$ channels in guard cells requires cytosolic NAD(P)H and is differentially disrupted upstream and downstream of reactive oxygen species production in *abi1-1* and *abi2-1* protein phosphatase 2C mutants, *Plant Cell*, **13**, 2513–2523.

Nakamura, R.L., Anderson, J.A. & Gaber, R.F. (1997) Determination of key structural requirements of a $K^+$ channel pore, *J. Biol. Chem.*, **272**, 1011–1018.

Nakamura, R.L., McKendree, W.L., Hirsch, R.E., Sedbrook, J.C., Gaber, R.F. & Sussman, M.R. (1995) Expression of an *Arabidopsis* potassium channel gene in guard-cells, *Plant Physiol.*, **109**, 371–374.

Neher, E. & Sakmann, B. (1976) Single-channel currents recorded from membrane of denervated frog muscle fibres, *Nature*, **260**, 799–802.

Ng, C.K., Carr, K., McAinsh, M.R., Powell, B. & Hetherington, A.M. (2001) Drought-induced guard cell signal transduction involves sphingosine-1-phosphate, *Nature*, **410**, 596–599.

Noda, M., Shimizu, S., Tanabe, T., *et al.* (1984) Primary structure of *Electrophorus electricus* sodium channel deduced from cDNA sequence, *Nature*, **312**, 121–127.

Pantoja, O., Gelli, A. & Blumwald, E. (1992a) Characterization of vacuolar malate and $K^+$ channels under physiological conditions, *Plant Physiol*, **100**, 1137–1141.

Pantoja, O., Gelli, A. & Blumwald, E. (1992b) Voltage-dependent calcium channels in plant vacuoles, *Science*, **255**, 1567–1570.

Pantoja, O. & Smith, J.A. (2002) Sensitivity of the plant vacuolar malate channel to pH, $Ca^{2+}$ and anion-channel blockers, *J. Membr. Biol.*, **186**, 31–42.

Pei, Z.M., Murata, Y., Benning, G., *et al.* (2000) Calcium channels activated by hydrogen peroxide mediate abscisic acid signalling in guard cells, *Nature*, **406**, 731–734.

Pei, Z.M., Ward, J.M., Harper, J.F. & Schroeder, J.I. (1996) A novel chloride channel in *Vicia faba* guard cell vacuoles activated by the serine/threonine kinase, CDPK, *EMBO J.*, **15**, 6564–6574.

Pei, Z.M., Ward, J.M. & Schroeder, J.I. (1999) Magnesium sensitizes slow vacuolar channels to physiological cytosolic calcium and inhibits fast vacuolar channels in *Fava* bean guard cell vacuoles, *Plant Physiol.*, **121**, 977–986.

Perozo, E., MacKinnon, R., Bezanilla, F. & Stefani, E. (1993) Gating currents from a nonconducting mutant reveal open-closed conformations in Shaker $K^+$ channels, *Neuron*, **11**, 353–358.

Philippar, K., Fuchs, I., Luthen, H., *et al.* (1999) Auxin-induced $K^+$ channel expression represents an essential step in coleoptile growth and gravitropism, *Proc. Natl. Acad. Sci. U.S.A.*, **96**, 12186–12191.

Pilot, G., Lacombe, B., Gaymard, F., *et al.* (2001) Guard cell inward $K^+$ channel activity in *Arabidopsis* involves expression of the twin channel subunits KAT1 and KAT2, *J. Biol. Chem.*, **276**, 3215–3221.

Pineros, M.A., Magalhaes, J.V., Carvalho, A.V. & Kochian, L.V. (2002) The physiology and biophysics of an aluminum tolerance mechanism based on root citrate exudation in maize, *Plant Physiol.*, **129**, 1194–1206.

Pineros, M. & Tester, M. (1997) Calcium channels in higher plant cells: selectivity, regulation and pharmacology, *J. Exp. Bot.*, **48**, 551–577.

Plant, P.J., Gelli, A. & Blumwald, E. (1994) Vacuolar chloride regulation of an anion-selective tonoplast channel, *J. Membr. Biol.*, **140**, 1–12.

Pottosin, I.I. & Martinez-Estevez, M. (2003) Regulation of the fast vacuolar channel by cytosolic and vacuolar potassium, *Biophys. J.*, **84**, 977–986.

Pottosin, I.I., Martinez-Estevez, M., Dobrovinskaya, O.R. & Muniz, J. (2003) Potassium-selective channel in the red beet vacuolar membrane, *J. Exp. Bot.*, **54**, 663–667.

Pottosin, I.I., Tikhonova, L.I., Hedrich, R. & Schonknecht, G. (1997) Slowly activating vacuolar channels can not mediate $Ca^{2+}$-induced $Ca^{2+}$ release, *Plant J.*, **12**, 1387–1398.

Pusch, M. (2004) Structural insights into chloride and proton-mediated gating of CLC chloride channels, *Biochemistry*, **43**, 1135–1144.

Raschke, K. (2003) Alternation of the slow with the quick anion conductance in whole guard cells effected by external malate, *Planta*, **217**, 651–657.

Raschke, K., Shabahang, M. & Wolf, R. (2003) The slow and the quick anion conductance in whole guard cells: their voltage-dependent alternation, and the modulation of their activities by abscisic acid and $CO_2$, *Planta*, **271**, 639–650.

Rea, P.A. & Sanders, D. (1987) Tonoplast energization: two pumps, one membrane, *Physiol. Plant.*, **71**, 131–141.

Reintanz, B., Szyroki, A., Ivashikina, N., *et al.* (2002) AtKC1, a silent *Arabidopsis* potassium channel alpha-subunit modulates root hair $K^+$ influx, *Proc. Natl. Acad. Sci. U.S.A.*, **99**, 4079–4084.

Roberts, M.R. (2003) 14-3-3 proteins find new partners in plant cell signalling, *Trends Plant Sci.*, **8**, 218–223.

Roelfsema, M.R. & Prins, H.B. (1998) The membrane potential of *Arabidopsis thaliana* guard cells: depolarizations induced by apoplastic acidification, *Planta*, **205**, 100–112.

Ryan, P.R., Skerrett, M., Findlay, G.P., Delhaize, E. & Tyerman, S.D. (1997) Aluminum activates an anion channel in the apical cells of wheat roots, *Proc. Natl. Acad. Sci. U.S.A.*, **94**, 6547–6552.

Saalbach, G., Schwerdel, M., Natura, G., Buschmann, P., Christov, V. & Dahse, I. (1997) Over-expression of plant 14-3-3 proteins in tobacco: enhancement of the plasmalemma $K^+$ conductance of mesophyll cells, *FEBS Lett.*, **413**, 294–298.

Sanders, D., Brownlee, C. & Harper, J.F. (1999) Communicating with calcium, *Plant Cell*, **11**, 691–706.

Sanders, D., Pelloux, J., Brownlee, C. & Harper, J.F. (2002) Calcium at the crossroads of signaling, *Plant Cell*, **14** (Suppl.), S401–S417.

Sather, W.A. & McCleskey, E.W. (2003) Permeation and selectivity in calcium channels, *Annu. Rev. Physiol.*, **65**, 133–159.

Schauf, C.L. & Wilson, K.J. (1987) Properties of single $K^+$ and $Cl^-$ channels in *Asclepias tuberosa* protoplasts, *Plant Physiol.*, **85**, 413–418.

Schmidt, C., Schelle, I., Liao, Y.J. & Schroeder, J.I. (1995) Strong regulation of slow anion channels and abscisic acid signaling in guard cells by phosphorylation and dephosphorylation events, *Proc. Natl. Acad. Sci. U.S.A.*, **92**, 9535–9539.

Schonknecht, G., Spoormaker, P., Steinmeyer, R., *et al.* (2002) KCO1 is a component of the slow-vacuolar (SV) ion channel, *FEBS Lett.*, **511**, 28–32.

Schroeder, J.I. (1988) $K^+$ transport properties of $K^+$ channels in the plasma membrane of *Vicia faba* guard cells, *J. Gen. Physiol.*, **92**, 667–683.

Schroeder, J.I., Allen, G.J., Hugouvieux, V., Kwak, J.M. & Waner, D. (2001) Guard cell signal transduction, *Annu. Rev. Plant Physiol. Plant Mol. Biol.*, **52**, 627–658.

Schroeder, J.I. & Hagiwara, S. (1989) Cytosolic calcium regulates ion channels in the plasma membrane of *Vicia faba* guard cells, *Nature*, **338**, 427–430.

Schroeder, J.I. & Hagiwara, S. (1990) Repetitive increases in cytosolic $Ca^{2+}$ of guard cells by abscisic acid activation of nonselective $Ca^{2+}$ permeable channels, *Proc. Natl. Acad. Sci. U.S.A.*, **87**, 9305–9309.

Schroeder, J.I., Hedrich, R. & Fernandez, J.M. (1984) Potassium-selective single channels in guard cell protoplasts of *Vicia faba*, *Nature*, **312**, 361–362.

Schroeder, J.I. & Keller, B.U. (1992) Two types of anion channel currents in guard cells with distinct voltage regulation, *Proc. Natl. Acad. Sci. U.S.A.*, **89**, 5025–5029.

Schroeder, J.I., Raschke, K. & Neher, E. (1987) Voltage dependence of $K^+$ channels in guard-cell protoplasts, *Proc. Natl. Acad. Sci. U.S.A.*, **84**, 4108–4112.

Schulz-Lessdorf, B. & Hedrich, R. (1995) Protons and calcium modulate SV-type channels in the vacuolar-lysosomal compartment – channel interaction with calmodulin inhibitors, *Planta*, **197**, 655–671.

Schulz-Lessdorf, B., Lohse, G. & Hedrich, R. (1996) GCAC1 recognizes the pH gradient across the plasma membrane: a pH-sensitive and ATP-dependent anion channel links guard cell membrane potential to acid and energy metabolism, *Plant J.*, **10**, 993–1004.

Sentenac, H., Bonneaud, N., Minet, M., *et al.* (1992) Cloning and expression in yeast of a plant potassium ion transport system, *Science*, **256**, 663–665.

Skerrett, M. & Tyerman, S.D. (1994) A channel that allows inwardly directed fluxes of anions in protoplasts derived from wheat roots, *Planta*, **192**, 295–305.

Spalding, E.P., Slayman, C.L., Goldsmith, M., Gradmann, D. & Bertl, A. (1992) Ion channels in *Arabidopsis* plasma membrane – transport characteristics and involvement in light-induced voltage changes, *Plant Physiol.*, **99**, 96–102.

Stevens, C.F. (1978) Interactions between intrinsic membrane protein and electric field. An approach to studying nerve excitability, *Biophys. J.*, **22**, 295–306.

Stoelzle, S., Kagawa, T., Wada, M., Hedrich, R. & Dietrich, P. (2003) Blue light activates calcium-permeable channels in *Arabidopsis* mesophyll cells via the phototropin signaling pathway, *Proc. Natl. Acad. Sci. U.S.A.*, **100**, 1456–1461.

Szyroki, A., Ivashikina, N., Dietrich, P., *et al.* (2001) KAT1 is not essential for stomatal opening, *Proc. Natl. Acad. Sci. U.S.A.*, **98**, 2917–2921.

Terry, B.R., Tyerman, S.D. & Findlay, G.P. (1991) Ion channels in the plasma membrane of *Amaranthus* protoplasts: one cation and one anion channel dominate the conductance, *J. Membr. Biol.*, **121**, 223–236.

Thiel, G. & Blatt, M.R. (1994) Phosphatase antagonist okadaic acid inhibits steady-state $K^+$ currents in guard-cells of *Vicia faba*, *Plant J.*, **5**, 727–733.

Thiel, G., Blatt, M.R., Fricker, M.D., White, I.R. & Millner, P. (1993) Modulation of $K^+$ channels in *Vicia* stomatal guard cells by peptide homologs to the auxin-binding protein C terminus, *Proc. Natl. Acad. Sci. U.S.A.*, **90**, 11493–11497.

Thiel, G., MacRobbie, E.A. & Blatt, M.R. (1992) Membrane transport in stomatal guard cells: the importance of voltage control, *J. Membr. Biol.*, **126**, 1–18.

Thomine, S., Guern, J. & Barbier-Brygoo, H. (1997) Voltage-dependent anion channel of *Arabidopsis* hypocotyls: nucleotide regulation and pharmacological properties, *J. Membr. Biol.*, **159**, 71–82.

Thomine, S., Zimmermann, S., Guern, J. & Barbier-Brygoo, H. (1995) ATP-dependent regulation of an anion channel at the plasma membrane of protoplasts from epidermal cells of *Arabidopsis* hypocotyls, *Plant Cell*, **7**, 2091–2100.

Thuleau, P., Moreau, M., Schroeder, J.I. & Ranjeva, R. (1994a) Recruitment of plasma membrane voltage-dependent calcium-permeable channels in carrot cells, *EMBO J.*, **13**, 5843–5847.

Thuleau, P., Ward, J.M., Ranjeva, R. & Schroeder, J.I. (1994b) Voltage-dependent calcium-permeable channels in the plasma membrane of a higher plant cell, *EMBO J.*, **13**, 2970–2975.

Tikhonova, L.I., Pottosin, I.I., Dietz, K.J. & Schonknecht, G. (1997) Fast-activating cation channel in barley mesophyll vacuoles. Inhibition by calcium, *Plant J.*, **11**, 1059–1070.

Tomita, M. (2001) Whole-cell simulation: a grand challenge of the 21st century, *Trends Biotechnol.*, **19**, 205–210.

Trewavas, A.J. & Malho, R. (1998) $Ca^{2+}$ signalling in plant cells: the big network!, *Curr. Opin. Plant Biol.*, **1**, 428–433.

Uozumi, N., Gassmann, W., Cao, Y. & Schroeder, J.I. (1995) Identification of strong modifications in cation selectivity in an *Arabidopsis* inward rectifying potassium channel by mutant selection in yeast, *J. Biol. Chem.*, **270**, 24276–24281.

Uozumi, N., Nakamura, T., Schroeder, J.I. & Muto, S. (1998) Determination of transmembrane topology of an inward-rectifying potassium channel from *Arabidopsis thaliana* based on functional expression in *Escherichia coli*, *Proc. Natl. Acad. Sci. U.S.A.*, **95**, 9773–9778.

van den Wijngaard, P.W., Bunney, T.D., Roobeek, I., Schonknecht, G. & de Boer, A.H. (2001) Slow vacuolar channels from barley mesophyll cells are regulated by 14-3-3 proteins, *FEBS Lett.*, **488**, 100–104.

Very, A.A. & Davies, J.M. (2000) Hyperpolarization-activated calcium channels at the tip of *Arabidopsis* root hairs, *Proc. Natl. Acad. Sci. U.S.A.*, **97**, 9801–9806.

Very, A.A. & Sentenac, H. (2002) Cation channels in the *Arabidopsis* plasma membrane, *Trends Plant Sci.*, **7**, 168–175.

Very, A.A. & Sentenac, H. (2003) Molecular mechanisms and regulation of $K^+$ transport in higher plants, *Annu. Rev. Plant Biol.*, **54**, 575–603.

Walker, D.J., Leigh, R.A. & Miller, A.J. (1996) Potassium homeostasis in vacuolate plant cells, *Proc. Natl. Acad. Sci. U.S.A.*, **93**, 10510–10514.

Wang, X.Q., Ullah, H., Jones, A.M. & Assmann, S.M. (2001) G protein regulation of ion channels and abscisic acid signaling in *Arabidopsis* guard cells, *Science*, **292**, 2070–2072.

Wang, X.Q., Wu, W.H. & Assmann, S.M. (1998) Differential responses of abaxial and adaxial guard cells of broad bean to abscisic acid and calcium, *Plant Physiol.*, **118**, 1421–1429.

Ward, J.M. & Schroeder, J.I. (1994) Calcium-activated $K^+$ channels and calcium-induced calcium release by slow vacuolar ion channels in guard cell vacuoles implicated in the control of stomatal closure, *Plant Cell*, **6**, 669–683.

Weber, H. (2002) Fatty acid-derived signals in plants, *Trends Plant Sci.*, **7**, 217–224.

Wegner, L.H. & de Boer, A.H. (1997) Two inward $K^+$ channels in the xylem parenchyma cells of barley roots are regulated by G-protein modulators through a membrane-delimited pathway, *Planta*, **203**, 506–516.

Weiser, T., Blum, W. & Bentrup, F.W. (1991) Calmodulin regulates the $Ca^{2+}$-dependent slow-vacuolar ion channel in the tonoplast of *Chenopodium rubrum* suspension cells, *Planta*, **185**, 440–442.

White, P.J. (1998) Calcium channels in the plasma membrane of root cells, *Ann. Bot. (Lond.)*, **81**, 173–183.

White, P.J. (2000) Calcium channels in higher plants, *Biochim. Biophys. Acta*, **1465**, 171–189.

White, P.J., Bowen, H., Demidchik, V., Nichols, C. & Davies, J. (2002) Genes for calcium-permeable channels in the plasma membrane of plant root cells, *Biochim. Biophys. Acta*, **1564**, 299.

White, P.J. & Broadley, M.R. (2001) Chloride in soils and its uptake and movement within the plant: a review, *Ann. Bot.*, **88**, 967–988.

Worrall, D., Ng, C.K. & Hetherington, A.M. (2003) Sphingolipids, new players in plant signaling, *Trends Plant Sci.*, **8**, 317–320.

Wu, W.H. & Assmann, S.M. (1994) A membrane-delimited pathway of G-protein regulation of the guard-cell inward $K^+$ channel, *Proc. Natl. Acad. Sci. U.S.A.*, **91**, 6310–6314.

Zhang, W.H., Skerrett, M., Walker, N.A., Patrick, J.W. & Tyerman, S.D. (2002) Nonselective currents and channels in plasma membranes of protoplasts from coats of developing seeds of bean, *Plant Physiol.*, **128**, 388–399.

Zhou, Y., Morais-Cabral, J.H., Kaufman, A. & MacKinnon, R. (2001) Chemistry of ion coordination and hydration revealed by a $K^+$ channel-Fab complex at 2.0 A resolution, *Nature*, **414**, 43–48.

Zimmermann, S., Thomine, S., Guern, J. & Barbier-Brygoo, H. (1994) An anion current at the plasma membrane of tobacco protoplasts shows ATP-dependent voltage regulation and is modulated by auxin, *Plant J.*, **6**, 707–716.

# 7    Ligand-gated ion channels

Frans Maathuis

## 7.1    Introduction

So far in this volume, the various aspects of *voltage-gated* ion channels have been discussed. Opening and closing (gating) of voltage-gated ion channels is strictly dependent on changes in membrane polarisation although many additional factors may modify this dependency. The *ligand-gated* ion channels constitute another main class of ion channel. In this type of transporter, gating occurs only after binding of specific compounds to the channel protein. These compounds are called 'ligands', i.e. they bind to another compound to form a complex (Fig. 7.1; Berg *et al.*, 2001). The term 'agonist' (literally meaning a 'competitor') is also frequently used, since often different ligands can have similar actions and compete for the same binding site. In contrast, 'antagonists' also bind at the same site but have an opposite action to ligands. Ligand binding to the channel protein causes a conformational change that switches the channel from the closed to the open state or vice versa. Binding of ligand is a minimum requirement for channel gating but, as for voltage-gated channels, overall gating properties of ligand-gated channels are often modulated by additional parameters, which can include membrane polarisation (Aidley & Stanfield, 1996; Hille, 2001).

So why do cells need ion channels that are regulated through binding of agonists rather than via changes in membrane voltage? There is a multitude of answers to this question depending on the function of particular ion channels. Most of our knowledge regarding voltage- and ligand-gated ion channels is derived from studies on animals and the overall generalised picture that emerges is that both voltage-gated and ligand-gated ion channels play indispensable roles in most signal transduction pathways, the regulation of the membrane potential and cellular ionic homeostasis. Nevertheless, in a number of cases it is clear why ion transport through channels under control of an agonist is more suitable than through channels that are modulated via changes in membrane polarisation: Ligand-gated channels are excellent molecular switches that can fulfil complex tasks in information processing. For example, in the brain where synaptic plasticity is believed to play a role in learning and memorising, moderate release of neurotransmitter from a synaptic end plate will activate both NMDA and AMPA receptors, two different types of glutamate receptors in the post-synaptic membrane (Bigge, 1999; Dingledine *et al.*, 1999). Of these, the NMDA receptors participate in long-term changes in synaptic efficacy. Depolarising current (in the form of $Na^+$ ions) will only occur through AMPA-type channels because NMDA receptors are blocked by $Mg^{2+}$. However, when there is

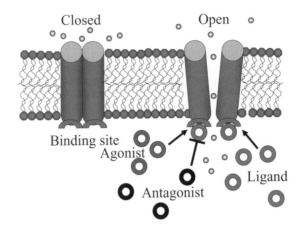

**Fig. 7.1** Schematic representation of ligand-activated channels. The channel protein remains in the closed configuration unless the binding sites are occupied by either *ligands* or *agonists*. Binding of *antagonists*, which compete for the same site, does not cause a transition to the open state.

coincident activity from other synapses in the vicinity, the amount of neurotransmitter is sufficient to cause post-synaptic depolarisation, which in turn relieves the $Mg^{2+}$ block. The resulting NMDA-mediated $Ca^{2+}$ influx contributes to the modification of synaptic function called 'long-term potentiation'. This arrangement of molecular switches forms a 'logical gate' where signals only perpetuate when more than one synapse is 'on' (O'Boyle *et al.*, 2003). It is hard to envisage how molecular switches such as these could function adequately without neurotransmitters that act as channel ligand.

In many biological processes, information has to be rapidly converted from one type of signal into another. Where signals from inter- or intracellular messengers have to be transduced into electrical signals, ligand-gated ion channels form the obvious mechanism. For example, in the mammalian eye, light perception in rod and cone cells of the retina occurs via a G-protein that stimulates phosphodiesterases to hydrolyse cGMP to GMP (Kaupp & Seifert, 2002). The reduction in cGMP levels leads to closure of a class of ligand-gated channels, the CNG channels (cyclic-nucleotide-gated channels) that keep the cell depolarised in the dark. In this way, an initial stimulus in the form of a light photon is translated into a chemical signal in the form of cGMP and, through the action of the ligand-gated channel, ultimately into an electric signal that is transmitted to the brain via the optic nerve. At the same time, a large degree of signal amplification is achieved (Berg *et al.*, 2001; Kaupp & Seifert, 2002).

Ligand-gated ion channels may also be crucial when ion fluxes and transmembrane signals need to be controlled at intracellular membranes where large deviations in membrane potential are either unlikely to occur or where control of the membrane potential may be difficult. For example, endomembranes like the endoplasmatic reticulum (ER) membrane, Golgi membrane or nuclear envelopes may not experience large swings in membrane potential, providing little scope for regulating ionic fluxes through voltage-gated channels. Indeed, the current picture suggests that the bulk of $Ca^{2+}$ transport across these membranes is carried through ligand- rather than voltage-gated ion channels. Nevertheless, most membranes, including those from plant cells, probably contain both voltage- and ligand-gated ion channels.

## 7.2   Acetylcholine receptors, the archetypal ligand-gated ion channels

Effective control of the activity of ligand-gated ion channels involves many aspects. These can include ligand specificity, ligand affinity, modulation by other binding factors, subunit composition and post-translational regulation such as phosphoryla- tion. The nicotinic acetylcholine receptor (nAcR), one of the most intensely studied ligand-gated channels, forms an excellent model system, since many of the princi- ples governing the functioning of nAcRs also pertain to other ligand-gated channels (Hille, 2001).

nAcRs were initially observed in neuromuscular junctions but were later found to be prevalent in other tissues also, particularly of the central nervous system. The electrical stimulation of a nerve supplying a muscle releases acetylcholine that opens the nAcRs to mediate post-synaptic $Na^+$ and $Ca^{2+}$ influx. The resulting depolarisation triggers opening of voltage-gated ($Ca^{2+}$) channels, ultimately leading to muscular contraction. The antagonistic action of highly specific snake venoms such as bungarotoxin that bind to nAcRs greatly facilitated research as did the existence of rich stores of nAcRs in the form of electric organs of *Torpedo* electric ray. It is now clear that the nAcR gene family is large with at least 17 genes coding for mammalian subunits, giving scope to a large array of different channel types. nAcRs consist of heteropentamers containing, $2\alpha$, $1\beta$, $1\gamma$ and either $1\delta$ or $1\eta$ subunit, with each subunit predicted to have four transmembrane segments M1–M4. The second transmembrane segment, M2, shapes the lumen of the pore and forms the gate of the closed channel. The functional protein has a total mass of around 290 kDa and forms a non-selective ion channel that can conduct monovalent and divalent cations.

How does the nAcR change from closed to open state? The five subunits are assembled in a funnel-like protein, wide at the synaptic side and narrow on the cytoplasmic side, with $\alpha$ helices from the M2 segments forming the pore that is the central ion conduction pathway. The pore is narrowest where it has traversed the membrane around 50% and this region constitutes the channel gate. Acetylcholine that has been released into the synaptic cleft can bind to the $\alpha$ subunits of the receptor. Binding of acetylcholine to the $\alpha$ subunits initially causes a rotation of the inner (pore facing) $\beta$-sheets, which is transferred to the $\alpha$ helices of the M2 segments (Miyazawa *et al.*, 2003). Movement of the pore lining $\alpha$ helices destabilises the gate, causing a conformational change that effectively widens the pore so it becomes conductive. Release of acetylcholine from its binding sites produces the reverse, i.e. channel closure. The kinetics of binding and release of ligand and the associated channel gating can be formulated as a simple two-state model with a closed and open state (C ⇌ O) in analogy to that discussed for voltage-gated channels (see Chapter 6). The only difference is the forward and reverse reaction rates that are determined by binding and release of ligand rather than by membrane voltage (Hille, 2001).

Kinetic experiments using different acetylcholine concentrations showed that more than one molecule of acetylcholine is needed to produce channel opening (Hille, 2001). If one ligand molecule would be sufficient for channel opening, membrane conductance would rise in a linear manner at low ligand concentration and saturate at high concentration, according to classical first-order Michaelis–Menten kinetics. However, the correlation between conductance and ligand concentration was found to be much steeper than expected from a first-order process, with conductance increasing fourfold when ligand concentration doubled, giving a slope of around 2 in a double log plot. This slope factor is usually expressed as the 'Hill coefficient', which in this case would equate to 2. Such a quadratic dependence can be formalised by assuming that the channel has at least two closed states (C ⇌ C ⇌ O) and that two ligand molecules have to bind consecutively before channel opening occurs.

In living tissues, the exposure of channel protein to ligand can be extremely brief, i.e. less than 1 ms in the case of nAcRs. This can be the result of a limited quantity of ligand, rapid diffusion of ligand or enzymatic breakdown of ligand as is the case for acetylcholine, which is quickly inactivated by acetylcholine esterases (Colquhoun, 1998[Q3]; Berg et al., 2001). Prolonged exposure to ligand can lead to channel 'desensitisation'. This process is analogous to inactivation of voltage-gated channels (see Chapter 6), that is, channel activity is reduced in spite of the presence of a constant stimulus. The physiological role of this phenomenon is not always clear but may form part of an adaptive response to excessive stimulation, for example in the constitutive activation of nAcRs by biological insecticides that inhibit acetylcholine esterases.

Ligand protein interactions are usually highly specific (Buchanan et al., 2000; Berg et al., 2001). However, other agonists and antagonists can bind to sites where the primary ligand binds (Fig. 7.1). Such compounds may be able to interact with binding sites because of similarities in physicochemical properties. For example, all mammalian CNGCs respond to both cGMP and cAMP (Kaupp & Seifert, 2002), and this also appears to be true for plant CNGCs (see below), which therefore function as competitive agonists. But while rod and cone CNGCs strongly discriminate between cyclic nucleotides and hence need approximately 500 times higher cAMP than cGMP for similar levels of activity, olfactory CNGCs hardly discriminate and show similar open probabilities when exposed to equal concentrations of either cyclic nucleotide (Kaupp & Seifert, 2002). In nAcRs, apart from acetylcholine, other agonists and antagonists can bind to the sites at the α subunits, affecting nAcR gating properties. The alkaloid nicotine from tobacco is a competitive agonist, and application of nicotine to a neuromuscular junction will produce muscle contraction just as well as acetylcholine. Antagonists, i.e. compounds that inhibit channel activation, can bind to the ligand site (competitive) or other sites of the channel protein (non-competitive). A well-known nAcR antagonist is curare, used on arrowheads in the Amazon to paralyse prey. Antagonists can play important regulatory roles, as is the case for calmodulin, which non-competitively binds to CNGCs and plays a crucial role in the adaptation of vision to varying light levels (see below). Other antagonists are evolutionary adaptations that play a role in defense and predation. The earlier mentioned nicotine is primarily an insecticide to discourage insect pests from munching away tobacco leaves. Many insect and snake venoms act as antagonists of both voltage- and ligand-gated ion channels: when you get bitten by a krait or a cobra and enough venom gets into the blood, neurotoxins like bungarotoxin inhibit nAcRs, paralysing the diaphragm muscles, and asphyxiation results.

Other factors influence channel activity. Subunit composition will affect many channel properties such as conductance, ion selectivity and often gating (e.g. Jones *et al.*, 1999). For example, some nAcRs in the central nervous system contain as many as five acetylcholine binding sites. Ligand dose–response curves of CNGCs will depend on subunit composition. In addition, post-translational modification such as phosphorylation, oxidation/reduction of disulfide bonds and binding of for example 14-3-3 proteins may all impact on channel activity (Hille *et al.*, 2001).

## 7.3   Techniques to study ligand-gated channels

High affinity binding of agonists and antagonists to channel proteins of interest can be exploited for biochemical studies. For example, the immobilisation of a binding compound to a high molecular substrate such as cellulose allows the construction of affinity chromatography columns where specific binding of the studied protein to its ligand is used to purify the protein (Buchanan *et al.*, 2000; Berg *et al.*, 2001). Pure protein can then be used for reconstitution studies, for example in lipid bilayers or oocytes, to reveal functional characteristics, or it can be sequenced to identify and clone encoding genes.

By using a radioactive form, ligands and antagonists can also be used to accurately determine channel expression patterns. Labelling of the neuromuscular junction with [$^{125}$I]bungarotoxin has been used to show dense packing (up to 20 000 per $\mu m^2$) of nAcRs in certain areas of the post-synaptic membrane but not in others, thus building a detailed picture of where and in what numbers specific channel receptors are expressed (Salpeter & Loring, 1985).

As with all current-mediating mechanisms, ligand-gated channels are best studied with voltage clamp methods (see Chapter 1). Before the advent of patch clamping, this took place using two-electrode voltage clamping in muscle cells, nerve cells and giant algal cells. Generally, two-electrode voltage clamp only allows the recording of macroscopic currents, i.e. a summation of currents through many different channel types. Patch clamping allows the resolution of microscopic currents, those mediated by a single channel protein. However, in contrast to voltage-gated channels, stimulation of ligand-gated channels cannot be accomplished by stepwise changes in the membrane potential, but requires the presence of ligand or agonist to the

appropriate site of the protein. Although agonist delivery itself is a minimal requirement, experimenters will ideally be able to control agonist concentration and also
the dynamics of agonist/antagonist delivery and removal. Such requirements can
be technically challenging, especially if binding sites are cytoplasmic and therefore
not easily accessible. Fortunately, the patch clamp technique allows manipulation
of the membrane preparation so that either side of the membrane is accessible (see
Chapter 1) for experimentation. Perfusion of the preparation is a simple and effective way to expose a preparation to ligands. If unstirred layer effects are negligible,
this method ensures adequate control of ligand concentration. However, to study the
kinetics of channel gating, rapid changes in ligand delivery are necessary. This can
be achieved using motorised or gravity-fed solution exchange systems that locally
change ligand concentrations, for example by rapidly moving a tube outlet that delivers the test solution in front of the membrane preparation. Time resolutions in the
milliseconds range can be achieved with such a setup. If charged agonists are used,
iontophoretic delivery can be made accurately and rapidly from a locally positioned
microelectrode that is filled with agonists (see Chapter 1).

Ligands and other compounds can also be chemically modified to an inactive
form. Inactivated ligands loaded into cells are activated by a strong flash of UV
light which cleaves the inactivating group. The usage of such 'caged' compounds
(e.g. Kaupp & Seifert, 2002) allows very rapid (within $\mu$s) localised application of
ligands and/or antagonists and is therefore ideal to study channel kinetics.

## 7.4   Plant ligand-gated ion channels

In comparison to what is known about animal ligand-gated channels, our knowledge regarding this class of transporter in plant cells is in its infancy. The first study
showing the activity of ligand-gated plant ion channels occurred in 1990 when it
was observed that the signalling molecule inositol 1,4,5-trisphosphate ($IP_3$) opened
$Ca^{2+}$ channels in red beet vacuoles (Alexandre et al., 1990), promoting $Ca^{2+}$ elevation and $Ca^{2+}$-regulated channel gating in guard cells (Blatt et al., 1990; Gilroy
et al., 1990). Subsequently, reports appeared regarding ion channels gated by other
ligands such as cADPR and NAADP (see Sanders et al., 2002, for a review). However, studies into the structure and function of these channels were hampered by
the fact that the encoding genes were unknown. This prevented cloning and heterologous expression for functional analysis and the use of reverse genetics to study
physiological roles. After completing the sequencing of the first plant genome from
Arabidopsis, sequence homology searches allowed the identification of the entire
complement of potential ligand-gated ion channels in a plant species. Surprisingly,
many of the ligand-gated channels for which functional data were obtained from
in planta studies, e.g. $IP_3$, cADPR- and NAADP-gated channels have still not been
identified at the gene level. Nevertheless, a large number of plant genes shows
high degrees of homology to animal ligand-gated channels. Two major families
of glutamate-receptor-like proteins (GLRs; see Davenport, 2002, for a review) and
cyclic-nucleotide-gated channels (CNGCs; see Talke et al., 2003, for a review) have

been identified in the *Arabidopsis* genome and are also represented in other plant genomes. Functional characterisation of these gene products has proved problematic but the few data available suggest that at least some of them function as genuine ligand-gated channels with specific physiological roles.

## 7.5   $Ca^{2+}$ release channels from endomembranes

Cells have to be able to respond to a plethora of internal and external stimuli including changes in light or temperature, pathogen attack and the availability of nutrients. Many of the signalling pathways implicated in these responses are based on a general scheme that includes primary receptors, non-protein (second) messengers, protein effectors such as kinases, and end-point targets such as transcription factors and transporters (Buchanan *et al.*, 2000). In plant cells, as in most eukaryotic cells, $Ca^{2+}$ plays essential roles as second messenger in the intermediate steps of transducing environmental and developmental clues to meaningful physiological responses (see Chapter 8). Normally, cytoplasmic $Ca^{2+}$ is maintained at around 100–200 nM through the activity of $H^+$-coupled $Ca^{2+}$ antiporters and $Ca^{2+}$-ATPases that extrude $Ca^{2+}$ (Fig. 7.2). The rise in the level of cytoplasmic $Ca^{2+}$, which forms the basis of

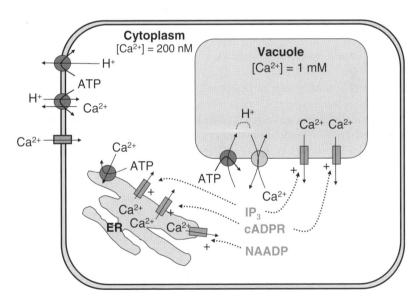

**Fig. 7.2** Some of the main transport mechanisms involved in cellular $Ca^{2+}$ signalling. Cytoplasmic $Ca^{2+}$ is typically maintained at a resting level of around 100–200 nM, which requires constant extrusion of $Ca^{2+}$ into the apoplast, vacuole and endoplasmic reticulum (ER). Extrusion is powered by primary $Ca^{2+}$ pumps that are fuelled by ATP hydrolysis and/or by antiport systems that use the electrochemical $H^+$ gradient as a driving force. During signalling events, cytoplasmic $Ca^{2+}$ can increase by opening various voltage- and/or ligand-gated $Ca^{2+}$-permeable ion channels situated in the membranes that separate the cytosol from internal and external $Ca^{2+}$ stores.

the signal, is achieved by allowing $Ca^{2+}$ entry from internal and external stores such as the apoplast, the vacuole and the ER, where concentrations can be up to 10 000 times higher. Thus, $Ca^{2+}$ release from these stores is down the electrochemical gradient and mediated by ion channels (Fig. 7.2). At the plasma membrane, tonoplast and ER membrane, voltage-gated ion channels have been observed that contribute to the entry of $Ca^{2+}$ into the cytoplasm (for a review, see White, 2000). Some of these respond to membrane hyperpolarisation whereas others only activate when the membrane depolarises and these different modes of gating may point to different functions. However, several types of ligand-gated ion channels have been found in plant endomembranes and it is believed that these play crucial roles in plant $Ca^{2+}$ signalling (White, 2000; Sanders *et al.*, 2002).

### 7.5.1   $IP_3$-gated channels

Several signalling pathways in animal cells are characterised by phosphoinositol intermediates and it appears that plants possess analogous pathways (see Drobak *et al.*, 1999, for a review). In a nutshell, these pathways start with some external trigger that binds to its appropriate receptor on the plasma membrane. Receptor binding results, via activation of a G-protein, in stimulation of phospholipase-C, which hydrolyses the membrane lipid phosphatidylinositol 4,5-bisphosphate to two second messengers: diacylglycerol (DAG) and myo-inositol 1,4,5-triphosphate or $IP_3$. $IP_3$ subsequently causes $Ca^{2+}$ release from intracellular stores, leading to further downstream $Ca^{2+}$-based signalling. Various reports showed that in plants too, $IP_3$ could release $Ca^{2+}$ from isolated vacuoles or vacuolar vesicles (e.g. Ranjeva *et al.*, 1988). However, the underlying mechanism was only revealed by applying the patch clamp technique directly to intact vacuoles from red beet (Alexandre *et al.*, 1990). Those experiments showed that application of ligand in the form of $IP_3$ led to opening of tonoplast ion channels that were later shown to be highly selective for $Ca^{2+}$ with a relative permeability for $K^+$ over $Ca^{2+}$, $P_{K:Ca}$, of 0.01 to 0.0013 (Allen & Sanders, 1994; Allen *et al.*, 1995). As expected for a true ligand-gated channel, both whole vacuole and single channel currents were strictly dependent on the presence of $IP_3$. Current magnitude was sensitive to $IP_3$ concentration in a first-order fashion (i.e. a Hill coefficient of 1), showing no evidence of cooperativity in ligand binding, as is the case for many animal $IP_3$ receptors. Therefore, plant $IP_3$ receptors are likely to possess only one ligand binding site per channel protein. Channel activation occurred in response to $IP_3$ concentrations in the nM–μM range with a $K_m$ of around 200 nM, well within the range of physiological $IP_3$ levels (e.g. Heilmann *et al.*, 1999). No or negligible currents were recorded after exposure to phosphoinositides other than $IP_3$, and currents could be blocked with various known $Ca^{2+}$ channel blockers such as verapamil. Single-channel recordings showed a unitary slope conductance of around 30 pS (Alexandre *et al.*, 1990).

Interestingly, $IP_3$-induced currents were steeply voltage–dependent, resulting in a strongly inward (cytoplasmic $Ca^{2+}$ influx) rectifying $Ca^{2+}$ current. It is not clear where this rectification originates and what its physiological relevance is. At all conceivable tonoplast potentials (typically 10–30 mV, vacuolar side positive, e.g., Walker et al., 1996) will the electrochemical $Ca^{2+}$ gradient be inward. So rectification is not necessary to ensure a 'one-way valve' like is the case for many voltage-dependent $K^+$ channels (see Chapter 6). However, it may be that these receptors, like the NMDA receptors in nerve cells, form logical switches that require two parameters to be in a 'permissive' state, i.e. the membrane voltage and $IP_3$ concentration. How voltage dependence is accomplished remains to be established. Does the tonoplast $IP_3$ receptor contain a voltage sensor similar to that found in voltage-gated channels? This will be hard to prove or disprove until sequence data are available. Alternatively, divalent ions such as $Mg^{2+}$ may bind to the pore in a voltage-dependent manner and so prevent outward current. However, removal of $Mg^{2+}$ from either side of the membrane had no effect on channel rectification properties (Alexandre et al., 1990). It is noteworthy, too, that inositol hexakisphosphate is around 100-fold more potent than $IP_3$ in modulating $K^+$ channel activity mediated by $Ca^{2+}$ (Lemtiri-Chlieh et al., 2000, 2003).

### 7.5.2 cADPR-gated channels

Apart from $IP_3$ receptors, the tonoplast contains a second class of ligand-gated $Ca^{2+}$ channel that functions in vacuolar $Ca^{2+}$ release (Fig. 7.2). These channels resemble the mammalian ryanodine receptor, a $Ca^{2+}$-selective channel named after its propensity to bind the plant alkaloid ryanodine (see Zucchi & RoncaTestoni, 1997, for a review). Mammalian ryanodine receptors are most prevalent in muscle sarcoplasmic reticulum (SR; Zucchi & Ronca-Testoni, 1997; Stokes & Wagenknecht, 2000) where they participate in $Ca^{2+}$ release that is necessary for muscle contraction. Gating characteristics of mammalian ryanodine receptors are complex and involve many factors: in isolated vesicles, $Ca^{2+}$ and ATP act synergistically as the main activators, leading to channel opening whereas $Mg^{2+}$ and calmodulin inhibit channel opening. Muscle contraction starts with a depolarisation of the plasma membrane which opens voltage-gated L-type $Ca^{2+}$ channels to allow small amounts of $Ca^{2+}$ to enter the cytoplasm. However, this is insufficient to cause contraction and mainly serves to open ryanodine receptors, leading to a much larger $Ca^{2+}$ influx and muscle contraction. Thus, the stimulating effect of $Ca^{2+}$ on the ryanodine receptor open probability forms a positive feedback loop that greatly amplifies the original signal and is known as calcium-induced calcium release (CICR; Yamamori et al., 2004). Although $Ca^{2+}$ is believed to be the main physiological agonist, there are endogenous and exogenous factors that influence gating (Zucchi & Ronca-Testoni, 1997). The NAD (nicotinamide–adenine–dinucleotide) metabolite cyclic ADP-ribose (cADPR) was found to release $Ca^{2+}$ from intracellular stores from many cell types. This release was in several cases potentiated by $Ca^{2+}$ and ryanodine. On the basis of these observations it was suggested that cADPR acts on ryanodine receptors, but this model has now been questioned (Zucchi & Ronca-Testoni, 1997). Ryanodine itself has a strong stimulating effect on the receptor open probability, possibly by sensitising the protein to $Ca^{2+}$ (Du et al., 2002). A similar effect is produced by the methylxanthine caffeine.

In plants too, cADPR has been postulated to be involved in many ($Ca^{2+}$-dependent) signalling events (e.g. Grabov & Blatt, 1998, 1999; Moyen *et al.*, 1998; Ng *et al.*, 2001; Sangwan *et al.*, 2001; Garcia-Mata *et al.*, 2003; Wu *et al.*, 2003). Both in microsomal vesicles and in intact red beet vacuoles, nanomolar amounts of cADPR but not ADPR give rise to $Ca^{2+}$ release, which can be inhibited by ruthenium red (Allen *et al.*, 1995; Navazio *et al.*, 2001). Patch clamp experiments showed cADPR-dependent currents that are mainly carried by $Ca^{2+}$ with a $P_{K:Ca}$ between 0.04 and 0.1 (Allen *et al.*, 1995). Like vacuolar $IP_3$-induced currents, cADPR-dependent currents are prevalent at physiological tonoplast potentials ($-10$ to $-40$ mV) and largely absent at positive potentials. The use of intact vacuoles further showed that subsequent addition of $IP_3$ and cADPR produces additive $Ca^{2+}$ currents, proving that both types of receptor are present at the same membrane (Allen *et al.*, 1995).

There is good evidence that cADPR not only releases vacuolar $Ca^{2+}$ but may also access $Ca^{2+}$ from the ER pool (Navazio *et al.*, 2001). Cauliflower inflorescence cells contain an extensive ER network and very small vacuoles and are therefore ideal in obtaining an ER-vesicle-enriched membrane fraction. Incubation with 1 $\mu$M cADPR of vesicle fractions preloaded with $Ca^{2+}$ resulted in $Ca^{2+}$ release from both vacuolar and ER-derived vesicles, which was specifically sensitive to cADPR, inhibited by ruthenium red but not heparin (a well-known blocker of $IP_3$ receptors). Interestingly, ER vesicles from which ribosomes were removed lost their ability to respond to cADPR. The reason for this is not entirely clear but it has been suggested that cADPR may not directly interact with the channel protein (Walseth *et al.*, 1993) but would require peripheral proteins. In the latter case, cADPR should not be regarded as a 'true' ligand of the ryanodine receptor.

### 7.5.3 NAADP-gated channels

An additional compound has been shown to elicit $Ca^{2+}$ release from the ER in plants. NAADP (nicotinic acid dinucleotide phosphate) is a NADP metabolite previously identified as a potent $Ca^{2+}$-mobilising agent in several animal cell types. Experiments with cauliflower and beet ER-enriched microsomes showed NAADP-dependent $Ca^{2+}$ release with a $K_{1/2}$ of around 100 nM (Navazio *et al.*, 2000). This mechanism exhibits a 'self-desensitising' response in the sense that pre-incubation with subthreshold levels of NAADP ($<3$ nM) inhibits $Ca^{2+}$ release by subsequent saturating doses of NAADP. Although it is postulated that NAADP-dependent $Ca^{2+}$ release functions through a (ligand-gated) ion channel, it has to be pointed out that no direct evidence (e.g. from patch clamping) is yet available for this notion.

Thus, apart from voltage-gated channels, plant endomembranes are likely to contain at least three different ligand-gated channel types that contribute to $Ca^{2+}$ mobilisation from internal stores. At the tonoplast, $IP_3$- and cADPR-gated pathways act in parallel, whereas the ER contains cADPR- and NAADP-dependent pathways and possibly also $IP_3$-gated channels (FranklinTong *et al.*, 1996; Muir & Sanders, 1997; Martinec *et al.*, 2000). An obvious question arising from these observations is why cells need such a plethora of different $Ca^{2+}$ release pathways and what their functions are. The relevance of multiple pathways may lie in providing specificity for $Ca^{2+}$ signalling (see Chapter 9). At least 20 stimuli have been described where a rise in $Ca^{2+}$ forms part of the response. Nevertheless, both stimuli and responses are usually highly specific and therefore cannot be mediated by a uniform $Ca^{2+}$ signal. Much recent research has shown that $Ca^{2+}$ signals subsume complicated amplitude and frequency modulations, and spatial variations that all contribute to the '$Ca^{2+}$ signature'. The presence of multiple mechanisms through which $Ca^{2+}$ can be released with different kinetics, at different locations and in response to a multitude of factors that impact on channel activity, ensures the capacity to generate different $Ca^{2+}$ signatures (e.g. Buchanan et al., 2000; Allen et al., 2001).

For example, CICR may depend on the physical proximity between systems that generate a primary $Ca^{2+}$ signal and those that are sensitive to small increases in $Ca^{2+}$ (Stokes & Wagenknecht, 2000). Such signal-amplifying mechanisms are examplified by the voltage-gated $Ca^{2+}$ channels and ryanodine receptors that act during the excitation/contraction process in muscle cells (Du et al., 2002). However, in contrast to animals, in plants neither the $IP_3$, the cADPR nor the NAADP receptor has been found to be sensitive to $Ca^{2+}$. Nevertheless, certain endomembrane voltage-gated channels such as the SV (slow vacuolar, Hedrich et al., 1988) channel are $Ca^{2+}$-permeable and greatly stimulated by a rise in $Ca^{2+}$ (Fig. 7.2). Thus, in plants, CICR may function in the opposite fashion where ligand-gated channels provide an initial rise in cytoplasmic $Ca^{2+}$, which subsequently activates voltage-gated channels such as the SV channel to increase $Ca^{2+}$ release (Ward & Schroeder, 1994; Bewell et al., 1999).

Indeed, many data are now emerging regarding the specific roles that some of the ligand-gated $Ca^{2+}$ pathways play in physiological responses to environmental and developmental stimuli. $IP_3$-based signalling is firmly implicated in cell turgor regulation. First, hyperosmotic stress and salinity stress both have been observed to cause elevated levels of $IP_3$ in a range of plant species (Heilmann et al., 1999; Drobak & Watkins, 2000; DeWald et al., 2001). The increase in $IP_3$ occurs in a dose-dependent fashion, suggesting that plants translate the severity of the (osmotic) stress into increasing $IP_3$ concentrations. The rise in $IP_3$ occurs within seconds after application of the stress condition and appears to precede the $Ca^{2+}$ signal. Second, release of caged $IP_3$ in guard cells leads to a rapid increase in $Ca^{2+}$ (Gilroy et al., 1990). Release of caged $IP_3$ and the subsequent increase in $Ca^{2+}$ has a direct inhibitory effect on inward $K^+$ channels at the guard cell plasma membrane and so promotes stomatal closure (Blatt, 1990; Blatt et al., 1990). Third, both hyperosmotic pretreatment and the application of an osmotic gradient during channel recordings greatly increased the $IP_3$-receptor-mediated $Ca^{2+}$ current in beet tissue vacuoles (Allen & Sanders, 1994). These observations combined infer that one of the earliest events during the perception of osmotic stress in plant cells is an increase in $IP_3$, which is translated into a specific $Ca^{2+}$ signal by tonoplast, and possibly ER, located $IP_3$ receptors. $IP_3$ levels also rise up to fivefold within seconds in response to a gravitropic stimulus (Perera et al., 1999). However, the role of $Ca^{2+}$ in gravitropic signalling is currently much debated, with some laboratories reporting a $Ca^{2+}$ transient whereas others were unable to detect any change in cytoplasmic $Ca^{2+}$ (see Chapter 9). It would be interesting if the gravitropic response constitutes a system where increases in $IP_3$ do not result in $Ca^{2+}$ signals.

Nitric oxide (NO) leads to induction of defence gene expression in tobacco plants (Durner et al., 1998). cADPR, which can act as second messenger in NO-triggered signalling pathways in animals, was also able to induce some defence genes in the same system, but not in the presence of ruthenium red. This suggests that both cADPR and $Ca^{2+}$ are intermediates in NO-triggered defence responses in plants, possibly through the action of ryanodine-type receptors at the tonoplast. cADPR has been implicated in ABA and voltage-evoked $Ca^{2+}$ oscillations (Grabov & Blatt, 1998, 1999) and it acts as a second messenger in $Ca^{2+}$ release by NO and ABA (Garcia-Mata et al., 2003). cADPR has also been implicated as part of an ABA-based signalling cascade in gene expression (Wu et al., 1997). Using expression of the cold- and drought-responsive genes *RD29A* and *KIN2* as reporters, it was shown that their expression was not only induced by ABA but that microinjection of 1 μM cADPR mimicked the effect of ABA. Furthermore, indirect evidence suggests that ABA leads to increased cADPR concentrations *in vivo* and that the cADPR-evoked gene expression relies on a $Ca^{2+}$ signal possibly mediated by ryanodine-type receptors (Wu et al., 1997).

The same authors show data suggesting that $IP_3$ is not involved in the primary ABA response. The latter appears to contradict the many observations that $IP_3$ signalling and $IP_3$-triggered $Ca^{2+}$ signalling are important in drought and osmotic stress. However, the $IP_3$-based signal and its concomitant $Ca^{2+}$ transient are extremely rapid (within seconds), whereas ABA production and its associated signalling cascades only come into effect at timescales of minutes and hours. Thus, osmotic stress may provide an example of a stimulus where the early response, such as regulation of transporters for salts and water, relies on phosphoinositide ($IP_3$) signalling, whereas during later stages ABA triggers long-term responses consisting of changes in gene expression that require cADPR as an intermediate. The different second messengers $IP_3$ and cADPR would evoke temporally diverse $Ca^{2+}$ signals. In addition, cADPR may be able to access spatially different $Ca^{2+}$ stores such as the ER (Navazio et al., 2001), adding a further dimension to the creation of specific $Ca^{2+}$ signatures.

## 7.6   Non-selective ligand-gated ion channels

The predominant type of ion channel found in plant membranes is $K^+$-selective and voltage-dependent (see Chapter 6). However, during many *in planta* studies currents were observed that did not show any clear time and/or voltage dependence and often no selectivity amongst cations. Initially, these currents were often regarded as artefactual and merely constituting a 'leak' without much physiological relevance.

However, it became clear that discrete single channel events underlie these currents and more recently researchers have come to recognise the importance of non-selective cation channels (NSCCs) or voltage-independent channels.

There are two areas of plant research where interest in NSCCs has been substantial. The first is the mechanism of $Ca^{2+}$ entry into plant cells. As discussed in the previous sections, there is good evidence for the presence of (ligand-gated) highly selective $Ca^{2+}$ channels at plant endomembranes. However, $Ca^{2+}$ currents recorded at the plasma membrane frequently appear to be mediated by ion channels that are not, or only weakly, selective for $Ca^{2+}$. These $Ca^{2+}$-permeable NSCCs may play roles in $Ca^{2+}$ acquisition and $Ca^{2+}$ signalling as examplified by studies on *Arabidopsis* root hairs (Very & Davies, 2000; Demidchik *et al.*, 2002a), on reconstituted channels (e.g. White, 1993) and on the effects of fungal elicitors (Zimmermann *et al.*, 1997).

The second area where NSSCs are likely to play a major role is in plant salinity stress, where a major question concerns the mechanism of $Na^+$ entry into the plant root. No $Na^+$-selective ion channel has been found in plants and the search for $Na^+$ entry pathways focused initially on voltage-dependent cation-selective ion channels since these constitute the dominant component of plasma membrane conductance (Schachtman *et al.*, 1991). However, both inward and outward rectifying (voltage-dependent) cation channels were generally found to be highly selective for $K^+$ with a negligible $Na^+$ permeability (see Chapter 6; Amtmann & Sanders, 1999; Tyerman & Skerrett, 1999). The subsequently characterised voltage-independent channels (White, 1993; Elzenga & Volkenburg, 1994; Tyerman *et al.*, 1997; Amtmann & Sanders, 1999; Maathuis & Sanders, 2001) showed little or no selectivity amongst monovalent cations and thus formed prime suspects as entry pathways for $Na^+$. Later studies showed a remarkable similarity between $Ca^{2+}$-dependent block of NSSC current (as recorded in patch clamp experiments) and $Na^+$ influx in intact tissue, measured with radiolabelled $Na^+$ (Tyerman & Skerett, 1999; Davenport & Tester, 2000). These combined observations have led to the current paradigm that NSSCs form a major pathway for $Na^+$ entry into plants.

*In planta* experiments have shown that gating of many NSCCs is not dependent on membrane voltage. A crucial question, therefore, is how the activity of such channels is regulated. If membrane polarisation is not, or not the major, gating control mechanism, it is likely that activity of voltage-independent NSCCs is modulated by ligands and/or post-translation modification.

One of the major obstacles in progressing our knowledge regarding gating characteristics of NSCCs is a lack of molecular insights into which genes encode these transporters. Although full genome sequences are available for several plant species, it has remained problematic to definitively ascribe *in vivo* obtained data for NSCCs to specific gene products. Nevertheless, sequencing has revealed two major gene families in genomes of *Arabidopsis* and other plants that bear the hallmarks of voltage-independent non-selective cation channels, namely the glutamate receptor family and the cyclic-nucleotide-gated channel family.

### 7.6.1  Glutamate receptors

The predominant excitatory neurotransmitter in the central nervous system of vertebrates is the amino acid glutamate, though other amino acids such as aspartate can also act as neurotransmitter (Berg *et al.*, 2001). Glutamate acts on (ionotropic) glutamate receptors (GluRs), which can be subdivided into NMDA (gated by glutamate or aspartate and requiring glycine as a cofactor), kainate (gated by glutamate and kainic acid) and AMPA (gated by glutamate and amino-hydroxy-methyl-isoazoleproprionate) types, according to their selective agonists (Bigge, 1999). AMPA and kainate receptors generally show limited permeability to $Ca^{2+}$. Gating is generally not sensitive to membrane voltage but rectification may occur through voltage-dependent block by $Mg^{2+}$. The functioning channel complex of GluRs is likely to consist of heterotetra- or pentamers, with each subunit containing three transmembrane segments and a hydrophobic loop that lines the pore region and includes the selectivity filter (Fig. 7.3; Dingledine *et al.*, 1999). Ligand binding is believed to take place at two extracellular sites. There is a large diversity in GluRs because of large gene families (e.g. more than 17 in mammals) encoding subunits, alternative splicing variants and post-transcriptional mRNA editing. This allows fine-tuning of channel characteristics for specific cell types. For example, some GluRs show exceptionally rapid desensitisation, with channel opening lasting for only a few milliseconds in spite of continuing presence of agonists (Bigge *et al.*, 1999).

The *Arabidopsis* genome sequence revealed a large gene family (20 members) encoding proteins with high degrees of homology to animal GluRs (Lam *et al.*, 1998; Davenport, 2002). These plant putative glutamate-receptor-like gene products

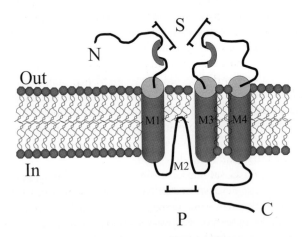

**Fig. 7.3** Glutamate receptor topology. From sequence data it is deduced that a plant glutamate receptor subunit is likely to contain four transmembrane spanning regions (M1–M4) with a pore region (P) that loops into the membrane and contains the selectivity filter. On the external side of the membrane, two putative glutamate-binding sites are located (S). Assembly of four or possibly five subunits is necessary to form a functional ion channel.

(GLRs) show similar membrane topology to their animal counterparts, with three membrane spanning regions, a pore between M1 and M3 and two putative extra-cellular glutamate binding sites (Fig. 7.3). However, the selectivity filter in the pore region of GLRs is substantially different from that of animal GluRs (Plate 1, see colour plate section). Cation-selective channels, including animal GluRs, typically contain anionic residues in the filter region, which are believed to form the basis of electrostatic interaction between the pore wall and permeating ion (Hille, 2001). Surprisingly, GLRs show an incidence for positive residues in these positions, which is reminiscent of anion selectivity filters (Dutzler *et al.*, 2002). This could impact on the type of ion that is transported by GLRs or indicate that many of the GLRs may not form functional ion channels.

The 20 *Arabidopsis* GLR isoforms are mostly homologous to mammalian NMDA receptor subunits and can be distinguished into three subfamilies (Davenport, 2002). For all but six of the GLRs, either ESTs (expressed sequence tags) or mRNA se-quences are available, with GLR3-4 appearing the most prevalent, being expressed in root, shoot, silique and flower tissues (http://mips.gsf.de). Indeed, isoforms from the third subgroup appear to be more frequently transcribed than those of the other two groups, possibly suggesting a greater physiological relevance.

Presently, very little is known regarding the physiological role of most GLRs, although glutamate and other amino acids have been implicated in plant signalling, particularly with respect to the nitrogen status and sensing of the nitrogen–carbon ratio (Hsieh *et al.*, 1998; Jeschke & Hartung, 2000). Application of the GluR antago-nist DNQX (6,7-dinitroquinoxaline-2,3-dione) to the growth substrate of *Arabidop-sis* seedlings produced plants that mimicked the 'long hypocotyl' or 'hy' phenotype (Lam *et al.*, 1998). This abnormal hypocotyl elongation occurred only when DNQX-treated plants were grown in the light. Furthermore, similar experiments showed that chlorophyll production was adversely affected. Assuming that DNQX blocks plant GLRs, these observations suggest a role of GLRs in light signal transduction in plants possibly by mediating $Ca^{2+}$ transport. Another study (Brenner *et al.*, 2000[Q5]) showed that the addition of BMAA ($S(+)$-β-methyl-α,β-diaminopropionic acid, a cycad-derived GLR agonist) to the medium of *Arabidopsis* seedlings led to a two- to threefold increase in hypocotyl elongation and inhibited cotyledon opening during early seedling development. The effects of BMAA, as those of DNQX, on hypocotyl elongation were light-specific, supporting the notion that GLRs are involved in light signal transduction in plants.

More specific characterisations have been carried out for some GLR isoforms. Overexpression of *AtGLR3.2* led to $Ca^{2+}$ deficiency symptoms in the transgenic plants, which were alleviated when extra $Ca^{2+}$ was supplied (Kim *et al.*, 2001). In addition, transgenic plants exhibited hypersensitivity to $Na^+$ and $K^+$ stress, and these symptoms were also alleviated by supplementation with $Ca^{2+}$. Surprisingly, overall $Ca^{2+}$ tissue levels were not different in overexpressing plants, showing that $Ca^{2+}$ uptake as such is probably not affected. Promoter-GUS studies showed expression patterns predominantly in the vicinity of phloem and xylem vessels in both roots and shoots. Further expression analysis (Turano *et al.*, 2002) showed that *AtGLR3.2* transcript accumulation also occurs in rapidly dividing regions such as developing flower buds. The prominent $Ca^{2+}$ phenotype and expression pattern may indicate that at least AtGLR3.2 is involved in $Ca^{2+}$ (re)distribution.

The possibility that glutamate gates plant GLRs was tested by measuring alterations in membrane potential after exposure of root cells to various agonists (Dennison & Spalding, 2001). These experiments showed that a large membrane depolarisation ensued when cells were exposed to 1 mM L-glutamate. By using aequorin expressing cells, changes in cytoplasmic $Ca^{2+}$ were recorded in parallel experiments and this assay was also used to test the effects of other agonists such as D-glutamate, arginine, NMDA and AMPA, showing that L-glutamate was by far the most potent inducer of a cytoplasmic $Ca^{2+}$ signal. The glutamate-induced $Ca^{2+}$ signal could be effectively eliminated by (pre-)incubation with the $Ca^{2+}$ channel blocker $La^{3+}$. A more recent study showed that glycine (a cosubstrate required for activation of mammalian NMDA receptors) can also cause a rise in cytoplasmic $[Ca^{2+}]$ and affects hypocotyl growth (Dubos et al., 2003); however, no membrane potential recordings were carried out to test the effect of glycine.

AtGLR1.1 antisense studies indicate that this particular isoform may be involved in sensing carbon and nitrogen metabolism via transcriptional regulation of distinct N- and C-metabolic enzymes (Kang & Turano, 2003). This interpretation should be treated with caution though, since the authors failed to show that their antisense construct does not affect expression of other GLRs.

So what do all these observations signify? First of all, most of this work is based on pharmacological studies that often extend across several days. Many of the used compounds may have (unknown) side effects that will confound interpretation of the data. However, the observations seem to show that external glutamate, and also glycine, induces a cytoplasmic $Ca^{2+}$ signal and that the $Ca^{2+}$ is likely to be of apoplastic origin. Still, there is no *direct* confirmation that a GLR-type ion channel is involved. Legion stimuli cause a $Ca^{2+}$ signal and it cannot be ruled out that glutamate interferes with receptors of unrelated stimuli. Alternatively, glutamate may act on a genuine glutamate receptor, but this receptor is not necessarily a GLR or not even an ion channel but activates a $Ca^{2+}$-permeable channel as a secondary response. More intriguing is the consideration why plants would have glutamate and/or glycine responsive receptors in the first place, especially with ligand-binding sites that would respond to external glutamate and/or glycine. Both these compounds have been found in root exudates (Paynel et al., 2001), with glutamate being released in certain nutritional conditions as are other carboxylic acids like malate and citrate. For example, phosphate depletion induces the extrusion of citrate, which acidifies the rhizosphere to release bound phosphate and can also act as phosphate chelator (Schachtman et al., 1998). Release of chelating organic acids is also believed to be important in response to $Al^{3+}$ toxicity where $Al^{3+}$ itself activates an anion conductance that may be involved in glutamate secretion. Subsequently, the glutamate-induced activation of a GLR would ensure $Ca^{2+}$ signalling and downstream responses (Sivaguru et al., 2003).

For unequivocal evidence that GLRs function as ionotropic signalling pathways, an electrophysiological approach is crucial. Only such techniques will allow us to establish whether glutamate or glycine are capable of directly opening plant ion channels, whether these ion channels are GLRs and what their characteristics are in terms of type of ion(s) that can permeate, voltage dependence, pharmacology, etc. Various groups have attempted the heterologous expression of GLRs for functional characterisation, but no glutamate-dependent currents have been reported. Indeed, it has been suggested that (some) GLRs may be constitutively active ion channels (Chiu *et al.*, 1999; Davenport, 2002) in spite of the conservation of ligand-binding sites. If the latter were the case, characterisation in heterologous expression systems would remain problematic and a major problem to solve would concern the gating mechanism that operates to control GLR conductance.

### 7.6.2 Cyclic-nucleotide-gated channels

Many reports have shown that the second messengers $3',5'$-cyclic AMP (cAMP) and $3',5'$-cyclic GMP (cGMP) play important roles in many aspects of growth and development of higher plants. Such roles include gibberellic acid induced signalling (Penson *et al.*, 1996), phytochrome signalling (Bowler *et al.*, 1994), stomatal movement (Garcia-Mata *et al.*, 2003; Garcia-Mata & Lamattina, 2003), pollen tube tip growth (Moutinho *et al.*, 2001), plant cell cycle progression (Ehsan *et al.*, 1998) and salt stress tolerance (Maathuis & Sanders, 2001).

In all cases, the signalling pathways remain mostly unknown, both at the biochemical level and regarding the genes that participate. A classic set of downstream effectors of cyclic nucleotide signalling in both animals and fungi are protein kinases (PKA and PKG), which in turn regulate downstream enzyme targets and the expression of many genes through phosphorylation of transcription factors such as the cAMP-responsive element-binding protein (CREB). CREB-type transcription factors have been reported in tobacco (Katagiri *et al.*, 1989) and in pollen tubes (Moutinho *et al.*, 2001).

Biochemical activity of the key enzymes for the generation and breakdown of cyclic nucleotides, adenylyl and guanylyl cyclases and phosphodiesterases, respectively, has been demonstrated in plants (Newton *et al.*, 1999). Furthermore, a cDNA named PSiP for ('pollen-signalling protein') was isolated recently from a maize pollen cDNA library (Moutinho *et al.*, 2001) that showed adenylyl cyclase activity when expressed in *Escherichia coli*. In addition, the identification and *in vitro* functional characterisation of the first plant guanylyl cyclase, AtGC1, was reported recently (Ludidi & Gehring, 2003).

In animals, non-selective ion channels function in the transduction of sensory input and in $Ca^{2+}$ signalling (e.g. Biel *et al.*, 1996). Many of the participating ion channels are gated via binding of cAMP or cGMP to a domain near the C-terminus. Additional control over gating is provided by a calmodulin (CaM) binding site at the N-terminus, which interacts with the cyclic-nucleotide-binding domain, thereby lowering the affinity for cAMP or cGMP. As in animal cells, plant cells too may contain ion channels that form cAMP or cGMP targets. Many $K^+$-selective plant channels contain putative cyclic-nucleotide-binding sites and in some cases their activity has been found to be modulated by cGMP: expressed in oocytes, 0.1 to

2 mM cGMP reduced the activity of *Arabidopsis thaliana* Shaker-type $K^+$ channels AKT1 and KAT1 with a time constant of several minutes (Hoshi, 1995; Gaymard *et al.*, 1996). On the basis of a large time constant and high nucleotide concentration, it is unlikely that modulation of activity involves direct binding of nucleotides to the channel protein. cAMP has also been shown to stimulate activity of the $K^+$-selective outward rectifier ion channel in *Vicia faba* mesophyll cells. In this case too, the data do not suggest direct channel–nucleotide interactions but the involvement of a PKA-like kinase (Li *et al.*, 1994). Thus, these channels do not constitute CNGCs, since the main mode of gating is probably through alteration of the membrane voltage. Other voltage-independent non-selective channels in *Arabidopsis* root plasma membranes contribute to $Na^+$ uptake (Maathuis & Sanders, 2001) and their activity is rapidly decreased in the presence of cyclic nucleotides. This type of channel appears to be directly modulated by cyclic nucleotides but, in contrast to animal CNGCs and the few characterised plant CNGCs, activity is downregulated by cyclic nucleotides.

Putative CNGCs have recently been cloned in plants. HvCBT1, cloned from barley on the basis of its capacity to bind calmodulin (Schuurink *et al.*, 1998), has a protein sequence that shows similarity to Shaker-like $K^+$ channels and animal CNGCs (Fig. 7.4). HvCBT homologues have now been identified in many species, including tobacco, rice and *Arabidopsis* (see Talke *et al.*, 2003, for a review), which contains a large gene family of 20 members that encodes CNGCs.

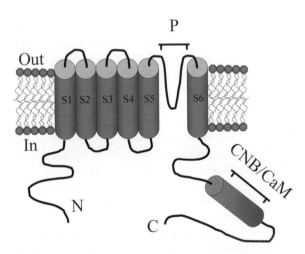

**Fig. 7.4** Cyclic-nucleotide-gated channel structure. Plant CNGCs are predicted to contain six transmembrane spanning regions (S1–S6) in analogy with voltage-gated channels from the Shaker family. The pore region (P) that loops into the membrane contains the selectivity filter, whereas the C-terminus includes a cyclic-nucleotide-binding (CNB) domain and a calmodulin-binding (CaM) domain. The latter do partly overlap and the possible competition between the cyclic nucleotides that function as ligand and calmodulin that acts as an antagonist may thus provide an additional mechanism to modulate channel activity. Assembly of four subunits forms a functional ion channel.

Like animal CNGCs, plant CNGCs have a predicted structure of six transmembrane domains, S1–S6, with a pore domain (P loop) between S5 and S6, and C-terminal cyclic-nucleotide-binding (CNB) and CaM-binding (CaMB) domains (see Fig. 7.4). However, caution has to be taken in interpreting hydrophobicity plots to define transmembrane domains. For example, the Institute for Genomic Research (TIGR: http://www.tigr.org) predicts anything between three and seven transmembrane domains for the 20 members of the *Arabidopsis* CNGC family.

In contrast to their animal counterparts, the CaMB region of plant CNGCs is located at the C-terminus and overlaps the CNB domain (Arazi *et al.*, 1999; Koehler & Neuhaus, 2000). It has been suggested that the binding of CaM to the C-terminus of plant CNGCs might interfere with cyclic nucleotide binding and thus channel activation (Arazi *et al.*, 2000). This implies a different mechanism for regulation of plant CNGCs by CaM than that described for olfactory and rod CNGCs of animals (Kaupp & Seifert, 2002). Differential regulation of CNGC isoforms is provided by CaMB domains with different CaM affinities. In addition, the CaMB domains of various CNGC isoforms are targeted by different CaM isoforms (Koehler & Neuhaus, 2000).

The P loop is another domain that sets plant CNGCs apart from other known ion channel subunits (Talke *et al.*, 2003). As illustrated in Plate 2 (see colour plate section), the region that is believed to endow ion selectivity differs significantly both from the pores of animal CNGCs, which are non-selective among cations, and from pores of $K^+$-selective channels.

$K^+$ channel selectivity is believed to originate from a short motif GYG (glycine, tyrosine, glycine) situated in the pore region that forms the selectivity filter and is responsible for discrimination against $Na^+$ permeation. In most plant and in all animal CNGCs, the first glycine (G) that forms part of the $K^+$-selective channel signature (GYG) is conserved (Plate 2), but the substitution of the Y and second G residue leading to GET (glycine, glutamate, threonine) or G_ET sequences in animal CNGC are believed to create non-selectivity among monovalent cations (Zagotta & Siegelbaum, 1996). Most plant CNGCs show GQN sequences in this region, and electrophysiological characterisation of AtCNGC1 (Leng *et al.*, 2002) and AtCNGC4 (Balague *et al.*, 2003) indicate that such selectivity filters may retain the ability to allow $Na^+$ permeation. However, AtCNGC2, AtCNGC20 and OsIIC-NGCa contain an AND (alanine, asparagine, aspartate) motif in the selectivity filter and recent work (Hua *et al.*, 2003) shows that such motifs may endow $K^+$ over $Na^+$ selectivity without the archetypal GYG signature.

Only limited information is available regarding how and where CNGCs function and what their physiological role would be. Figure 7.5 shows a summary of the currently available data on *Arabidopsis* CNGC expression. It appears that particularly members of subgroup I are relatively highly expressed in all tissues. In contrast, several members, especially from subgroup III and IVA family are not represented in the MPSS (massively parallel signature sequence) database. These isoforms might either be expressed at very low levels, in specific conditions, or they might be pseudogenes. In addition, the various CNGC isoforms may be expressed

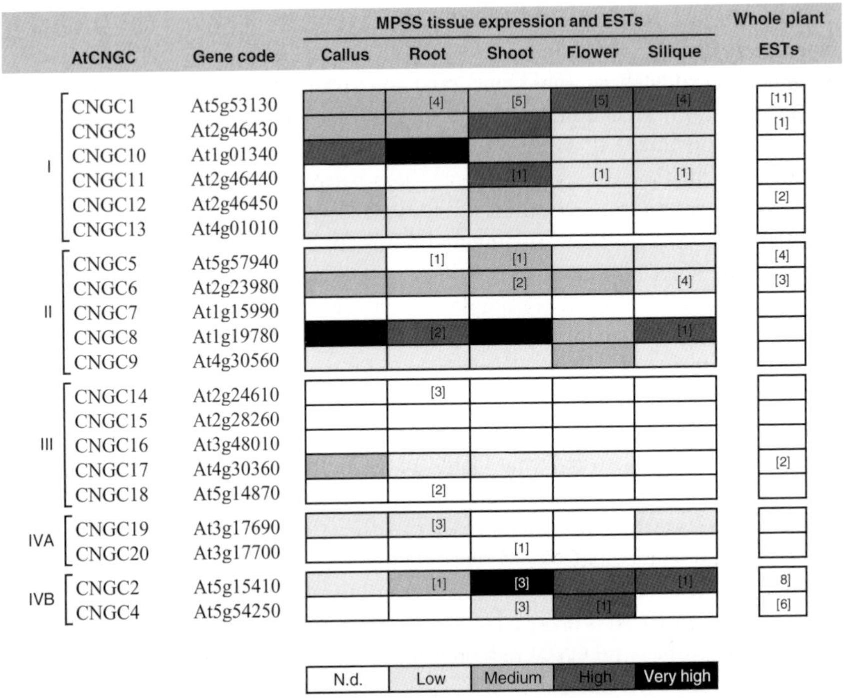

| | AtCNGC | Gene code | MPSS tissue expression and ESTs | | | | | Whole plant ESTs |
|---|---|---|---|---|---|---|---|---|
| | | | Callus | Root | Shoot | Flower | Silique | |
| I | CNGC1 | At5g53130 | | [4] | [5] | [5] | [4] | [11] |
| | CNGC3 | At2g46430 | | | | | | [1] |
| | CNGC10 | At1g01340 | | | | | | |
| | CNGC11 | At2g46440 | | | [1] | [1] | [1] | |
| | CNGC12 | At2g46450 | | | | | | [2] |
| | CNGC13 | At4g01010 | | | | | | |
| II | CNGC5 | At5g57940 | | [1] | [1] | | | [4] |
| | CNGC6 | At2g23980 | | | [2] | | [4] | [3] |
| | CNGC7 | At1g15990 | | | | | | |
| | CNGC8 | At1g19780 | | [2] | | | [1] | |
| | CNGC9 | At4g30560 | | | | | | |
| III | CNGC14 | At2g24610 | | [3] | | | | |
| | CNGC15 | At2g28260 | | | | | | |
| | CNGC16 | At3g48010 | | | | | | |
| | CNGC17 | At4g30360 | | | | | | [2] |
| | CNGC18 | At5g14870 | | [2] | | | | |
| IVA | CNGC19 | At3g17690 | | [3] | | | | |
| | CNGC20 | At3g17700 | | | [1] | | | |
| IVB | CNGC2 | At5g15410 | | [1] | [3] | | [1] | 8 |
| | CNGC4 | At5g54250 | | | [3] | [1] | | [6] |

| N.d. | Low | Medium | High | Very high |
|---|---|---|---|---|

**Fig. 7.5** Expression pattern of *Arabidopsis* CNGCs. Relative expression of each CNGC gene was graded from 'low' to 'very high', on the basis of the expression data in the MPSS (massively parallel signature sequence) database (http://mpss.ucdavis.edu). The normalised abundance data available for a given gene were classified as follows: N.d., not detected; Low, 1–19 ppm; Medium, 20–49 ppm; High, 50–100 ppm; Very high, > 100 ppm. Numbers in brackets indicate the number of expressed sequence tags (ESTs) found for a given gene in tissue-specific or whole plant cDNA libraries. Information on ESTs was obtained from The *Arabidopsis* Information Resource database (http://arabidopsis.org). Figure reproduced from Talke *et al.* (2003), by permission of Elsevier Science.

in a tightly regulated cell-specific expression pattern. Two CNGCs, HvCBT1 and NtCBP4 (Schuurink *et al.*, 1998; Arazi *et al.*, 1999), were shown to localise to the plasma membrane. However, whether this is also the case for other members of the CNGC family remains to be seen and so far it cannot be excluded that some CNGCs are targeted to internal membranes.

An efficient way to functionally characterise plant transporters is the complementation of clearly defined phenotypes in systems such as yeast. However, transformation with various plant CNGCs of yeast deficient in $K^+$ uptake, $Na^+$ efflux or $Ca^{2+}$ uptake frequently failed or was only partially successful (Schuurink *et al.*, 1998; Arazi *et al.*, 1999; Leng *et al.*, 1999). So far only one report demonstrating cAMP-dependent full complementation of a yeast mutant deficient in $K^+$ uptake (Leng *et al.*, 1999) has been published, suggesting that plant CNGCs are capable of transporting $K^+$.

As was mentioned above regarding GLRs, it has also proven frustrating to achieve functional expression of plant CNGCs in heterologous expression systems for electrophysiological analysis. Functional mammalian CNGCs have long been believed to be heterotetramers, based on studies comparing the functional properties of CNGCs in native tissues with those of CNGC subunits expressed in heterologous systems as either homomers or heteromers (Kaupp & Seifert, 2002). Indeed, homomeric expression of certain (previously called β-) subunits did not lead to functional channels, whereas so-called α-subunits did. Thus, some of the plant CNGC subunits may be regulatory rather than being capable of forming functioning channels. Nevertheless, the use of two-electrode voltage-clamp and patch-clamp approaches have led to the recording of cAMP- and cGMP-induced cation currents in oocytes expressing AtCNGC1, AtCNGC2, AtCNGC4 and NtCBP4 (Leng *et al.*, 1999; Leng *et al.*, 2002; Balague *et al.*, 2003). However, the very low incidence of functional channel expression means that only a rudimentary description of plant CNGCs is available. From these limited experimental data some surprising characteristics have emanated. For example, AtCNGC2 exhibited a high degree of $K^+$ selectivity with regard to $Na^+$ (see above) a feature that is unknown in animal CNGCs (Hua *et al.*, 2003). In contrast, AtCNGC4 does not discriminate between $K^+$ and $Na^+$ but is blocked in a voltage-dependent manner by $Cs^+$. Interestingly, AtCNGC1, AtCNGC2 and NtCBP4 all show considerable voltage dependence resulting in a strong inward rectification. It remains to be revealed whether the rectification is the result of a 'genuine' voltage sensor (see Chapter 6) that regulates gating, is caused by channel block from cytoplasmic impermeable ions such as $Mg^{2+}$, or originates in asymmetrical channel architecture. No voltage dependence was observed for AtCNGC4 (Balague *et al.*, 2003).

Expression of AtCNGC2 in human embryonic kidney (HEK) cells loaded with $Ca^{2+}$ reporter dye showed that cAMP could induce an increase in cytosolic $Ca^{2+}$. In brief, the heterologous expression data suggest that plant CNGCs are activated by cyclic nucleotides, all isoforms studied conduct $K^+$, whereas some appear to discriminate against $Na^+$. CNGCs may be blocked by $Cs^+$ and divalent cations such as $Mg^{2+}$, and indirect evidence shows they may form pathways for $Ca^{2+}$ influx. Thus, *CNGC* genes may encode the non-selective channel proteins that mediate cation currents including $Ca^{2+}$ currents that have been observed by electrophysiological methods *in planta* (Demidchik *et al.*, 2002b; Very & Sentenac, 2002).

Several mutations in CNGCs have been reported and characterised. Truncation and null mutants in *NtCBP4* and *AtCNGC1* showed $Pb^{2+}$ and $Ni^{2+}$ associated phenotypes, indicating that CNGCs may be involved in heavy metal transport (Sunkar *et al.*, 2000). AtCNGC2 (Clough *et al.*, 2000) and AtCNGC4 (Balague *et al.*, 2003) mutants both exhibit altered responses to pathogen attack. It is well documented that $Ca^{2+}$ as well as $K^+$ and $Cl^-$ fluxes occur during the early phase of plant defence responses and in addition, cyclic nucleotides have been implicated in plant defence-related signalling. For example, the production of reactive oxygen species (ROS) (which have antimicrobial activity) requires cAMP and $Ca^{2+}$ in French bean cells (Bindschedler *et al.*, 2001). Treatment of tobacco suspension cells with nitric oxide causes a transient increase in cGMP levels (Durner *et al.*, 1998). Thus, both $Ca^{2+}$ and cyclic-nucleotide-based signalling form part of defence responses, which in CNGC2 null mutants lack the 'hypersensitive response', i.e. programmed cell death, whereas CNGC4 null mutants show a 'lesion mimic' phenotype where spontaneous cell death occurs without the presence of a pathogen. In both instances, ion fluxes that form part of the early signalling response to pathogen attack may be disturbed because of non-functional CNGCs. Interestingly, mutations in AtCNGC2 also affect homeostasis of ions such as $Ca^{2+}$ (Chan *et al.*, 2003).

The parallel occurrence and/or requirement of cyclic nucleotide and $Ca^{2+}$ signalling is a recurring theme, not only in defence-related responses but also in guard cell signalling (Garcia-Mata *et al.*, 2003), pollen tube tip growth (Moutinho *et al.*, 2001; Bindschedler *et al.*, 2001) and aspects of phytochrome phototransduction (Bowler *et al.*, 1994). Indeed, there is good evidence for a direct link between increased levels of cyclic nucleotides and a rise in cytoplasmic $Ca^{2+}$ (Volotovski *et al.*, 1998; Leng *et al.*, 1999). From these observations and the occurrence of $Ca^{2+}$-related phenotypes in CNGC null mutants, it is tempting to hypothesise that a main function of CNGCs may be to form intermediates where cyclic nucleotide signals are translated into $Ca^{2+}$ signals for downstream applications.

## 7.7   Concluding remarks

When a protein–protein homology search is carried out with plant or animal ligand-gated channels as template, nil or only very few hits are obtained from prokaryotic or unicellular eukaryotic organisms. One putative glutamate receptor, probably originating from an amino acid binding protein (Chen *et al.*, 1999), has been found in bacteria, but prokaryotes do not appear to contain ryanodine receptors, $IP_3$ receptors, CNGCs or other specialised ligand-gated channels. Most ligand-gated ion channels appeared relatively late in evolution (around 800 million years ago), after eukaryotic life was established, and most of these transporters are found only in multicellular organisms where they play specific roles in inter- and intracellular communication. Voltage-gated channels, on the other hand, can be traced back at least 2.5 billion years.

In plants, as in animals, it is likely that ligand-gated channels are crucial when rapid signalling is required, particularly when signals need conversion from one form into another. Examples of the latter include the $IP_3$-induced opening of $Ca^{2+}$ channels or the cyclic-nucleotide-dependent gating of non-selective cation channels during pathogen attack. Thus, the main function of plant ligand-gated ion channels is likely to be in signalling events, although participation in heavy metal or $Na^+$ transport may have evolved as secondary functions or simply occur as side effects.

Future research will have to reveal why plants need such large families of putative ligand-gated channels such as the CNGC and GLR families. Are all members of these families functioning channels or are some subunits merely regulatory subunits or not expressed at all? Large size gene families would allow an enormous diversity in physiological functioning where specific isoforms are expressed in specific cell types or during particular stages. Since many channels have been found to function as heteromers with two, three or even four different subunits, the number of permutations is seemingly endless. An additional complicating mechanism is the occurrence of post-transcriptional editing of pre-mRNAs. Development-dependent residue substitutions have been shown to alter animal channel characteristics such as selectivity (Hanrahan *et al.*, 2000). Similar systems may be present in plants to fine-tune channel functioning according to varying physiological demands in space and time.

We are only beginning to answer questions regarding physiological roles of plant ligand-gated channels. Expression studies and the use of reverse genetics ('knockout mutants') can greatly help in revealing putative roles of proteins. In the case of several plant ligand channels, such as the IP$_3$-, cADPR- and NAADP-gated channels, such approaches are frustrated by the fact that the encoding genes are not known even though their animal counterparts have long been identified. Such apparent lack of homology between animal and plant orthologues is remarkable, pointing to an early evolutionary divergence, but is not unprecedented. Until recently, plant adenyl and guanyl cyclases had been detected only at a biochemical level but now several genes whose products show cyclase activity (e.g. Ludidi & Gehring, 2003) have been identified and some of these show only low levels of homology to animal cyclases.

In addition, successful heterologous expression of many plant ligand-gated channels has proved very difficult, thereby greatly hampering progress in establishing fundamental properties such as channel selectivity, ligand affinity, etc.

Thus, gaining better insights into the exact roles of plant ligand-gated channels may be a challenge but by applying new techniques such as proteomics of plant membrane proteins (Kubis *et al.*, 2003) and single molecule labelling of ion channels (Dahan *et al.*, 2003), we should be able to progress significantly in the near future.

# References

Aidley, D.J. & Stanfield, P.R. (1996) *Ion Channels, Molecules in Action*, Cambridge University Press, Cambridge, UK.

Alexandre, J., Lassalles, J.P. & Kado, R.T. (1990) Opening of $Ca^{2+}$ channels in isolated red beet root vacuole membrane by inositol 1,4,5-triphosphate, *Nature*, **343**, 567–570.

Allen, G.J., Chu, S.P., Harrington, C.L., *et al.* (2001) A defined range of guard cell calcium oscillation parameters encodes stomatal movements, *Nature*, **411**, 1053–1057.

Allen, G.J., Muir, S.R. & Sanders, D. (1995) Release of $Ca^{2+}$ from individual plant vacuoles by both Insp(3) and cyclic ADP-ribose, *Science*, **268**, 735–737.

Allen, G.J. & Sanders, D. (1994) Osmotic-stress enhances the competence of *Beta vulgaris* vacuoles to respond to inositol 1,4,5-trisphosphate, *Plant J.*, **6**, 687–695.

Amtmann, A. & Sanders, D. (1999) Mechanisms of $Na^+$ uptake by plant cells, *Adv. Bot. Res. Incorporating Adv. Plant Pathol.*, **29**, 75–112.

Arazi, T., Kaplan, B. & Fromm, H. (2000) A high-affinity calmodulin-binding site in a tobacco plasma-membrane channel protein coincides with a characteristic element of cyclic nucleotide-binding domains, *Plant Mol. Biol.*, **42**, 591–601.

Arazi, T., Sunkar, R., Kaplan, B. & Fromm, H. (1999) A tobacco plasma membrane calmodulin-binding transporter confers $Ni^{2+}$ tolerance and $Pb^{2+}$ hypersensitivity in transgenic plants, *Plant J.*, **20**, 171–182.

Balague, C., Lin, B.Q., Alcon, C., *et al.* (2003) HLM1, an essential signaling component in the hypersensitive response, is a member of the cyclic nucleotide-gated channel ion channel family, *Plant Cell*, **15**, 365–379.

Berg, J.M., Tymockzko, J.L. & Stryer, L. (2001) *Biochemistry*, 5th edn, Freeman and Co., New York.

Bewell, M.A., Maathuis, F.J.M., Allen, G.J. & Sanders, D. (1999) Calcium-induced calcium release mediated by a voltage-activated cation channel in vacuolar vesicles from red beet, *FEBS Lett.*, **458**, 41–44.

Biel, M., Zong, X.G. & Hofmann, F. (1996) Cyclic nucleotide-gated cation channels – molecular diversity, structure, and cellular functions, *Trends Cardiovasc. Med.*, **6**, 274–280.

Bigge, C.F. (1999) Ionotropic glutamate receptors, *Curr. Opin. Chem. Biol.*, **3**, 441–447.

Bindschedler, L.V., Minibayeva, F., Gardner, S.L., Gerrish, C., Davies, D.R. & Bolwell, G.P. (2001) Early signalling events in the apoplastic oxidative burst in suspension cultured French bean cells involve cAMP and $Ca^{2+}$, *New Phytol.*, **151**, 185–194.

Blatt, M.R. (1990) Potassium channel currents in intact stomatal guard cells: rapid enhancement by abscisic acid, *Planta*, **180**, 445–455.

Blatt, M.R., Thiel, G. & Trentham, D. (1990) Reversible inactivation of $K^+$ channels of *Vicia* stomatal guard cells following the photolysis of caged inositol 1,4,5-trisphosphate, *Nature*, **346**, 766–769.

Bowler, C., Neuhaus, G., Yamagata, H. & Chua, N.H. (1994) Cyclic GMP and calcium mediate phytochrome phototransduction, *Cell*, **77**, 73–81.

Brenner, E.D., Martinez-Barboza, N., Clark, A.P., Liang, Q.S., Stevenson, D.W. & Coruzzi, G.M. (2000) *Arabidopsis* mutants resistant to $S(+)$-beta-methyl-alpha, beta-diaminopropionic acid, a cycad-derived glutamate receptor agonist, *Plant Physiol.*, **124**, 1615–1624.

Buchanan, B.B., Gruissem, W. & Jones, R.L. (2000) *Biochemistry and Molecular Biology of Plants*, ASPP, Rockville.

Chan, C.W.M., Schorrak, L.M., Smith, R.K., Bent, A.F. & Sussman, M.R. (2003) A cyclic nucleotide-gated ion channel, CNGC2, is crucial for plant development and adaptation to calcium stress, *Plant Physiol.*, **132**, 728–731.

Chen, G.Q., Cui, C.H., Mayer, M.L. & Gouaux, E. (1999) Functional characterization of a potassium-selective prokaryotic glutamate receptor, *Nature*, **402**, 817–821.

Chiu, J., DeSalle, R., Lam, H.M., Meisel, L. & Coruzzi, G. (1999) Molecular evolution of glutamate receptors: a primitive signaling mechanism that existed before plants and animals diverged, *Mol. Biol. Evol.*, **16**, 826–838.

Clough, S.J., Fengler, K.A., Yu, I., Lippok, B., Smith, R.K. & Bent, A. (2000) The *Arabidopsis dnd1* 'defense, no death' gene encodes a mutated cyclic nucleotide-gated ion channel, *Proc. Natl. Acad. Sci. U.S.A.*, **97**, 9323–9328.

Colquhoun, D. (1998) Binding, gating, affinity and efficacy: the interpretation of structure–activity relationships for agonists and of the effects of mutating receptors, *Brit. J. Pharmacol.*, **125**, 924–947.

Dahan, M., Humbert, M., Hanus, C., Levi, S., Vannier, C. & Triller, A. (2003) Dynamics of glycine and GABA receptors in spinal cord neurons studied by single molecule fluorescence imaging, *Biophys. J.*, **84** (Pt 2, Suppl.), 123A.

Davenport, R. (2002) Glutamate receptors in plants, *Ann. Bot.*, **90**, 549–557.

Davenport, R.J. & Tester, M. (2000) A weakly voltage-dependent, nonselective cation channel mediates toxic sodium influx in wheat, *Plant Physiol.*, **122**, 823–834.

Demidchik, V., Bowen, H.C., Maathuis, F.J.M., *et al.* (2002a) *Arabidopsis thaliana* root nonselective cation channels mediate calcium uptake and are involved in growth, *Plant J.*, **32**, 799–808.

Demidchik, V., Davenport, R.J. & Tester, M. (2002b) Nonselective cation channels in plants, *Ann. Rev. Plant Biol.*, **53**, 67–107.

Dennison, K.L. & Spalding, E.P. (2001) Glutamate-gated calcium fluxes in *Arabidopsis*, *Plant Physiol.*, **125**, 2203–2203.

DeWald, D.B., Torabinejad, J., Jones, C.A., *et al.* (2001) Rapid accumulation of phosphatidylinositol 4,5-bisphosphate and inositol 1,4,5-trisphosphate correlates with calcium mobilization in salt-stressed *Arabidopsis*, *Plant Physiol.*, **126**, 759–769.

Dingledine, R., Borges, K., Bowie, D. & Traynelis, S.F. (1999) The glutamate receptor ion channels, *Pharmacol. Rev.*, **51**, 7–61.

Drobak, B.K., Dewey, R.E. & Boss, W.F. (1999) Phosphoinositide kinases and the synthesis of polyphosphoinositides in higher plant cells, *Int. Rev. Cytol.*, **189**, 95–130.

Drobak, B.K. & Watkins, P.A.C. (2000) Inositol(1,4,5) trisphosphate production in plant cells: an early response to salinity and hyperosmotic stress, *FEBS Lett.*, **481**, 240–244.

Du, G.G., Guo, X.H., Khanna, V.K. & MacLennan, D.H. (2002) Ryanodine sensitizes the cardiac ryanodine receptor (RyR2) to $Ca^{2+}$ activation and dissociates as the channel is closed by $Ca^{2+}$ depletion, *Biophys. J.*, **82**, 359.

Dubos, C., Huggins, D., Grant, G.H., Knight, M.R. & Campbell, M.M. (2003) A role for glycine in the gating of plant NMDA-like receptors, *Plant J.*, **35**, 800–810.

Durner, J., Wendehenne, D. & Klessig, D.F. (1998) Defense gene induction in tobacco by nitric oxide, cyclic GMP, and cyclic ADP-ribose, *Proc. Natl. Acad. Sci. U.S.A.*, **95**, 10328–10333.

Dutzler, R., Campbell, E.B., Cadene, M., Chait, B.T. & MacKinnon, R. (2002) X-ray structure of a ClC chloride channel at 3.0 angstrom reveals the molecular basis of anion selectivity, *Nature*, **415**, 287–294.

Ehsan, H., Reichheld, J.P., Roef, L., *et al.* (1998) Effect of indomethacin on cell cycle dependent cyclic AMP fluxes in tobacco BY-2 cells, *FEBS Lett.*, **422**, 165–169.

Elzenga, J.T.M. & Volkenburg, V.E. (1994) Characterization of ion channels in the plasma-membrane of epidermal cells of expanding pea (*Pisum Sativum Arg*) leaves, *J. Membr. Biol.*, **137**, 229–235.

FranklinTong, V.E., Drobak, B.K., Allan, A.C., Watkins, P.A.C. & Trewavas, A.J. (1996) Growth of pollen tubes of *Papaver rhoeas* is regulated by a slow-moving calcium wave propagated by inositol 1,4,5- trisphosphate, *Plant Cell*, **8**, 1305–1321.

Garcia-Mata, C., Gay, R., Sokolovski, S., Hills, A., Lamattina, L. & Blatt, M.R. (2003) Nitric oxide regulates $K^+$ and $Cl^-$ channels in guard cells through a subset of abscisic acid-evoked signaling pathways, *Proc. Natl. Acad. Sci. U.S.A.*, **100**, 11116–11121.

Garcia-Mata, C. & Lamattina, L. (2003) Abscisic acid, nitric oxide and stomatal closure – is nitrate reductase one of the missing links? *Trends Plant Sci.*, **8**, 20–26.

Gaymard, F., Cerutti, M., Horeau, C., *et al.* (1996) The baculovirus/insect cell system as an alternative to *Xenopus* oocytes, *J. Biochem.*, **271**, 22863–22870.

Gilroy, S., Read, N.D. & Trewavas, A.J. (1990) Elevation of cytoplasmic calcium by caged calcium or caged inositol trisphosphate initiates stomatal closure, *Nature*, **346**, 769–771.

Grabov, A. & Blatt, M.R. (1998) Membrane voltage initiates $Ca^{2+}$ waves and potentiates $Ca^{2+}$ increases with abscisic acid in stomatal guard cells, *Proc. Natl. Acad. Sci. U.S.A.*, **95**, 4778–4783.

Grabov, A. & Blatt, M.R. (1999) A steep dependence of inward-rectifying potassium channels on cytosolic free calcium concentration increase evoked by hyperpolarization in guard cells, *Plant Physiol.*, **119**, 277–287.

Hanrahan, C.J., Palladino, M.J., Ganetzky, B. & Reenan, R.A. (2000) RNA editing of the *Drosophila* para $Na^+$ channel transcript: evolutionary conservation and developmental regulation, *Genetics*, **155**, 1149–1160.

Hedrich, R., Barbier-Brygoo, H., Felle, H.H., *et al.* (1988) General mechanisms for solute transport across the tonoplast of plant vacuoles: a patch clamp survey of ion channels and proton pumps, *Botanica Acta*, **101**, 7–13.

Heilmann, I., Perera, I.Y., Gross, W. & Boss, W.F. (1999) Changes in phosphoinositide metabolism with days in culture affect signal transduction pathways in *Galdieria sulphuraria, Plant Physiol.*, **119**, 1331–1339.

Hille, B. (2001) *Ion Channels of Excitable Membranes*, Sinauer Associates, Inc., Sunderland, MA.

Hoshi, T. (1995) Regulation of voltage dependence of the KAT1 channel by intracellular factors, *J. Gen. Physiol.*, **105**, 309–328.

Hsieh, M.H., Lam, H.M., van de Loo, F.J. & Coruzzi, G. (1998) A PII-like protein in *Arabidopsis*: putative role in nitrogen sensing, *Proc. Natl. Acad. Sci. U.S.A.*, **95**, 13965–13970.

Hua, B.G., Mercier, R.W., Leng, Q. & Berkowitz, G.A. (2003) Plants do it differently. A new basis for potassium/sodium selectivity in the pore of an ion channel, *Plant Physiol.*, **132**, 1353–1361.

Jeschke, W.D. & Hartung, W. (2000) Root–shoot interactions in mineral nutrition, *Plant Soil*, **226**, 57–69.

Jones, S., Sudweeks, S. & Yakel, J.L. (1999) Nicotinic receptors in the brain: correlating physiology with function, *Trends Neurosci.*, **22**, 555–561.

Kang, J.M. & Turano, F.J. (2003) The putative glutamate receptor 1.1 (AtGLR1.1) functions as a regulator of carbon and nitrogen metabolism in *Arabidopsis thaliana*, *Proc. Natl. Acad. Sci. U.S.A.*, **100**, 6872–6877.

Katagiri, F., Lam, E. & Chua, N.H. (1989) Two tobacco DNA-binding proteins with homology to the nuclear factor CREB, *Nature*, **340**, 727–730.

Kaupp, U.B. & Seifert, R. (2002) Cyclic nucleotide-gated ion channels, *Physiol. Rev.*, **82**, 769–824.

Kim, S.A., Kwak, J.M., Jae, S.K., Wang, M.H. & Nam, H.G. (2001) Overexpression of the AtGluR2 gene encoding an *Arabidopsis* homolog of mammalian glutamate receptors impairs calcium utilization and sensitivity to ionic stress in transgenic plants, *Plant Cell Physiol.*, **42**, 74–84.

Koehler, C. & Neuhaus, G. (2000) Characterisation of calmodulin binding to cyclic nucleotide-gated ion channels from *Arabidopsis thaliana*, *FEBS Lett.*, **471**, 133–136.

Kubis, S., Baldwin, A., Patel, R., *et al.* (2003) The *Arabidopsis* ppi1 mutant is specifically defective in the expression, chloroplast import, and accumulation of photosynthetic proteins, *Plant Cell*, **15**, 1859–1871.

Lam, H.M., Chiu, J., Hsieh, M.H., *et al.* (1998) Glutamate-receptor genes in plants, *Nature*, **396**, 125–126.

Lemtiri-Chlieh, F., MacRobbie, E.A. & Brearley, C.A. (2000) Inositol hexakisphosphate is a physiological signal regulating the $K^+$-inward rectifying conductance in guard cells, *Proc. Natl. Acad. Sci. U.S.A.*, **97**, 8687–8692.

Lemtiri-Chlieh, F., MacRobbie, E.A., Webb, A.A., *et al.* (2003) Inositol hexakisphosphate mobilizes an endomembrane store of calcium in guard cells, *Proc. Natl. Acad. Sci. U.S.A.*, **100**, 10091–10095.

Leng, Q., Mercier, R.W., Hua, B.G., Fromm, H. & Berkowitz, G.A. (2002) Electrophysiological analysis of cloned cyclic nucleotide-gated ion channels, *Plant Physiol.*, **128**, 400–410.

Leng, Q., Mercier, R.W., Yao, W.Z. & Berkowitz, G.A. (1999) Cloning and first functional characterization of a plant cyclic nucleotide-gated cation channel, *Plant Physiol.*, **121**, 753–761.

Li, W., Luan, S., Schreiber, S.L. & Assmann, S.M. (1994) Cyclic AMP stimulates $K^+$ channel activity in mesophyll cells of *Vicia faba* L., *Plant Physiol.*, **106**, 957–961.

Ludidi, N. & Gehring, C. (2003) Identification of a novel protein with guanylyl cyclase activity in *Arabidopsis thaliana*, *J. Biol. Chem.*, **278**, 6490–6494.

Maathuis, F.J.M. & Sanders, D. (2001) Sodium uptake in *Arabidopsis thaliana* roots is regulated by cyclic nucleotides, *Plant Physiol.*, **127**, 1617–1625.

Martinec, J., Feltl, T., Scanlon, C.H., Lumsden, P.J. & Machackova, I. (2000) Subcellular localization of a high affinity binding site for D-myo-inositol 1,4,5-trisphosphate from *Chenopodium rubrum*, *Plant Physiol.*, **124**, 475–483.

Miyazawa, A., Fujiyoshi, Y. & Unwin, N. (2003) Structure and gating mechanism of the acetylcholine receptor pore, *Nature*, **423**, 949–955.

Moutinho, A., Hussey, P.J., Trewavas, A.J. & Malho, R. (2001) cAMP acts as a second messenger in pollen tube growth and reorientation, *Proc. Natl. Acad. Sci. U.S.A.*, **98**, 10481–10486.

Moyen, C., Hammond-Kosack, K.E., Jones, J., Knight, M.R. & Johannes, E. (1998) Systemin triggers an increase of cytoplasmic calcium in tomato mesophyll cells: $Ca^{2+}$ mobilization from intra- and extracellular compartments, *Plant Cell Environ.*, **21**, 1101–1111.

Muir, S.R. & Sanders, D. (1997) Inositol 1,4,5-trisphosphate-sensitive $Ca^{2+}$ release across nonvacuolar membranes in cauliflower, *Plant Physiol.*, **114**, 1511–1521.

Navazio, L., Bewell, M.A., Siddiqua, A., Dickinson, G.D., Galione, A. & Sanders, D. (2000) Calcium release from the endoplasmic reticulum of higher plants elicited by the NADP metabolite nicotinic acid adenine dinucleotide phosphate, *Proc. Natl. Acad. Sci. U.S.A.*, **97**, 8693–8698.

Navazio, L., Mariani, P. & Sanders, D. (2001) Mobilization of $Ca^{2+}$ by cyclic ADP-ribose from the endoplasmic reticulum of cauliflower florets, *Plant Physiol.*, **125**, 2129–2138.

Newton, R.P., Roef, L., Witters, E. & VanOnckelen, H. (1999) Tansley review no. 106 – cyclic nucleotides in higher plants: the enduring paradox, *New Phytol.*, **143**, 427–455.

Ng, C.K.Y., McAinsh, M.R., Gray, J.E., *et al.* (2001) Calcium-based signalling systems in guard cells, *New Phytol.*, **151**, 109–120.

O'Boyle, M.P., Do, V., Derrick, B.E. & Claiborne, B.J. (2003) In vivo recordings of long-term potentiation and long-term depression in the dentate gyrus of the neonatal rat, *J. Neurophysiol.*, **91**, 613–622.

Paynel, F., Murray, P.J. & Cliquet, J.B. (2001) Root exudates: a pathway for short-term N transfer from clover and ryegrass, *Plant Soil*, **229**, 235–243.

Penson, S.P., Schuurink, R.C., Fath, A., Gubler, F., Jacobsen, J.V. & Jones, R.L. (1996) cGMP is required for gibberellic acid-induced gene expression in barley aleurone, *Plant Cell*, **8**, 2325–2333.

Perera, I., Heilmann, I. & Boss, W.F. (1999) Transient and sustained increases in inositol 1,4,5-bisphosphate precede the differential growth response in gravistimulated maize pulvini, *Proc. Natl. Acad. Sci. U.S.A.*, **96**, 5838–5843.

Ranjeva, R., Carrasco, A. & Boudet, A.M. (1988) Inositol trisphosphate stimulates the release of calcium from intact vacuoles isolated from acer cells, *FEBS Lett.*, **230**, 137–141.

Salpeter, M.M. & Loring, R.H. (1985) Nicotinic acetylcholine-receptors in vertebrate muscle – properties, distribution and neural control, *Prog. Neurobiol.*, **25**, 297–325.

Sanders, D., Pelloux, J., Brownlee, C. & Harper, J.F. (2002) Calcium at the crossroads of signaling, *Plant Cell*, **14**, S401–S417.

Sangwan, V., Foulds, I., Singh, J. & Dhindsa, R.S. (2001) Cold-activation of *Brassica napus* BN115 promoter is mediated by structural changes in membranes and cytoskeleton, and requires $Ca^{2+}$ influx, *Plant J.*, **27**, 1–12.

Schachtman, D.P., Reid, J.D. & Ayling, S.M. (1998) Phosphorus uptake by plants: from soil to cell, *Plant Physiol.*, **116**, 447–453.

Schachtman, D.P., Tyerman, S.D. & Yerry, B.R. (1991) The $K^+/Na^+$ selectivity of a cation channel in the plasma membrane of root cells does not differ in salt-tolerant and salt-sensitive wheat species, *Plant Physiol.*, **97**, 598–605.

Schuurink, R.C., Shartzer, S.F., Fath, A. & Jones, R.L. (1998) Characterization of a calmodulin-binding transporter from the plasma membrane of barley aleurone, *Proc. Natl. Acad. Sci. U.S.A.*, **95**, 1944–1949.

Sivaguru, M., Pike, S., Gassmann, W. & Baskin, T.I. (2003) Aluminum rapidly depolymerizes cortical microtubules and depolarizes the plasma membrane: evidence that these responses are mediated by a glutamate receptor, *Plant Cell Physiol.*, **44**, 667–675.

Stokes, D.L. & Wagenknecht, T. (2000) Calcium transport across the sarcoplasmic reticulum – Structure and function of $Ca^{2+}$-ATPase and the ryanodine receptor, *Eur. J. Biochem.*, **267**, 5274–5279.

Sunkar, R., Kaplan, B., Bouche, N., *et al.* (2000) Expression of a truncated tobacco *NtCBP4* channel in transgenic plants and disruption of the homologous *Arabidopsis CNGC1* gene confer $Pb^{2+}$ tolerance, *Plant J.*, **24**, 533–542.

Talke, I.N., Blaudez, D., Maathuis, F.J.M. & Sanders, D. (2003) CNGCs: prime targets of plant cyclic nucleotide signalling? *Trends Plant Sci.*, **8**, 286–293.

Turano, F.J., Muhitch, M.J., Felker, F.C. & McMahon, M.B. (2002) The putative glutamate receptor 3.2 from *Arabidopsis thaliana* (AtGLR3.2) is an integral membrane peptide that accumulates in rapidly growing tissues and persists in vascular-associated tissues, *Plant Sci.*, **163**, 43–51.

Tyerman, S.D. & Skerrett, I.M. (1999) Root ion channels and salinity, *Sci. Horticulturae*, **78**, 175–235.

Tyerman, S.D., Skerrett, M., Garrill, A., Findlay, G.P. & Leigh, R.A. (1997) Pathways for the permeation of $Na^+$ and $Cl^-$ into protoplasts derived from the cortex of wheat roots, *J. Exp. Bot.*, **48**, 459–480.

Very, A.A. & Davies, J.M. (2000) Hyperpolarization-activated calcium channels at the tip of *Arabidopsis* root hairs, *Proc. Natl. Acad. Sci. U.S.A.*, **97**, 9801–9806.

Very, A.A. & Sentenac, H. (2002) Cation channels in the *Arabidopsis* plasma membrane, *Trends Plant Sci.*, **7**, 168–175.

Volotovski, I.D., Sokolovski, S.G., Molchan, O.V. & Knight, M.R. (1998) Second messengers mediate increases in cytosolic calcium in tobacco protoplasts, *Plant Physiol.*, **117**, 1023–1030.

Walker, D.J., Leigh, R.A. & Miller, A.J. (1996) Potassium homeostasis in vacuolate plant cells, *Proc. Natl. Acad. Sci. U.S.A.*, **93**, 10510–10514.

Walseth, T.F., Aarhus, R., Kerr, J.A. & Lee, H.C. (1993) Identification of cyclic ADP-ribose-binding proteins by photoaffinity-labeling, *J. Biol. Chem.*, **268**, 26686–26691.

Ward, J.M. & Schroeder, J.I. (1994) Calcium activated $K^+$ channels and calcium-induced calcium release by slow vacuolar ion channels in guard cell vacuoles implicated in the control of stomatal closure, *Plant Cell*, **6**, 669–683.

White, P.J. (1993) Characterization of a high-conductance, voltage-dependent cation channel from the plasma membrane of rye roots in planar lipid bilayers, *Planta*, **191**, 541–551.

White, P.J. (2000) Calcium channels in higher plants, *Biochim. Biophys. Acta*, **1465**, 171–189.

Wu, Y., Kuzma, J., Marechal, E., *et al.* (1997) Abscisic acid signaling through cyclic ADP-ribose in plants, *Science*, **278**, 2126–2130.

Wu, Y., Sanchez, J.P., Lopez-Molina, L., Himmelbach, A., Grill, E. & Chua, N.H. (2003) The abi1-1 mutation blocks ABA signaling downstream of cADPR action, *Plant J.*, **34**, 307–315.

Yamamori, E., Iwasaki, Y., Oki, Y., *et al.* (2004) Possible involvement of ryanodine receptor-mediated intracellular calcium release in the effect of corticotropin-releasing factor on adrenocorticotropin secretion, *Endocrinology*, **145**, 36–38.

Zagotta, W.N. & Siegelbaum, S.A. (1996) Structure and function of cyclic nucleotide-gated channels, *Annu. Rev. Neurosci.*, **19**, 235–263.

Zimmermann, S., Nurnberger, T., Frachisse, J.M., *et al.* (1997) Receptor-mediated activation of a plant $Ca^{2+}$-permeable ion channel involved in pathogen defense, *Proc. Natl. Acad. Sci. U.S.A.*, **94**, 2751–2755.

Zucchi, R. & Ronca-Testoni, S. (1997) The sarcoplasmic reticulum $Ca^{2+}$ channel/ryanodine receptor: modulation by endogenous effectors, drugs and disease states, *Pharmacol. Rev.*, **49**, 1–51.

# 8    Aquaporins in plants

Clare Vander Willigen, Lionel Verdoucq, Yann Boursiac and Christophe Maurel

## 8.1   Introduction

The regulation of water movement across plant membranes is fundamental to both the maintenance of a dynamic water balance within the plant, and to volume flow of water through the whole body. Understanding these processes is central to plant biology, as they critically determine the performance of crops under varying environmental or culturing conditions.

The era of molecular studies on water transport, in plants and animals, was sparked quite recently, in 1992, by the incidental discovery of CHIP28, an abundant intrinsic membrane protein in erythrocytes and kidney tubules (Preston *et al.*, 1992). CHIP28 turned out to be one of the long-sought-after water channels, which had been hypothesised in certain animal cell membranes. In the years following the discovery of CHIP28, homologous water channel proteins could be identified in nearly all mammalian tissues, but also in all living kingdoms (Chrispeels & Agre, 1994). These proteins belong to a superfamily of membrane channels named after its founding member the Major Intrinsic Protein (MIP) of lens fibres. While some MIPs, such as bacterial glycerol facilitator GlpF, function as solute channels, a large majority of MIPs can transport substantial amounts of water and were thus designated as aquaporins (Chrispeels & Agre, 1994).

Within the last decade, studies on prototypic MIP channels, i.e. human CHIP28 renamed Aquaporin1 (AQP1) and *Escherichia coli* GlpF, have provided a fairly good understanding of water and solute permeation mechanisms through the aquaporin channel (Murata *et al.*, 2000; Sui *et al.*, 2001). However, the integrated function of aquaporins in higher organisms, including plants, is far from being fully understood. In plants, pharmacological and genetic approaches have revealed that water channels contribute a large part (50–90%) of water uptake by roots (reviewed by Javot & Maurel, 2002). Aquaporins may also play a critical role in other aspects of plant water relations, such as water stress avoidance or extension growth, but data are scarce and their mechanistic interpretation is still unclear.

Present research in the field aims at linking molecular and cellular knowledge on aquaporins with early reports illustrating an alteration of hydraulic conductivity of cells and tissues, during development or upon exposure to such environmental stimuli as salinity or anoxia (Clarkson *et al.*, 2000; Javot & Maurel, 2002). We now realise that much of this regulation is mediated by aquaporins present in cellular membranes. Recent investigations are beginning to describe in detail how their abundance and dynamic permeability is highly regulated, thus enabling the fine control observed in cellular water relations. However, the signalling mechanisms that lead to water channel regulation by developmental and environmental cues are largely unknown. Thus, it becomes increasingly clear that aquaporins provide a unique molecular entry into multiple aspects of plant water relations (Maurel & Chrispeels, 2001).

Several facets of aquaporin function, including transport selectivity, molecular and cellular mechanisms of regulation or integrated function, will be described in Sections 8.3 and 8.4. The underlying studies all rely on measurements of water permeability at the level of isolated membranes, single cells or whole organs. Because these measurements can be the object of numerous artefacts or misinterpretation, it is critical that the reader first understands the methodologies used and their founding physical principles. These principles and methods, as have been enounced and developed by earlier plant and animal physiologists, are described briefly in Section 8.2 below.

## 8.2 Water transport measurements: principles and methods

### 8.2.1 Theory

Water potential ($\Psi$, in energy per unit volume, or pressure) is a practical measure of the free energy status of water. It is defined, relative to free, pure water at the same temperature as that of the system, and at atmospheric pressure. For living plant cells, $\Psi$ is classically less than zero, and split into hydrostatic pressure ($P$) and osmotic pressure ($\Pi$) components:

$$\Psi = P - \Pi \tag{8.1}$$

Note that the matric component, which can play an important role in substrates such as soils or xylem vessels, is considered as negligible in living cells.

The theory of irreversible thermodynamics predicts that the net flow of water ($J_v$, in $m^3\ s^{-1}$) between two membrane compartments is proportional to the exchange surface $A$ and the driving force, i.e. the gradient in water potential ($\Delta\Psi$):

$$J_v = A \cdot L_p \cdot \Delta\Psi = A \cdot L_p \cdot (\Delta P - \sigma\Delta\Pi) \tag{8.2}$$

where $L_p$ is the hydraulic conductivity and $\sigma$ is the reflection coefficient (Dainty, 1963; Steudle, 1989). This coefficient balances the osmotic driving force according to the permeability properties of the membrane. The value of $\sigma$ is generally comprised between 1 (semipermeable membrane) and 0 (no selectivity between water and solutes). Because Equation 8.2 was derived through certain approximations (Dainty, 1963), it is critical to note that it can be used only when the volume (water) flow is low and may not fully apply in certain drastic osmotic swelling assays.

For practical reasons and direct comparison to diffusional water permeability ($P_d$) (see below), an osmotic water permeability coefficient ($P_f$, also noted $P_{os}$ by certain authors) can be derived from $L_p$ according to the following equation:

$$P_f = L_p \cdot R \cdot T / V_w \qquad (8.3)$$

where $R$ is the universal gas constant, $T$ is absolute temperature and $V_w$ is the molar volume of water.

Although Equation 8.2 primarily applies to membranes or individual cells, many authors have considered whole roots as equivalent to a single osmotic barrier between the soil solution and the xylem vessels. This assumption has proved to be acceptable in many species and, although some deviation from linearity may exist at low pressure and/or flow (see Passioura, 1988, for discussion), root water transport can also be analysed according to Equation 8.2.

Thus, most water transport assays that aim at establishing the hydraulic properties of a membrane or cell barrier rely on Equation 8.2. In practice, water flow intensity is measured for a single or a range of imposed driving forces, these being osmotic and/or hydrostatic. However, a bias may exist between the apparent and the real driving forces. For instance, unstirred layers, i.e. local changes in osmotic (solute) gradients at the vicinity of a membrane, may result from rapid permeation of the solute across the membrane ('gradient dissipation effect') or from solute accumulation or depletion induced by high-intensity water flows ('sweep away effect') (Steudle, 1989).

A critical aspect of these measurements can also be to estimate properly the exchange surface $A$. As will be discussed below, membrane water permeability of cells or vesicles can be conveniently derived from the kinetics of volume adjustment in response to a rapid osmotic challenge. For any object, the kinetics are determined by both its intrinsic water permeability and its surface-to-volume ratio. The latter is inversely proportional to the object diameter and can span over four orders of magnitude from Xenopus oocytes (diameter $\sim 10^{-3}$ m) to membrane vesicles (diameter $\sim 10^{-7}$ m). For oocytes, measurements over several tens of seconds are sufficient to monitor the initial swelling phase, whereas for membrane vesicles the kinetics of volume adjustment can be completed over times as low as 100 ms. This of course results in strikingly different experimental constraints.

In contrast to the water transport assays mentioned above, those assays in which a unidirectional flow of water molecules is followed, in the absence of any driving force, yield a diffusional water permeability value ($P_d$). This parameter may reflect genuine permeability properties of the membrane, and is indeed dependent on aquaporin activity. In red blood cells for instance, the predominant aquaporin (AQP1) contributes 64% of the membrane $P_d$ (Mathai et al., 1996). However, $P_d$ has less biological significance than the hydraulic parameters $L_p$ and $P_f$, since only the osmotically and hydrostatically driven net flows of water play a physiological role. Note also that measurements of $P_d$ are subjected to numerous pitfalls, in particular because unstirred layers create significant barriers for water diffusion (Steudle, 1989).

Although aquaporins create a path for passive (downstream) water transport along water potential gradients, attention has been drawn by certain authors to other membrane systems that may achieve upstream transport of water. These ideas originate from measurements made in animal epithelial and *Xenopus* oocytes (some of them expressing a plant amino acid transporter) (Loo *et al.*, 1996). It was suggested that direct coupling of solute and water transport may occur within certain membrane transporters and lead to a significant secondary active transport of water molecules against a water potential gradient. The significance of this process in plants has been theoretically discussed by Morillon *et al.* (2001), to account for some rapid movements of plant organs. However, there is, to date, no experimental data that would support the existence of such a mechanism in plants. Also, the principle itself of this apparent pumping of water has been the object of a very vivid debate as it may simply result from unstirred layer effects (Schultz, 2001).

### 8.2.2  Stopped-flow techniques

These techniques typically apply to small membrane objects (vesicles) for which rapid flow kinetics (in the 0.1–1 s range) can be expected. They allow accurate physical measurements, provided that vesicle size can be determined accurately using independent techniques (dynamic light scattering, electron microscopy).

In practice, improved procedures for efficient solution mixing in a stopped-flow apparatus are used to impose an abrupt solute concentration gradient to a membrane vesicle suspension, and subsequent changes (dead time <5 ms) in vesicle optical properties are recorded (Verkman, 2000) (Fig. 8.1A). Mean vesicle volume can typically be monitored through the scattering of incident light by the vesicle suspension (Fig. 8.1B) or through concentration-dependent quenching of a vesicle-entrapped fluorophore (Verkman, 2000). Using these approaches, the osmotic water permeability of membranes can then be determined, primarily from osmotic experiments in the presence of an impermeant solute (mannitol or sucrose). Stopped-flow techniques have also proved useful to elucidate various aspects of aquaporin transport selectivity. For instance, iso- or hyperosmotic challenges in the presence of a concentration gradient for small neutral solutes, such as glycerol or urea, allowed the determination of the membrane permeability to these solutes (Gerbeau *et al.*, 1999). In addition, proton and $CO_2$ transport can be followed by means of pH-dependent probes loaded within the vesicles, but in the case of $CO_2$, exogenous carbonic anhydrase also has to be present within the vesicles (Prasad *et al.*, 1998).

Although stopped-flow measurements exhibit a great biophysical accuracy, there have been concerns that water transport properties (aquaporin activity) may be altered during membrane isolation from plant material. In studies on plasma membranes purified from *Arabidopsis* cell suspensions, Gerbeau *et al.* (2002) suggested that protection of aquaporins from dephosphorylation by protein phosphatases or from inhibition by divalent ions may be critical to maintain purified plasma membranes with their native water permeability.

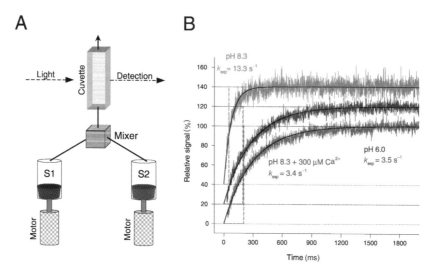

**Fig. 8.1** Measurement of water permeability of membrane vesicles by stopped-flow techniques. (A) Schematic stopped-flow apparatus. Motor-driven syringes allow the efficient mixing of a hyperosmotic solution (S1) with a membrane vesicle suspension (S2). The resulting mixture passes through a cuvette where its time-dependent optical properties are followed (see text). (B) Typical time course of shrinking of purified *Arabidopsis* plasma membrane vesicles, as followed by an increase in light scattering intensity. A hyperosmotic challenge was imposed to the vesicle suspension at time 0, and the subsequent changes in light signal were recorded. These changes can be fitted with an exponential function. The deduced rate constant ($k_{exp}$) together with the size (surface and volume) of the vesicles are used to calculate the osmotic water permeability coefficient ($P_f$) of the vesicles. The figure illustrates that, when compared with measurements in control conditions (pH 8.3; $k_{exp} = 13.3 \text{ s}^{-1}$), the addition of calcium (pH 8.3 + 300 µM $Ca^{2+}$; $k_{exp} = 3.4 \text{ s}^{-1}$) or the acidification of the bathing solution (pH 6.0; $k_{exp} = 3.3 \text{ s}^{-1}$) induces a 75% inhibition of $k_{exp}$.

Stopped-flow techniques have also been instrumental in analysing the activity of aquaporins heterogeneously expressed in yeast or the activity of purified aquaporins reconstituted in proteoliposomes (Laizé *et al.*, 1995; Karlsson *et al.*, 2003).

### 8.2.3  Swelling of isolated cells, protoplasts and vacuoles

Isolated cells, plant protoplasts and vacuoles are much more fragile than membrane vesicles, and measurements on such objects can be problematic because the rapid imposition of an osmotic gradient may be destructive. Ramahaleo *et al.* (1999) developed a device in which isolated protoplasts or vacuoles were held by a pipette through which a gentle suction was applied. The pipette was used for quick transfer of the object in solutions of varying osmolalities, and early changes in protoplast/vacuole volume were recorded by video-microscopy. A slightly different approach for selection of protoplasts from well-defined tissues and individual transfer into a hypotonic solution has recently been developed by Suga *et al.* (2003). Techniques for continuous protoplast perfusion and immobilisation in a microscopic observation chamber have been improved by Moshelion *et al.* (2002) but a lag of several seconds remains before homogeneous perfusion of the protoplast is achieved.

As discussed for purified membrane vesicles, the possibility remains that water transport properties of native cells are altered during protoplast isolation. Indeed, water permeability values reported by certain authors (Moshelion *et al.*, 2002; Kaldenhoff *et al.*, 1998) are extremely low ($P_f \leq 10 \ \mu m\,s^{-1}$) as compared with those values reported in plant cells by other authors and techniques (Maurel, 1997; Steudle, 1989). In certain cases, however, protoplasts have proved to be relevant materials to address certain aspects of plant aquaporin function and regulation. For instance, Morillon and Chrispeels (2001) observed a consistent relationship between the $P_f$ of leaf protoplasts and the transpiration regime of wild-type and ABA-deficient *Arabidopsis* plants. This relationship may reflect an exquisite balance between apoplastic and cell-to-cell paths for water transfer across intact leaf tissues. In other studies, measurements of protoplast $P_f$ were used to demonstrate the occurrence of aquaporin downregulation in plants expressing an aquaporin antisense transgene (Kaldenhoff *et al.*, 1998; Martre *et al.*, 2002; Siefritz *et al.*, 2002).

*Xenopus* oocytes represent an extremely favourable system for assaying the water and solute transport of cloned aquaporins. First, native oocytes are virtually devoid of endogenous water channel activity and have thus a low basal water permeability. Second, oocytes can efficiently express exogenous membrane proteins upon intracellular injection of *in vitro* transcribed complementary RNAs. In addition, because of the large size of oocytes, the increase in $P_f$ conferred by aquaporin expression can be easily followed by video-microscopy and the capacity of aquaporins to transport solutes such as glycerol or urea can be monitored by the uptake of radiolabelled molecules (Maurel *et al.*, 1993). In plant cells, aquaporins are specifically targeted to intracellular or plasma membranes. Although plant aquaporins of the PIP1 subfamily have been recalcitrant to oocyte expression, in part because of a deficient targeting to the cell surface (Fetter *et al.*, 2004), many plant aquaporins are at least in part routed to the oocyte surface and can be functionally characterised in oocytes.

### 8.2.4  The pressure probe technique

Pressure probes allow measurements of water relation parameters of intact walled plant cells, in native cell suspensions or tissues (Steudle, 1993; Tomos & Leigh, 1999). Therefore, this technique has been instrumental to reveal *in situ* the regulation of water transport (aquaporin activity) by environmental or developmental cues. It requires that an oil-filled micropipette connected to a pressure transducer be inserted into the cell. A meniscus is formed at the tip of the pipette by the contact between the oil (which fills the pipette) and the aqueous bathing solution or cell sap. Displacement of the meniscus, as observed under a microscope, can be used to monitor bidirectional volume exchanges between the probe and the cell. Because a hydraulic connection is established between the cell and the probe, impalement of a turgid plant cell results in a dramatic shift of the meniscus, which can be compensated by applying a counter-pressure within the probe (Fig. 8.2). Thus, stabilisation of the meniscus at its initial position allows one to measure a steady-state cell turgor. Furthermore, osmotic or hydrostatic perturbations can be imposed by altering the cell bathing solution or by displacing the meniscus itself. These manoeuvres result in cell

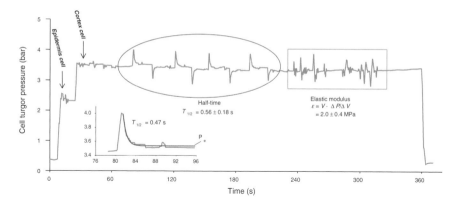

**Fig. 8.2** Measurement of cell hydraulic conductivity ($L_p$) in the root of *Arabidopsis thaliana* using a cell pressure probe. After insertion of the pressure probe micropipette into the first cell layer (epidermis) of the root, pressure was equilibrated and a first stationary turgor was measured. The next cell layer (cortex) was reached upon further impalement and a second, higher stationary turgor was measured ($t > 30$ s). A first series of hydrostatic relaxations was triggered by displacement of the pressure probe meniscus (see text) and the half-time of cell relaxation ($T_{1/2}$) was determined (see inset for exponential fit of a single relaxation). In the second sequence ($t > 240$ s), rapid and reversible changes in pressure (before any cell relaxation occurs) were used to determine the elastic properties of the cell wall (elastic modulus, $\varepsilon$). Both $T_{1/2}$ and $\varepsilon$ are used, together with cell surface and volume, to calculate $L_p$.

pressure relaxations, which provide crucial information on the mechanical properties of the cell (cell wall elastic modulus) and on its hydraulic conductivity (Fig. 8.2). In addition, the membrane reflection coefficients for different solutes can also be determined using this technique. Thus, all key parameters of the plant cell water relations can be measured. It is important to note that the pressure probe technique allows one to monitor both endo-osmotic (inward) and exo-osmotic (outward) water movements and that changes in cell pressure provide a strong signal for reduced exchange of water and/or changes in cell volume (usually $<5\%$). However, the pipette may induce cell damages and disturb fine cell regulations. It is also difficult to determine in which compartment (cytosol or vacuole) the pipette tip is inserted.

### 8.2.5   Water transport measurements on excised organs

Crude water transport assays have been developed in leaf or stem fragments, based on osmotic challenges imposed to the whole tissue (see, for instance, Cosgrove & Steudle, 1981; Terashima & Ono, 2002). However, the epidermis of the material may have to be peeled off or abraded and it is extremely difficult to model the propagation of an osmotic driving force within the tissue. These constraints make the calculation of a genuine tissue hydraulic conductivity very challenging.

Excised roots are in contrast much more amenable to integrated water transport assays. One of the most classical techniques to measure root hydraulic conductivity ($L_{p_r}$) requires the excised root to be inserted within a chamber, with a hermetic seal maintaining the stem section outside. Application of a hydrostatic pressure within the chamber results in a flow of sap exuding from the stem section. Within certain pressure limits, this flow is usually proportional to the hydrostatic pressure gradient present between the root solution (at the imposed pressure chamber) and the xylem vessels (at constant, atmospheric pressure) and $Lp_r$ is easily derived from the pressure-to-flow relationship (Passioura, 1988). High-pressure flow meters allow simultaneous imposition of a hydrostatic driving force and measurements of the resulting volume flow, provided that a tight seal between the measuring device and the root section is achieved. Interestingly, this device can be applied to root systems maintained in soil, but in this case a reverse flow, from root to soil, is generated (Siefritz $et\ al.$, 2002). High-pressure flow meters can also be used to measure the $L_p$ of stem fragments. In this case, measurements on a complete set of sectioned parts allow the hydraulic architecture of the whole plant to be modelled as an equivalent circuit of series or derivatized hydraulic resistances. Since axial rather than radial resistances are measured, these analyses mostly concern water transport along xylem vessels and are not directly related to aquaporin function (Tyree, 1997).

Pressure probe techniques have also been extended to measure water transport in excised roots (Steudle, 1989). This approach requires a pressure probe to be hydraulically connected to the xylem vessels of a sectioned root, by tight sealing of the probe onto the root section. As discussed above, the tissues between the xylem vessels and the root bathing solution can be assimilated to a single cell barrier, and osmotic and hydrostatic relaxations are used to calculate equivalent $L_p$ or $\sigma$ values.

### 8.2.6  Nuclear magnetic resonance techniques

Proton nuclear magnetic resonance (NMR) and associated magnetic resonance imaging (MRI) techniques are non-invasive techniques to investigate the water status of intact plants. Classical approaches typically provide measurements of $P_d$ and are thus of reduced relevance for physiological studies. They are subject to numerous unstirred layer artefacts (see above) and rely on assumptions on the diffusion and compartmentalisation of paramagnetic ions such as $Mn^{2+}$, which are intended to dope the proton signal in extracellular spaces (for discussion, see Maurel, 1997). However, these approaches have been recently refined and current developments in dynamic MRI give access to information on the molecular displacement of water molecules in cell compartments and across membranes (Scheenen $et\ al.$, 2000). Although more precise connections need to be established between spectroscopic parameters as measured by MRI and membrane water permeability values, MRI may become a useful approach to study adaptation of membrane water transport, or exchange of water within tissues during the process of growth, or stress responses (van der Weerd $et\ al.$, 2002).

Finally, a large variety of techniques, based for instance on the use of potometers or infrared $H_2O/CO_2$ analysers, have been developed to measure water transport throughout whole plants. These techniques primarily determine the dynamic status of transpiration, which is in large part determined by the stomatal conductance. Thus, they do not give direct access to the hydraulic conductivity of cell and tissues and will not be reviewed in the present chapter. We note however that these techniques can provide crucial information on the water relations of plants with genetically altered functions (Kaldenhoff *et al.*, 1998; Martre *et al.*, 2002; Siefritz *et al.*, 2002).

## 8.3  Aquaporins at the level of molecules, cells and tissues

### 8.3.1  Classification of plant aquaporins

Phylogenetic analyses suggested that the MIP family, with members in all living organisms, has evolved from two prokaryotic ancestors (Zardoya & Villalba, 2001). These ancestors delineate two main classes of homologues, with conserved sequence motifs and somehow similar transport selectivity: aquaporins, as represented by *E. coli* AqpZ, mostly function as water-selective channels whereas aquaglyceroporins, whose archetype is *E. coli* GlpF, can also transport small neutral solutes such as glycerol or urea (Agre *et al.*, 1998). Genomic and EST sequences revealed a high isoform multiplicity for *MIP* genes in plants, but surprisingly no close homologue of GlpF, i.e. no aquaglyceroporin has been identified so far. In contrast to mammals that have 11 *MIP* genes (named *AQP0* to *AQP10*), the model species *Arabidopsis thaliana* has up to 35 MIPs (Johanson *et al.*, 2001; Quigley *et al.*, 2001). Other plant genomes encode a similar multiplicity, with more than 30 distinct MIP cDNAs in maize (Chaumont *et al.*, 2001), and at least 14 full-length or partial MIP cDNAs in the halophytic species *Mesembryanthemum crystallinum* (Kirch *et al.*, 2000). Plant MIPs can be further subdivided into four subclasses, based on their sequence homology (Fig. 8.3). Within the same subclass, plant MIPs can share 27–97% amino acid sequence identity, whereas this homology can drop to ∼20% when members of two distinct subclasses are considered. More specifically, *Arabidopsis* MIPs can be distributed as follows (Fig. 8.3):

(a)  13 plasma membrane intrinsic proteins (PIPs), with some members known to be localised on the plasma membrane, and further subdivided into two groups: PIP1 and PIP2 (Fig. 8.3) (Schäffner, 1998);

(b)  10 tonoplast intrinsic proteins (TIPs), with some members known to be localised on the vacuolar membrane (tonoplast) (Höfte *et al.*, 1992);

(c)  9 NOD26-like intrinsic proteins (NIPs) corresponding to close homologues of soybean Nodulin 26 (NOD26), an aquaporin that transports water and glycerol in the peribacteroid membrane of $N_2$-fixing symbiotic root nodules (Dean *et al.*, 1999);

(d)  3 small basic intrinsic proteins (SIPs), forming the most diverged subclass of MIPs, with no data available on their function or subcellular localisation (Johanson & Gustavsson, 2002).

We stress, however, that this classification is purely based on sequence homologies and does not allow any firm prediction as to MIP functions or subcellular localisations.

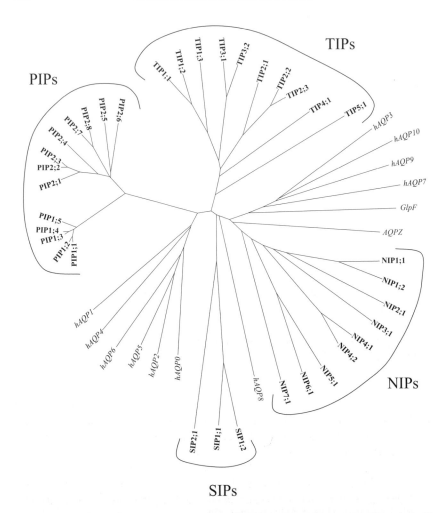

**Fig. 8.3** Sequence relationship between aquaporin homologues of *Arabidopsis*, human and *E. coli*. The amino acid sequences were aligned with ClustalW and a phylogenetic tree was constructed with TreeExplorer. The tree illustrates the subdivision of the aquaporin family of *Arabidopsis thaliana* (35 members) into four subfamilies: the PIPs, TIPs, NIPs and SIPs (see text). *E. coli* has two aquaporin homologues: GlpF and AqpZ. The 11 human aquaporins (hAQP0 to hAQP10) can be grouped into two subfamilies. The branch containing GlpF corresponds to the aquaglyceroporins. Note that there is no plant homologue in this branch.

Although it has raised significant experimental difficulties in aquaporin research, the high diversity of MIP homologues in plants suggest that, in these organisms, aquaporins with various subcellular localisation, selectivity profile or regulatory properties are needed to accomplish a large variety of functions. In the present section, we will emphasise how the high isoform multiplicity of plant aquaporins indeed provides an interesting angle to address basic aspects of aquaporins at the level of molecules, cells and tissues.

## 8.3.2 Molecular level: a variety of selectivity profiles

### 8.3.2.1 Transport selectivity

The transport activity of plant MIP homologues has been mostly characterised after heterologous expression in *Xenopus* oocytes (see Section 8.2) and, more recently, by functional expression in yeast (Weig & Jakob, 2000). Altogether, these studies revealed a large array of transport selectivity profiles. For instance, extensive characterisation of *Arabidopsis* γ-TIP (TIP1;1) in oocytes showed this aquaporin to be permeable only to water (Maurel *et al.*, 1993). In contrast, its tobacco homologue NtTIPa transported urea and glycerol in addition to water when expressed in the same cells (Gerbeau *et al.*, 1999). Other TIP homologues have recently been identified as permeable to urea, through complementation of a yeast mutant defective in its endogenous urea transporter (Klebl *et al.*, 2003; Liu *et al.*, 2003). Solute transport is, however, not restricted to TIPs and other plant aquaporins of the PIP (Biela *et al.*, 1999) and NIP subfamilies (Rivers *et al.*, 1997; Weig & Jakob, 2000) have been shown to transport glycerol in addition to water. Glycerol facilitators characterised in microorganisms (bacteria, yeast) play a key role in sugar uptake and cell osmoregulation (Luyten *et al.*, 1995), but the physiological significance of neutral solute transport by plant aquaporins still remains unclear.

The capacity of certain aquaporins to transport gaseous compounds has recently raised great interest. Similar to human AQP1, tobacco NtAQP1 was found to facilitate $CO_2$ diffusion across the oocyte membrane (Uehlein *et al.*, 2003). Human AQP1 also facilitates ammonia transport in oocytes (Nakhoul *et al.*, 2001) and such activity has been hypothesised for NOD26 (Niemietz & Tyerman, 2000) in the peribacteroid membrane of symbiotic root nodules. Other molecules relevant to plant physiology have also been shown to be transported by MIPs (for review, see Tyerman & Niemietz, 2002). These putative substrates include boron (Dordas *et al.*, 2000), antimonite (Sanders *et al.*, 1997) and hydrogen peroxide (Henzler & Steudle, 2000). Interestingly, some human aquaporins such as AQP1 and AQP6 can function as ionic channels provided that they are placed under very specific conditions, i.e. acidic extracellular pH (Yasui *et al.*, 1999) or stimulation by cGMP (Anthony *et al.*, 2000). Similar behaviour has not yet been identified in plant aquaporins.

This variety of selectivity profiles raises the intriguing possibility that plant aquaporins fulfil novel, multiple functions, besides water transport. The variety of selectivity profiles of aquaporins also point to very specific molecular properties of their permeating pore. In these respects, extraordinary insights have been brought about by structural studies on two MIP prototypes differing in their transport selectivity, i.e. mammalian AQP1 and *E. coli* glycerol facilitator GlpF.

### 8.3.2.2 Aquaporin structure and molecular basis of aquaporin selectivity

Primary sequence analyses indicate that, with the exception of a few homologues in yeast and *Drosophila*, all MIPs consist of small (25–34 kDa) hydrophobic proteins (Zardoya & Villalba, 2001). MIPs also exhibit a typical symmetrical organisation comprising six transmembrane domains, the N- and C-terminal extremities of the peptidic chain being located in the cytoplasm. In addition, MIPs contain two highly conserved loops, located on either side of the membrane, which both carry a highly conserved Asn-Pro-Ala (NPA) motif (Chrispeels & Agre, 1994; Agre *et al.*, 1998). Early structure function analyses of AQP1 in oocytes suggested an hourglass model for this aquaporin, the two NPA loops dipping into the membrane to pair their two NPA motifs (Jung *et al.*, 1994).

Biochemical and functional studies also suggested that aquaporins fold into homotetramers of independent monomeric channels. This design was directly visualised by electron microscopy of two-dimensional aquaporin crystals. Such low-resolution structures were first produced with human AQP1 (Walz *et al.*, 1994), but similar results at 7–8 Å resolution obtained with *E. coli* AqpZ (Ringler *et al.*, 1999), common bean α-TIP (Daniels *et al.*, 1999) and two spinach PIP homologues (Fotiadis *et al.*, 2001) confirmed that aquaporins from all organisms exhibit a conserved tetrameric organisation.

A major breakthrough was achieved recently with the resolution by cryoelectron microscopy of a 3.8 Å atomic structure for human AQP1 (Murata *et al.*, 2000), and by X-ray crystallography of 2.2 Å atomic structures for bacterial GlpF (Fu *et al.*, 2000) and bovine AQP1 (Sui *et al.*, 2001). These studies confirmed the hourglass model and showed that aquaporin monomers fold with a right-hand twisted arrangement of six membrane-spanning α-helices tilted along the plane of the membrane.

Two main structural motifs were identified as critical to determine the transport properties of aquaporins (Plate 3). A first arrangement named Ar/R because it is comprised of aromatic (H180, F56 in AQP1) and polar (R195 in AQP1) residues forms a constriction of 2.8 Å in diameter in the extracellular half of the pore. The positive charge of Arg195 would generate an electrostatic repulsion which, together with size restriction, would act as an efficient filter for protons and ions (for review, see Fujiyoshi *et al.*, 2002). A second constriction must be formed in the centre of the membrane, by the gathering of the two NPA motifs. This constriction contains Phe24-Leu149 in AQP1, while most glycerol facilitators have a pair of Leu-Leu, leading to a larger pore and allowing the permeation of small solutes (Heymann & Engel, 2000).

To explain polyol permeation in GlpF, Fu *et al.* (2000) also proposed that the carbon backbone of glycerol would move against a hydrophobic wedge in the pore whereas its OH groups would form hydrogen bonds with residues on the opposite side of the pore. This model implied that substrate stereo-specificity, in addition to molecular size, must be critical for permeation. It was checked indeed that linear polyol sugars that have their OH groups lined up in the same direction with respect to the carbon backbone (i.e. glycerol, ribitol) can efficiently permeate GlpF whereas unsymmetrical stereoisomers such as xylitol or D-arabitol are poorly transported (Fu *et al.*, 2000).

Molecular structures of AQP1 and GlpF also provided a solid basis to molecular dynamics (MD) simulations of glycerol and water transport in proteins of distinct selectivity (de Groot and Grubmüller, 2001; Tajkhorshid *et al.*, 2002). It has been known for a long time that membrane pores filled with a file of water molecules joined by hydrogen bonds conduct protons very efficiently (de Groot *et al.*, 2003). Therefore, one crucial question has been to understand how the aqueous pore of aquaporins can restrict proton flux to a rate that is at least 1000-fold slower than the flux of water molecules ($3 \times 10^9$ molecules s$^{-1}$). MD studies revealed, in particular, that the NPA constriction functions as an important barrier for proton exclusion, by imposing a 180° dipole reorientation of the water molecule upon its passage. This creates a break in the single file of water molecules and abrogates proton conductance through the channel. Protons and hydroxide ions are also excluded by a combination of other electrostatic barriers and by entropic effects (de Groot *et al.*, 2003).

The structure of AQP1 also provided a clear mechanistic explanation to its blockade by mercurials. The target of these compounds had been identified as a Cys residue, which turned out to be located right at the Ar/R constriction, in front of H180 and R195 (Sui *et al.*, 2001). Plant aquaporins can also be blocked by mercury but the structural basis for this inhibition is unclear since there is no conserved cysteine residue in the pore of these proteins (Daniels *et al.*, 1996). Silver and gold salts have recently been described as extremely efficient blockers of aquaporins in membranes from roots, soybean nodules and human red blood cells (Niemietz & Tyerman, 2002) but the molecular basis of this inhibition is still unknown.

Following structural work on AQP1 and GlpF, the overall molecular structure of plant aquaporins could be predicted (Plate 3, see colour plate section) and it was shown that the transport selectivity of some plant aquaporins can be changed by a single amino acid substitution. In particular, a conserved tryptophane in helix 2 of a NIP protein is a major determinant of its glycerol versus water selectivity (Wallace *et al.*, 2002). This approach is reminiscent of previous work on insect AQPcic, in which a two amino acid substitution was sufficient to switch this water-selective aquaporin to a glycerol channel (Lagrée *et al.*, 1999). It is interesting to note that the only aquaporin present in *Toxoplasma gondii*, which shows 47% similarity with TIP1;1 (γ-TIP) and conservation of typical pore-forming residues, is a water/glycerol/urea facilitator and is permeable to the hydroxyurea drug (Pavlovic-Djuranovic *et al.*, 2003), while TIP1;1 only transports water (Maurel *et al.*, 1993). However, the lack of structural data for TIP proteins, and plant aquaporins in general, is hampering a more precise characterisation of the specificity determinants. Although it might be tempting, based on existing structures, to predict the transport selectivity of plant aquaporins from their primary sequence, structural knowledge of these proteins is insufficient to come up with reliable modelling.

In conclusion, structural data gathered on AQP1 and GlpF have proved critical for understanding molecular mechanisms of water and solute permeation in aquaporins. However, the next challenge in this field will be understanding fundamental structural aspects involved in channel gating. As was illustrated by the work of Tournaire-Roux *et al.* (2003) on pH gating of plant aquaporins (see Section 8.4), these studies may have a strong impact on our understanding of aquaporin function in plants.

### 8.3.2.3   Significance of $CO_2$ transport

Although water transport undoubtedly remains the primary function of aquaporins in plants, the diversity of transport profiles uncovered in these proteins suggested that they may fulfil important subsidiary functions. Among these, $CO_2$ transport may be one of most significant.

Similar to water, $CO_2$ can pass freely across lipid membranes and this process has long been considered as sufficient to account for the diffusion of the gas across the mesophyll, from the stomatal chamber to the chloroplast. However, the drop in soluble $CO_2$ concentration between the two compartments suggested that a significant resistance to $CO_2$ transport can be encountered within the mesophyll. The idea that leaf aquaporins may contribute to $CO_2$ diffusion was first examined by Terashima and Ono (2002). These authors based their study on the inhibition properties of aquaporins by mercury, and investigated in two plant species (*Vicia faba* and *Phaseolus vulgaris*), the effects of the blocker on the $CO_2$ dependency of photosynthesis. The latter was characterised by a combination of infrared gas analyses and fluorescence measurements. At moderate mercury concentrations, the authors observed that the stimulation of photosynthesis by limiting intercellular $CO_2$ was reduced whereas the dependency of photosynthesis on chloroplastic $CO_2$ was unchanged. This was interpreted to mean that mercury blocked $CO_2$ diffusion into the chloroplast without affecting the $CO_2$ fixation machinery. More recently, Uehlein *et al.* (2003) investigated the specific role of aquaporin NtAQP1 for $CO_2$ transport in tobacco leaves. Functional expression in oocytes had shown that indeed this aquaporin significantly transports gaseous $CO_2$. A set of tobacco plants altered in the expression level of NtAQP1 was obtained, by expression of an antisense or a tetracycline-inducible sense transgene. In these plants, the rate of photosynthesis, as measured by $CO_2$ fixation, and also the rate of growth were both well correlated to NtAQP1 expression, whereas the intercellular $CO_2$ concentration was unchanged. The authors proposed that $CO_2$ transport was limiting in both antisense and wild-type tobacco plants. Intriguingly, wild-type and transgenic plants also showed significantly different stomatal conductances. A similar phenotype has been observed by Aharon *et al.* (2003) upon ectopic expression of AtPIP1;2 in tobacco. This means that pleiotropic effects of aquaporin expression on stomatal function (and therefore water status) may be present in transgenic plants. Therefore, a thorough characterisation of $CO_2$ dependency of photosynthesis in these transgenic materials will be useful to distinguish between all possible effects of aquaporin expression.

### 8.3.3   Cell level: subcellular targeting

#### 8.3.3.1   Pattern of aquaporin expression within the cell

The diversity of plant aquaporin isoforms can be explained in part by specific subcellular localisations, on the plasma membrane or on intracellular membranes. These suggest complementary functions for aquaporins in cell osmoregulation and control of transcellular transport. Subcellular localisation of MIPs in plants was mostly achieved using antibodies raised against members of the two main subclasses of MIPs, the PIPs and the TIPs. These studies revealed that representative members of the two classes are mostly localised in the plasma membrane and the tonoplast, respectively, and hence their names.

For instance, immunocytolocalisation or biochemical purification of tonoplast fractions followed by immunodetection has confirmed the localisation of a number of TIPs to the tonoplast in a variety of plant species and organs, including bean seeds (Johnson *et al.*, 1990), radish roots (Maeshima, 1992) and spinach leaves (Karlsson *et al.*, 2000). In many of these studies, antibodies were raised against the entire protein, but later even more intricate detail of TIP localisation has been revealed using antipeptide antibodies that are more specific to particular TIP isoforms. This advancement has lead to the realisation that specific TIP isoforms may be associated with distinct vacuolar subtypes, which perform different functions within the cell. Studies by Jauh *et al.* (1998, 1999) in various plant materials including barley and pea root tips, and potato tubers showed for instance that expression of two isoforms, δ-TIP (TIP2) and γ-TIP (TIP1), was associated with vegetative storage protein and lytic vacuoles, respectively. In contrast, α-TIP (TIP3) was localised to the protein storage vacuoles only in conjunction with δ-TIP, whereas the expression of the α-TIP isoform alone was associated with autophagic vacuoles. From these studies, it remains unclear whether TIPs can just be taken as vacuolar markers or whether they are actively involved in the functional specialisation of vacuolar subtypes.

The first localisation studies of PIPs were performed using biochemically purified membrane fractions obtained using aqueous two-phase partitioning (Daniels *et al.*, 1994; Kammerloher *et al.*, 1994) and later confirmed by immunolocalisation studies (Kaldenhoff *et al.*, 1995; Robinson *et al.*, 1996). Because of the close homology of many of the PIP isoforms, cross-reactivity of PIP antibodies has remained a limitation in detailing their localisation further. However, some localisation studies have been undertaken including detailed investigations of three PIP isoforms from the ice plant *M. crystallinum*, MIP-A, MIP-B and MIP-C (Barkla *et al.*, 1999, Kirch *et al.*, 2000). Separation of the membrane fractions by discontinuous sucrose density gradients and subsequent immunodetection revealed surprising patterns in the localisation of the three PIP isoforms. MIP-C was found mostly in the plasma membrane fraction but also in a fraction of intermediate density, whereas MIP-A seemed to be associated with the tonoplastic fraction only. MIP-B was found to be only weakly associated with the plasma membrane and was present in all other fractions (Kirch *et al.*, 2000). These observations have led these and other authors to caution that the sequence-based classification of plant aquaporins does not necessarily confirm the subcellular localisation. It was also hypothesised that the subcellular localisation of plant aquaporins may be regulated upon cell stimulation, as has been established for vasopressin-regulated AQP2 in mammalian kidney (Brown, 2003) (see discussion in Section 8.4). Using *Xenopus* oocytes, Fetter *et al.* (2004) recently unravelled a novel mechanism, in which physical interaction between aquaporins of the PIP1 and PIP2 subgroups were necessary for the latter to be expressed at the cell surface.

These observations, which still need to be extended to plant cells, may correspond to a central mechanism for controlling the expression of aquaporins in the plant plasma membrane.

To circumvent the use of anti-aquaporin polyclonal antibodies and their limited specificity, new methods have been used to ascertain the subcellular specificity of plant aquaporins. For instance, Chaumont *et al.* (2000) characterised the localisation of two PIPs to the cell surface in maize using fusions of these aquaporins with the green fluorescent protein (GFP). Interestingly, labelling was also noticeable in a perinuclear compartment, which probably corresponded to the ER. In contrast, fusions of TIPs with GFP yielded specific labelling of the vacuole as shown by Cutler *et al.* (2000) in a study where random fusions between plant proteins and GFP were made with the aim of identifying labelling markers of specific subcellular structures.

Recently, Santoni *et al.* (2003) demonstrated that 5 out of the 13 *Arabidopsis* PIPs are expressed in root plasma membrane extracts, and characterised these using a proteomic approach coupling 2D gel electrophoresis and mass spectrometry. Three other isoforms might also be present, but in a lower extent. Only one peptide from a TIP was found, probably because of a slight contamination of the plasma membrane preparation (Santoni *et al.*, 2003).

While we possess a lot of information on the subcellular localisation of PIPs and TIPs, information for other MIPs families are scarce. In the case of the NIP subfamily, nothing is known except that the founding member, the soybean NOD26, is specifically expressed in the symbiosome membrane surrounding nitrogen-fixing bacteria (Miao *et al.*, 1992). A preliminary report on localisation of a SIP homologue in *Arabidopsis* plasma membrane has been presented recently (Gustavsson & Johanson, 2003).

### 8.3.3.2  *Role of aquaporins in cell osmoregulation*

The cytoplasm of a plant cell is the medium in which most metabolic activity occurs and the necessity for fine regulation of both the volume and osmotic potential of this compartment is crucial. The plasma membrane and the tonoplast form the borders between the exterior and the vacuole respectively and thus play a major role in maintaining this equilibrium. Several studies have pointed to strikingly different solute and water transport properties in the plasma membrane and the tonoplast of various plant materials including tobacco suspension cells (Maurel *et al.*, 1997; Gerbeau *et al.*, 1999), wheat roots (Niemietz & Tyerman, 1997) and radish taproot (Ohshima *et al.*, 2001). In tobacco for instance, the tonoplast had high permeability values to water ($P_f \sim 690 \, \mu m \, s^{-1}$) and urea ($P_{urea} \sim 75 \times 10^{-3} \mu m \, s^{-1}$), as compared to low values in the plasma membrane ($P_f \sim 6 \, \mu m \, s^{-1}$; $P_{urea} \sim 1 \times 10^{-3} \mu m \, s^{-1}$) (Maurel *et al.*, 1997; Gerbeau *et al.*, 1999). These striking differences can be accounted for by a higher level of aquaporin activity in the former membrane (Maurel *et al.*, 1997; Gerbeau *et al.*, 1999) (Fig. 8.4). Morillon and Lassalles (1999) also observed that vacuoles isolated from various plant materials had remarkably high water permeabilities (200–1000 $\mu m \, s^{-1}$). We and others have proposed that by not limiting water and solute transport across the tonoplast plant cells may be able to use the vacuolar

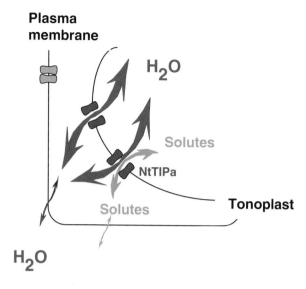

**Plasma membrane**

$H_2O$

Solutes

NtTIPa

Solutes

**Tonoplast**

$H_2O$

**Fig. 8.4** Putative model for water and solute transport by aquaporins in tobacco suspension cells. This model was derived from permeability measurements made in purified tonoplast and plasma membrane vesicles, the former having a 75- to 100-fold higher permeability to water and urea than the latter (see text) (Maurel *et al.*, 1997; Gerbeau *et al.*, 1999). The scheme shows how aquaporins can mediate efficient transport of water or neutral solutes across the tonoplast. NtTIPa, an aquaporin permeable to both water and solutes, has been characterised in detail by Gerbeau *et al.* (1999). The high permeability properties of the tonoplast may provide a means for rapid osmotic adjustment of the cytosol (see text). Aquaporins may be present on the plasma membrane but they are proposed to be inactive.

space for efficient osmoregulation of the cytosol (Maurel *et al.*, 1997; Niemietz & Tyerman, 1997) (Fig. 8.4). The significance of these properties for the kinetics of volume and osmotic potential equilibration in response to a sudden change in extracellular water potential has been examined in detail elsewhere (Tyerman *et al.*, 1999; Maurel *et al.*, 2002). For this model to be generalised, it will be crucial to characterise water transport in subcellular membranes of novel plant materials. The CAM plant *Graptopetalum paraguayense* appears to have peculiar water transport properties, since purified plasma membrane and tonoplast both had extremely low $P_f$ (Ohshima *et al.*, 2001).

The model discussed above for efficient osmotic equilibration of the cytosol and vacuole can surely be extended to other systems in which large membrane surfaces delineate small-sized subcellular compartments. It may apply in particular to the case of arbuscular mycorrhiza that express specific aquaporins (Krajinski *et al.*, 2000). The solute and water transport properties of NOD26 in the symbiotic membrane of soybean root nodules (Rivers *et al.*, 1997; Dean *et al.*, 1999) may also be critical for fine osmoregulation of the peribacteroid space. Accordingly, expression and phosphorylation-dependent activity of NOD26 is tightly regulated during nodule development and as a function of nodule water status (Guenther *et al.*, 2003).

### 8.3.4   Tissue level: the role of aquaporins in root water uptake

#### 8.3.4.1   Cell-specific expression patterns

The diversity of aquaporin isoforms in plants can surely be linked to a large variety of tissue-specific expression patterns. There is indeed a wealth of data reporting on expression properties of aquaporins, in response to environmental stimuli or during development. *In situ* hybridization studies or the use of reporter genes driven by aquaporin promoters in transgenic plants have revealed exquisite cell- and tissue-specific gene expression patterns. These studies have been extensively analysed in a previous review article (Maurel *et al.*, 2002). Altogether, they suggested that aquaporins may be critically involved in most physiological and developmental processes of plants, including root water uptake, fluid transport in vascular tissues, seed maturation and germination, extension growth and flower development. However, it has proved difficult to establish a link between aquaporin expression data and direct evidence for a role in tissue water transport. In this respect, most studies have focused on aquaporin functions in roots.

#### 8.3.4.2   Role of cell membranes and aquaporins in water uptake

First attempts to probe for the role of water channels in root water uptake were based on the use of mercury, a general blocker of aquaporins (reviewed in Javot & Maurel, 2002). Because of its toxicity and lack of specificity, mercury remains a very poor pharmacological compound, which must be used at concentrations and with exposure times as low as possible. Despite these restrictions, a consistent set of data gathered from various plant species has accumulated over the years suggesting that aquaporins, and therefore cell membranes, contribute a large part (50–90%) to root $L_{pr}$ (Javot & Maurel, 2002). This means that parallel water transport across cell walls (apoplastic path) or cytoplasmic continuities (symplastic path) may not be as important as was initially hypothesised. In recent years, reverse genetics has provided a more reliable approach than mercury treatment to explore the integrated function of aquaporins. For instance, antisense inhibition of PIP1 and PIP2 aquaporin gene expression in *Arabidopsis* and tobacco has revealed the role of these two groups of aquaporins for water transport in roots (and water stress attenuation in leaves) (Kaldenhoff *et al.*, 1998; Martre *et al.*, 2002; Siefritz *et al.*, 2002). In particular, expression in transgenic *Arabidopsis* of antisense genes for *pip1;2*, *pip2;3* or both reduced $L_{pr}$ by 60%, 47% and 68%, respectively (Martre *et al.*, 2002). Using aquaporin PIP knockout mutants in *Arabidopsis*, our group recently determined, for the first time, the contribution of a single aquaporin gene to root water uptake (Javot *et al.*, 2003). Cell pressure probe measurements in normal and PIP2;2 mutant lines revealed a high cell hydraulic conductivity in the cortex ($L_p \sim 2$–$3 \times 10^{-6}$ m s$^{-1}$ MPa) and showed that this aquaporin contributed to at least 25% of this conductivity. These altered water transport properties did not result in any significant change in hydrostatic $L_{pr}$ of whole roots, as measured with a pressure chamber. In contrast, free exudation by excised roots yielded sap of greater osmolality in mutant than in wild-type plants. A reduced osmotic $L_{pr}$ could be calculated in the former plants, showing that PIP2;2 functions as an aquaporin specialised in osmotic fluid transport

(Javot *et al.*, 2003). Interestingly, PIP2;2 has a close homologue named PIP2;3. The two aquaporins share >97% amino acid identity and have evolved by a recent gene duplication. The identification of phenotype in PIP2;2 knockout plants provides evidence that, despite their high isoform multiplicity, plant aquaporins do not have fully overlapping functions. Thus, this study demonstrates that, in principle, reverse genetics analysis using single gene knockouts allows to dissect the diversity of plant aquaporin functions.

Besides their role in soil water uptake, an interesting feature of roots is that $Lp_r$ can vary in response to many hormonal and environmental factors (reviewed in Javot & Maurel, 2002, and Clarkson *et al.*, 2000). Thus, root water transport has emerged as a paradigm, in which to investigate aquaporin regulation. It is now used by many laboratories to investigate the early signalling events and the downstream molecular and cellular mechanisms leading to water transport regulation (see Section 8.4).

## 8.4   Mechanisms of regulation

### 8.4.1   Levels of regulation

It has been known for a long time that the lipid composition of membranes can determine to some extent their water and solute permeability properties (Lande *et al.*, 1995). Also, extreme and sustained stress conditions (cold acclimatation and desiccation) can induce significant changes in plant lipid membranes (Lynch & Steponkus, 1987). Therefore, a central role for lipids in membrane water transport regulation has been much favoured in the past (reviewed in Maurel, 1997). However, novel evidence has emerged over the last decade, showing that the course and fine regulation of membrane water permeability is, in fact, mostly due to a multistorey regulation of aquaporins. This is an area of intense interest, and significant advances have been made. However, many crucial aspects, both in terms of the stimuli involved as well as the responses themselves, are still unclear.

In theory, both the density of aquaporins within the membrane and their intrinsic water permeability can contribute to regulating membrane water transport. Evidence for both types of mechanisms exists and will be discussed below. The density of aquaporins within the membrane is generally under the regulation of the gene and protein expression responses within the cell, allowing regulation on a rather long-term scale (hours to days). Dynamic changes in the permeability of each pore can rapidly be induced through changes in protein conformation and are generally involved in shorter term responses to stimuli. In most cases, the regulation of aquaporins is an integrated process involving a combination of the various levels of control. Nevertheless, and for clarity, we will consider and exemplify each of these levels individually.

### 8.4.2   Regulation of gene expression

There is a considerable body of evidence demonstrating that many aquaporins are under transcriptional regulation. Recent advances in transcriptome analyses have contributed significantly to the understanding of regulation of many genes, and aquaporins are no exception. Both large-scale microarray studies (Kawasaki *et al.*, 2001; Klok *et al.*, 2002; Seki *et al.*, 2002; Maathuis *et al.*, 2003) as well as smaller scale in-depth investigations have demonstrated that transcription of aquaporins can be controlled by environmental stimuli including temperature (Kammerloher *et al.*, 1994), drought (Yamaguchi-Shinozaki *et al.*, 1992), salinity (Yamada *et al.*, 1995) and anoxia (Klok *et al.*, 2002). Blue light as well as hormones such as abscisic acid and gibberellic acid have also been shown to induce aquaporin expression (Kaldenhoff *et al.*, 1993; Phillips & Huttly, 1994).

In some studies, a correlation between the abundance of aquaporin transcripts and membrane water permeability has been observed. The largest body of evidence concerns the daily cycles in water permeability that can be observed in various plant organs. One of the first aquaporin transcripts observed to follow a daily cycle was a δ-TIP homologue in sunflower guard cells (Sarda *et al.*, 1997). Daily rhythms in aquaporin transcripts have also been observed in the roots of a number of species including *Raphanus sativus* (radish) (Higuchi *et al.*, 1998), *Lotus japonicus* (Henzler *et al.*, 1999), *Nicotiana excelsior* (Yamada *et al.*, 1997), *Arabidopsis* (Harmer *et al.*, 2000) and *Zea mays* (maize) (Gaspar *et al.*, 2003). In the latter species, the diurnal regulation of a *PIP* gene (*ZmPIP1-5b*) was found to be tissue-specific, since gene expression was constitutive in the root epidermal layer but fluctuating in the stele and cortex (Gaspar *et al.*, 2003). On the other hand, diurnal fluctuations in root hydraulic conductivity are well known as they were first observed in cotton 30 years ago (Parsons & Kramer, 1974). In *L. japonicus*, such rhythms could directly be correlated with PIP1 mRNA expression (Henzler *et al.*, 1999). In another study, Moshelion *et al.* (2002) found that the daily movements of the leaflets of *Samanea saman* can be correlated with daily changes in mRNA expression of a PIP homologue, SsAQP1. In that study the authors demonstrated that motor cell volume changes and permeability fluctuated in parallel with both the aquaporin transcript and protein expression in those tissues. Another striking example was revealed in the petioles of tobacco plants by Siefritz *et al.* (2004). It was shown that cyclic expression of NtAQP1 in this organ correlated with both circadian and diurnal light cycles. The authors suggested that the observed cyclic movements of leaves might be attributed to rapid and differential changes in cell elongation in petioles, as governed by NtAQP1 expression (Siefritz *et al.*, 2004).

### 8.4.3 *Protein translation and degradation*

The examples mentioned above imply a direct coordination between transcript abundance and tissue water permeability. However, there are many levels of possible control, subsequent to mRNA accumulation. Consequently, there should be extreme caution in interpreting transcriptional patterns of aquaporins. Studies by Suga *et al.* (2001, 2002) in radish have demonstrated that the transcript and protein levels of a variety of PIPs and TIPs were not always correlated with each other, suggesting the possible occurrence of post-transcriptional regulations. In the later study for instance (Suga *et al.*, 2002), these authors have noted that although some TIP isoforms are constitutively expressed, the transcript expression of some PIP2 isoforms increased transiently in response to osmotic stress but the protein levels of these isoforms remained constant.

As with transcript accumulation, the translation of aquaporin proteins has been correlated with various environmental stresses and, for example, the abundance of a TIP homologue was downregulated during the response of ice plant to drought or salt exposure (Kirch et al., 2000). The accumulation of aquaporins can also be developmentally regulated (Melroy & Herman, 1991). During seed maturation and germination, significant changes in the vacuolar apparatus occur, and consequent changes in the protein expression of various TIP isoforms have been identified. The accumulation of α-TIP-like aquaporins on the tonoplast of protein storage vacuoles is thought to be important during the early phases of imbibition in *P. vulgaris* (Johnson et al., 1989) and soybean (Melroy & Herman, 1991) seeds. During germination, this isoform is progressively substituted by a γ-TIP homologue (Maeshima et al., 1994), which is usually associated with lytic vacuoles (Jauh et al., 1999). It is thought that regulated expression of these various TIP isoforms plays a role in controlling the volume changes and dramatic fluctuations in osmotic potentials associated with seed maturation and germination.

The molecular mechanisms that govern aquaporin translation and stability in plant cells are largely unknown. In maturing pumpkin seeds, a MIP found in protein storage vacuoles is initially transcribed as a 29 kDa precursor, and subsequently cleaved to a 23 kDa protein (Inoue et al., 1995). This post-translational degradation suggests a possible means of aquaporin regulation but as yet no functional role has been demonstrated. A pathway for aquaporin degradation in mammalian cells has recently been unravelled by Leitch et al. (2001). These authors demonstrated that the increased expression of human AQP1 during a hyperosmotic stress (4–24 h) was the consequence of an increased stability of the protein owing to a decrease of ubiquitin-mediated AQP1 degradation.

## 8.4.4  *Protein targeting*

In animals there are a number of reports suggesting that protein targeting may allow for rapid changes in membrane permeability. Using immunohistochemistry, it has been demonstrated that some mammalian aquaporins shuttle from intracellular vesicles to the plasma membrane in response to various stimuli. Some examples include AQP2 trafficking to apical membranes in kidney collecting ducts under vasopression regulation (Brown, 2003), AQP1 relocation to apical plasma membranes of cholangiocytes upon secretin induction (Marinelli et al., 1999), AQP5 shuttling to apical plasma membranes of parotid acinar cells through nitric oxide/cGMP transduction (Ishikawa et al., 2002) and AQP8 redistribution to canalicular membranes of hepatocytes upon cAMP stimulation (Huebert et al., 2002). In the case of AQP2, it was shown that stimulus-induced subcellular targeting is dependent on exquisite molecular mechanisms. In particular, phosphorylation of AQP2 takes place on Ser256 and a critical balance between phosphorylated and dephosphorylated subunits in each tetramer determines its subcellular fate (Kamsteeg et al., 2000). This process is controlled in part by vasopressin-activated protein kinase A, thereby providing a control for apical membrane targeting. Recent studies showed that transition of AQP2 in the Golgi apparatus was also controlled by phosphorylation of Ser256, but by another stimulus-independent protein kinase, possibly a Golgi-casein kinase (Procino *et al.*, 2003). Similar mechanisms, though less documented, have also been described for mammalian aquaporins AQP1 and AQP4 (Han & Patil, 2000; Madrid *et al.*, 2001).

In plants, no direct evidence for aquaporin trafficking has been reported yet. Robinson *et al.* (1996) suggested, however, that apparent invaginations of the plasma membrane (called plasmalemmasomes), which exhibit a high density of PIPs, may act in a way similar to those intracellular vesicles described in animals. The presence of PIPs in the endomembrane fractions of *M. crystallinum* was also interpreted as evidence of their passage to the plasma membrane via the secretory system, this being possibly regulated in response to stresses (Barkla *et al.*, 1999; Kirch *et al.*, 2000). However, these studies were performed using immunolocalisation with PIP antibodies and hence only provide a static view of the system. Thus dynamic shuttling of aquaporins has yet to be conclusively demonstrated in plants.

### 8.4.5   *Molecular mechanisms of aquaporin gating*

#### 8.4.5.1   *Regulation by phosphorylation*

In contrast to phosphorylation of mammalian aquaporins, which is generally associated with signalling processes and membrane targeting, the reversible phosphorylation reported in plant aquaporins is thought to mediate direct gating of the pore itself. This process was recently demonstrated in peribacteroid membrane vesicles purified from symbiotic root nodules. The water permeability of these vesicles was shown to be controlled by phosphorylation of NOD26 (Guenther *et al.*, 2003), this being modulated through an exogenous protein kinase. In previous studies, several groups have used *Xenopus* oocyte expression to prove that phosphorylation regulates the activity of a few other plant aquaporin isoforms (Maurel *et al.*, 1995; Johansson *et al.*, 1998). An α-TIP from *P. vulgaris*, for instance, has been shown to be reversibly activated through serine phosphorylation at three sites, two of them being located on the N-terminal tail of the protein, whereas the last one is on the NPA-containing loop B (Maurel *et al.*, 1995). Johansson *et al.* (1998) showed that a similar site in loop B of spinach PIP2 homologue PM28A, together with phosphorylation of a Ser in the C-terminal tail, regulate this aquaporin in oocytes.

As with the phosphorylation of all plant aquaporins observed thus far, phosphorylation of PM28A in spinach leaves is mediated by a calcium-dependent protein kinase (Johansson *et al.*, 1996). In addition, an increase in the osmolarity of the apoplasm (simulating drought conditions) was found to cause a decrease in phosphorylation of this aquaporin. In contrast, phosphorylation of NOD26 in soybean nodules was enhanced by both water stress and salinity (Guenther *et al.*, 2003). Thus, regulation of plant aquaporin phosphorylation by osmotic stresses seems quite common and may form part of a coordinated adaptive response to regulate cell turgor and/or redirect water flows.

### 8.4.5.2 *Regulation by protons*

Evidence that plant aquaporins are regulated by pH was initially obtained in purified membrane vesicles or organelles. For instance, Amodeo *et al.* (2002) observed pH sensitivity of tonoplast water channels in *Beta vulgaris* storage roots, using swelling assays on isolated vacuoles. In another study, Gerbeau *et al.* (2002) used stopped-flow spectrophotometry to show that protons reversibly inhibit water channels in plasma membrane vesicles purified from *Arabidopsis* suspension cells (Fig. 8.1B). Half inhibition occurred at pH 7.2–7.5, suggesting that protons were acting on the cytosolic side of the water channels.

The subsequent use of *Xenopus* oocyte expression provided further mechanistic insights into this regulation. It was first shown that all members of the PIP subfamily of plasma membrane aquaporins in *Arabidopsis* were specifically blocked by cytosplasmic acidosis (Tournaire-Roux *et al.*, 2003) (Fig. 8.5A). Furthermore, the gating mechanism was explained by the presence on the cytosolic side of these proteins of a specifically conserved, pH-sensing His residue (Tournaire-Roux *et al.*, 2003) (Fig. 8.5B). Using aquaporin PIP2;2 as a model, it was shown that point mutations at this site, to a Asp residue for instance (PIP2;2 H197D), result in active aquaporins that are insensitive to cytosolic pH (Fig. 8.5C). The physiological significance of this regulation has been investigated by parallel measurements of cytosolic pH (using $^{31}$P *in vivo* NMR) and water transport in excised roots and root cortical cells. These studies revealed a strong parallel between cytoplasmic pH and $Lp_r$, which both dropped within minutes during anoxia (Tournaire-Roux *et al.*, 2003). Thus, variations in cytoplasmic pH provide an important, novel mechanism for dynamic short-term regulation of transcellular water transport mediated by the PIP aquaporins. For vacuoles, the general significance of pH-dependent water permeability is less clear: pH gating has been demonstrated in root storage vacuoles (Amodeo *et al.*, 2002) but TIP1;1 from *Arabidopsis* has been found to be pH-insensitive (Tournaire-Roux *et al.*, 2003). Finally, $Ca^{2+}$ and other divalent ions have been shown to regulate water channels in purified plasma membrane vesicles but the underlying mechanisms remain as yet unknown (Gerbeau *et al.*, 2002) (Fig. 8.1B).

## 8.5  Conclusion

Recent physiological and genetic evidence has provided compelling evidence that aquaporins are central players in key aspects of plant water relations, such as regulation of root water uptake (Tournaire-Roux *et al.*, 2003), or water stress recovery (Martre *et al.*, 2002; Siefritz *et al.*, 2002). However, much remains to be learned about the integrated function of aquaporins.

In particular, the molecular and cellular mechanisms that lead to aquaporin regulation will be very important fields for future investigations (Fig. 8.6). The elucidation of novel molecular mechanisms for aquaporin gating or trafficking, and their in-depth analysis will undoubtedly provide new tools to molecular physiologists. For instance, transgenic plants that overexpress mutated, pH-insensitive aquaporins

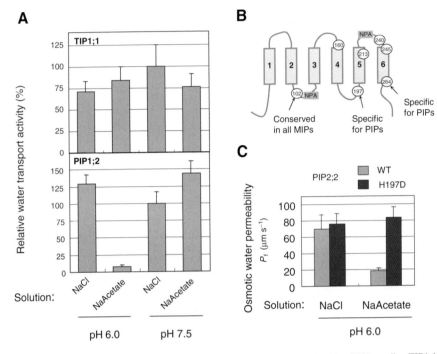

**Fig. 8.5** Regulation of PIPs by intracellular pH. (A) Oocytes were injected with mRNAs coding TIP1;1 or PIP1;2 aquaporins, incubated for 3 days to allow exogenous aquaporin expression, and then transferred in a solution at pH 6.0 or 7.5, containing either 50 mM NaCl or 50 mM NaAcetate. In contrast to NaCl, NaAcetate can efficiently diffuse across the cell membrane under its acidic form, thereby equilibrating external and internal pH. Note that the water permeability of oocytes expressing TIP1;1 is not significantly changed by any of the treatments whereas the water permeability of oocytes expressing PIP1;2 is dramatically decreased by an acid load at pH 6. (B) Schematic representation of the structure of PIP aquaporins, with the six transmembrane helices and the highly conserved NPA motifs. The histidine (H) residues conserved in PIPs and their position in the PIP2;2 sequence are circled. Two residues are specific for the PIP subfamily: H197 and H264. (C) Osmotic water permeability of oocytes injected with mRNA coding wild-type PIP2;2 (WT) or a mutant PIP2;2 (H197D) where the histidine at position 197 was replaced by an aspartate. Under acid load conditions (NaAcetate, pH 6.0), WT but not H197D exhibits a marked decrease in water transport activity. This suggests the involvement of histidine at position 197 in cytosolic pH sensing and controlled-gating of the aquaporin.

are currently used in our laboratory to investigate the adaptive significance of water transport inhibition by cytosol acidosis. A central objective will also be to integrate the disconnected regulatory mechanisms unravelled so far into a comprehensive picture, from initial stimulus (stress) perception to terminal aquaporin and water status regulation (Fig. 8.6).

Finally, the diversity of aquaporin gene expression patterns and of transport selectivities, which underlies the high isoform multiplicity of plant aquaporins, has pointed to previously unpredicted functions for water channel proteins. The provocative idea that aquaporins play a role in $CO_2$ diffusion in leaves has recently found some experimental support (Terashima & Ono, 2002; Uehlein *et al.*, 2003). Another

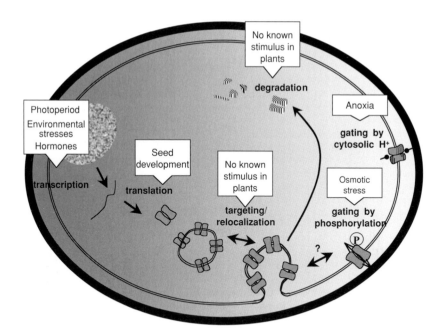

**Fig. 8.6** Simplistic representation of a few of the multiple signals and their respective points of regulation affecting the activity of aquaporins. Evidence for regulation of aquaporin expression at the transcription and translation levels have been obtained in various species including *Arabidopsis thaliana*, *Nicotiana excelsior*, *Lotus japonicus*, *Phaseolus vulgaris* and *Zea mays* (see text). Evidence for regulation of both relocalisation/targeting and protein degradation still remains unknown in plants but such mechanisms have been discovered in mammals (see text). Similarly, a link between aquaporin phosphorylation and aquaporin trafficking has been established in animals and may exist in plants. To date, both phosphorylation and cytosolic pH have been demonstrated to result in gating of some plant aquaporins.

important question concerns the role of aquaporins and membrane water permeability in controlling cell expansion (Siefritz *et al.*, 2004). These two difficult issues still need careful examination and will surely trigger lively scientific discussions in the coming years.

# References

Agre, P., Bonhivers, M. & Borgnia, M.J. (1998) The aquaporins, blueprints for cellular plumbing systems, *J. Biol. Chem.*, **273**, 14659–14662.

Aharon, R., Shahak, Y., Wininger, S., Bendov, R., Kapulnik, Y. & Galili, G. (2003) Overexpression of a plasma membrane aquaporin in transgenic tobacco improves plant vigor under favorable growth conditions but not under drought or salt stress, *Plant Cell*, **15**, 439–447.

Amodeo, G., Sutka, M., Dorr, R. & Parisi, M. (2002) Protoplasmic pH modifies water and solute transfer in *Beta vulgaris* root vacuoles, *J. Membr. Biol.*, **187**, 175–184.

Anthony, T.L., Brooks, H.L., Boassa, D., *et al.* (2000) Cloned human aquaporin-1 is a cyclic GMP-gated ion channel, *Mol. Pharmacol.* **57**, 576–588.

Barkla, B.J., Vera-Estrella, R., Pantoja, O., Kirch, H.-H. & Bohnert, H.J. (1999) Aquaporin localization – how valid are the TIP and PIP labels. *Trends Plant Sci.*, **4**, 86–88.

Biela, A., Grote, K., Otto, B., Hoth, S., Hedrich, R. & Kaldenhoff, R. (1999) The *Nicotiana tabacum* plasma membrane aquaporin NtAQP1 is mercury-insensitive and permeable for glycerol, *Plant J.*, **18**, 565–570.

Brown, D. (2003) The ins and outs of aquaporin-2 trafficking, *Am. J. Physiol. Renal. Physiol.*, **284**, F893–F901.

Chaumont, F., Barrieu, F., Jung, R. & Chrispeels, M.J. (2000) Plasma membrane intrinsic proteins from maize cluster in two sequence subgroups with differential aquaporin activity, *Plant Physiol.*, **122**, 1025–1034.

Chaumont, F., Barrieu, F., Wojcik, E., Chrispeels, M.J. & Jung, R. (2001) Aquaporins constitute a large and highly divergent protein family in maize, *Plant Physiol.*, **125**, 1206–1215.

Chrispeels, M.J. & Agre, P. (1994) Aquaporins: water channel proteins of plant and animal cells, *Trends Biochem. Sci.*, **19**, 421–425.

Clarkson, D.T., Carvajal, M., Henzler, T., *et al.* (2000) Root hydraulic conductance: diurnal aquaporin expression and the effects of nutrient stress, *J. Exp. Bot.*, **51**, 61–70.

Cosgrove, D. & Steudle, E. (1981) Water relations of growing pea epicotyl segments, *Planta*, **153**, 343–350.

Cutler, S.R., Ehrhardt, D.W., Griffitts, J.S. & Somerville, C.R. (2000) Random GFP::cDNA fusions enable visualization of subcellular structures in cells of *Arabidopsis* at high frequency, *Proc. Natl. Acad. Sci. U.S.A.*, **97**, 3718–3723.

Dainty, J. (1963) Water relations of plant cells, *Adv. Bot. Res.*, **1**, 279–326.

Daniels, M.J., Chaumont, F., Mirkov, T.E. & Chrispeels, M.J. (1996) Characterization of a new vacuolar membrane aquaporin sensitive to mercury at a unique site, *Plant Cell*, **8**, 587–599.

Daniels, M.J., Chrispeels, M.J. & Yeager, M. (1999) Projection structure of a plant vacuole membrane aquaporin by electron cryo-crystallography, *J. Mol. Biol.*, **294**, 1337–1349.

Daniels, M.J., Mirkov, T.E. & Chrispeels, M.J. (1994) The plasma membrane of *Arabidopsis thaliana* contains a mercury-insensitive aquaporin that is a homolog of the tonoplast water channel protein TIP, *Plant Physiol.*, **106**, 1325–1333.

Dean, R.M., Rivers, R.L., Zeidel, M.L. & Roberts, D.M. (1999) Purification and functional reconstitution of soybean Nodulin 26. An aquaporin with water and glycerol transport properties, *Biochemistry*, **38**, 347–353.

de Groot, B.L., Frigato, T., Helms, V. & Grubmuller, H. (2003) The mechanism of proton exclusion in the aquaporin-1 water channel, *J. Mol. Biol.*, **333**, 279–293.

de Groot, B. & Grubmüller, H. (2001) Water permeation across biological membranes: mechanism and dynamics of Aquaporin-1 and GlpF, *Science*, **294**, 2353–2357.

Dordas, C., Chrispeels, M.J. & Brown, P.H. (2000) Permeability and channel-mediated transport of boric acid across membrane vesicles isolated from squash roots, *Plant Physiol.*, **124**, 1349–1361.

Fetter, K., Van Wilder, V., Moshelion, M. & Chaumont, F. (2004) Interactions between plasma membrane aquaporins modulate their water channel activity, *Plant Cell*, **16**, 215–228.

Fotiadis, D., Jeno, P., Mini, T., *et al.* (2001) Structural characterization of two aquaporins isolated from native spinach leaf plasma membranes, *J. Biol. Chem.*, **276**, 1707–1714.

Fu, D., Libson, A., Miercke, L.J.W., *et al.* (2000) Structure of a glycerol-conducting channel and the basis for its selectivity, *Science*, **290**, 481–486.

Fujiyoshi, Y., Mitsuoka, K., de Groot, B.L., *et al.* (2002) Structure and function of water channels, *Curr. Opin. Struct. Biol.*, **12**, 509–515.

Gaspar, M., Bousser, A., Sissoëff, I., Roche, O., Hoarau, J. & Mahe, A. (2003) Cloning and characterization of *ZmPIP1-5b*, an aquaporin transporting water and urea, *Plant Sci.*, **165**, 21–31.

Gerbeau, P., Amodeo, G., Henzler, T., Santoni, V., Ripoche, P. & Maurel, C. (2002) The water permeability of *Arabidopsis* plasma membrane is regulated by divalent cations and pH, *Plant J.*, **30**, 71–81.

Gerbeau, P., Guclu, J., Ripoche, P. & Maurel, C. (1999) Aquaporin Nt-TIPa can account for the high permeability of tobacco cell vacuolar membrane to small neutral solutes, *Plant J.*, **18**, 577–587.

Guenther, J.F., Chanmanivone, N., Galetovic, M.P., Wallace, I.S., Cobb, J.A. & Roberts, D.M. (2003) Phosphorylation of soybean nodulin 26 on serine 262 enhances water permeability and is regulated developmentally and by osmotic signals, *Plant Cell*, **15**, 981–991.

Gustavsson, S. & Johanson, U. (2003) A new subfamily of water channel-like proteins in plants, in *7th International Congress of Plant Molecular Biology*, Barcelona, Spain, Abstract S24–10.

Han, Z. & Patil, R.V. (2000) Protein kinase A-dependent phosphorylation of aquaporin-1, *Biochem. Biophys. Res. Commun.*, **273**, 328–332.

Harmer, S.L., Hogenesch, J.B., Straume, M., *et al.* (2000) Orchestrated transcription of key pathways in *Arabidopsis* by the circadian clock, *Science*, **290**, 2110–2113.

Henzler, T. & Steudle, E. (2000) Transport and metabolic degradation of hydrogen peroxide in *Chara corallina*. Model calculations and measurements with the pressure probe suggest transport of $H_2O_2$ across water channels, *J. Exp. Bot.*, **51**, 2053–2066.

Henzler, T., Waterhouse, R.N., Smyth, A.J., *et al.* (1999) Diurnal variations in hydraulic conductivity and root pressure can be correlated with the expression of putative aquaporins in the roots of *Lotus japonicus*, *Planta*, **210**, 50–60.

Heymann, J.B. & Engel, A. (2000) Structural clues in the sequences of the aquaporins, *J. Mol. Biol.*, **295**, 1039–1053.

Higuchi, T., Suga, S., Tsuchiya, T., *et al.* (1998) Molecular cloning, water channel activity and tissue specific expression of two isoforms of radish vacuolar aquaporin, *Plant Cell Physiol.*, **39**, 905–913.

Höfte, H., Hubbard, L., Reizer, J., Ludevid, D., Herman, E.M. & Chrispeels, M.J. (1992) Vegetative and seed-specific forms of tonoplast intrinsic protein in the vacuolar membrane of *Arabidopsis thaliana*, *Plant Physiol.*, **99**, 561–570.

Huebert, R., Splinter, P., Garcia, F., Marinelli, R. & LaRusso, N. (2002) Expression and localization of aquaporin water channels in rat hepatocytes. Evidence for a role in canalicular bile secretion, *J. Biol. Chem.*, **277**, 22710–22717.

Inoue, K., Takeuchi, Y., Nishimura, M. & Hara-Nishimura, I. (1995) Molecular characterization of proteins in protein-body membrane that disappear most rapidly during transformation of protein bodies into vacuoles, *Plant Mol. Biol.*, **28**, 1089–1101.

Ishikawa, Y., Iida, H. & Ishida, H. (2002) The muscarinic acetylcholine receptor-stimulated increase in aquaporin-5 levels in the apical plasma membrane in rat parotid acinar cells is coupled with activation of nitric oxide/cGMP signal transduction, *Mol. Pharmacol.*, **61**, 1423–1434.

Jauh, G.Y., Fischer, A.M., Grimes, H.D., Ryan, C.A. & Rogers, J.C. (1998) Delta-Tonoplast intrinsic protein defines unique plant vacuole functions, *Proc. Natl. Acad. Sci. U.S.A.*, **95**, 12995–12999.

Jauh, G.Y., Phillips, T.E. & Rogers, J.C. (1999) Tonoplast intrinsic protein isoforms as markers for vacuolar functions, *Plant Cell*, **11**, 1867–1882.

Javot, H., Lauvergeat, V., Santoni, V., *et al.* (2003) Role of a single aquaporin isoform in root water uptake, *Plant Cell*, **15**, 509–522.

Javot, H. & Maurel, C. (2002) The role of aquaporins in root water uptake, *Ann. Bot.*, **90**, 301–313.

Johanson, U. & Gustavsson, S. (2002) A new subfamily of major intrinsic proteins in plants, *Mol. Biol. Evol.*, **19**, 456–461.

Johanson, U., Karlsson, M., Gustavsson, S., *et al.* (2001) The complete set of genes encoding Major Intrinsic Proteins in *Arabidopsis* provides a framework for a new nomenclature for Major Intrinsic Proteins in plants, *Plant Physiol.*, **126**, 1358–1369.

Johansson, I., Karlsson, M., Shukla, V.K., Chrispeels, M.J., Larsson, C. & Kjellbom, P. (1998) Water transport activity of the plasma membrane aquaporin PM28A is regulated by phosphorylation, *Plant Cell*, **10**, 451–459.

Johansson, I., Larsson, C., Ek, B. & Kjellbom, P. (1996) The major integral proteins of spinach leaf plasma membranes are putative aquaporins and are phosphorylated in response to $Ca^{2+}$ and apoplastic water potential, *Plant Cell*, **8**, 1181–1191.

Johnson, K.D., Herman, E.M. & Chrispeels, M.J. (1989) An abundant, highly conserved tonoplast protein in seeds, *Plant Physiol.*, **91**, 1006–1013.

Johnson, K.D., Höfte, H. & Chrispeels, M.J. (1990) An intrinsic tonoplast protein of protein storage vacuoles in seeds is structurally related to a bacterial solute transporter (GlpF), *Plant Cell*, **2**, 525–532.

Jung, J.S., Preston, G.M., Smith, B.L., Guggino, W.B. & Agre, P. (1994) Molecular structure of the water channel through aquaporin CHIP. The hourglass model, *J. Biol. Chem.* **269**, 14648–14654.

Kaldenhoff, R., Grote, K., Zhu, J.-J. & Zimmermann, U. (1998) Significance of plasmalemma aquaporins for water-transport in *Arabidopsis thaliana*, *Plant J.*, **14**, 121–128.

Kaldenhoff, R., Kölling, A., Meyers, J., Karmann, U., Ruppel, G. & Richter, G. (1995) The blue light-responsive *AthH2* gene of *Arabidopsis thaliana* is primarily expressed in expanding as well as in differentiating cells and encodes a putative channel protein of the plasmalemma, *Plant J.*, **7**, 87–95.

Kaldenhoff, R., Kölling, A. & Richter, G. (1993) A novel blue light- and abscisic acid-inducible gene of *Arabidopsis thaliana* encoding an intrinsic membrane protein, *Plant Mol. Biol.*, **23**, 1187–1198.

Kammerloher, W., Fischer, U., Piechottka, G.P. & Schäffner, A.R. (1994) Water channels in the plant plasma membrane cloned by immunoselection from a mammalian expression system, *Plant J.*, **6**, 187–199.

Kamsteeg, E.J., Heijnen, I., van Os, C.H. & Deen, P.M.T. (2000) The subcellular localization of an aquaporin-2 tetramer depends on the stoichiometry of phosphorylated and nonphosphorylated monomers, *J. Cell Biol.*, **151**, 919–930.

Karlsson, M., Fotiadis, D., Sjovall, S., *et al.* (2003) Reconstitution of water channel function of an aquaporin overexpressed and purified from *Pichia pastoris*, *FEBS Lett.*, **537**, 68–72.

Karlsson, M., Johansson, I., Bush, M., *et al.* (2000) An abundant TIP expressed in mature highly vacuolated cells, *Plant J.*, **21**, 83–90.

Kawasaki, S., Borchert, C., Deyholos, M., *et al.* (2001) Gene expression profiles during the initial phase of salt stress in rice, *Plant Cell*, **13**, 889–905.

Kirch, H.-H., Vera-Estrella, R., Golldack, D., *et al.* (2000) Expression of water channel proteins in *Mesembryanthemum crystallinum*, *Plant Physiol.*, **123**, 111–124.

Klebl, F., Wolf, M. & Sauer, N. (2003) A defect in the yeast plasma membrane urea transporter Dur3p is complemented by CpNIP1, a Nod26-like protein from zucchini (*Cucurbita pepo* L.), and by *Arabidopsis thaliana* delta-TIP or gamma-TIP, *FEBS Lett.*, **547**, 69–74.

Klok, E.J., Wilson, I.W., Wilson, D., *et al.* (2002) Expression profile analysis of the low-oxygen response in *Arabidopsis* root cultures, *Plant Cell*, **14**, 2481–2494.

Krajinski, F., Biela, A., Schubert, D., Gianinazzi-Pearson, V., Kaldenhoff, R. & Franken, P. (2000) Arbuscular mycorrhiza development regulates the mRNA abundance of *Mtaqp1* encoding a mercury-insensitive aquaporin of *Medicago truncatula*, *Planta*, **211**, 85–90.

Lagrée, V., Froger, A., Deschamps, S., *et al.* (1999) Switch from an aquaporin to a glycerol channel by two amino acids substitution, *J. Biol. Chem.*, **274**, 6817–6819.

Laizé, V., Rousselet, G., Verbavatz, J.-M., *et al.* (1995) Functional expression of the human CHIP28 water channel in a yeast secretory mutant, *FEBS Lett.*, **373**, 269–274.

Lande, M.B., Donovan, J.M. & Zeidel, M.L. (1995) The relationship between membrane fluidity and permeabilities to water, solutes, ammonia, and protons, *J. Gen. Physiol.*, **106**, 67–84.

Leitch, V., Agre, P. & King, L.S. (2001) Altered ubiquitination and stability of aquaporin-1 in hypertonic stress, *Proc. Natl. Acad. Sci. U.S.A.*, **98**, 2894–2898.

Liu, L.H., Ludewig, U., Gassert, B., Frommer, W.B. & von Wiren, N. (2003) Urea transport by nitrogen-regulated tonoplast intrinsic proteins in *Arabidopsis*, *Plant Physiol.*, **133**, 1220–1228.

Loo, D.D.F., Zeuthen, T., Chandy, G. & Wright, E.M. (1996) Cotransport of water by the $Na^+$/glucose cotransporter, *Proc. Natl. Acad. Sci. U.S.A.*, **93**, 13367–13370.

Luyten, K., Albertyn, J., Skibbe, W.F., *et al.* (1995) Fps1, a yeast member of the MIP family of channel proteins, is a facilitator for glycerol uptake and efflux and is inactive under osmotic stress, *EMBO J.*, **14**, 1360–1371.

Lynch, D.V. & Steponkus, P.L. (1987) Plasma membrane lipid alterations associated with cold acclimation of winter rye seedlings (*Secale cereale* L. cv Puma), *Plant Physiol.*, **83**, 761–767.

Maathuis, F.J.M., Filatov, V., Herzyk, P., *et al.* (2003) Transcriptome analysis of root transporters reveals participation of multiple gene families in the response to cation stress, *Plant J.*, **35**, 675–692.

Madrid, R., Le Maout, S., Barrault, M.B., Janvier, K., Benichou, S. & Merot, J. (2001) Polarized trafficking and surface expression of the AQP4 water channel are coordinated by serial and regulated interactions with different clathrin-adaptor complexes, *EMBO J.*, **20**, 7008–7021.

Maeshima, M. (1992) Characterization of the major integral protein of vacuolar membrane, *Plant Physiol.*, **98**, 1248–1254.

Maeshima, M., Hara-Nishimura, I., Takeuchi, Y. & Nishimura, M. (1994) Accumulation of vacuolar H$^+$-pyrophosphatase and H$^+$-ATPase during reformation of the central vacuole in germinating pumpkin seeds, *Plant Physiol.*, **106**, 61–69.

Marinelli, R.A., Tietz, P.S., Pham, L.D., Rueckert, L., Agre, P. & LaRusso, N.F. (1999) Secretin induces the apical insertion of aquaporin-1 water channels in rat cholangiocytes, *Am. J. Physiol. (Gastrointest Liver Physiol.)*, **39**, G280–G286.

Martre, P., Morillon, R., Barrieu, F., North, G.B., Nobel, P.S. & Chrispeels, M.J. (2002) Plasma membrane aquaporins play a significant role during recovery from water deficit, *Plant Physiol.*, **130**, 2101–2110.

Mathai, J.C., Mori, S., Smith, B.L., *et al.* (1996) Functional analysis of aquaporin-1 deficient red cells. The Colton-null phenotype, *J. Biol. Chem.*, **271**, 1309–1313.

Maurel, C. (1997) Aquaporins and water permeability of plant membranes, *Annu. Rev. Plant Physiol. Plant Mol. Biol.*, **48**, 399–429.

Maurel, C. & Chrispeels, M.J. (2001) Aquaporins. A molecular entry into plant water relations, *Plant Physiol.*, **125**, 135–138.

Maurel, C., Javot, H., Lauvergeat, V., *et al.* (2002) Molecular physiology of aquaporins in plants, *Int. Rev. Cytol.*, **215**, 105–148.

Maurel, C., Kado, R.T., Guern, J. & Chrispeels, M.J. (1995) Phosphorylation regulates the water channel activity of the seed-specific aquaporin a-TIP, *EMBO J.*, **14**, 3028–3035.

Maurel, C., Reizer, J., Schroeder, J.I. & Chrispeels, M.J. (1993) The vacuolar membrane protein g-TIP creates water specific channels in *Xenopus* oocytes, *EMBO J.*, **12**, 2241–2247.

Maurel, C., Tacnet, F., Güclü, J., Guern, J. & Ripoche, P. (1997) Purified vesicles of tobacco cell vacuolar and plasma membranes exhibit dramatically different water permeability and water channel activity, *Proc. Natl. Acad. Sci. U.S.A.*, **94**, 7103–7108.

Melroy, D.L. & Herman, E.M. (1991) TIP, an integral membrane protein of the protein-storage vacuoles of the soybean cotyledon undergoes developmentally regulated membrane accumulation and removal, *Planta*, **184**, 113–122.

Miao, G.-H., Hong, Z. & Verma, D.P.S. (1992) Topology and phosphorylation of soybean nodulin-26, an intrinsic protein of the peribacteroid membrane, *J. Cell Biol.*, **118**, 481–490.

Morillon, R. & Chrispeels, M.J. (2001) The role of ABA and the transpiration stream in the regulation of the osmotic water permeability of leaf cells, *Proc. Natl. Acad. Sci. U.S.A.*, **98**, 14138–14143.

Morillon, R. & Lassalles, J.P. (1999) Osmotic water permeability of isolated vacuoles, *Planta*, **210**, 80–84.

Morillon, R., Lienard, D., Chrispeels, M.J. & Lassalles, J.-P. (2001) Rapid movements of plants organs require solute-water cotransporters or contractile proteins, *Plant Physiol.*, **127**, 720–723.

Moshelion, M., Becker, D., Biela, A., *et al.* (2002) Plasma membrane aquaporins in the motor cells of *Samanea saman*: diurnal and circadian regulation, *Plant Cell*, **14**, 727–739.

Murata, K., Mitsuoka, K., Hirai, T., *et al.* (2000) Structural determinants of water permeation through aquaporin-1, *Nature*, **407**, 599–605.

Nakhoul, N.L., Hering-Smith, K.S., Abdulnour-Nakhoul, S.M. & Hamm, L.L. (2001) Transport of NH$_3$/NH$_4$$^+$ in oocytes expressing aquaporin-1, *Am. J. Physiol. Renal Physiol.*, **281**, F255–F263.

Niemietz, C.M. & Tyerman, S.D. (1997) Characterization of water channels in wheat root membrane vesicles, *Plant Physiol.*, **115**, 561–567.

Niemietz, C.M. & Tyerman, S.D. (2000) Channel-mediated permeation of ammonia gas through the peribacteroid membrane of soybean nodules, *FEBS Lett.*, **465**, 110–114.

Niemietz, C.M. & Tyerman, S.D. (2002) New potent inhibitors of aquaporins: silver and gold compounds inhibit aquaporins of plant and human origin, *FEBS Lett.*, **531**, 443–447.

Ohshima, Y., Iwasaki, I., Suga, S., Murakami, M., Inoue, K. & Maeshima, M. (2001) Low aquaporin content and low osmotic water permeability of the plasma and vacular membranes of a CAM plant *Graptopetalum paraguayense*: comparison with radish, *Plant Cell Physiol.*, **42**, 1119–1129.

Parsons, L.R. & Kramer, P.J. (1974) Diurnal cycling in root resistance to water movement, *Physiol. Plant.*, **30**, 19–23.

Passioura, J.B. (1988) Water transport in and to roots, *Annu. Rev. Plant Physiol. Plant Mol. Biol.*, **39**, 245–265.

Pavlovic-Djuranovic, S., Schultz, J.E. & Beitz, E. (2003) A single aquaporin gene encodes a water/glycerol/urea facilitator in *Toxoplasma gondii* with similarity to plant tonoplast intrinsic proteins, *FEBS Lett.*, **555**, 500–504.

Phillips, A.L. & Huttly, A.K. (1994) Cloning of two gibberellin-regulated cDNAs from *Arabidopsis thaliana* by subtractive hybridization: expression of the tonoplast water channel, $\gamma$-TIP, is increased by GA3, *Plant Mol. Biol.*, **24**, 603–615.

Prasad, G.V.R., Coury, L.A., Finn, F. & Zeidel, M.L. (1998) Reconstituted aquaporin 1 water channels transport CO2 across membranes, *J. Biol. Chem.*, **273**, 33123–33126.

Preston, G.M., Carroll, T.P., Guggino, W.B. & Agre, P. (1992) Appearance of water channels in *Xenopus* oocytes expressing red cell CHIP28 protein, *Science*, **256**, 385–387.

Procino, G., Carmosino, M., Marin, O., *et al.* (2003) Ser-256 phosphorylation dynamics of Aquaporin 2 during maturation from the ER to the vesicular compartment in renal cells, *FASEB J.*, **17**, 1886–1888.

Quigley, F., Rosenberg, J.M., Shachar-Hill, Y. & Bohnert, H.J. (2001) From genome to function: the *Arabidopsis* aquaporins. *Genome Biol.*, **3**, 1–17.

Ramahaleo, T., Morillon, R., Alexandre, J. & Lassalles, J.-P. (1999) Osmotic water permeability of isolated protoplasts. Modifications during development, *Plant Physiol.*, **119**, 885–896.

Ringler, P., Borgnia, M.J., Stahlberg, H., Maloney, P.C., Agre, P. & Engel, A. (1999) Structure of the water channel AqpZ from *Escherichia coli* revealed by electron crystallography, *J. Mol. Biol.*, **291**, 1181–1190.

Rivers, R.L., Dean, R.M., Chandy, G., Hall, J.E., Roberts, D.M. & Zeidel, M.L. (1997) Functional analysis of Nodulin 26, an aquaporin in soybean root symbiosomes, *J. Biol. Chem.*, **272**, 16256–16261.

Robinson, D.G., Sieber, H., Kammerloher, W. & Schäffner, A.R. (1996) PIP1 aquaporins are concentrated in plasmalemmasomes of *Arabidopsis thaliana* mesophyll, *Plant Physiol.*, **111**, 645–649.

Sanders, O.I., Rensing, C., Kuroda, M., Mitra, B. & Rosen, B.P. (1997) Antimonite is accumulated by the glycerol facilitator GlpF in *Escherichia coli*, *J. Bacteriol.*, **179**, 3365–3367.

Santoni, V., Vinh, J., Pflieger, D., Sommerer, N. & Maurel, C. (2003) A proteomic study reveals novel insights into the diversity of aquaporin forms expressed in the plasma membrane of plant roots, *Biochem. J.*, **372**, 289–296.

Sarda, X., Tousch, D., Ferrare, K., *et al.* (1997) Two TIP-like genes encoding aquaporins are expressed in sunflower guard cells, *Plant J.*, **12**, 1103–1111.

Schäffner, A.R. (1998) Aquaporin function, structure, and expression: are there more surprises to surface in water relations? *Planta*, **204**, 131–139.

Scheenen, T.W., van Dusschoten, D., de Jager, P.A. & Van As, H. (2000) Quantification of water transport in plants with NMR imaging, *J. Exp. Bot.*, **51**, 1751–1759.

Schultz, S.G. (2001) Epithelial water absorption: osmosis or cotransport? *Proc. Natl. Acad. Sci. U.S.A.*, **98**, 3628–3630.

Seki, M., Narusaka, M., Ishida, J., *et al.* (2002) Monitoring the expression profiles of 7000 *Arabidopsis* genes under drought, cold and high-salinity stresses using a full-length cDNA microarray, *Plant J.*, **31**, 279–292.

Siefritz, F., Otto, B., Bienert, G.P., Van Der Krol, A. & Kaldenhoff, R. (2004) The plasma membrane aquaporin NtAQP1 is a key component of the leaf unfolding mechanism in tobacco, *Plant J.*, **37**, 147–155.

Siefritz, F., Tyree, M.T., Lovisolo, C., Schubert, A. & Kaldenhoff, R. (2002) PIP1 plasma membrane aquaporins in tobacco: from cellular effects to function in plants, *Plant Cell*, **14**, 869–876.

Steudle, E. (1989) Water flow in plants and its coupling to other processes: an overview, *Methods Enzymol.*, **174**, 183–225.

Steudle, E. (1993) Pressure probe techniques: basic principles and application to studies of water and solute relations at the cell, tissue and organ level, in *Water Deficits: Plant Responses from Cell to Community* (eds J.A.C. Smith & H. Griffiths), Bios Scientific Publishers Ltd, Oxford, UK, pp. 5–36.

Suga, S., Imagawa, S. & Maeshima, M. (2001) Specificity of the accumulation of mRNAs and proteins of the plasma membrane and tonoplast aquaporins in radish organs, *Planta*, **212**, 294–304.

Suga, S., Komatsu, S. & Maeshima, M. (2002) Aquaporin isoforms responsive to salt and water stresses and phytohormones in radish seedlings, *Plant Cell Physiol.*, **43**, 1229–1237.

Suga, S., Murai, M., Kuwagata, T. & Maeshima, M. (2003) Differences in aquaporin levels among cell types of radish and measurement of osmotic water permeability of individual protoplasts, *Plant Cell Physiol.*, **44**, 277–286.

Sui, H., Han, B., Lee, J., Walian, P. & Jap, B. (2001) Structural basis of water-specific transport through the AQP1 water channel, *Nature*, **414**, 872–878.

Tajkhorshid, E., Nollert, P., Jensen, M., *et al.* (2002) Control of the selectivity of the aquaporin water channel family by global orientational tuning, *Science*, **296**, 525–530.

Terashima, I. & Ono, K. (2002) Effects of $HgCl_2$ on $CO_2$ dependence of leaf photosynthesis: evidence indicating involvement of aquaporins in $CO_2$ diffusion across the plasma membrane, *Plant Cell Physiol.*, **43**, 70–78.

Tomos, A.D. & Leigh, R.A. (1999) The pressure probe: a versatile tool in plant cell physiology, *Annu. Rev. Plant Physiol. Plant Mol. Biol.*, **50**, 447–472.

Tournaire-Roux, C., Sutka, M., Javot, H., *et al.* (2003) Cytosolic pH regulates root water transport during anoxic stress through gating of aquaporins, *Nature*, **425**, 393–397.

Tyerman, S.D., Bohnert, H.J., Maurel, C., Steudle, E. & Smith, J.A. (1999) Plant aquaporins: their molecular biology, biophysics and significance for plant water relations, *J. Exp. Bot.*, **50**, 1055–1071.

Tyerman, S.D. & Niemietz, C.M. (2002) Plant aquaporins: multifunctional water and solute channels with expanding roles, *Plant Cell Environ.*, **25**, 173–194.

Tyree, M.T. (1997) The cohesion-tension theory of sap ascent: current controversies, *J. Exp. Bot.*, **48**, 1753–1765.

Uehlein, N., Lovisolo, C., Siefritz, F. & Kaldenhoff, R. (2003) The tobacco aquaporin NtAQP1 is a membrane $CO_2$ pore with physiological functions, *Nature*, **425**, 734–737.

van der Weerd, L., Claessens, M.M.A.E., Efdé, C. & Van As, H. (2002) Nuclear magnetic resonance imaging of membrane permeability changes in plants during osmotic stress, *Plant Cell Environ.*, **25**, 1539–1549.

Verkman, A.S. (2000) Water permeability measurement in living cells and complex tissues, *J. Membr. Biol.*, **173**, 73–87.

Wallace, I.S., Wills, D.M., Guenther, J.F. & Roberts, D.M. (2002) Functional selectivity for glycerol of the nodulin 26 subfamily of plant membrane intrinsic proteins, *FEBS Lett.*, **523**, 109–112.

Walz, T., Smith, B.L., Zeidel, M.L., Engel, A. & Agre, P. (1994) Biologically active 2-dimensional crystals of aquaporin CHIP, *J. Biol. Chem.*, **269**, 1583–1586.

Weig, A.R. & Jakob, C. (2000) Functional identification of the glycerol permease activity of *Arabidopsis thaliana* NLM1 and NLM2 proteins by heterologous expression in *Saccharomyces cerevisiae*, *FEBS Lett.*, **481**, 293–298.

Yamada, S., Katsuhara, M., Kelly, W.B., Michalowski, C.B. & Bohnert, H.J. (1995) A family of transcripts encoding water channel proteins: tissue-specific expression in the common ice plant, *Plant Cell*, **7**, 1129–1142.

Yamada, S., Komori, T., Myers, P.N., Kuwata, S., Kubo, T. & Imaseki, H. (1997) Expression of plasma membrane water channel genes under water stress in *Nicotiana excelsior*, *Plant Cell Physiol.*, **38**, 1226–1231.

Yamaguchi-Shinozaki, K., Koizumi, M., Urao, S. & Shinozaki, K. (1992) Molecular cloning and characterization of 9 cDNAs for genes that are responsive to desiccation in *Arabidopsis thaliana*: sequence analysis of one cDNA clone that encodes a putative transmembrane channel protein, *Plant Cell Physiol.*, **33**, 217–224.

Yasui, M., Hazama, A., Kwon, T.-H., Nielsen, S., Guggino, W.B. & Agre, P. (1999) Rapid gating and anion permeability of an intracellular aquaporin, *Nature*, **402**, 184–187.

Zardoya, R. & Villalba, S. (2001) A phylogenetic framework for the aquaporin family in eukaryotes, *J. Mol. Evol.*, **52**, 391–404.

# 9 $Ca^{2+}$ and pH as integrating signals in transport control

Tatiana N. Bibikova, Sarah M. Assmann and Simon Gilroy

## 9.1 Introduction

Plant transporters are often placed in the context of the functional machinery that maintains the plant. For example, $K^+$ channels are essential for regulating membrane potential and for $K^+$ nutrition, and sucrose transporters are part of the symplastic photosynthate highway. However, it is becoming increasingly clear that, in addition to these 'maintenance' functions, transporters are essential elements of environmental sensing, response and developmental control within the plant body. Our understanding of these regulatory roles has advanced considerably in the last 5 years with the advent of genomic tools that are allowing us to define these transport activities at the molecular level. In this chapter, we will explore the themes of how transporters are involved in developmental processes and control and how environmental signals are processed through transport phenomena.

## 9.2 Transport and the control of development

There is now a wealth of information implicating the precise control of membrane transport phenomena in the regulation of cellular growth. Nowhere is this theme seen more clearly than in the precisely localized growth that characterizes the elongation of tip-growing plant cells such as pollen tubes and root hairs. Tip growth represents a highly localized exocytosis that deposits new membrane and new wall material at the growing apex of the elongating cell. One general observation seen to accompany such growth is a highly localized tip-focused $Ca^{2+}$ gradient centered on the point of elongation that seems essential to localize and promote cell expansion. This $Ca^{2+}$-driven tip-growth system is seen in diverse cells ranging from fungal hyphae to algal rhizoids and even extending neurons. Work on these non-plant systems has revealed that the localized gradient is maintained via restricting $Ca^{2+}$ influx to the apex either through localizing $Ca^{2+}$ channels to the tip or locally gating these channels.

For example, in fungi, cytosolic $Ca^{2+}$ is highest at the very tip of growing hyphae of *Saprolegnia* and *Neurospora* (Garrill *et al.*, 1993; Levina *et al.*, 1994, 1995) and the presence of this gradient correlates well with growth (Garrill *et al.*, 1993; Levina *et al.*, 1995). $Ca^{2+}$-activated $Ca^{2+}$ channels are found throughout the plasma membrane of *Neurospora* hyphae (Levina *et al.*, 1995). Because these channels are not concentrated at the apex, some localized gating must be responsible for the preferential opening at the tip. It has been suggested that these $Ca^{2+}$ channels maintain the elevated $Ca^{2+}$ levels by allowing localized influx of $Ca^{2+}$, which then reinforces itself through preferentially opening the apical channels (Garrill *et al.*, 1993). The initial localized $Ca^{2+}$ influx that triggers this self-reinforcing system could be via the release of $Ca^{2+}$ from internal stores towards the tip (Levina *et al.*, 1995). Indeed, it has been proposed that in *Neurospora*, inositol-1,4,5-trisphosphate ($InsP_3$), produced by a stretch-activated phospholipase C, causes $Ca^{2+}$ release from internal stores at the tip of the tube and that subapical endoplasmatic reticulum (ER) then sequesters $Ca^{2+}$, leading to the tip-focused $Ca^{2+}$ gradient (Silverman-Gavrila & Lew, 2002). Such a mechanism is consistent with the observation that unlike most other tip-growing cells, it has proven difficult to detect the $Ca^{2+}$ influx across the *Neurospora* hypha apical plasma membrane that would be expected if direct gating of plasma membrane $Ca^{2+}$ influx channels were generating the tip-focused gradient (Lew, 1999; Silverman-Gavrila & Lew, 2001). The putative $InsP_3$-gated channel has been characterized as a 13 pS conductance localized to the plasma membrane and ER. Inhibition of this activity blocked formation of the tip-focused $Ca^{2+}$ gradient and tip growth (Silverman-Gavrila & Lew, 2002). However, the precise molecular identity of the relevant channel has yet to be defined.

An alternative localized $Ca^{2+}$ channel gating mechanism has been reported in the fungus *Saprolegnia ferax*, where tip growth is associated with $Ca^{2+}$ influx at the apical plasma membrane (Lew, 1999). In these hyphae, stretch-activated, $Ca^{2+}$-permeable channels are concentrated in the plasma membrane at the growing tip of the hypha via an actin-dependent mechanism (Garrill *et al.*, 1993). Thus, the local membrane tension that is greatest in the enlarging dome at the tip may gate these channels, leading to a local influx, i.e. localized gating is generated by the inherent biophysical forces of tip growth. Such a self-reinforcing, self-sustaining $Ca^{2+}$ gradient bears striking similarity to that found at the tip of *Arabidopsis* root hairs (Bibikova *et al.*, 1998, and see below). However, a note of caution must be added when applying models derived from studies of fungal hyphae to the mechanisms of tip growth in plant cells. In fungal and some algal cells, secretory vesicles accumulate in the hyphal tip to create an apical body called the Spitzenkorper (Girbardt, 1969). Although the structure of the apex of tip-growing plant cells resembles the tips of these fungal hyphae and algal rhizoids, a structure equivalent to a Spitzenkorper has not been seen in plant cells. However, these studies on fungal hyphae do highlight how growth is driven by a tip-focused $Ca^{2+}$ gradient generated by either localized $Ca^{2+}$ channel positioning, localized gating and/or a possible role of $Ca^{2+}$ fluxes from organelles. The question now arises as to how far these insights translate into revealing how plant cells undergo localized growth?

## 9.3   Plant and algal transporters and tip-growth control

The cytoplasmic $Ca^{2+}$ concentration of plant cells is usually maintained at about 100 nM, but at the growing tip of pollen tubes, root hairs and algal rhizoids, this level is elevated to several micromolars (e.g. Brownlee *et al.*, 1998; Bibikova & Gilroy, 2002; Holdaway-Clarke & Hepler, 2003; Plate 4A, see colour plate section). Faster growing pollen tubes and root hairs appear to have a more pronounced gradient, whereas cessation of apical growth is associated with complete dissipation of tip-focused gradient. This $Ca^{2+}$ gradient does appear to be important for driving growth as experimental dissipation of the gradient arrests elongation (Pierson *et al.*, 1994; Wymer *et al.*, 1997) and artificial redirection of the gradient redirects growth in both pollen tubes and root hairs (Bibikova *et al.*, 1998). Measurements with an extracellular self-referencing (vibrating-reed) microelectrode have shown $Ca^{2+}$ influx localized to the tip of pollen tubes, root hairs and rhizoids (Pierson *et al.*, 1994; Jones *et al.*, 1995; Brownlee *et al.*, 1998; Plate 4B, see colour plate section), implying that localized plasma membrane $Ca^{2+}$ channel activity may be responsible for this tip-focused gradient.

In addition to localized $Ca^{2+}$ fluxes, cytosolic pH has also been proposed as an important regulator of cell growth. pH is a second messenger that may control many cellular activities; for example, changes in the intracellular pH can trigger developmental (Friedman *et al.*, 1990; Sater *et al.*, 1994) and morphological (Roncal *et al.*, 1993) transitions. Differences in the intracellular pH within a cell may also serve as markers of the polar axis (Kropf, 1994). Direct evidence for a role of cytoplasmic pH in apical growth is not strong at present. Influx of protons, localized to the tip of growing root hairs (Weisenseel *et al.*, 1979; Jones *et al.*, 1995) and pollen tubes (Messerli & Robinson, 1998; Feijo *et al.*, 1999; Messerli *et al.*, 1999), has been reported, but the extent to which these fluxes represent $H^+$ entering the cell, as opposed to, for example, $H^+$ incorporation into wall polymers at the growing apex, remains to be determined. Localized $H^+$ fluxes may result in a cytoplasmic pH gradient; however, considering the high mobility of $H^+$ in the cytosol, there would have to be a considerable $H^+$ flux to sustain the localized change versus its dissipation by diffusion. Alternatively, a cytoplasmic structure, such as an accumulation of organelle membranes, could effectively make a localized and separate domain of cytosol where differences in pH could be maintained. Indeed, in emerging root hairs this appears to be the case (Bibikova *et al.*, 1998). Confocal ratio imaging of cytoplasmic pH in growing pollen tubes (Fricker *et al.*, 1997; Parton *et al.*, 1997; Messerli & Robinson, 1998) or root hairs (Bibikova *et al.*, 1998) did not detect consistent pH gradients. However, Feijo *et al.* (1999) have reported that lily pollen tubes possess a constitutive alkaline band at the base of the vesicle-rich apical clear zone and an 'acidic' domain at the apex.

Although disruption of cytosolic pH can inhibit subsequent tip growth in both root hairs and pollen tubes (Herrmann & Felle, 1995; Bibikova *et al.*, 1998; Feijo *et al.*, 1999), this effect may well be through the dramatic effects on cellular metabolism that shifts in cytosolic pH can exert, rather than a specific effect on the localized pH gradient at the tip. On the other hand, it is important to note that there are some tip-growth-related activities that could be specific targets for a pH signaling system. For example, inwardly and outwardly rectifying $K^+$ channels have been identified in pollen tubes and shown to be important for tube growth and development (Fan *et al.*,

1999, 2001, 2003). The outwardly rectifying $K^+$ channel appears to be directly regulated by internal and external pH fluctuations (Fan *et al.*, 2003). In *Arabidopsis*, the AKT and AtKT/AtKUP/HAK families of $K^+$ transporters also appear important for root hair growth (Rigas *et al.*, 2001; Desbrosses *et al.*, 2003) although how pH might regulate these growth-related activities remains to be determined. The possibility that such $K^+$ transporters represent an integration point for both $Ca^{2+}$ and pH regulation will be discussed in greater detail below in the context of guard cell function.

## 9.4   Tip growth shows oscillations in fluxes and growth

Apical growth is known to be pulsatile in nature in many different types of tip-growing cells. In pollen tubes, the tip-focused $Ca^{2+}$ gradient oscillates in phase with these bursts of growth (Pierson *et al.*, 1996; Holdaway-Clarke *et al.*, 1997; Messerli & Robinson, 1997; Messerli *et al.*, 2000), suggesting it may play a role in driving the pulses of cell elongation. However, the oscillations in $Ca^{2+}$ are actually preceded by the oscillations in growth rate by about 4 s (Messerli *et al.*, 2000). In addition, $Ca^{2+}$ oscillations have been observed in the absence of detectable growth (Messerli & Robinson, 2003; Parton *et al.*, 2003). It is possible that an intermediate $Ca^{2+}$ store, either an organelle at the tip or the wall itself, may act as a buffer, causing the temporal separation of maximal $Ca^{2+}$ influx, growth and the elevation in cytosolic $Ca^{2+}$. However, these observations do indicate that periodic $Ca^{2+}$ increases might not be tightly coupled with growth and that the growth process itself is unlikely to generate them.

Oscillations in cytoplasmic pH, with pH being highest when the growth rate is the slowest, were observed in the apex of the pollen tube (Messerli & Robinson, 1998; Feijo *et al.*, 1999) but again their link to growth control remains obscure. The significance of these oscillations may well lie in increasing the control over the spatial localization of the ion fluxes and/or encoding information about the growth process. Thus, a pulsatile ion influx may help restrict the extent of the gradient the ions form by allowing pumps and buffer systems time to efficiently remove those ions that diffuse into the basal part of the cytoplasm. Similarly, pulsatile ion changes have been shown to encode information to plant cells, e.g. in ABA signaling in the stomatal guard cell (see below). Thus, the oscillations seen in tip-growing cells could well be encoding regulatory information about the tip-growth process.

Even though the precise role of these oscillations in ion levels in mediating tip growth remains to be determined, it seems clear that they are linked somehow to regulating the tip-growth machinery. What then might these local changes in $Ca^{2+}$ and pH be regulating and how might they be generated? Apical growth is maintained by continuous exocytosis of cell wall material containing vesicles at the growth point. Elevated levels of $Ca^{2+}$ stimulate exocytosis in both plant and animal cells (Zorec & Tester, 1992; Battey *et al.*, 1999; Thiel *et al.*, 2000, see Chapter 10), and in apically growing plant cells, the tip-focused $Ca^{2+}$ gradient, the site of exocytosis and the site of fastest growth are colocalized. Calcium may play a direct role in stimulating membrane fusion and regulating the $Ca^{2+}$-dependent proteins, such as the annexins and related proteins (e.g. Hawkins et al., 2000), involved in secretory vesicle targeting and fusion. However, ionic regulation of the actin cytoskeleton has also recently emerged as likely an essential control element for tip growth and a possible integration point where many regulatory factors might combine to spatially restrict secretion to the growing apex of the cell (Wasteneys & Galway, 2003). For example, the Arabidopsis mutant deformed root hairs 1 (der1) that fails to control apical growth is deficient in the major vegetative actin ACTIN2 (Ringli et al., 2002; Nishimura et al., 2003). The tip-focused cytosolic $Ca^{2+}$ gradient, and possibly the alkaline band, have both been implicated in the spatially defined regulation of actin assembly. Thus, micromolar cytosolic $Ca^{2+}$, such as found at the growing tip, induces actin fragmentation. In addition, an array of actin-binding proteins, including profilins, actin-depolymerizing factor (ADF)/cofilin, villin, EF-1α and capping proteins are in some way regulated by $Ca^{2+}$ and/or pH (Wasteneys & Galway, 2003). The actin cytoskeleton is thought to mediate transport of secretory vesicles (e.g. Bick et al., 2001) and thus the apical $Ca^{2+}$ gradient may well help localize exocytosis to the growing tip. Consistent with this idea, artificially imposed elevation of cytosolic $Ca^{2+}$ stimulates exocytosis in pollen tubes (Camacho & Malho, 2003). However, this enhancement of secretion does not actually affect the growth rate of the pollen tubes. This observation suggests that other factors, possibly turgor or cell wall yield, might be a limiting factor for apical growth (Camacho & Malho, 2003). Interestingly, wall modifying enzymes thought to be important for tip growth such as pectin methyl esterase (Li et al., 1994, 1995) and exo-α-glucanase (Kotake et al., 2000) show pH-dependent regulation, providing a possible link between the proton fluxes seen at the growing tip and modulation of the wall's contribution to growth control.

## 9.5   How are local $Ca^{2+}$ gradients formed?

It seems clear from the above discussion that localized ion fluxes are important determinants of cell morphogenesis. In addition, we now have many targets such as the cytoskeleton, wall dynamics and the secretory pathway whereby these fluxes might exert their spatial control on growth. One obvious question is therefore how are such fluxes generated and regulated? The intracellular $Ca^{2+}$ gradient that is characteristic of growing pollen tubes and root hairs must arise either from influx of $Ca^{2+}$ at the tip via $Ca^{2+}$-permeable channels, or via localized release of $Ca^{2+}$ from intracellular stores, such as the ER or vacuole. As we have seen, in fungi both mechanisms are operative. In plants it has long been known that extracellularly supplied radioactive $^{45}Ca$ is taken up by lily pollen tubes and concentrated in the growing tip (Jaffe et al., 1975), strongly suggesting that open $Ca^{2+}$-permeable channels are localized at the pollen tube tip. In similar studies, $Mn^{2+}$ supplied to the medium preferentially quenched fluorescence of a cytoplasmically localized fluorescent dye in the apical cytoplasm of both pollen tubes (Malho et al., 1995) and root hairs (Wymer *et al.*, 1997). As $Mn^{2+}$ is thought to enter cells through $Ca^{2+}$-permeable channels, these results also suggest increased $Ca^{2+}$ channel activity at the tips of the pollen tubes and growing hairs. Again, measurements with an extracellular (vibrating-reed) $Ca^{2+}$-selective microelectrode confirm an influx of $Ca^{2+}$ at the apex of pollen tube (Pierson *et al.*, 1994) and root hairs (reviewed in Gilroy & Jones, 2000).

It has therefore been suggested that these cells contain a plasma-membrane-localized, stretch-activated $Ca^{2+}$ channel that is either concentrated or activated at the very tip (Pierson *et al.*, 1994; Feijo *et al.*, 1995). Mechanosensory $Ca^{2+}$-permeable channels have been detected in the plasma membranes of guard cells (Cosgrove & Hedrich, 1991) and onion epidermal cells (Ding & Pickard, 1993). Similarly, *Fucus* rhizoids exhibit a stretch-activated $Ca^{2+}$ conductance throughout the rhizoid and thallus, suggestive of the presence of a ubiquitously expressed stretch-activated channel (Taylor *et al.*, 1996). However, to date, no stretch-activated channels have been characterized in apically growing plant cells.

The drugs that block voltage-gated $Ca^{2+}$ channels in animals inhibit apical growth in pollen tubes, root hairs and algal rhizoids (e.g. Shaw & Quatrano, 1996; Wymer *et al.*, 1997; Geitmann & Cresti, 1998), and fluorescently labeled dihydropyridine drugs that bind to animal $Ca^{2+}$ channels accumulate at the apex of algal rhizoids (Shaw & Quatrano, 1996; Braun & Richter, 1999) and root hairs (Bibikova & Gilroy, 2000), suggesting that perhaps this class of channels may be active in these tip-growing cells. Indeed, a hyperpolarization-activated, inwardly rectifying $Ca^{2+}$-selective channel has been electrophysiologically identified in root hairs of *Arabidopsis*. The channel is most prevalent in protoplasts arising from the apex of the root hair, suggesting that the channel may be concentrated at the growing tip (Very & Davies, 2000). The calculated macroscopic current through this channel (3–5 pmol cm² s⁻¹) would be sufficient to support the apical $Ca^{2+}$ fluxes measured in root hairs *in vivo*. Therefore, it is likely that the activity of this channel is intimately involved in the apical influx of $Ca^{2+}$ into growing root hairs. In addition, reactive oxygen species (ROS) were shown to stimulate the activity of the root hair channel, suggesting that ROS might regulate root hair apical growth through the localized activation of $Ca^{2+}$ channels (Foreman *et al.*, 2003). In *Fucus*, a nonselective, plasma-membrane-localized cation channel is also activated by ROS and may trigger the cytosolic $Ca^{2+}$ increases seen in these cells in response to membrane deformation (Coelho *et al.*, 2002). A similar theme of ROS regulation of channel activity will be discussed below in the context of guard cell function, suggesting that ROS may play an important role in regulating transporter coordination in a range of plant cells and response systems. The critical outstanding challenge is the molecular characterization of these voltage/ROS-sensitive $Ca^{2+}$ channels in root hairs and rhizoids and investigation of whether a similar system is in operation in pollen tubes.

Although from the discussion above there seems to be clear evidence that $Ca^{2+}$ influx through $Ca^{2+}$ channels in the apical plasma membrane is strongly associated with apical growth, there are some indications that this is not the only $Ca^{2+}$ transport activity in operation. For example, it is possible to grow roots in a humid chamber where they are not in contact with any substrate. These roots generate root hairs despite lacking a medium to supply more external $Ca^{2+}$ for influx above what is secreted from the cell during wall polymer exocytosis. The alternative to a growth-coupled trans-plasma-membrane influx would be spatially restricted $Ca^{2+}$ release from intracellular stores driving the $Ca^{2+}$ gradient. Release of InsP3 can induce $Ca^{2+}$ release and growth inhibition in *Papaver rhoeas* (Franklin-Tong *et al.*, 1996) and *Agapanthus* (Malho, 1998) pollen tubes, suggesting that a phosphoinositide signaling system might be involved in the regulation of pollen tube growth. However, the InsP3-induced $Ca^{2+}$ increases were predominately in the cytoplasm around the nucleus rather than at the tube apex (Franklin-Tong *et al.*, 1996; Malho, 1998). Thus, their relationship to growth remains to be determined. In animal cells, InsP3 stimulates release of $Ca^{2+}$ from internal stores such as the ER, through interaction with an InsP3 receptor (Mikoshiba, 2002). In plants, the vacuole is a strong candidate for possessing these InsP3-sensitive channels (Brosnan & Sanders, 1993) but as the vacuole does not protrude to the growing tip, it is unlikely to serve this role in tip-growing cells. There are indications of sparse ER fragments at the growing apex (Holdaway-Clarke & Hepler, 2003) but in general the Golgi and secretory vesicles represent the major membranous components of the apical cytoplasm.

It is important to note here that there are many components needed to shape a spatially localized ion gradient. The elements that sequester the ion and so attenuate the extent of the gradient are just as important as the influx channels that generate the elevated ion level. Thus, it is perhaps not surprising that intracellular stores are likely to be critical elements in maintaining successful localized growth. Interestingly, a T-DNA mutant in the *Arabidopsis* putative plasma membrane $Ca^{2+}$ transporter gene *ACA9* appears to have a lesion in pollen tube growth (J. Harper, personal communication, 2004), further highlighting that $Ca^{2+}$ stores and efflux pumps are an essential component generating the apical $Ca^{2+}$ gradient. However, the possibility that internal release of $Ca^{2+}$ contributes to the apical calcium gradient in growing pollen tubes and root hairs cannot be ruled out at present.

## 9.6   G-proteins regulating ion fluxes at the apex

The plant-specific group of Rho-like monomeric GTP-binding proteins (members of the Ras superfamily of small GTPases), the Rops, have recently emerged as impor-tant regulators of plant development (Yang, 2002). Rops have been implicated in the control of a vast range of processes including defense responses, disease resistance, hydrogen peroxide production (Agrawal *et al.*, 2003), oxidative reactions involved in secondary wall formation (Potikha *et al.*, 1999), parasite entry (Schultheiss *et al.*, 2003), hormone responses (Lemichez *et al.*, 2001; Li *et al.*, 2001) and possibly even meristem development (Trotochaud *et al.*, 1999). These G-proteins also seem to be integral to the control of tip growth by controlling both cytoskeletal dynamics and the $Ca^{2+}$ gradient. Genetically altering the activity of Rops expressed in *Arabidop-sis* pollen disrupts the localization of growth leading to pollen tube swelling and delocalization of the apical $Ca^{2+}$ gradient (Fu et al., 2001), suggesting that Rops may regulate formation and localization of this gradient. In addition, overexpressing the CRIB domain Rop interacting proteins (RICs) also disrupts growth, $Ca^{2+}$ gradient and actin cytoskeletal dynamics (Wu et al., 2001).

In root hairs, the Rops also seem to be involved in apical growth regulation and $Ca^{2+}$ gradient formation. Thus, Rop2 and Rop4 are localized to the root hair apex (Molendijk et al., 2001; Jones et al., 2002) and changes in their activity affect not only root hair growth rate but also the $Ca^{2+}$ gradient, and the directionality and localization of growth. These data suggest that in plants, Rops are involved in both regulation of actin dynamics and stabilization of the apical $Ca^{2+}$ gradient required for tip growth (Molendijk et al., 2001; Jones et al., 2002). How Rops mediate this localized activity is still unclear but it seems likely that Rops are involved either directly or indirectly in gating the influx channels responsible for maintaining the apical $Ca^{2+}$ gradient. Some Rop phenotypes can be suppressed by low concentrations of actin-fragmenting drugs (Fu et al., 2001) or overexpression of ADF (Chen et al., 2003), strongly linking Rop action to actin dynamics. Rop localization appears to represent a feed forward system where RICs recruit activated Rop to the growing apex. Clearly there are still many unanswered questions about how Rops maintain their apical localization but they are promising candidates for key components of the $Ca^{2+}$ channel complex located at the tube apex.

Rops are also involved in the early phase of cell expansion and patterning in non-tip-growing cells such as leaf epidermal cells. Thus, for example, expressing a dominant negative version of ROP2 inhibited the early polar expansion of epidermal cells, whereas a constitutively active mutant caused isotropic growth (Fu et al., 2002). These disruptions in polarity were associated with a disruption of fine F-actin bundles. Rop2 was also seen to be localized to the cortex of the cell, where growth control would be expected to be exerted. Unfortunately, at present there is no evidence for localized $Ca^{2+}$ gradients in these diffusely growing cells which correlates with their localized growth patterns. In addition, diffuse cell expansion in Micrasterias has been shown to proceed without localized $Ca^{2+}$ gradients (Holzinger et al., 1995). Clearly, more studies of the ionic requirements for localized growth in these diffuse growing cells will be required to assess how universal the link is between localized growth, ROP activity and $Ca^{2+}$ fluxes.

In addition to these monomeric G-proteins, heterotrimeric G-proteins have also been tentatively implicated in the control of polar growth. The G-protein effectors pertussis and cholera toxin alter pollen tube growth (Ma et al., 1999) and a 41-kDa pollen plasma membrane protein from lily was recognized by an antibody against a mammalian G-protein α-subunit. However, the GPA1 mutant of Arabidopsis, which is defective in the sole prototypical α-subunit of the heterotrimeric G-proteins found in the Arabidopsis genome, shows apparently normal root hair elongation and is not impaired in reproduction, as might be expected if pollen tube growth was compromised. Thus, although this evidence is far from placing heterotrimeric G-proteins in the tip-growth machinery, considering the wealth of data from stomatal guard cells that this class of G-proteins regulates ion channel activity (see below), it is a tantalizing possibility they may also play a role in regulating the ionic fluxes seen in plant growth.

## 9.7   Regulation of $H^+$ fluxes

In contrast to the wealth of data on possible mechanisms regulating the tip-focused $Ca^{2+}$ gradient, the regulation of the growth-related $H^+$ fluxes remains less well-defined. Housekeeping maintenance of cytosolic pH is accomplished by a range of $H^+$ buffers and transporters including $H^+$-ATPases, $H^+$-pyrophosphatases and numerous $H^+$ symporters and antiporters. Although we do now have molecularly characterized pumps expressed in pollen (e.g. AHA9 in *Arabidopsis*, PMA4 in tobacco and LILHA1 in lily pollen), we must await functional characterization of their role(s) in tip growth. At present it is also unclear to what extent the localized pH fluxes associated with tip growth represent contributions from exocytosis, $H^+$ production/consumption by processes in the wall, coupled transport activities or direct action of $H^+$ pumps at the apical plasma membrane. Similarly, how such proton fluxes might be being regulated remains largely unknown. The plasma membrane proton pumping ATPases form a large multigene family, with up to 12 members depending on species, raising the possibility of control at the transcriptional/isoform level (Arango *et al.*, 2003). Activity is also known to be regulated via a C-terminal autoinhibitory domain, by phosphorylation and through allosteric activators such as auxin-binding proteins (Arango *et al.*, 2003). These pumps are also activated by 14-3-3 proteins and this interaction itself can be regulated by protein kinases (Pertl *et al.*, 2001). This regulatory system might also be operating in pollen tubes. Lily pollen contains 14-3-3 proteins and the level of this regulator increased upon germination and tip growth (Pertl *et al.*, 2001). However, whether these 14-3-3 proteins are functionally regulating the ATPase *in vivo* remains to be determined.

## 9.8   Transport and the reversible control of cell volume

In order for plant cells to grow, adequate turgor pressure must be maintained to drive cell expansion. Control of cellular turgor is also central to the reversible changes in cell volume that occur in specialized motor cells such as guard cells and the pulvinars cells of legumes. The above discussion has highlighted the central roles of $Ca^{2+}$ and pH in the control of irreversible cell volume changes during polarized growth. Below, in parallel format, the roles of these two important ions in the control of reversible volume changes are discussed. Such a comparison reveals many parallels but also significant differences between irreversible and reversible volume changes (Fig. 9.1B). The following discussion focuses primarily on the guard cell system, which is better-studied, but brings in information on the pulvinar system where such data are available.

## A. Tip-growing cell

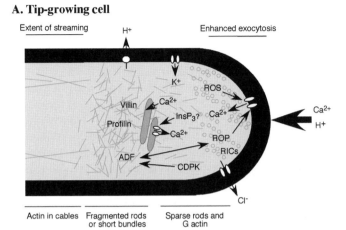

## B. Motor cell

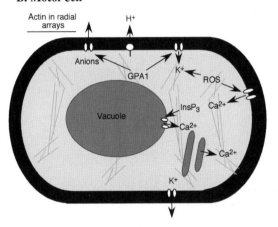

**Fig. 9.1** (A) Spatial relationships and potential interactions between ion fluxes, cytoskeletal structure and putative regulators of tip growth. (B) Components implicated in the reversible volume changes of motor cells. Ion transporters illustrated have been implicated in both guard cells and pulvinar cells either by direct electrophysiological analysis or by pharmacological tools. Regulatory elements discussed in this chapter are also indicated; of these, InsP$_3$ has been implicated in both types of motor cells, while the other components illustrated have, to date, only been identified in guard cells.

## 9.9  The mechanistic basis of reversible cell volume change

In both guard cells and pulvinar cells, changes in cell solute potential drive trans-membrane water fluxes that effect changes in cell volume. By such osmotically driven swelling or shrinking, pairs of guard cells in the aerial epidermis of terrestrial plants open or close stomatal pores. Alterations in stomatal aperture mediated by guard-cell volume changes can occur quite rapidly, presumably without changes in gene expression. For example, under certain conditions, blue-light-stimulated stomatal opening in sugarcane can be detected within seconds after imposition of the blue light signal (Assmann & Grantz, 1990). In grasses such as sugarcane and rice (Fig. 9.2A), swelling of the bulbous ends of the guard cells separates the two cells from one another. In nongraminaceous angisperms, the radial pattern of re-inforcement imposed by cellulose microfibrils in the walls of the kidney-shaped guard cells constrains the cells such that cell swelling causes the two guard cells to bow apart, thus widening the stomatal pore (Fig. 9.2A). Conversely, stomatal clo-sure results from osmotically driven reductions in guard cell volume that cause the two guard cells to deflate against each other, thereby narrowing the pore. The ma-jor osmotically active ions that redistribute across the guard cell membrane during stomatal movements are $K^+$, $Cl^-$ and malate$^{2-}$. Under certain conditions, uptake of nitrate ($NO_3^-$) or sucrose, or photosynthetic production of sugars within the guard cells can also contribute significantly to osmotic buildup (Poffenroth et al., 1992; Talbott & Zeiger, 1993, 1996; Guo et al., 2003; Outlaw, 2004).

The pulvinus contains two specialized cell types, flexors and extensors, the swelling and shrinking of which are coordinately and oppositely regulated, resulting in leaf movement (Fig. 9.2B). Swelling of the abaxially located extensor cells and shrinking of the adaxially located flexor cells result in elevation of the leaf or leaflet opening, whereas the converse processes result in a lowering of the organ or leaflet closure (Satter et al., 1990). It is worth noting that changes in pulvinar cell volume

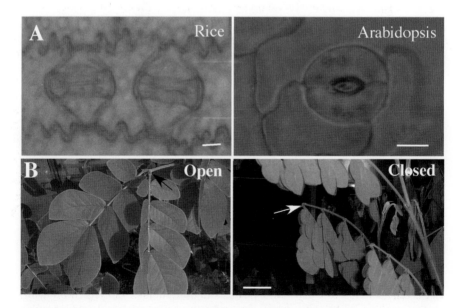

**Fig. 9.2** (A) Epidermal peels of the model species *Oryza sativa* and *Arabidopsis thaliana* showing stomatal complexes and pavement cells. Scale bar indicates 10 μm. (B) Leaves of the leguminous tree *Samanea saman* with leaflets in the open and closed position. The arrow indicates the location of the pulvinus, *S. saman*. Pictures courtesy of Dr Nava Moran. Scale bar indicates 10 mm.

can also be involved in gravitropic bending of stems in graminaceous species (e.g. Perera *et al.*, 1999); however, this is a nonreversible growth, and is not the focus of this section.

Early X-ray microanalysis studies of the pulvini of the leguminous tree *Samanea saman* showed that in the closed state the flexor cells are high in $K^+$ as well as $Cl^-$, while only low levels of these ions are present in the extensor cells (Satter *et al.*, 1982). These results indicate that in pulvinar cells, as in guard cells, $K^+$ and $Cl^-$ contribute significantly to osmotic buildup. The roles of $NO_3^-$ and organic solutes have yet to be studied in the pulvinar system.

In both guard cells and pulvinar cells, the $K^+$ uptake that contributes to cell swelling occurs through sets of inwardly rectifying $K^+$ channels that are voltage-regulated such that they open when the membrane is hyperpolarized, i.e. under conditions where the electrochemical driving force favors passive $K^+$ influx (Schroeder *et al.*, 1987; Schroeder, 1988; Blatt, 1992; Roelfsema *et al.*, 2001; Yu *et al.*, 2001). *KAT1, KAT2, AKT1* and *AKT2/3* genes encoding inward $K^+$ channels are expressed in *Arabidopsis* guard cells (Szyroki *et al.*, 2001). In *S. saman*, genes *SPICK1* and *SPICK2*, encoding inwardly rectifying $K^+$ channels with highest homology to *Arabidopsis* AKT2, are expressed in the pulvinus (Moshelion *et al.*, 2002). Conversely, $K^+$ efflux during shrinkage of guard cells, flexor cells and extensor cells occurs through a molecularly distinct set of $K^+$ channels that are activated by membrane depolarization (Schroeder *et al.*, 1987, 1988; Blatt 1988; Moran *et al.*, 1988; Blatt and Armstrong 1993; Moshelion & Moran, 2000). The outward channel GORK1 plays a significant role in $K^+$ efflux from *Arabidopsis* guard cells (Hosy *et al.*, 2003). Two genes encoding homologues to outwardly rectifying $K^+$ channels also have been cloned from a pulvinar cDNA library (Moshelion *et al.*, 2002): *SPORK1* is homologous to the above-mentioned *GORK*, while *SPOCK1* is similar to *KCO1*, the vacuolar *Arabidopsis* two-pore-domain $K^+$-conducting SV channel (Czempinski *et al.*, 2002; Schonknecht *et al.*, 2002).

Genes encoding anion channels expressed in guard cells and the pulvinus have yet to be cloned, but electrophysiological observations indicate the existence of two types or modes of anion channels in guard cells, called S-type or R-type by virtue of their deactivation kinetics (Hedrich *et al.*, 1990; Schroeder & Keller, 1992; Raschke, 2003). Pulvinar anion channels have not been characterized electrophysiologically, although the observation that anion channel inhibitors block shrinkage of flexor and extensor cells of *Phaseolus vulgaris* (Iino *et al.*, 2001; Wang *et al.*, 2001) suggests anion channel function in the pulvinus.

## 9.10   Calcium and volume change in motor cells

Localized elevation of $Ca^{2+}$ at the growth site is a hallmark of tip-growing cells, as discussed previously. Motor cells are not known to display comparable standing asymmetries in $Ca^{2+}$ concentration. However, guard cells do display elevations in cytosolic $Ca^{2+}$ in response to the plant hormones ABA and auxin (Gilroy *et al.*, 1991; Irving *et al.*, 1992; McAinsh *et al.*, 1992; Fricker *et al.*, 1994). One paper has reported spatially discernable waves of cytosolic $Ca^{2+}$ in response to ABA (Grabov & Blatt, 1998), while several reports illustrate the importance to signaling of temporal $Ca^{2+}$ oscillations (reviewed in Blatt, 2000a,b; McAinsh *et al.*, 2000; Hetherington, 2001; Schroeder *et al.*, 2001a,b). Information contained in the frequency and waveform of these oscillations influences the extent of stomatal closure (McAinsh *et al.*, 1995; Allen *et al.*, 1999; Staxen *et al.*, 1999; Allen *et al.*, 2000; Allen *et al.*, 2001). In wild-type guard cells, experimental abrogation of $Ca^{2+}$ oscillations interferes with ABA-induced stomatal closure (Allen *et al.*, 2000). In ABA-insensitive mutants such as *abi1-1*, *abi2-1*, *rcn1* and *gca2*, abberant $Ca^{2+}$ oscillations in response to ABA are seen (Allen *et al.*, 1999, 2001; Kwak *et al.*, 2002). In *det3* and *gca2* mutants, experimental manipulations that mimic wild-type cytosolic $Ca^{2+}$ oscillations trigger stomatal closure (Allen *et al.*, 2000), clearly illustrating the importance of this signal. However, $Ca^{2+}$ elevation is not invariably evoked in response to ABA (e.g. Gilroy *et al.*, 1991; Romano *et al.*, 2000). $Ca^{2+}$ oscillations may be more important for maintaining, rather than generating, the ABA-induced state, since they often proceed after stomatal closure has reached steady-state levels (Allen *et al.*, 2001; Evans *et al.*, 2001).

ABA is a potent inhibitor of stomatal opening and promoter of stomatal closure via its effects on ion transport activity. Many (but not all) of these ABA effects are recapitulated by $Ca^{2+}$ application, thus implicating $Ca^{2+}$ as a proximate regulator of these transporters. ABA activates the plasma membrane $Ca^{2+}$-permeable channel (Hamilton *et al.*, 2000; Pei *et al.*, 2000), and ABA and cytosolic $Ca^{2+}$ both inhibit the inwardly rectifying $K^+$ channels responsible for $K^+$ uptake, and increase the activation state of anion channels that mediate anion efflux during stomatal closure (reviewed in Grabov *et al.*, 1997; MacRobbie, 1998; Blatt, 2000b; Assmann & Wang, 2001; Schroeder *et al.*, 2001b). However, while ABA also promotes activation of outward $K^+$ channels, this effect is not mediated by cytosolic $Ca^{2+}$ increases but instead is modulated independently of $Ca^{2+}$ by cytosolic pH (Blatt & Armstrong, 1993; Miedema & Assmann, 1996; Assmann & Armstrong, 1999; see Section 9.7).

Auxin stimulates stomatal opening at low concentrations and inhibits opening at higher concentrations in *Vicia faba* (Marten *et al.*, 1991; Lohse & Hedrich, 1992), and promotes stomatal opening in *Arabidopsis* (Klein *et al.*, 2003). Consistent with this observation, auxin enhances $K^+$ uptake currents in *V. faba* at low concentrations but inhibits them at high concentrations (Blatt & Thiel, 1994). High auxin concentrations also enhance outward $K^+$ currents and anion currents in *V. faba* (Marten *et al.*, 1991; Blatt & Thiel, 1994). However, the specific role of auxin-induced cytosolic $Ca^{2+}$ elevations in regulating guard cell channels and transporters remains unknown, except by inference from the experimental $Ca^{2+}$ manipulations already discussed above.

Similarly to guard cells, in *P. vulgaris* pulvini, ABA causes shrinkage of both flexor and extensor cells, while auxin has the reverse effect (Iino *et al.*, 2001). Cytosolic $Ca^{2+}$ changes have not been measured in pulvinar cells, but may be involved in ion channel regulation, because $Ca^{2+}$ or $Ca^{2+}$ionophore treatment of excised intact legume leaflets enhances leaflet movement (Moysset & Simon, 1989), while

EGTA application has the opposite effect (Roblin & Fleurat-Lessard, 1984). In electrophysiological assays, experimental elevation of cytosolic $Ca^{2+}$ concentrations in flexor cells of *S. saman* results in small effects on maximum open channel probability and voltage-sensitivity of outward $K^+$ channels of flexor cells, but has no effect on these channels in extensor cells (Moshelion & Moran, 2000). Given the guard cell paradigm, inward $K^+$ channels may be more likely targets of $Ca^{2+}$ modulation in pulvinar cells.

## 9.11   $Ca^{2+}$, secretion and the cytoskeleton

Guard cell surface area changes significantly during stomatal responses (Shope *et al.*, 2003). Exocytotic events during guard cell swelling triggered by hypoosmotic conditions, and endocytotic events accompanying guard cell shrinkage in response to hyperosmotic conditions have been detected (Homann & Thiel, 1999; Bick *et al.*, 2001) as step changes in membrane capacitance (capacitance is proportional to membrane surface area) and, for endocytosis, by internalization of fluorescently labeled membrane (Kubitscheck *et al.*, 2000). Evidence that vesicle trafficking is important for the ABA response comes from the study of syntaxins, integral membrane proteins that are important for vesicle movement and fusion (Geelen *et al.*, 2002). The lethal neurotoxin, botulinum C, inhibits syntaxin-dependent vesicle trafficking. Botulinum C interferes with ABA regulation of inward and outward $K^+$ channels and slow anion channels of guard cells when microinjected into guard cells. A truncated version of a tobacco syntaxin, Nt-Syr1, similarly impedes ABA responses (Leyman *et al.*, 1999). Given that ABA influences cytosolic $Ca^{2+}$ status, as described above, one can surmise that cytosolic $Ca^{2+}$ is involved in ABA modulation of vesicle trafficking in guard cells.

Vesicles traffic along cytoskeletal components (e.g. Bick *et al.*, 2001). Just as is the case for tip-growing cells, so too for guard cells the actin cytoskeleton appears to be important in the control of reversible changes in cell volume. ABA-induced disruption of the radially organized actin filaments of guard cells occurs consistently in several different species, and elevated cytosolic $Ca^{2+}$ is an important trigger for actin fragmentation (Eun & Lee, 1997; Eun *et al.*, 2001; Hwang & Lee, 2001; Lemichez *et al.*, 2001). However, pharmacological agents that act as actin antagonists inhibit both stomatal opening and stomatal closure, so the cause–effect relationship between ABA and cytoskeletal reorganization is currently murky. Possibly, depolymerization of actin is simply required for changes in stomatal aperture to occur (Kim *et al.*, 1995).

## 9.12   How are $Ca^{2+}$ oscillations generated?

In tip-growing plant cells, evidence points to $Ca^{2+}$ uptake across the plasma membrane as a critical component for the production of a tip-focused $Ca^{2+}$ gradient. By contrast, in guard cells and motor cells, both internal and external sources contribute to stimulus-evoked changes in cytosolic $Ca^{2+}$ concentrations (Stoeckel & Takeda, 1995; Kim et al., 1996; McAinsh et al., 2000; Assmann & Wang, 2001; Hetherington, 2001; Schroeder et al., 2001a).

The guard cell plasma membrane contains $Ca^{2+}$-permeable, voltage-activated channels (Hamilton et al., 2000; Pei et al., 2000), as well as stretch-activated channels that exhibit permeability to a number of ions, including $Ca^{2+}$ (Cosgrove & Hedrich, 1991). While the role of the latter channel type in guard cell signaling remains unknown, the former channel type is implicated in the production of $Ca^{2+}$ oscillations. Upon transition of the guard cell membrane potential to a strongly hyperpolarized state, these voltage-regulated $Ca^{2+}$ channels are activated, allowing $Ca^{2+}$ influx and cytosolic $Ca^{2+}$ elevation to occur. Because these $Ca^{2+}$ channels are inhibited by cytosolic $Ca^{2+}$ increases, negative feedback of $Ca^{2+}$ channel activity will occur, a crucial component in establishment of an oscillatory system (Blatt, 2000a,b; Hamilton et al., 2000). In support of this model, experimentally imposed oscillations in guard cell membrane potential result in parallel oscillations in cytosolic $Ca^{2+}$ levels (Allen et al., 2000).

Secondary messengers also participate in modulation of the voltage-regulated $Ca^{2+}$ channels. $Ca^{2+}$ channel current is stimulated by ROS (Pei et al., 2000; Murata et al., 2001; Kohler et al., 2003), similar to the scenario in root hairs and Fucus zygotes, where ROS application triggers cytosolic $Ca^{2+}$ increases. ABA does stimulate increases in endogenous ROS in guard cells (Murata et al., 2001; Zhang et al., 2001), and recent evaluation of knockout mutants of NADPH oxidase implicate AtrbohD and AtrbohF in guard cell ROS generation during this response (Kwak et al., 2003; Desikan et al., 2004).

Regarding the other mechanism of cytosolic $Ca^{2+}$ elevation, namely release of $Ca^{2+}$ from internal stores, a number of secondary messengers, many of them lipids or lipid metabolites, initiate this process in guard cells. Phospholipase D (PLD), which produces phosphatidic acid and a headgroup upon phospholipid hydrolysis, is activated in response to ABA in guard cells (Jacob et al., 1999), and phosphatidic acid mimics ABA effects on stomatal opening and closure, and inward $K^+$ channel inhibition. However, PLD inhibitors only partially block ABA response, indicating that other pathways are operating in parallel. In addition, phosphatidic acid does not elicit a rise in cytosolic $Ca^{2+}$, so $Ca^{2+}$ apparently functions upstream of or independently of PLD activation. A second phospholipase whose activity is ABA-stimulated in guard cells is phospholipase C (Lee et al., 1996; Hunt et al., 2003), which catalyses production of $InsP_3$ and diacyl glycerol. Microinjection of $InsP_3$ into guard cells results in cytosolic $Ca^{2+}$ elevation (Gilroy et al., 1990) and inhibits inward $K^+$ channels (Blatt et al., 1990). Based on mammalian responses, this effect is thought to be mediated by $InsP_3$-gated $Ca^{2+}$ channels located in organellar membranes (Hetherington, 2001; Sanders et al., 2002). Other phosphorylated inositides have also been implicated in ABA and $Ca^{2+}$-related events. Pharmacological inhibitors of PI 3-kinase (PI3K) and PI 4-kinase (PI4K) decrease the proportion of guard cells exhibiting ABA stimulation of $Ca^{2+}$ transients and ROS generation, and inhibit ABA-induced stomatal closure (Jung *et al.*, 2002; Park *et al.*, 2003). Myo-inositol-hexakisphosphate ($InsP_6$) produced in response to ABA triggers $Ca^{2+}$ release from guard cell internal stores and inhibits inward $K^+$ channels in a $Ca^{2+}$-dependent manner (Lemtiri-Chlieh *et al.*, 2000, 2003).

ABA activation of the enzyme sphingosine kinase, and resultant production of the phosphorylated long-chain amino alcohol, sphingosine-1-phosphate (S1P), also occurs in guard cells (Coursol *et al.*, 2003). S1P elevates cytosolic $Ca^{2+}$ in guard cells (Ng *et al.*, 2001); in mammalian cells this effect results from internal $Ca^{2+}$ release, although this mechanism has not been directly investigated in guard cells. In addition, nitric oxide promotes release of $Ca^{2+}$ from guard cell internal stores, and studies with pharmacological inhibitors implicate guanylate cyclase and cyclic ADPribose in this phenomenon (Garcia-Mata *et al.*, 2003); cyclic ADPribose is known to activate vacuolar $Ca^{2+}$-release channels in guard cells (Leckie *et al.*, 1998), and also acts on the plant endoplasmic reticulum (Sanders *et al.*, 2002).

While most of these signaling elements remain to be studied in pulvinar cells, in both extensor and flexor cells of *S. saman*, $InsP_3$ plays a role analogous to its function in guard cells; $InsP_3$ is produced in response to signals that cause both cell shrinkage and closure of the channels that mediate $K^+$ uptake (Kim *et al.*, 1996).

## 9.13   G-proteins regulating signaling in guard cells

Two ROPs, ROP6 and ROP10, have been implicated in guard cell ABA response. Overexpression of a constitutively inactive form of ROP6 mimics ABA effects on actin disruption and stomatal closure, suggesting that ROP6 functions as a negative regulator of ABA action (Lemichez *et al.*, 2001). Null mutants of ROP10 show ABA hypersensitivity in stomatal closure, seed germination and root elongation; whether these effects are related to cytoskeletal integrity is currently unknown (Zheng *et al.*, 2002). Heterotrimeric G-proteins also have been implicated in ABA action. Null mutants in the G-protein α-subunit gene, *GPA1*, show insensitivity to inhibition of stomatal opening by ABA or S1P and are also insensitive to S1P-promotion of stomatal closure, suggesting that following ABA-stimulated S1P production, the S1P signal is transduced via a G-protein-based signaling cascade (Wang *et al.*, 2001; Coursol *et al.*, 2003; Worrall *et al.*, 2003).

## 9.14   Regulation of $H^+$ fluxes

Regulation of plasma membrane $H^+$ fluxes in motor cells plays several important roles. First, as already mentioned above, $H^+$ extrusion is a critical effector of membrane hyperpolarization, which in turn regulates fluxes of osmotically active ions across the cell membrane. Second, regulation of $H^+$-ATPase activity can influence both apoplastic and cytosolic pH and, as in tip-growing cells, these parameters can function as signaling intermediates. Thus, experimentally imposed decreases in extracellular pH stimulate inward $K^+$ currents of guard cells (Blatt, 1992; Ilan *et al.*, 1996) and inhibit outward $K^+$ currents (Ilan *et al.*, 1994) consistent with the well-defined coupling between $H^+$ extrusion and stomatal opening. Surprisingly, apoplastic acidification inhibits inward $K^+$ currents in flexor cells as well as in extensor cells (Yu *et al.*, 2001). On the cytosolic side, experimental elevation of internal pH has no appreciable effect on guard-cell inward $K^+$ currents (Blatt, 1992; Grabov & Blatt, 1997) but enhances outward $K^+$ currents by increasing the availability of activatable outward $K^+$ channels (Blatt & Armstrong, 1993; Miedema & Assmann, 1996).

Signals that regulate $H^+$-ATPase activity in guard cells include ABA and $Ca^{2+}$, both of which inhibit plasma membrane $H^+$-ATPase activity (Kinoshita *et al.*, 1995; Goh *et al.*, 1996), and auxin, which stimulates $H^+$-ATPase activity (Lohse & Hedrich, 1992; Blatt & Thiel, 1994). Even though ABA and auxin, at least at low concentrations, have opposite effects on stomatal apertures and $H^+$-ATPase activity, they both somehow induce elevation of cytosolic pH (Irving *et al.*, 1992; Fricker *et al.*, 1994). In the *P. vulgaris* pulvinus, auxin-induced swelling of motor cells is inhibited by the P-type ATPase inhibitor, vanadate, suggesting that auxin also activates the $H^+$-ATPase in these cells (Iino *et al.*, 2001).

Blue light also stimulates guard-cell $H^+$-ATPase activity, contributing to the membrane hyperpolarization that drives $K^+$ uptake and stomatal opening (Assmann *et al.*, 1985; Shimazaki *et al.*, 1986; Roelfsema *et al.*, 2001). In *Arabidopsis* guard cells, the blue light receptors for phototropism, phot1 and phot2, function redundantly as blue light sensors. *phot1 phot2* double mutants fail to exhibit blue-light-stimulated stomatal opening and $H^+$ extrusion (Kinoshita *et al.*, 2001). Blue light stimulates binding of 14-3-3 protein to guard cell phototropins that are phosphorylated in response to blue light. Blue light also stimulates 14-3-3 binding to phosphorylated $H^+$-ATPase, consistent with a role of 14-3-3 proteins as scaffolding proteins. 14-3-3 binding is necessary for $H^+$-ATPase activation (Kinoshita & Shimazaki, 1999, 2002; Kinoshita *et al.*, 2003).

Blue light opens leaflets in *S. saman*, stimulating swelling of extensor cells (i.e. analogous to the guard cell response) and shrinkage of flexor cells (Satter *et al.*, 1981). In contrast to guard cells, where there is no evidence that blue light regulates $K^+$ channel activity, in *S. saman* flexor cells, blue light not only depolarizes the membrane potential via apparent inhibition of $H^+$-ATPase activity (Suh *et al.*, 2000), but also activates outward $K^+$ channels, as assessed by monitoring apoplastic $K^+$ activity (Lowen & Satter, 1989) and patch clamping (Suh *et al.*, 2000). Less direct assays also suggest that blue light also closes inward $K^+$ channels in *S. saman* flexor protoplasts and opens them in extensor protoplasts (Kim *et al.*, 1992, 1993, 1996), consistent with the whole-plant phenomenon of light-induced leaflet opening. Blue light also stimulates $InsP_3$ production in flexor protoplasts of *S. saman*, implicating $InsP_3$ in blue-light-induced cell shrinkage of this cell type (Kim *et al.*, 1996).

## 9.15   Roles of extracellular Ca²⁺ and pH in wall structure/activity of guard cells and pulvinar cells

In the pulvinus, promotion of leaflet opening by light or the biological clock (Satter *et al.*, 1981) is accompanied by extracellular acidification in extensor tissue and extracellular alkalinization in flexor tissue (Iglesias & Satter, 1983; Lee & Satter, 1989), consistent with regulation of H⁺-ATPase activity that drives extensor cell swelling and flexor cell shrinkage under these conditions. In the substomatal cavity of the intact *V. faba* leaf, apoplastic Ca²⁺ and H⁺ concentrations transiently decrease following ABA application (Starrach & Mayer, 1989; Felle, 2000), plausibly reflecting ABA-stimulated Ca²⁺ uptake and H⁺-ATPase inhibition. A protein functioning as an apoplastic Ca²⁺ sensor, 'CAS', has recently been identified in *Arabidopsis* (Han *et al.*, 2003). Genetic abrogation of CAS impairs stomatal closure elicited by extracellular Ca²⁺ elevation but not that elicited by extracellular ABA application (Han *et al.*, 2003). These results suggest that apoplastic Ca²⁺ may be a signal in its own right, rather than simply signaling via its induction of cytosolic Ca²⁺ increases. Consistent with this idea, external ABA application and external Ca²⁺ application result in different oscillatory patterns in cytosolic Ca²⁺ (Allen *et al.*, 2000). Extracellular Ca²⁺ also blocks K⁺ permeation through inward K⁺ channels (Fairley-Grenot & Assmann, 1992; Dietrich *et al.*, 1998).

Whether, and to what extent, extracellular pH and Ca²⁺ influence the biophysical properties of guard cell and pulvinar cell walls remains to be determined. The guard cell wall has a unique ultrastructure and the walls of motor cells in general are unusual in their ability to undergo elastic (i.e. reversible) extension. Thus, one can speculate that the biophysical parameters of the motor cell wall may be unique, and uniquely influenced by Ca²⁺ and pH changes. Biochemical analysis of the guard cell wall is just beginning, but one report shows that cell wall arabinan is crucial to the elasticity of the guard cell wall (Jones *et al.*, 2003). The effects of apoplastic Ca²⁺ and H⁺ concentrations on the wall elasticity of motor cells appear a fruitful area for future investigation.

## 9.16   Conclusions and perspectives

From the above discussion, there is clearly now a wealth of data showing that ion fluxes are critical regulators of plant growth, development and environmental sensing. Indeed, we have only touched the surface of the studies in this area and we must apologize to the many researchers whose equally important work we have had to omit because of space constraints. Ca²⁺ and H⁺ have emerged as key controlling factors in plant growth and environmental response but we can anticipate that as more information is gathered on the dynamics of other ions, these too will emerge as imposing regulation on cellular activities. In comparing tip growth to stomatal and pulvinar responses, some striking parallels emerge in how these apparently disparate response systems control and utilize their ionic signaling networks. Thus,

ROS and the ROP family of G-proteins seem to play parallel roles in regulating $Ca^{2+}$ channel activity and so modulating cellular response. In addition, oscillations in regulatory ion fluxes seem to underlie both tip growth and stomatal movements, perhaps highlighting a conserved role for encoding information in $Ca^{2+}$ oscillation for plant cells. Alternatively, the oscillatory phenomena may reflect the limits imposed on ion transport systems by the cell's requirement to maintain control over its internal ionic environment, i.e. oscillations are easier to precisely regulate than large-scale steady-state changes in ion flux.

With the advent of genomics and genomic tools, molecular identification of signaling-related transporters and the proteins that respond to ionic signals at a molecular level is gathering pace. Progress in this area continues to be a critical goal for defining how these membrane transporters generate signals and how these are then turned into cellular response. Our current understanding of the $Ca^{2+}$ fluxes responsible for growth control in pollen tubes and root hairs and in signaling in guard cells highlights this need. We have a very detailed knowledge of the spatial and temporal dynamics of $Ca^{2+}$ in these systems and a wealth of data defining what these fluxes are likely doing. However, we still have not identified the channels and pumps responsible for these membrane transport phenomena. Molecular characterization of the channels would allow testing of, for example, how ROS activates channel activity and whether ROPs directly gate the channels. Membrane proteins are inherently difficult to functionally analyze but as can be seen from the other chapters in this book, rapid advances are being made in going from cloning a transporter to its functional characterization. We can predict that the next few years will yield some very exciting insights into the proteins that make ionic regulation work.

## Acknowledgements

The authors would like to thank Drs Nava Moran and Sarah Swanson for critical reading of the manuscript, and Ms Anne Gibson for assistance with figure preparation. This work was supported by grants to SG from NSF and NASA and to SMA from NSF and USDA.

## References

Agrawal, G.K., Iwahashi, H. & Rakwal, R. (2003) Small GTPase 'Rop': molecular switch for plant defense responses, *FEBS Lett.*, **546**, 173–180.

Allen, G.J., Chu, S.P., Harrington, C.L., *et al.* (2001) A defined range of guard cell calcium oscillation parameters encodes stomatal movements, *Nature*, **411**, 1053–1057.

Allen, G.J., Chu, S.P., Schumacher, K., *et al.* (2000) Alteration of stimulus-specific guard cell calcium oscillations and stomatal closing in *Arabidopsis* det3 mutant, *Science*, **289**, 2338–2342.

Allen, G.J., Kuchitsu, K., Chu, S.P., Murata, Y. & Schroeder, J.I. (1999) *Arabidopsis* abi1-1 and abi2-1 phosphatase mutations reduce abscisic acid-induced cytoplasmic calcium rises in guard cells, *Plant Cell*, **11**, 1785–1798.

Arango, M., Gevaudant, F., Oufattole, M. & Boutry, M. (2003) The plasma membrane proton pump ATPase: the significance of gene subfamilies, *Planta*, **216**, 355–365.

Assmann, S. & Armstrong, F. (1999) Hormonal regulation of ion transporters: the guard cell system in *Biochemistry and Molecular Biology of Plant Hormones* (eds P.J.J. Hooykass, M.A. Hall & K.R. Libbonga), Elsevier, Amsterdam, The Netherlands.

Assmann, S. & Grantz, D. (1990) Stomatal response to humidity in sugarcane and soybean: effect of vapour pressure difference on the kinetics of the blue light response, *Plant Cell Environ.*, **13**, 163–169.

Assmann, S., Simoncini, L. & Schroeder, J. (1985) Blue light activates electrogenic ion pumping in guard cell protoplasts of *Vicia faba*, *Nature*, **318**, 285–287.

Assmann, S. & Wang, X. (2001) From milliseconds to millions of years: guard cells and environmental responses, *Curr. Opin. Plant Biol.*, **4**, 421–428.

Battey, N.H., James, N.C., Greenland, A.J. & Brownlee, C. (1999) Exocytosis and endocytosis, *Plant Cell*, **11**, 643–660.

Bibikova, T. & Gilroy, S. (2000) Calcium in root hair growth and development, in *The Cellular and Molecular Biology of Root Hairs* (eds R. Ridge & A.M. Emons), Springer Verlag, Berlin, pp. 141–163.

Bibikova, T. & Gilroy, S. (2002) Root hair development, *J. Plant Growth Regul.*, **21**, 383–415.

Bibikova, T.N., Jacob, T., Dahse, I. & Gilroy, S. (1998) Localized changes in apoplastic and cytoplasmic pH are associated with root hair development in *Arabidopsis thaliana*, *Development*, **125**, 2925–2934.

Bick, I., Thiel, G. & Homann, U. (2001) Cytochalasin D attenuates the desensitisation of pressure-stimulated vesicle fusion in guard cell protoplasts, *Eur. J. Cell Biol.*, **80**, 521–526.

Blatt, M.R. (1988) Potassium-dependent bipolar gating of potassium channels in guard cells, *J. Membr. Biol.*, **102**, 235–246.

Blatt, M.R. (1992) K$^+$ channels of stomatal guard cells. Characteristics of the inward rectifier and its control by pH, *J. Gen. Physiol.*, **99**, 615–644.

Blatt, M.R. (2000a). Ca$^{2+}$ signalling and control of guard-cell volume in stomatal movements, *Curr. Opin. Plant Biol.*, **3**, 196–204.

Blatt, M.R. (2000b). Cellular signalling and volume control in stomatal movements in plants, *Ann. Rev. Cell Develop. Biol.*, **16**, 221–241.

Blatt, M. & Armstrong, F. (1993) K$^+$ channels of stomatal guard cells: abscisic-acid-evoked control of the outward rectifier mediated by cytoplasmic pH, *Planta*, **191**, 330–341.

Blatt, M. & Thiel, G. (1994) K$^+$ channels of stomatal guard cells: bimodal control of the K$^+$ inward-rectifier evoked by auxin, *Plant J.*, **5**, 55–68.

Blatt, M., Thiel, G. & Trentham, D. (1990) Reversible inactivation of K$^+$ channels of *Vicia* stomatal guard cells following the photolysis of caged inositol 1,4,5-trisphosphate, *Nature*, **346**, 766–769.

Braun, M. & Richter, P. (1999) Relocalization of the calcium gradient and a dihydropyridine receptor is involved in upward bending by bulging of *Chara* protonemata, but not in downward bending by bowing of *Chara* rhizoids, *Planta*, **209**, 414–423.

Brosnan, J.M. & Sanders, D. (1993) Identification and characterization of high-affinity binding sites for inositol trisphosphate in red beet, *Plant Cell*, **5**, 931–940.

Brownlee, C., Manison, N. & Anning, R. (1998) Calcium, polarity and osmoregulation in *Fucus* embryos: one messenger, multiple messages, *Exp. Biol. Online*, **3**, 11.

Camacho, L. & Malho, R. (2003) Endo/exocytosis in the pollen tube apex is differentially regulated by Ca$^{2+}$ and GTPases, *J. Exp. Bot.*, **54**, 83–92.

Chen, C.Y., Cheung, A.Y. & Wu, H.M. (2003) Actin-depolymerizing factor mediates Rac/Rop GTPase-regulated pollen tube growth, *Plant Cell*, **15**, 237–249.

Coelho, S.M., Taylor, A.R., Ryan, K.P., Sousa-Pinto, I., Brown, M.T. & Brownlee, C. (2002) Spatiotemporal patterning of reactive oxygen production and Ca$^{2+}$ wave propagation in *Fucus* rhizoid cells, *Plant Cell*, **14**, 2369–2381.

Cosgrove, D.J. & Hedrich, R. (1991) Stretch-activated chloride, potassium, and calcium channels coexisting in plasma membranes of guard cells of *Vicia faba* L., *Planta*, **186**, 143–153.

Coursol, S., Fan, L.M., Le Stunff, H., Spiegel, S., Gilroy, S. & Assmann, S.M. (2003) Sphingolipid signalling in *Arabidopsis* guard cells involves heterotrimeric G proteins, *Nature*, **423**, 651–654.

Czempinski, R., Frachisse, J., Maurel, C., Barbier-Brygoo, H. & Mueller-Roeber, B. (2002) Vacuolar membrane localization of the *Arabidopsis* 'two-pore' $K^+$ channel KCO1, *Plant J.*, **29**, 809–820.

Desbrosses, G., Josefsson, C., Rigas, S., Hatzopoulos, P. & Dolan, L. (2003) AKT1 and TRH1 are required during root hair elongation in *Arabidopsis*, *J. Exp. Bot.*, **54**, 781–788.

Desikan, R., Cheung, M.K., Bright, J., Henson, D., Hancock, J.T. & Neill, S.J. (2004) ABA, hydrogen peroxide and nitric oxide signalling in stomatal guard cells, *J. Exp. Bot.*, **55**, 205–212.

Dietrich, P., Dreyer, I., Wiesner, P. & Hedrich R. (1998) Cation sensitivity and kinetics of guard-cell potassium channels differ among species, *Plants*, **205**, 277–287.

Ding, J.P. & Pickard, B.G. (1993) Mechanosensory calcium-selective cation channels in epidermal cells, *Plant J.*, **3**, 83–110.

Eun, S.O., Bae, S.H. & Lee, Y. (2001) Cortical actin filaments in guard cells respond differently to abscisic acid in wild-type and abi1-1 mutant *Arabidopsis*, *Planta*, **212**, 466–469.

Eun, S.O. & Lee, Y. (1997) Actin filaments of guard cells are reorganized in response to light and abscisic acid, *Plant Physiol.*, **115**, 1491–1498.

Evans, N.H., McAinsh, M.R. & Hetherington, A.M. (2001) Calcium oscillations in higher plants, *Curr. Opin. Plant Biol.*, **4**, 415–420.

Fairley-Grenot, K.A. & Assmann, S.M. (1992) Permeation of $Ca^{2+}$ through $K^+$ channels in the plasma membrane of *Vicia faba* guard cells, *J. Membr. Biol.*, **128**, 103–113.

Fan, L.M., Wang, Y.F., Wang, H. & Wu, W.H. (2001) In vitro *Arabidopsis* pollen germination and characterization of the inward potassium currents in *Arabidopsis* pollen grain protoplasts, *J. Exp. Bot.*, **52**, 1603–1614.

Fan, L.M., Wang, Y.F. & Wu, W.H. (2003) Outward $K^+$ channels in *Brassica chinensis* pollen protoplasts are regulated by external and internal pH, *Protoplasma*, **220**, 143–152.

Fan, L.M., Wu, W.H. & Yang, H.Y. (1999) Identification and characterization of the inward $K^+$ channel in the plasma membrane of *Brassica* pollen protoplasts, *Plant Cell Physiol.*, **40**, 859–865.

Feijo, J., Malho, R. & Obermeyer, G. (1995) Ion dynamics and its possible role during in vitro pollen germination and tube growth, *Protoplasma*, **187**, 155–167.

Feijo, J.A., Sainhas, J., Hackett, G.R., Kunkel, J.G. & Hepler, P.K. (1999) Growing pollen tubes possess a constitutive alkaline band in the clear zone and a growth-dependent acidic tip, *J. Cell. Biol.*, **144**, 483–496.

Foreman, J., Demidchik, V., Bothwell, J.H., *et al.* (2003) Reactive oxygen species produced by NADPH oxidase regulate plant cell growth, *Nature*, **422**, 442–446.

Franklin-Tong, V.E., Drobak, B.K., Allan, A.C., Watkins, P. & Trewavas, A.J. (1996) Growth of pollen tubes of *Papaver rhoeas* is regulated by a slow-moving calcium wave propagated by inositol 1,4,5-trisphosphate, *Plant Cell*, **8**, 1305–1321.

Fricker, M.D., White, N.S. & Obermeyer, G. (1997) pH gradients are not associated with tip growth in pollen tubes of *Lilium longiflorum*, *J. Cell. Sci.*, **110**, 1729–1740.

Fricker, M., White, N., Thiel, G., Millner, P. & Blatt, M. (1994) Peptides derived from the auxin binding protein elevate $Ca^{2+}$ and pH in stomatal guard cells of *Vicia faba*: a confocal fluorescence ratio imaging study, *Symp. Soc. Exp. Biol.*, **48**, 215–228.

Friedman, H., Spiegelstein, H., Goldschmidt, E. & Halevy, A. (1990) Flowering response of Pharbitis-nil to agents affecting cytoplasmic pH, *Plant Physiol.*, **94**, 114–119.

Fu, Y., Li, H. & Yang, Z. (2002) The ROP2 GTPase controls the formation of cortical fine F-actin and the early phase of directional cell expansion during *Arabidopsis* organogenesis, *Plant Cell*, **14**, 777–794.

Fu, Y., Wu, G. & Yang, Z. (2001) Rop GTPase-dependent dynamics of tip-localized F-actin controls tip growth in pollen tubes, *J. Cell. Biol.*, **152**, 1019–1032.

Garcia-Mata, C., Gay, R., Sokolovski, S., Hills, A., Lamattina, L. & Blatt, M.R. (2003) Nitric oxide regulates $K^+$ and $Cl^-$ channels in guard cells through a subset of abscisic acid-evoked signaling pathways, *Proc. Natl. Acad. Sci. U.S.A.*, **100**, 11116–11121.

Garrill, A., Jackson, S.L., Lew, R.R. & Heath, I.B. (1993) Ion channel activity and tip growth: tip-localized stretch-activated channels generate an essential Ca$^{2+}$ gradient in the oomycete *Saprolegnia ferax*, *Eur. J. Cell. Biol.*, **60**, 358–365.

Geelen, D., Leyman, B., Batoko, H., Di Sansebastiano, G.P., Moore, I. & Blatt, M.R. (2002) The abscisic acid-related SNARE homolog NtSyr1 contributes to secretion and growth: evidence from competition with its cytosolic domain, *Plant Cell*, **14**, 387–406.

Geitmann, A. & Cresti, M. (1998) Ca$^{2+}$ channels control the rapid expansions in pulsating growth of Petunia hybrida pollen tubes, *J. Plant Physiol.*, **152**, 439–447.

Gilroy, S., Fricker, M.D., Read, N.D. & Trewavas, A.J. (1991) Role of calcium in signal transduction of *Commelina* guard cells, *Plant Cell*, **3**, 333–344.

Gilroy, S. & Jones, D.L. (2000) Through form to function: root hair development and nutrient uptake, *Trends Plant Sci.*, **5**, 56–60.

Gilroy, S., Read, N.D. & Trewavas, A.J. (1990) Elevation of cytoplasmic calcium by caged calcium or caged inositol triphosphate initiates stomatal closure, *Nature*, **346**, 769–771.

Girbardt, M. (1969) Die Uitrastruktur der Apikalregion Pilzhyphen, *Protoplasma*, **67**, 413–441.

Goh, C.H., Kinoshita, T., Oku, T. & Shimazaki, K. (1996) Inhibition of blue light-dependent H$^+$ pumping by abscisic acid in *Vicia* guard-cell protoplasts, *Plant Physiol.*, **111**, 433–440.

Grabov, A. & Blatt, M.R. (1997) Parallel control of the inward-rectifier K$^+$ channel by cytosolic-free Ca$^{2+}$ and pH in *Vicia* guard cells, *Planta*, **201**, 84–95.

Grabov, A. & Blatt, M.R. (1998) Membrane voltage initiates Ca$^{2+}$ waves and potentiates Ca$^{2+}$ increases with abscisic acid in stomatal guard cells, *Proc. Natl. Acad. Sci. U.S.A.*, **95**, 4778–4783.

Grabov, A., Leung, J., Giraudat, J. & Blatt, M.R. (1997) Alteration of anion channel kinetics in wild-type and *abi1-1* transgenic *Nicotiana benthamiana* guard cells by abscisic acid, *Plant J.*, **12**, 203–213.

Guo, F.Q., Young, J. & Crawford, N.M. (2003) The nitrate transporter AtNRT1.1 (CHL1) functions in stomatal opening and contributes to drought susceptibility in *Arabidopsis*, *Plant Cell*, **15**, 107–117.

Hamilton, D.W., Hills, A., Kohler, B. & Blatt, M.R. (2000) Ca$^{2+}$ channels at the plasma membrane of stomatal guard cells are activated by hyperpolarization and abscisic acid, *Proc. Natl. Acad. Sci. U.S.A.*, **97**, 4967–4972.

Han, S., Tang, R., Anderson, L.K., Woerner, T.E. & Pei, Z.M. (2003) A cell surface receptor mediates extracellular Ca$^{2+}$ sensing in guard cells, *Nature*, **425**, 196–200.

Hedrich, R., Busch, H. & Raschke, K. (1990) Ca$^{2+}$ and nucleotide dependent regulation of voltage dependent anion channels in the plasma membrane of guard cells, *Embo J.*, **9**, 3889–3892.

Herrmann, A. & Felle, H. (1995) Tip growth in root hair cells of *Sinapis alba* L.: significance of internal and external Ca$^{2+}$ and pH, *New Phytol.*, **129**, 523–533.

Hetherington, A.M. (2001) Guard cell signaling, *Cell*, **107**, 711–714.

Holdaway-Clarke, T.L., Feijo, J.A., Hackett, G.R., Kunkel, J.G. & Hepler, P.K. (1997) Pollen tube growth and the intracellular cytosolic calcium gradient oscillate in phase while extracellular calcium influx is delayed, *Plant Cell*, **9**, 1999–2010.

Holdaway-Clarke, T. & Hepler, P. (2003) Control of pollen tube growth: role of ion gradients and fluxes, *New Phytol.*, **159**, 539–563.

Holzinger, A., Callaham, D.A., Hepler, P.K. & Meindl, U. (1995) Free calcium in *Micrasterias*: local gradients are not detected in growing lobes, *Eur. J. Cell. Biol.*, **67**, 363–371.

Homann, U. & Thiel, G. (1999) Unitary exocytotic and endocytotic events in guard-cell protoplasts during osmotically driven volume changes, *FEBS Lett.*, **460**, 495–499.

Hosy, E., Vavasseur, A., Mouline, K., *et al.* (2003) The *Arabidopsis* outward K$^+$ channel GORK is involved in regulation of stomatal movements and plant transpiration, *Proc. Natl. Acad. Sci. U.S.A.*, **100**, 5549–5554.

Hunt, L., Mills, L.N., Pical, C., *et al.* (2003) Phospholipase C is required for the control of stomatal aperture by ABA, *Plant J.*, **34**, 47–55.

Hwang, J.U. & Lee, Y. (2001) Abscisic acid-induced actin reorganization in guard cells of dayflower is mediated by cytosolic calcium levels and by protein kinase and protein phosphatase activities, *Plant Physiol.*, **125**, 2120–2128.

Iglesias, A. & Satter, R. (1983) H$^+$ fluxes in excised *Samanea* motor tissue, I: promotion by light, *Plant Physiol.*, **72**, 564–569.

Iino, M., Long, C. & Wang, X. (2001) Auxin- and abscisic acid-dependent osmoregulation in protoplasts of *Phaseolus vulgaris* pulvini, *Plant Cell Physiol.*, **42**, 1219–1227.

Ilan, N., Schwartz, A. & Moran, N. (1994) External pH effects on the depolarization-activated K channels in guard cell protoplasts of *Vicia faba*, *J. Gen. Physiol.*, **103**, 807–831.

Ilan, N., Schwartz, A. & Moran, N. (1996) External protons enhance the activity of the hyperpolarization-activated K channels in guard cell protoplasts of *Vicia faba*, *J. Membr. Biol.*, **154**, 169–181.

Irving, H.R., Gehring, C.A. & Parish, R.W. (1992) Changes in cytosolic pH and calcium of guard cells precede stomatal movements, *Proc. Natl. Acad. Sci. U.S.A.*, **89**, 1790–1794.

Jacob, T., Ritchie, S., Assmann, S.M. & Gilroy, S. (1999) Abscisic acid signal transduction in guard cells is mediated by phospholipase D activity, *Proc. Natl. Acad. Sci. U.S.A.*, **96**, 12192–12197.

Jaffe, L.A., Weisenseel, M.H. & Jaffe, L.F. (1975) Calcium accumulations within the growing tips of pollen tubes, *J. Cell. Biol.*, **67**, 488–492.

Jones, D., Shaff, J. & Kochian, L. (1995) Role of calcium and other ions in directing root hair tip growth in *Limnobium stoloniferum*, I: Inhibition of tip growth by aluminum, *Planta*, **197**, 672–680.

Jones, L., Milne, J.L., Ashford, D. & McQueen-Mason, S.J. (2003) Cell wall arabinan is essential for guard cell function, *Proc. Natl. Acad. Sci. U.S.A.*, **100**, 11783–11788.

Jones, M.A., Shen, J.J., Fu, Y., Li, H., Yang, Z. & Grierson, C.S. (2002) The *Arabidopsis* Rop2 GTPase is a positive regulator of both root hair initiation and tip growth, *Plant Cell*, **14**, 763–776.

Jung, J.Y., Kim, Y.W., Kwak, J.M., *et al.* (2002) Phosphatidylinositol 3- and 4-phosphate are required for normal stomatal movements, *Plant Cell*, **14**, 2399–2412.

Kim, H., Coté, G. & Crain, R. (1992) Effects of light on the membrane potential of protoplasts from *Samanea saman* pulvini. Involvement of K$^+$ channels and the H$^+$ ATPase, *Plant Physiol.*, **99**, 1532–1539.

Kim, H., Coté, G. & Crain, R. (1996) Inositol 1,4,5-trisphosphate may mediate closure of K$^+$ channels by light and darkness in *Samanea saman* motor cells, *Planta*, **198**, 279–287.

Kim, M., Hepler, P.K., Eun, S.O., Ha, K.S. & Lee, Y. (1995) Actin filaments in mature guard cells are radially distributed and involved in stomatal movement, *Plant Physiol.*, **109**, 1077–1084.

Kim, S.R., Kim, Y. & An, G. (1993) Molecular cloning and characterization of anther-preferential cDNA encoding a putative actin-depolymerizing factor, *Plant Mol. Biol.*, **21**, 39–45.

Kinoshita, T., Doi, M., Suetsugu, N., Kagawa, T., Wada, M. & Shimazaki, K. (2001) Phot1 and phot2 mediate blue light regulation of stomatal opening, *Nature*, **414**, 656–660.

Kinoshita, T., Emi, T., Tominaga, M., *et al.* (2003) Blue-light- and phosphorylation-dependent binding of a 14-3-3 protein to phototropins in stomatal guard cells of broad bean, *Plant Physiol.*, **133**, 1453–1463.

Kinoshita, T., Nishimura, M. & Shimazaki, K. (1995) Cytosolic concentration of Ca$^{2+}$ regulates the plasma membrane H$^+$-ATPase in guard cells of *Fava* bean, *Plant Cell*, **7**, 1333–1342.

Kinoshita, T. & Shimazaki, K. (1999) Blue light activates the plasma membrane H(+)-ATPase by phosphorylation of the C-terminus in stomatal guard cells, *Embo J.*, **18**, 5548–5558.

Kinoshita, T. & Shimazaki, K. (2002) Biochemical evidence for the requirement of 14-3-3 protein binding in activation of the guard-cell plasma membrane H$^+$-ATPase by blue light, *Plant Cell Physiol.*, **43**, 1359–1365.

Klein, M., Perfus-Barbeoch, L., Frelet, A., *et al.* (2003) The plant multidrug resistance ABC transporter AtMRP5 is involved in guard cell hormonal signalling and water use, *Plant J.*, **33**, 119–129.

Kohler, B., Hills, A. & Blatt, M.R. (2003) Control of guard cell ion channels by hydrogen peroxide and abscisic acid indicates their action through alternate signaling pathways, *Plant Physiol.*, **131**, 385–388.

Kotake, T., Nakagawa, N., Takeda, K. & Sakurai, N. (2000) Auxin-induced elongation growth and expressions of cell wall-bound exo- and endo-beta-glucanases in barley coleoptiles, *Plant Cell Physiol.*, **41**, 1272–1278.

Kropf, D.L. (1994) Cytoskeletal control of cell polarity in a plant zygote, *Dev. Biol.*, **165**, 361–371.

Kubitscheck, U., Homann, U. & Thiel, G. (2000) Osmotically evoked shrinking of guard-cell protoplasts causes vesicular retrieval of plasma membrane into the cytoplasm, *Planta*, **210**, 423–431.

Kwak, J.M., Moon, J.H., Murata, Y., *et al.* (2002) Disruption of a guard cell-expressed protein phosphatase 2A regulatory subunit, RCN1, confers abscisic acid insensitivity in *Arabidopsis*, *Plant Cell*, **14**, 2849–2861.

Kwak, J.M., Mori, I.C., Pei, Z.M., *et al.* (2003) NADPH oxidase AtrbohD and AtrbohF genes function in ROS-dependent ABA signaling in *Arabidopsis*, *Embo J.*, **22**, 2623–2633.

Leckie, C.P., McAinsh, M.R., Allen, G.J., Sanders, D. & Hetherington, A.M. (1998) Abscisic acid-induced stomatal closure mediated by cyclic ADP-ribose, *Proc. Natl. Acad. Sci. U.S.A.*, **95**, 15837–15842.

Lee, Y., Choi, Y.B., Suh, S., Lee, J., *et al.* (1996) Abscisic acid-induced phosphoinositide turnover in guard cell protoplasts of *Vicia faba*, *Plant Physiol.*, **110**, 987–996.

Lee, Y. & Satter, R. (1989) Effects of white, blue, red light and darkness on pH of the apoplast in the *Samanea* pulvinus, *Planta*, **353**, 31–40.

Lemichez, E., Wu, Y., Sanchez, J.P., Mettouchi, A., Mathur, J. & Chua, N.H. (2001) Inactivation of AtRac1 by abscisic acid is essential for stomatal closure, *Genes Dev.*, **15**, 1808–1816.

Lemtiri-Chlieh, F., MacRobbie, E.A. & Brearley, C.A. (2000) Inositol hexakisphosphate is a physiological signal regulating the K$^+$-inward rectifying conductance in guard cells, *Proc. Natl. Acad. Sci. U.S.A.*, **97**, 8687–8692.

Lemtiri-Chlieh, F., MacRobbie, E.A., Webb, A.A., *et al.* (2003) Inositol hexakisphosphate mobilizes an endomembrane store of calcium in guard cells, *Proc. Natl. Acad. Sci. U.S.A.*, **100**, 10091–10095.

Levina, N.N., Lew, R.R. & Heath, I.B. (1994) Cytoskeletal regulation of ion channel distribution in the tip-growing organism *Saprolegnia ferax*, *J. Cell Sci.*, **107**, 127–134.

Levina, N.N., Lew, R.R., Hyde, G.J. & Heath, I.B. (1995) The roles of Ca$^{2+}$ and plasma membrane ion channels in hyphal tip growth of *Neurospora crassa*, *J. Cell Sci.*, **108**, 3405–3417.

Lew, R.R. (1999) Comparative analysis of Ca$^{2+}$ and H$^+$ flux magnitude and location along growing hyphae of *Saprolegnia ferax* and *Neurospora crassa*, *Eur J. Cell. Biol.*, **78**, 892–902.

Leyman, B., Geelen, D., Quintero, F.J. & Blatt, M.R. (1999) A tobacco syntaxin with a role in hormonal control of guard cell ion channels, *Science*, **283**, 537–540.

Li, Y., Chen, F., Linskens, H. & Cresti, M. (1994) Distribution of unesterified and esterified pectins in cell walls of pollen tubes of flowering plants, *Sex. Plant Reprod.*, **7**, 145–152.

Li, Y., Faleri, C., Geitmann, A., Zhang, H. & Cresti, M. (1995) Immunogold localization of arabino-galactan proteins, unesterified and esterified pectins in pollen grains and pollen tubes of *Nicotiana tabacum* L., *Protoplasma*, **189**, 26–36.

Li, H., Shen, J.J., Zheng, Z.L., Lin, Y. & Yang, Z. (2001) The Rop GTPase switch controls multiple developmental processes in *Arabidopsis*, *Plant Physiol.*, **126**, 670–684.

Lohse, G. & Hedrich, R. (1992) Characteristics of the plasma membrane H$^+$ ATPase from *Vicia faba* guard cells; modulation by extracellular factors and seasonal changes, *Planta*, **188**, 206–214.

Lowen, C. & Satter, R. (1989) Light-promoted changes in apoplastic K$^+$ activity in the *Samanea saman* pulvinus, monitored with liquid membrane microelectrodes, *Planta*, **179**, 421–427.

Ma, L., Xu, X., Cui, S. & Sun, D. (1999) The presence of a heterotrimeric G protein and its role in signal transduction of extracellular calmodulin in pollen germination and tube growth, *Plant Cell*, **11**, 1351–1364.

MacRobbie, E.A. (1998) Signal transduction and ion channels in guard cells, *Philos. Trans. R. Soc. Lond. B, Biol. Sci.*, **353**, 1475–1488.

Malho, R. (1998) Role of 1,4,5-inositol triphosphate-induced Ca$^{2+}$ release in pollen tube orientation, *Sex. Plant Reprod.*, **11**, 231–235.

Malho, R., Read, N.D., Trewavas, A.J. & Pais, M.S. (1995) Calcium channel activity during pollen tube growth and reorientation, *Plant Cell*, **7**, 1173–1184.

Marten, I., Lohse, G. & Hedrich R. (1991) Plant growth hormones control voltage-dependent activity of anion channels in plasma membrane of guard cells, *Nature*, **353**, 758–762.

McAinsh, M.R., Brownlee, C. & Hetherington, A.M. (1992) Visualizing changes in cytosolic-free $Ca^{2+}$ during the response of stomatal guard cells to abscisic acid, *Plant Cell*, **4**, 1113–1122.

McAinsh, M.R., Gray, J.E., Hetherington, A.M., Leckie, C.P. & Ng, C. (2000) $Ca^{2+}$ signalling in stomatal guard cells, *Biochem. Soc. Trans.*, **28**, 476–481.

McAinsh, M.R., Webb, A., Taylor, J.E. & Hetherington, A.M. (1995) Stimulus-induced oscillations in guard cell cytosolic free calcium, *Plant Cell*, **7**, 1207–1219.

Messerli, M.A., Creton, R., Jaffe, L.F. & Robinson, K.R. (2000) Periodic increases in elongation rate precede increases in cytosolic $Ca^{2+}$ during pollen tube growth, *Dev. Biol.*, **222**, 84–98.

Messerli, M.A., Danuser, G. & Robinson, K.R. (1999) Pulsatile influxes of $H^+$, $K^+$ and $Ca^{2+}$ lag growth pulses of *Lilium longiflorum* pollen tubes, *J. Cell. Sci.*, **112**, 1497–1509.

Messerli, M. & Robinson, K.R. (1997) Tip localized $Ca^{2+}$ pulses are coincident with peak pulsatile growth rates in pollen tubes of *Lilium longiflorum*, *J. Cell. Sci.*, **110** (Pt 11), 1269–1278.

Messerli, M. & Robinson, K. (1998) Cytoplasmic acidification and current influx follow growth pulses of *Lilium longiflorum* pollen tubes, *Plant J.*, **16**, 87–91.

Messerli, M.A. & Robinson, K.R. (2003) Ionic and osmotic disruptions of the lily pollen tube oscillator: testing proposed models, *Planta*, **217**, 147–157.

Miedema, H. & Assmann, S.M. (1996) A membrane-delimited effect of internal pH on the $K^+$ outward rectifier of *Vicia faba* guard cells, *J. Membr. Biol.*, **154**, 227–237.

Mikoshiba, K. (2002) IP3 receptor, a $Ca^{2+}$ oscillator – role of $IP_3$ receptor in development and neural plasticity, *Nippon Yakurigaku Zasshi*, **120**, 6–10.

Molendijk, A.J., Bischoff, F., Rajendrakumar, C.S., Friml, J., Braun, M., Gilroy, S. & Palme, K. (2001) *Arabidopsis thaliana* Rop GTPases are localized to tips of root hairs and control polar growth, *EMBO J.*, **20**, 2779–2788.

Moshelion, M., Becker, D., Czempinski, K., *et al.* (2002) Diurnal and circadian regulation of putative potassium channels in a leaf moving organ, *Plant Physiol.*, **128**, 634–642.

Moshelion, M. & Moran, N. (2000) Potassium-efflux channels in extensor and flexor cells of the motor organ of *Samanea saman* are not identical. Effects of cytosolic calcium, *Plant Physiol.*, **124**, 911–919.

Moysset, L. & Simon, E. (1989) Role of calcium in phytochrome-controlled nyctinastic movements of *Albizzia lophantha* leaflets, *Plant Physiol.*, **90**, 1108–1114.

Murata, Y., Pei, Z.M., Mori, I.C. & Schroeder, J. (2001) Abscisic acid activation of plasma membrane $Ca^{2+}$ channels in guard cells requires cytosolic NAD(P)H and is differentially disrupted upstream and downstream of reactive oxygen species production in abi1-1 and abi2-1 protein phosphatase 2C mutants, *Plant Cell*, **13**, 2513–2523.

Nishimura, T., Yokota, E., Wada, T., Shimmen, T. & Okada, K. (2003) An *Arabidopsis* ACT2 dominant-negative mutation, which disturbs F-actin polymerization, reveals its distinctive function in root development, *Plant Cell Physiol.*, **44**, 1131–1140.

Outlaw, W.J. (2003) Integration of cellular and physiological functions of guard cells, *Crit. Rev. Plant Sci.*, **22**(6), 503–529.

Park, K.Y., Jung, J.Y., Park, J., *et al.* (2003) A role for phosphatidylinositol, 3-phosphate in abscisic acid-induced reactive oxygen species generation in guard cells, *Plant Physiol.*, **132**, 92–98.

Parton, R.M., Fischer, S., Malho, R., *et al.* (1997) Pronounced cytoplasmic pH gradients are not required for tip growth in plant and fungal cells, *J. Cell. Sci.*, **110**, 1187–1198.

Parton, R.M., Fischer-Parton, S., Trewavas, A.J. & Watahiki, M.K. (2003) Pollen tubes exhibit regular periodic membrane trafficking events in the absence of apical extension, *J. Cell. Sci.*, **116**, 2707–2719.

Pei, Z.M., Murata, Y., Benning, G., *et al.* (2000) Calcium channels activated by hydrogen peroxide mediate abscisic acid signalling in guard cells, *Nature*, **406**, 731–734.

Perera, I., Heilmann, I. & Boss, W. (1999) Transient and sustained increases in inositol 1,4,5-trisphosphate precede the differential growth response in gravistimulated maize pulvini, *PNAS*, **96**, 5838–5843.

Pierson, E.S., Miller, D.D., Callaham, D.A., *et al.* (1994) Pollen tube growth is coupled to the extracellular calcium ion flux and the intracellular calcium gradient: effect of BAPTA-type buffers and hypertonic media, *Plant Cell*, **6**, 1815–1828.

Pierson, E.S., Miller, D.D., Callaham, D.A., van Aken, J., Hackett, G. & Hepler, P.K. (1996) Tip-localized calcium entry fluctuates during pollen tube growth, *Dev. Biol.*, **174**, 160–173.

Poffenroth, M., Green, D. & Tallman, G. (1992) Sugar concentrations in guard cells of *Vicia faba* illuminated with red or blue light, *Plant Physiol.*, **98**, 1460–1471.

Potikha, T.S., Collins, C.C., Johnson, D.I., Delmer, D.P. & Levine, A. (1999) The involvement of hydrogen peroxide in the differentiation of secondary walls in cotton fibers, *Plant Physiol.*, **119**, 849–858.

Rigas, S., Debrosses, G., Haralampidis, K., *et al.* (2001) TRH1 encodes a potassium transporter required for tip growth in *Arabidopsis* root hairs, *Plant Cell*, **13**, 139–151.

Ringli, C., Baumberger, N., Diet, A., Frey, B. & Keller, B. (2002) ACTIN2 is essential for bulge site selection and tip growth during root hair development of *Arabidopsis*, *Plant Physiol.*, **129**, 1464–1472.

Roblin, G. & Fleurat-Lessard, P. (1984) A possible mode of calcium involvement in dark- and light-induced leaflet movements in *Cassia fasciculate* Michx, *Plant Cell Physiol.*, **25**, 1495–1499.

Roelfsema, M.R., Steinmeyer, R., Staal, M. & Hedrich, R. (2001) Single guard cell recordings in intact plants: light-induced hyperpolarization of the plasma membrane, *Plant J.*, **26**, 1–13.

Romano, L.A., Jacob, T., Gilroy, S. & Assmann, S.M. (2000) Increases in cytosolic Ca²⁺ are not required for abscisic acid-inhibition of inward K⁺ currents in guard cells of *Vicia faba* L., *Planta*, **211**, 209–217.

Roncal, T., Ugalde, U.O. & Irastorza, A. (1993) Calcium-induced conidiation in *Penicillium cyclopium*: calcium triggers cytosolic alkalinization at the hyphal tip, *J. Bacteriol.*, **175**, 879–886.

Sanders, D., Pelloux, J., Brownlee, C. & Harper, J.F. (2002) Calcium at the crossroads of signaling, *Plant Cell*, **14**, S401–417.

Sater, A.K., Alderton, J.M. & Steinhardt, R.A. (1994) An increase in intracellular pH during neural induction in *Xenopus*, *Development*, **120**, 433–442.

Satter, R., Garber, R., Khairallah, L. & Cheng, Y. (1982) Elemental analysis of freeze-dried thin sections of *Samanea* motor organs; barriers to ion diffusion through the apoplast, *J. Cell Biol.*, **95**, 893–902.

Satter, R., Gorton, H. & Vogelmann, T. (eds) (1990) *The Pulvinus: Motor Organ for Leaf Movement*, American Society of Plant Physiologists, Rockville, MD.

Satter, R., Guggino, S., Lonergan, T. & Galston, A. (1981) The effects of blue and far red light on rhythmic leaflet movements in *Samanea* and *Albizzia*, *Plant Physiol.*, **67**, 965–968.

Schonknecht, G., Spoormaker, P., Steinmeyer, R., *et al.* (2002) KCO1 is a component of the slow-vacuolar (SV) ion channel, *FEBS Lett.*, **511**, 28–32.

Schroeder, J.I. (1988) K⁺ transport properties of K⁺ channels in the plasma membrane of *Vicia faba* guard cells, *J. Gen. Physiol.*, **92**, 667–683.

Schroeder, J.I., Allen, G.J., Hugouvieux, V., Kwak, J.M. & Waner, D. (2001a). Guard cell signal transduction, *Annu. Rev. Plant Physiol., Plant Mol. Biol.*, **52**, 627–658.

Schroeder, J.I. & Keller, B.U. (1992) Two types of anion channel currents in guard cells with distinct voltage regulation, *Proc. Natl. Acad. Sci. U.S.A.*, **89**, 5025–5029.

Schroeder, J.I., Kwak, J.M. & Allen, G.J. (2001b). Guard cell abscisic acid signalling and engineering drought hardiness in plants. *Nature*, **410**, 327–330.

Schroeder, J., Raschke, K. & Neher, E. (1987) Voltage dependence of K⁺ channels in guard cell protoplasts, *Proc. Natl. Acad. Sci. U.S.A.*, **84**, 4108–4112.

Schultheiss, H., Dechert, C., Kogel, K.H. & Huckelhoven, R. (2003) Functional analysis of barley RAC/ROP G-protein family members in susceptibility to the powdery mildew fungus, *Plant J.*, **36**, 589–601.

Shaw, S.L. & Quatrano, R.S. (1996) The role of targeted secretion in the establishment of cell polarity and the orientation of the division plane in *Fucus* zygotes, *Development*, **122**, 2623–2630.

Shimazaki, K., Iino, M. & Zeiger E. (1986) Blue light-dependent proton extrusion of guard cell protoplasts of *Vicia faba*, *Nature*, **319**, 324–326.

Shope, J.C., DeWald, D.B. & Mott, K.A. (2003) Changes in surface area of intact guard cells are correlated with membrane internalization, *Plant Physiol.*, **133**, 1314–1321.

Silverman-Gavrila, L.B. & Lew, R.R. (2001) Regulation of the tip-high [$Ca^{2+}$] gradient in growing hyphae of the fungus *Neurospora crassa*, *Eur J. Cell. Biol.*, **80**, 379–390.

Silverman-Gavrila, L.B. & Lew, R.R. (2002) An $IP_3$-activated $Ca^{2+}$ channel regulates fungal tip growth, *J. Cell. Sci.*, **115**, 5013–5025.

Starrach, N. & Mayer, W. (1989) Changes of the apoplastic pH and $K^+$ concentration in the *Phaseolus* pulvinus in situ in relation to rhythmic leaf movements, *J. Exp. Biol.*, **217**, 865–873.

Staxen, I.I., Pical, C., Montgomery, L.T., Gray, J.E., Hetherington, A.M. & McAinsh, M.R. (1999) Abscisic acid induces oscillations in guard-cell cytosolic free calcium that involve phosphoinositide-specific phospholipase C, *Proc. Natl. Acad. Sci. U.S.A.*, **96**, 1779–1784.

Stoeckel, H. & Takeda, K. (1995) Calcium-sensitivity of the plasmalemmal delayed rectifier potassium current suggests that calcium influx in pulvinar protoplasts from *Mimosa pudica* L. can be revealed by hyperpolarization, *J. Membr. Biol.*, **146**, 201–209.

Suh, S., Moran, N. & Lee, Y. (2000) Blue light activates potassium-efflux channels in flexor cells from *Samanea saman* motor organs via two mechanisms, *Plant Physiol.*, **123**, 833–843.

Szyroki, A., Ivashikina, N., Dietrich, P., *et al.* (2001) KAT1 is not essential for stomatal opening, *Proc. Natl. Acad. Sci. U.S.A.*, **98**, 2917–2921.

Talbott, L.D. & Zeiger, E. (1993) Sugar and organic acid accumulation in guard cells of *Vicia faba* in response to red and blue light, *Plant Physiol.*, **102**, 1163–1169.

Talbott, L.D. & Zeiger, E. (1996) Central roles for potassium and sucrose in guard-cell osmoregulation, *Plant Physiol.*, **111**, 1051–1057.

Taylor, A.R., Manison, N., Fernandez, C., Wood, J. & Brownlee, C. (1996) Spatial organization of calcium signaling involved in cell volume control in the *Fucus* rhizoid, *Plant Cell*, **8**, 2015–2031.

Thiel, G., Sutter, J.U. & Homann, U. (2000) $Ca^{2+}$-sensitive and $Ca^{2+}$-insensitive exocytosis in maize coleoptile protoplasts, *Pflugers Arch.*, **439**, R152–153.

Trotochaud, A.E., Hao, T., Wu, G., Yang, Z. & Clark, S.E. (1999) The CLAVATA1 receptor-like kinase requires CLAVATA3 for its assembly into a signaling complex that includes KAPP and a Rho-related protein, *Plant Cell*, **11**, 393–406.

Very, A. & Davies, J. (2000) Hyperpolarization-activated calcium channels at the tip of *Arabidopsis* root hairs, in *Proceedings of the National Academy of Sciences of the United States of America*, Vol. 97, pp. 9801–9806.

Wang, X.Q., Ullah, H., Jones, A.M. & Assmann, S.M. (2001) G protein regulation of ion channels and abscisic acid signaling in *Arabidopsis* guard cells, *Science*, **292**, 2070–2072.

Wasteneys, G.O. & Galway, M.E. (2003) Remodeling the cytoskeleton for growth and form: an overview with some new views, *Annu. Rev. Plant Biol.*, **54**, 691–722.

Weisenseel, M., Dorn, A. & Jaffe, L.F. (1979) Natural $H^+$ currents traverse growing roots and root hairs of barley (*Hordeum vulgare* L.), *Plant Physiol.*, **64**, 512–518.

Worrall, D., Ng, C.K. & Hetherington, A.M. (2003) Sphingolipids, new players in plant signaling, *Trends Plant Sci.*, **8**, 317–320.

Wu, G., Gu, Y., Li, S. & Yang, Z. (2001) A genome-wide analysis of *Arabidopsis* Rop-interactive CRIB motif-containing proteins that act as Rop GTPase targets, *Plant Cell*, **13**, 2841–2856.

Wymer, C.L., Bibikova, T.N. & Gilroy, S. (1997) Cytoplasmic free calcium distributions during the development of root hairs of *Arabidopsis thaliana*, *Plant J.*, **12**, 427–439.

Yang, Z. (2002) Small GTPases: versatile signaling switches in plants, *Plant Cell*, **14**, S375–388.

Yu, L., Moshelion, M. & Moran, N. (2001) Extracellular protons inhibit the activity of inward-rectifying potassium channels in the motor cells of *Samanea saman* pulvini, *Plant Physiol.*, **127**, 1310–1322.

Zhang, X., Zhang, L., Dong, F., Gao, J., Galbraith, D.W. & Song, C.P. (2001) Hydrogen peroxide is involved in abscisic acid-induced stomatal closure in *Vicia faba*, *Plant Physiol.*, **126**, 1438–1448.

Zheng, Z.L., Nafisi, M., Tam, A., *et al.* (2002) Plasma membrane-associated ROP10 small GTPase is a specific negative regulator of abscisic acid responses in *Arabidopsis*, *Plant Cell*, **14**, 2787–2797.

Zorec, R. & Tester, M. (1992) Cytoplasmic calcium stimulates exocytosis in a plant secretory cell, *Biophys. J.*, **63**, 864–867.

# 10 Vesicle traffic and plasma membrane transport

Annette C. Hurst, Gerhard Thiel and Ulrike Homann

## 10.1 Introduction

Until recently, cell biology and ion transport physiology have been two different disciplines in the plant sciences, with very little overlap. On the assumption that the plasma membrane of plant cells is a rather static system, transport physiologists found little need to consider processes such as exo- and endocytosis for understanding ion transport properties. This simple view of a static plasma membrane has in recent years been fundamentally questioned. It is now clear that essential transport processes in the plasma membrane of plants are coupled to dynamic changes of this membrane. This chapter reviews recent data, which show that there is a causal dependency in the activity of membrane transport proteins on the dynamics of the plasma membrane.

## 10.2 Membrane turnover in plants

Cell biologists long ago found evidence that the plasma membrane, at least in fast-growing cells such as those in pollen tubes and grass coleoptiles, is subject to a high turnover (Steer, 1988). Analysis of electron micrographs of, for example, coleoptile cells revealed that the membrane area provided by exocytotic vesicles exceeds that required for cell elongation by a factor of about 2 (Quaite *et al.*, 1983; Phillips *et al.*, 1988). This implies that a large part of the membrane delivered to the plasma membrane must be recycled via endocytosis. On a quantitative basis, these data revealed turnover rates in these cells of up to 200 min$^{-1}$ (Phillips *et al.*, 1988). A similar value was obtained from the direct recording of endo- and exocytotic activity in intact maize coleoptile protoplasts, using patch-clamp capacitance measurements (Thiel *et al.*, 1998). In these measurements, capacitance steps corresponding to fusion/fission of exo- and endocytotic vesicles with a mean diameter of about 120 nm were recorded (Thiel *et al.*, 1998). The size distribution of these vesicles was similar to that of vesicular structures visible in electron micrographs (Thiel *et al.*, 1998). The quantitative similarity in the estimated vesicle sizes and membrane turnover rates obtained by two completely different methods underlined the significance of a high plasma membrane turnover.

Fast and reversible addition of membrane material to the plasma membrane is also an essential process during osmotically driven changes in cell volume. Large volume changes occur, for example, during the movement of guard cells in the process of opening and closing of the stomatal pore (Blatt, 2000). Reversible volume changes also take place in desiccation and in cold-tolerant plants during loss of cell water and during rehydration (e.g. Wolfe *et al.*, 1985; Tetteroo *et al.*, 1996).

Some quantitative data are available for guard cells. Swelling and shrinking, which is due to the uptake and release of $K^+$ salts and water, correlates with changes in cell volume and a concomitant increase and decrease in the plasma membrane surface area of 40% (Raschke, 1979; Shope *et al.*, 2003). To accomplish these large positive and negative excursions in surface area, vesicular membrane needs to be incorporated or retrieved to and from the plasma membrane. Indeed, high-resolution capacitance measurements of guard cell protoplasts have revealed fusion and fission of single vesicles with a diameter of around 300 nm during osmotically driven changes in surface area (Homann & Thiel, 1999).

More evidence for a fast membrane turnover in plants also comes from recent studies in which the internalization of fluorescent membrane markers was followed in protoplasts (Carroll *et al.*, 1998; Kubitscheck *et al.*, 2000) and intact plant cells (Parton *et al.*, 2001; Emans *et al.*, 2002; Shope *et al.*, 2003; Meckel *et al.*, 2004).

On the assumption that fluorescent membrane markers such as the styryl dyes are not membrane-permeable (Betz *et al.*, 1996), all the fluorescence, which accumulates inside a cell during incubation, is interpreted to be a result of endocytosis (Thiel *et al.*, 2001). With no perceptible decrease in the dimensions of the whole cell, such an endocytotic activity must be part of a constitutive membrane turnover. In some cases large rates of dye internalization have been reported in either non-stimulated protoplasts or intact plant cells, suggesting extremely high rates of membrane turnover. More recent studies however question the experimental basis of this approach, namely the assumption that the plasma membrane of plant cells is completely impermeable to the styryl dyes (Meckel *et al.*, 2004). In particular, with the less hydrophobic dyes FM1-43 and FM2-10, even a staining of the mitochondria was detectable, a process which is unlikely to be related to membrane cycling (Meckel *et al.*, 2004). This observation suggests that in a plant cell, not all the internalized fluorescence of styryl dyes should be seen as evidence for endocytosis and membrane turnover. In intact guard cells the best results were obtained with FM4-64, the dye with the highest hydrophobicity and the lowest toxicity. Using this marker it was possible to detect a constitutive retrieval of small vesicles ($\leq$250 nm) from the plasma membrane in intact turgid guard cells (Meckel *et al.*, 2004). These vesicles were abundant in the cortical region in the vicinity of the plasma membrane. They most likely reflect endocytotic vesicles, because some of them are found to retrieve not only membrane but also GFP-labelled $K^+$ channels from the plasma membrane (Meckel *et al.*, 2004). Unfortunately, it is not yet possible to estimate the rate of membrane turnover from these data.

## 10.3  Turnover of membrane proteins

While the aforementioned studies provided some information about the order of magnitude at which membrane turnover takes place, they give little or no information on the 'quality' of the membrane, which is inserted or retrieved from the plasma membrane. In other words it was not clear whether exo- and endocytotic activity affected the protein composition of the plasma membrane. However, from the above-mentioned retrieval of $K^+$ channels from the plasma membrane and a number of other recent studies, a picture is now emerging in which many, if not all, membrane proteins undergo rapid cycling and/or turnover. Also in some cases changes in the activity of plasma membrane transport proteins seem to be not only the result of a regulation of the existing proteins in the membrane, but also the consequence of an incorporation and retrieval of active proteins via the exo- and endocytotic pathway.

### 10.3.1   Cycling and redistribution of PIN

A process, which has recently received much attention with respect to membrane dynamics, has been the observation of PIN cycling. Various processes in plant growth and development are mediated by polar transport of the phytohormone auxin. This is thought to be accomplished by the asymmetric distribution of an efflux carrier, probably proteins of the PIN family (Galweiler *et al.*, 1998; Palme & Galweiler, 1999). The importance of membrane trafficking in this process became evident when it was observed that inhibitors of polar auxin transport resulted in a cytoplasmic accumulation of the PIN protein (Geldner *et al.*, 2001). Hence what appears at first glance as a stationary polar localization of a membrane protein is really only the result of a continuous cycling between the membrane and a cytoplasmic pool. This dynamic redistribution however is neither a specific property of the auxin efflux carrier nor is it only sensitive to inhibitors of polar transport (Geldner *et al.*, 2001). Brefeldin A, a generic inhibitor of vesicle trafficking, was also able to impose the same kind of protein redistribution. Inhibitor-sensitive cycling of membrane proteins in the same cells has also been reported for other membrane proteins such as $H^+$-ATPase. These observations imply that the protein composition of the plasma membrane is a dynamic flow continuum.

### 10.3.2   Cycling of $K^+$ channels in guard cells

As described above, the large excursions in surface area of guard cells during opening and closing of the stomatal pore are accomplished by addition or removal of membrane material to and from the plasma membrane. To examine the 'quality' of the membrane, which is inserted or retrieved, membrane capacitance and membrane conductance were recorded in parallel in *Vicia faba* guard cell protoplasts during swelling and shrinking (Homann & Thiel, 2002). These measurements revealed a good correlation between the two parameters, showing that changes in surface area are associated with changes in the number of active $K^+$ channels in the plasma membrane. In detail, it was observed that an increase in surface area was associated with a parallel rise in the number of the two dominant $K^+$ channels, i.e. the inward and the outward rectifier (Homann & Thiel, 2002). This process was reversed during shrinking such that both types of channels were retrieved from the plasma membrane together with endocytotic vesicles.

Although work on trafficking of plasma membrane channels associated with surface area changes is still at an early stage, we can already extract some valuable information from these data.

One piece of information comes from the observation that the rise in surface area and the incorporation of $K^+$ channels occur immediately (i.e. within seconds) upon pressure stimulation (Homann & Thiel, 2002). The lag time is much too short for any *de novo* synthesis of channel proteins. This means that the vesicles containing the active $K^+$ channels are already present in the cell, probably in a pool close to the plasma membrane. Although not yet examined in detail, it seems that the membrane containing the $K^+$ channels might be retrieved back into this pool upon shrinking.

More insight into details of the process was provided by a transient overexpression of the KAT1 channel in *Vicia faba* guard cells (Hurst *et al.*, 2004). It turned out that upon pressure stimulation these cells showed about the same increase in surface area as wild-type cells. But the number of inward-rectifying $K^+$ channels per unit of membrane area was greatly increased. These data suggest that the size of the pool of vesicular membrane that is available for insertion into the plasma membrane is about the same but gains a large increase in channel density because of an overexpression of the $K^+$ channel.

In the wild-type cell the density of both $K^+$ channels (inward and outward rectifier) in the vesicular membrane was found to be higher by a factor of about 10 compared to the channel density of the plasma membrane (Homann & Thiel, 2002). This imbalance in channel density was also visible in cells transfected with KAT1. In these cells the vesicular membrane had a channel density of the $K^+$ inward rectifier, which was about five times higher than that of the plasma membrane (Hurst *et al.*, 2004).

These observations give valuable insights into the molecular details of this process. First it implies that the channels are inserted into the plasma membrane in a clustered manner. They appear to remain in these clusters, because the channel density of the membrane, which is retrieved during endocytosis is also higher than that of the global plasma membrane. Since it is unlikely that channels are first concentrated in small areas before they are retrieved by endocytosis, the data are a good indication for stable clusters of $K^+$ channels in the plasma membrane. This concept of the formation of channel clusters accords well with the observation that clustering of the $K^+$ inward rectifier KST1 expressed in insect cells depends on the conserved $K_{HA}$ domain (Erhardt *et al.*, 1997). And electrophysiological recordings of $K^+$ channels often report a clustering of activity (Draber *et al.*, 1993).

The available data also have some implications for the physiology of guard cell movement. It is inherent in the function of guard cells that they favour either the influx or the efflux of $K^+$ for opening or closing of the stomatal pore (Thiel & Wolf, 1997). This is reflected in a general anti-parallel regulation of the inward and outward rectifiers (Blatt, 1991, 2000). In the case of a vesicle-mediated insertion or retrieval of $K^+$ channels however, the inward and outward rectifier change in parallel. This implies that this process has nothing to do with the physiological regulation of $K^+$-channel activity in the context of guard cell functioning. Instead, insertion and retrieval of ion channels may prevent large fluctuations in channel density of the plasma membrane during changes in surface area, and may thus allow a separation of cell size from regulation of current density. It would not be a surprise if all plant cells have the same ability to adapt the $K^+$-channel density of their plasma membrane during excursions in surface area. In fact, a similar mechanism of a concerted increase in membrane surface area and membrane conductance has also been reported for protoplasts from barley aleurone cells (Zorec & Tester, 1993).

### 10.3.3   Auxin-induced channel expression in elongating cells

The observation that $K^+$-channel inhibitors are able to block growth of *Zea mays* coleoptiles has indicated a rate-limiting role of $K^+$-channel activity in cell elongation (Thiel *et al.*, 1996; Claussen *et al.*, 1997; Phillippar *et al.*, 1999). This view has been further supported by the observation that auxins increase the $K^+$ conductance of the plasma membrane in coleoptile cells (Nelles & Müller, 1975) by stimulating the activity of a $K^+$ inward rectifier (Thiel & Weise, 1999). Indirect evidence reveals that the mechanism underlying this hormonal stimulation of channel activity is not due to an upregulation of channels, which already exist in the plasma membrane (Thiel & Weise, 1999). Stimulation of $K^+$-channel activity seems rather to rely on a *de novo* synthesis of the respective $K^+$ channel and an insertion into the plasma membrane via the exocytotic pathway. Such a hypothesis is based on the observation that a stimulation of corn coleoptiles, as well as of *Arabidopsis* seedlings, by auxins augments the transcription of genes coding for inward-rectifying $K^+$-channel proteins (ZMK1 in corn; KAT1 and KAT2 in *Arabidopsis*) (Philippar *et al.*, 1999, 2004). The resulting rapid translation of these messages into active channels was inferred from patch-clamp recordings. There it was found that an application of auxin to protoplasts from maize coleoptiles augmented the activity of the endogenous $K^+$ inward rectifier (Thiel & Weise, 1999). Also in protoplasts from etiolated *Arabidopsis* hypocotyls, the current density of the respective inward rectifier could be correlated with the expression profile of the $K^+$-channel genes. From these data the authors speculate that the auxin-induced stimulation of $K^+$-channel conductance in the plasma membrane results from a signal transduction pathway, leading to the fusion of newly synthesized $K^+$ channels into the plasma membrane.

A similar scenario to that of the hormonal activation of $K^+$ channels has also been reported for a stimulation of the $H^+$-ATPase in the plasma membrane of coleoptile cells. Hager *et al.* (1991) used a quantitative Western blots approach to detect the concentration of ATPases in the plasma membrane of these cells. They found that, after a lag phase of $\leq 10$ min, auxin augmented the concentration of the $H^+$-ATPase proteins in the plasma membrane. The dynamics of this increase in protein concentration was well correlated in time with the onset of auxin-stimulated $H^+$ extrusion and enhanced coleoptile elongation (Hager *et al.*, 1991). This fostered the hypothesis that the rise in $H^+$ transport in auxin-treated coleoptiles is the result of a *de novo* synthesis of ATPase, followed by its insertion into the plasma membrane via the secretory pathway. The proposed mechanisms show a nice parallel to the model for $K^+$-channel activation (Philippar *et al.*, 1999, 2004). It may further explain the dependency on cell elongation on any kind of inhibition of vesicle transport (Cunninghame & Hall, 1986; Schindler *et al.*, 1994; Cho & Hong, 1995). However, the question as to whether this entire process from transcription to insertion of active transport proteins into the plasma membrane is fast enough to account for the observed rise in channel and ATPase activity found only about 15 min after stimulation with auxin needs to be explored. It should also be mentioned that others were not able to detect the same kind of auxin-stimulated rise in ATPase concentration in the plasma membrane of cells from elongating tissue (Cho & Hong, 1995; Jahn *et al.*, 1996).

## 10.4   Parallels to mechanisms in animal cells

A tight coupling between a regulation of membrane trafficking and control of transport proteins is not unique to plants. There are also reports of similar processes in animal cells. A very interesting system is the regulation of water channels. Because most aquaporins are constitutively open, it has been a puzzling question as to how cells regulate the conductance of these channels in the context of the physiological requirements for water transport. Recent data now show that the concentration of active water channels in the plasma membrane is regulated by inserting and retrieving the proteins from the plasma membrane. Inhibition of clathrin-mediated endocytosis results in an accumulation of plasma membrane aquaporin-2 (AQP2) in transfected epithelial cells (Lu *et al.*, 2004). Similar regulation of AQP2, based on controlled vesicle trafficking, was described for the same protein in renal collecting duct cells (Deen *et al.*, 2000). Binding of the anti-diuretic hormone vasopressin to its receptor triggers an intracellular signalling cascade, which leads to the docking and fusion of AQP2 containing vesicles with the apical membrane. In pathological conditions, where urine fails to be concentrated, expression or apical targeting of AQP2 is reduced or absent, while in conditions of increased water uptake, total and apical expression of AQP2 is also elevated. More recent studies on transfected Madin-Derby canine kidney cells revealed that the cytoplasmic termini of AQP2 are important for the trafficking of AQP2 to the apical membrane via sorting into forskolin-sensitive vesicles (Van Balkom *et al.*, 2004). It is tempting to speculate that similar mechanisms to regulate water channel density in plant cells (Suga *et al.*, 2003) are also present, especially since the molecular structure of plant aquaporins shows high homologies to those present in other organisms.

In animal cells there are also reports on a regulation of ion channel activity in the plasma membrane via membrane trafficking. One example is the regulation by insulin. In animal cells it has been shown that insulin stimulates the movement of ions across the cell membrane by increasing the exocytotic transfer of transporters from intracellular pools to the surface membrane. An investigation carried out on apical membranes of cultured kidney cells revealed that the increase in membrane area of about 15% by insulin was accompanied by doubling the number of active $Na^+$ channels in the plasma membrane (Erlij *et al.*, 1994). This was determined by means of parallel measurements of capacitance and $Na^+$ conductance, without detectable changes in the open probability or effects on single-channel current because of treatment with insulin. These data suggest that either the additional membranes contain a considerably higher density of $Na^+$ channels than the resting apical membrane or in case insulin may stimulate both exo- and endocytosis, the transferred channels are left in the plasma membrane during endocytotic retrieval of excess membrane material. These findings suggest an important role for membrane traffic in the regulation of transepithelial $Na^+$ transport in animal cells.

More evidence for channel regulation via its density in the plasma membrane can be drawn from studies on polarized and regulated trafficking of the CFTR (cystic fibrosis transmembrane conductance regulator) chloride channel (Kleizen *et al.*, 2000). The CFTR protein is highly expressed in the apical membrane of epithelial cells, where it fulfils a major role in the vectorial transport of electrolytes and water. Its mislocalization or dysfunctioning results in the characteristic symptoms of cystic fibrosis. Most likely, CFTR is efficiently sorted to the apical membrane by cytoplasmic domains (PDZ-binding motifs) interacting with a protein machinery that is associated with the cytosolic surface of the trans-Golgi network, allowing proper cellular localization and polarization of target proteins. Upon cAMP stimulation, exocytotic insertion of CFTR during increases of the apical membrane area could be monitored, measuring membrane conductance and capacitance in parallel. Localization studies in stably transfected Chinese hamster ovary (CHO) cells showed approximately 50% of the total pool of CFTR protein localized to the plasma membrane and the remaining 50% localized intracellularly, while constitutive internalization of CFTR from the cell surface ranged between 5 and 16% per minute. Although native CFTR recycles from early endosomes back to the cell surface, misfolding of this protein prevented recycling and facilitated lysosomal targeting (Sharma *et al.*, 2004). The CFTR $Cl^-$ channel belongs to the ATP-binding cassette superfamily of transporters, and not only comprises a $Cl^-$ conducting pore, but also modifies the activity of a multitude of other channels. The cAMP stimulated exocytotic insertion of CFTR from sub-plasma-membrane vesicles could potentially serve as an additional mechanism for the upregulation of other transporters by facilitating their recruitment to the plasma membrane.

## 10.5  Regulatory mechanisms in membrane trafficking and their implications for activity of ion transport proteins

Surface density of membrane proteins is determined jointly by the exo- and endocytotic pathways. The understanding of the role of transport proteins at the plasma membrane therefore requires an integrated understanding of the trafficking of these proteins in plants. Because insights into the regulation of membrane trafficking in plants is still at an early stage, a discussion of this issue will only scratch the surface. Nonetheless, the following sections summarize some of the information known about general regulation strategies of the secretory pathways in plants, which could be extrapolated to the trafficking of transport proteins. In the specific case of surface expression of ion channels, some data from the animal literature are also reviewed with the intention of demonstrating principles, which will for sure also be relevant in plants.

### 10.5.1   ER export as control step in surface expression of ion channels

The ER (endoplasmic reticulum) is the first step in protein synthesis, folding and assembling. Several protein motifs are involved in ER retention and retrieval, as part of the quality control machinery. New data from animal cells underline that an interference with these steps can affect the surface density of ion channel proteins.

It has been generally assumed that proteins, once folded and assembled, reach the plasma membrane by default with no need for specific traffic signals. Studies on the trafficking of membrane proteins in animal cells have revealed the participation of diacidic sequence motifs in the efficient export from the ER. Potassium channels among the Kir family, such as Kir2.1, contain the ER export signal FCYENE at the C-terminal (Ma *et al.*, 2001). Mutations in this motif greatly reduced the steady-state density of channels in the plasma membrane. When fused to several unrelated proteins, FCYENE also promoted ER export and thus the steady-state surface density of the membrane proteins. Distinct ER export signals present in other Kir family members revealed that these motifs can also have a function for specifying the subunit composition of multimeric membrane protein complexes (Ma & Jan, 2002). Furthermore, many other membrane proteins that are efficiently exported from the ER contain diaromatic or dihydrophobic residues instead of diacidic motifs (Barlowe, 2003). While the yeast membrane protein Sys1p depends on its diacidic motif for direct binding to Sec23-Sec24, there is evidence that diaromatic or dihydrophobic residues also have a role in binding to the COPII subunits, as shown for ERGIC53 and its homologues in yeast. This form of regulation in the density of membrane transporters is also conceivable in plants, assuming that distinct motifs will be discovered.

A further potential mechanism for the control of surface expression of ion channels is an interaction with the 14-3-3 proteins. In animal cells, 14-3-3 proteins not only affected the assembly of the channel tetramers but also influenced the targeting of active channels to the plasma membrane (Yuan *et al.*, 2003). Because 14-3-3 proteins have also been reported in plants to increase $K^+$-channel activity (Saalbach *et al.*, 1997; De Boer, 2002), it is conceivable that this regulatory protein is also participating in the trafficking of plant channels.

### 10.5.2   $Ca^{2+}$ and exocytosis

For many years $Ca^{2+}$ has been recognized as an important factor in the regulation of growth and secretion in plants (Battey & Blackbourn, 1993). In tip-growing cells such as in pollen tubes or root hairs, cell growth has been found to be dependent on a tip-focused $Ca^{2+}$ gradient and an influx of $Ca^{2+}$ at the tip (Felle & Hepler,

1997; Wymer *et al.*, 1997; Franklin-Tong, 1999). These results have been interpreted as an indication that tip growth is achieved via $Ca^{2+}$-stimulated exocytosis. $Ca^{2+}$-dependent secretion has also been implicated for $\alpha$-amylase secretion from aleurone cells (Gilroy, 1996). In these cells, secretion occurs only in the presence of external $Ca^{2+}$ and is associated with an increase in intracellular $Ca^{2+}$. However, these previous investigations on the regulation of secretion in plants only provided indirect evidence for the role of $Ca^{2+}$ in exocytosis and did not allow a quantification of the effects. With new techniques these limitations have now been overcome.

Patch-clamp capacitance measurements have been conducted on protoplasts from cells with a high secretory activity such as aleurone and root caps. These recordings demonstrated an increase in exocytotic activity when the free cytosolic $Ca^{2+}$ concentration was increased over the physiological range from about 100 nm to $\geq 700$ nM (Zorec & Tester, 1992; Homann & Tester, 1997; Carroll *et al.*, 1998). An increase in exocytotic activity upon a rise in cytosolic $Ca^{2+}$ has also been demonstrated in capacitance measurements on maize coleoptile protoplasts (Thiel *et al.*, 1994; Sutter *et al.*, 2000). In coleoptile cells, the auxin-stimulated increase in cytosolic $Ca^{2+}$ (Felle, 1988) is considered an important step in the signal transduction pathway leading to cell growth. The capacitance measurements therefore imply that stimulation of secretory activity by $Ca^{2+}$ may be a central process in integrating vesicle trafficking and cell growth. The secretory vesicles may not only deliver cell wall material but may also contain proteins for cell wall loosening or the machinery required for acid-induced cell wall elongation. It remains to be examined whether this signalling step is also involved in the proposed trafficking and insertion of $K^+$ channels and/or $H^+$-ATPases into the plasma membrane of these cells.

### 10.5.3  Membrane tension and exo- and endocytosis

Large changes in surface area can be achieved only by the addition of new membrane material to the plasma membrane, because stretching of a biological membrane is limited to about 5% (Wolfe & Steponkus, 1983; Wolfe *et al.*, 1985). Membrane capacitance measurements of guard cell protoplasts demonstrated that osmotically driven swelling and shrinking is indeed associated with incorporation and removal of membrane material (Homann, 1998). Large osmotically driven changes in exocytotic activity have also been reported for barley aleurone and maize coleoptile protoplasts (Zorec & Tester, 1993; Sutter *et al.*, 2000). In both cases the process was independent of cytosolic $Ca^{2+}$.

So far, the underlying mechanisms have not been resolved at the molecular level. However, plant as well as animal cells appear to possess a safety mechanism whereby increase in membrane tension stimulates the insertion of additional membrane material into the plasma membrane in order to prevent fatal rupturing of the plasma membrane under tension (Morris & Homann, 2001). Tension-sensitive exo- and endocytosis may indeed be a common mechanism underlying the regulation of surface area in plant and animal cells. This is supported by the observation that insertion and retrieval of new membrane material can be modulated by application of hydrostatic pressure (Zorec & Tester, 1993; Bick *et al.*, 2001). In guard cell protoplasts, exocytotic activity was stimulated when the plasma membrane tension was increased above 1.2 mN/m (Bick *et al.*, 2001). Step-like changes in positive and negative pressure resulted in an immediate increase and decrease, respectively, in membrane capacitance (Bick *et al.*, 2001). The immediate responses of the cell to pressure changes implies that the membrane, which is incorporated in the process of exocytosis, must be readily available in the cell. Experiments using cytochalasin D, which depolymerizes actin filaments, furthermore suggested that the cytoskeleton is an important factor in the modulation of exocytotic activity by membrane tension (Bick *et al.*, 2001).

### 10.5.4   SNARE proteins and their possible role in ion channel trafficking and gating

Soluble NSF-attachment protein receptor (SNARE) proteins comprise a family of structurally related proteins in animals and plants, which catalyse the fusion between vesicles and target membranes (Blatt & Thiel, 2003). Current progress is uncovering the role of these proteins in numerous physiological processes in plants all related to vesicle trafficking (Sanderfoot & Raikhel, 1999; Ueda & Nakano, 2002; Blatt & Thiel, 2003; Pratelli *et al.*, 2004). Considering the significance of SNARE proteins in trafficking and surface expression of ion channels in animal cells (e.g. Atlas, 2001; Chang *et al.*, 2002; Peters *et al.*, 2001; Leung *et al.*, 2003) it is most likely that similar functions will also be discovered in plants. Indeed investigations on the role of Nt-Syr1, a SNARE protein from the plasma membrane of tobacco, have already demonstrated effects of this protein both in secretory activity and in regulating plasma membrane ion channel function (Leyman *et al.*, 1999, 2000; Geelen *et al.*, 2002). In a secretion assay, Geelen *et al.* (2002) found that the constitutive secretion of a fluorescent marker protein (secGFP) was abolished when Nt-Syr1 activity was inhibited via competition with the cytosolic domain of the protein. Inhibition of the same protein, either by proteolytic cleavage with *Clostridium botulinum* type C toxin or by competition with the cytosolic domain of Nt-Syr1, eliminated the typical response of $K^+$ and $Cl^-$ channels to ABA in guard cells. Whether these data relate to SNARE-regulated surface expression of the ion channels or to a direct role for the SNARE in signalling in guard cells remains to be seen. Significantly, for some channels in animal cells, SNARE proteins have been found not only to affect their trafficking but also to directly modulate the gating of the channels (e.g. Condliffe *et al.*, 2003, 2004; Leung *et al.*, 2003; Michaelevski *et al.*, 2003; Zhang *et al.*, 2003; Pratelli *et al.*, 2004).

### Acknowledgements

This work was supported by the Deutsche Forschungsgemeinschaft (Schwerpunkt 1108). We are grateful to Prof. Jack Dainty (Norwich) for help with the manuscript.

# References

Atlas, D. (2001) Functional and physical coupling of voltage-sensitive calcium channels with exocytotic proteins: ramifications for the secretion mechanism, *J. Neurochem.*, **77**, 972–985.

Barlowe, C. (2003) Signals from the COPII-dependent export from the ER: What's the ticket out, *Trends Cell Biol.*, **13**, 295–300.

Battey, N.H. & Blackbourn, H.D. (1993) The control of exocytosis in plant cells. Transley review No. 57, *New Phytol.*, **125**, 307–338.

Betz, W.J., Mao, F. & Smith, C.B. (1996) Imaging exocytosis and endocytosis, *Curr. Opin. Neurobiol.*, **6**, 365–371.

Bick, I., Thiel, G. & Homann, U. (2001) Cytochalasin D attenuates the desensitisation of pressure-stimulated vesicle fusion in guard cell protoplasts, *Eur. J. Cell Biol.*, **80**, 521–526.

Blatt, M.R. (1991) Ion channel gating in plants: physiological implications and integration for stomatal function, *J. Membr. Biol.*, **124**, 95–112.

Blatt, M.R. (2000) Cellular signalling and volume control in stomatal movements in plants, *Ann. Rev. Cell Dev. Biol.*, **16**, 221–241.

Blatt, M.R. & Thiel, G. (2003) Molecular components and mechanisms of exocytosis in plants, in *Golgi Apparatus and the Plant Secretory Pathway* (ed. D.G. Robinson), Blackwell Publishing, Oxford, UK.

Carroll, A.D., Moyen, C., Van Kesteren, P., Tooke, F., Battey, N.H. & Brownlee, C. (1998) $Ca^{2+}$, annexins, and GTP modulate exocytosis from maize root cap protoplasts, *Plant Cell*, **10**, 1267–1276.

Chang, S.Y., Di, A., Naren, A.P., Palfrey, H.C., Kirk, K.L. & Nelson, D.J. (2002) Mechanisms of CFTR regulation by syntaxin 1A and PKA, *J. Cell Sci.*, **115**, 783–791.

Cho, H.-T. & Hong, Y.-N. (1995) Effect of IAA on synthesis and activity of the plasma membrane $H^{+}$-ATPase of sunflower hypocotyls, in reaction to IAA-induced cell elongation and $H^{+}$ extrusion, *J. Plant Physiol.*, **145**, 717–725.

Claussen, M., Lüthen, H., Blatt, M. & Böttger, M. (1997) Auxin-induced growth and its linkage to potassium channels, *Planta*, **201**, 227–234.

Condliffe, S.B., Carattino, M.D., Frizzell, R.A. & Zhang, H. (2003) Syntaxin 1A regulates ENaC via domain-specific interactions, *J. Biol. Chem.*, **278**, 12796–12804.

Condliffe, S.B., Zhang, H. & Frizzell, R.A. (2004) Syntaxin 1A regulates ENaC channel activity, *J. Biol. Chem.*, **279**, 10085–10092.

Cunninghame, M.E. & Hall, J.L. (1986) The effect of calcium antagonists and inhibitors of secretory processes on auxin-induced elongation and fine structure of *pisum sativum* stem segments, *Protoplasma*, **133**, 149–159.

De Boer, A.H. (2002) Plant 14-3-3 proteins assist ion channels and pumps, *Biochem. Soc. Trans.*, **30**, 416–421.

Deen, P.M.T., van Balkom, B.W.M. & Kamsteeg, E.-J. (2000) Routing of the aquaporin-2 water channel in health and disease, *Eur. J. Cell Biol.*, **79**, 523–530.

Draber, S., Schultze, R. & Hansen, U.P. (1993) Cooperative behavior of $K^{+}$ channels in the tonoplast of *Chara corallina*, *Biophys. J.*, **65**, 1553–1559.

Emans, N., Zimmermann, S. & Fischer, R. (2002) Uptake of a fluorescent marker in plant cells is sensitive to Brefeldin A and wortmannin, *Plant Cell*, **14**, 71–86.

Erhardt, T., Zimmermann, S. & Müller-Röber, B. (1997) Association of plant $K_{in}^{+}$ channels is mediated by conserved C-termini and does not affect subunit assembly, *FEBS Lett.*, **409**, 166–170.

Erlij, D., De Smet, P. & Van Driessche, W. (1994) Effect of insulin on area and $Na^{+}$ channel density of apical membrane of cultured kidney cells, *J. Physiol.*, **481**, 533–542.

Felle, H. (1988) Auxin causes oscillations of cytosolic free calcium and pH in *Zea mays* coleoptiles, *Planta*, **174**, 495–499.

Felle, H.H. & Hepler, P. (1997) The cytosolic $Ca^{2+}$ concentration gradient of *Sinapis alba* root hairs as revealed by $Ca^{2+}$-selective microelectrode tests and Fura-dextran ratio imaging, *Plant Physiol.*, **114**, 39–45.

Franklin-Tong, V. (1999) Signaling and modulation of pollen tube growth, *Plant Cell*, **11**, 727–738.

Galweiler, L., Guan, C., Muller, A., *et al.* (1998) Regulation of polar auxin transport by AtPIN1 in *Arabidopsis* vascular tissue, *Science*, **282**, 2226–2230.

Geelen, D., Leyman, B., Batoko, H., Di Sansebastiano, G.P., Moore, I. & Blatt, M.R. (2002) The abscisic acid-related SNARE homolog NtSyr1 contributes to secretion and growth: evidence from competition with its cytosolic domain. *Plant Cell*, **14**, 387–406.

Geldner, N., Friml, J., Stierhof, Y.-D., Jürgens, G. & Palme, K. (2001) Auxin transport inhibitors block PIN1 cycling and trafficking, *Nature*, **413**, 425–428.

Gilroy, S. (1996) Signal transduction in barley aleurone protoplasts is calcium dependent and independent, *Plant Cell*, **8**, 2193–2209.

Hager, A., Debus, G., Edel, H.-G., Stransky, H. & Serrano, R. (1991) Auxin induces exocytosis and rapid synthesis of a high turnover-pool of plasma-membrane $H^+$-ATPase, *Plants*, **185**, 527–537.

Homann, U. (1998) Fusion and fission of plasma-membrane material accommodates for osmotically induced changes in the surface area of guard-cell protoplasts, *Planta*, **206**, 329–333.

Homann, U. & Tester, M. (1997) $Ca^{2+}$-independent and $Ca^{2+}$/GTP-binding protein-controlled exocytosis in a plant cell, *Proc. Natl. Acad. Sci. U.S.A.*, **94**, 6565–6570.

Homann, U. & Thiel, G. (1999) Unitary exocytotic and endocytotic events in guard-cell protoplasts during osmotic-driven volume changes, *FEBS Lett.*, **460**, 495–499.

Homann, U. & Thiel, G. (2002) Alteration of $K^+$-channel numbers in the plasma membrane of guard cell protoplasts during surface area changes, *Proc. Natl. Acad. Sci. U.S.A.*, **99**, 10215–10220.

Hurst, A.C., Meckel, T., Tayefeh, S., Thiel, G. & Homann, U. (2004) Trafficking of the plant potassium inward rectifier KAT1 in guard cells of *Vicia faba*, *Plant J.*, **37**, 391–397.

Jahn, T., Johansson, F., Lüthen, H., Volkmann, D. & Larsson, C. (1996) Reinvestigation of auxin and fusicoccin stimulation of the plasma membrane $H^+$-ATPase activity, *Planta*, **199**, 359–365.

Kleizen, B., Braakmann, I. & de Jonge, H.R. (2000) Regulated trafficking of the CFTR chloride channel, *Eur. J. Cell Biol.*, **79**, 544–556.

Kubitscheck, U., Homann, U. & Thiel, G. (2000) Osmotic evoked shrinking of guard cell protoplasts causes retrieval of plasma membrane into the cytoplasm, *Planta*, **210**, 423–431.

Leung, Y.M., Kang, Y., Gao, X., *et al.* (2003) Syntaxin 1A binds to the cytoplasmic C terminus of Kv2.1 to regulate channel gating and trafficking, *J. Biol. Chem.*, **278**, 17532–17538.

Leyman, B., Geelen, D. & Blatt, M.R. (2000) Localization and control of expression of Nt-Syr1, a tobacco SNARE protein, *Plant J.*, **24**, 369–381.

Leyman, B., Geelen, D., Quintero, F.J. & Blatt, M.R. (1999) A tobacco syntaxin with a role in hormonal control of guard cell ion channels, *Science*, **283**, 537–540.

Lu, H., Sun, T.X., Bouley, R., Blackburn, K., McLaughlin, M. & Brown, D. (2004) Inhibition of endocytosis causes phosphorylation (S256)-independent plasma membrane accumulation of AQP2, *Am. J. Physiol. Renal. Physiol.*, **286**, 233–243.

Ma, D. & Jan, L.Y. (2002) ER transport signals and trafficking of potassium channels and receptors, *Curr. Opin. Neurobiol.*, **12**, 287–292.

Ma, D., Zerangue, N., Lin, Y.-F., *et al.* (2001) Role of ER export signals in controlling surface potassium channel numbers, *Science*, **291**, 316–331.

Meckel, M., Hurst, A.C., Thiel, G. & Homann, U. (2004) Endocytosis against high turgor: intact guard cells of *Vicia faba* constitutively endocytose fluorescently labelled plasma membrane and GFP tagged $K^+$-channel KAT1. *Plant J.*, **39**, 182–193.

Michaelevski, I., Chikvashvili, D., Tsuk, S., *et al.* (2003) Direct interaction of target SNAREs with the Kv2.1 channel. Modal regulation of channel activation and inactivation gating. *J. Biol. Chem.*, **278**, 34320–34330.

Morris, C. & Homann, U. (2001) Cell surface area regulation and membrane tension. *J. Membr. Biol.*, **179**, 79–102.

Nelles, A. & Müller, E. (1975) Ioneninduzierte Depolarisation der Zellen von Maiskoleoptilen unter Einfluß von Kalzium und β-Indolllessigsäure (IES), *Biochem. Physiol. Pflanzen*, **167**, 253–260.

Palme, K. & Galweiler, L. (1999) PIN-pointing the molecular basis of auxin transport, *Curr. Opin. Plant Biol.*, **2**, 375–381.

Parton, R.M., Fischer-Parton, S., Watahiki, M.K. & Trewavas, A.J. (2001) Dynamics of the apical vesicle accumulation and the rate of growth are related in individual pollen tubes, *J. Cell Sci.*, **114**, 2685–2695.

Peters, K.W., Qi, J., Johnson, J.P., Watkins, S.C. & Frizzell, R.A. (2001) Role of snare proteins in CFTR and ENaC trafficking, *Pflügers Arch.*, **443** (Suppl. 1), S65–69.

Philippar, K., Fuchs, I., Lüthen, H., *et al.* (1999) Auxin induced $K^+$ channel expression represents an essential step in coleoptile growth and gravitropism, *PNAS*, **96**, 12186–12191.

Philippar, K., Ivashikina, N., Ache, P., *et al.* (2004) Auxin activates KAT1 and KAT2, two $K^+$-channel genes expressed in seedlings of *Arabidopsis thaliana*, *Plant J.*, **37**, 815–827.

Phillips, G.D., Preshaw, C. & Steer, M.W. (1988) Dictyosome vesicle production and plasma membrane turnover in auxin-stimulated outer epidermal cells of coleoptile segments from *Avena sativa* (L.), *Protoplasma*, **145**, 59–65.

Pratelli, R., Sutter, J.-U. & Blatt, M.R. (2004) A new catch to the SNARE, *Trends Plant Sci.*, **9**, 187–195.

Quaite, E., Parker, R.E. & Steer, M.W. (1983) Plant cell extension: structural implications for the origin of the plasma membrane plant, *Cell Environ.*, **6**, 429–432.

Raschke, K. (1979) Movements of stomata, in *Encyclopaedia of Plant Physiology* (eds W. Haupt & E. Feinlieb), Vol. 7, Springer Verlag, Berlin, pp. 383–441.

Saalbach, G., Schwerdel, M., Natura, G., Buschmann, P., Christov, V. & Dahse, I. (1997) Over-expression of plant 14-3-3 proteins in tobacco: enhancement of the plasmalemma $K^+$ conductance of mesophyll cells, *FEBS Lett.*, **413**, 294–298.

Sanderfoot, A.A. & Raikhel, N.V. (1999) The specificity of vesicle trafficking: coat proteins and SNAREs, *Plant Cell*, **11**, 629–642.

Schindler, T., Bergfeld, R., Hohl, M. & Schopfer, P. (1994) Inhibition of Golgi-apparatus function by Brefeldin A mediated growth, cell–wall extensibility and secretion of cell wall proteins, *Planta*, **192**, 404–413.

Sharma, M., Pampinella, F., Nemes, C., *et al.* (2004) Misfolding diverts CFTR from recycling to degradation: quality control at early endosomes, *J. Cell Biol.*, **164**, 923–933.

Shope, J.C., DeWald, D.B. & Mott, K.A. (2003) Changes in surface area of intact guard cells are correlated with membrane internalization, *Plant Physiol.*, **133**, 1–8.

Steer, M.W. (1988) Plasma membrane turnover in plant cells, *J. Exp. Bot.*, **39**, 987–996.

Suga, S., Murai, M., Kuwagata, T. & Maeshima, M. (2003) Differences in aquaporin levels among cell types of radish and measurement of osmotic water permeability of individual protoplasts, *Plant Cell Physiol.*, **44**, 277–286.

Sutter, J.-U., Homann, U. & Thiel, G. (2000) $Ca^{2+}$-stimulated exocytosis in maize coleoptile cells, *Plant Cell*, **12**, 1–11.

Tetteroo, F., De Bruijn, A.Y., Henselmans, R., Wolkers, W.F., Van Aelst, A.C. & Hoekstra, F.A. (1996) Characterization of membrane properties in desiccation-tolerant and -intolerant carrot somatic embryos, *Plant Physiol.*, **111**, 403–412.

Thiel, G., Brüdern, A. & Gradmann, D. (1996) Small inward rectifying $K^+$ channels in coleoptiles: inhibition by external $Ca^{2+}$ and function in cell elongation, *J. Membr. Biol.*, **149**, 9–20.

Thiel, G., Kreft, M. & Zorec, R. (1998) Unitary exocytotic and endocytotic events in *Zea mays* L. coleoptile protoplasts, *Plant J.*, **13**, 117–120.

Thiel, G., Rupnik, M. & Zorec, R. (1994) Raising cytosolic $Ca^{2+}$ concentration increases the membrane capacitance of maize coleoptile protoplasts: evidence for $Ca^{2+}$-stimulated exocytosis, *Planta*, **195**, 305–308.

Thiel, G., Sutter, J.U. & Homann, U. (2001). Monitoring exo- and endocytosis in real time with electrophysiological methods, in *Practical Approach Series* (eds C. Hawes & B. Satiat-jeunemaitre), Oxford University Press, Oxford, UK, pp. 17–1187.

Thiel, G. & Weise, R. (1999) Auxin augments $K^+$ inward rectifier in coleoptiles, *Planta*, **208**, 38–45.

Thiel, G. & Wolf, A.H. (1997). Operation of $K^+$-channels in stomatal movement, *Trends Plant Sci.*, **2**, 339–345.

Ueda, T. & Nakano, A. (2002) Vesicular traffic: an integral part of plant life, *Curr. Opin. Plant Biol.*, **5**, 513–527.

Van Balkom, B.W., Graat, M.P., van Raak, M., Hofman, E., van der Sluijs, P. & Deen, P.M. (2004) Role of cytoplasmic termini in sorting and shuttling of the aquaporin-2 water channel, *Am. J. Physiol. Cell Physiol.*, **286**, C372–3729.

Wolfe, J., Dowgert, M.F. & Steponkus, P.L. (1985) Dynamics of membrane exchange of the plasma membrane and the lysis of isolated protoplasts during rapid expansion in area, *J. Membr. Biol.*, **86**, 127–138.

Wolfe, J. & Steponkus, P.L. (1983) Mechanical properties of the plasma membrane of isolated plant protoplasts, *Plant Physiol.*, **71**, 276–285.

Wymer, C.L., Bibikova, T.N. & Gilroy, S. (1997) Cytoplasmic free calcium distributions during the development of root hairs of *Arabidopsis thaliana*, *Plant J.*, **12**, 427–439.

Yuan, H., Michelsen, K. & Schwappach, B. (2003) 14-3-3 Dimers probe the assembly status of multimeric membrane proteins, *Curr. Biol.*, **13**, 638–646.

Zorec, R. & Tester, M. (1992) Cytosolic calcium stimulates exocytosis in plant secretory cells, *Biophys. J.*, **63**, 864–867.

Zorec, R. & Tester, M. (1993) Rapid pressure driven exocytosis–endocytosis cycle in a single plant cell, *FEBS Lett.*, **333**, 283–286.

# 11 Potassium nutrition and salt stress

Anna Amtmann, Patrick Armengaud and Vadim Volkov

## 11.1 The physiology of potassium nutrition and salt stress

### 11.1.1 The physiology of potassium nutrition

#### 11.1.1.1 Roles of potassium in the plant

Potassium ($K^+$) is the most abundant inorganic cation in plants, comprising up to 10% of a plant's dry weight (Leigh & Jones, 1984). Potassium is an important macronutrient for plants, as it carries out vital functions in metabolism, growth and stress adaptation. These functions can be classified into those that rely on high and relatively stable concentrations of potassium in certain cellular compartments or tissues and those that rely on potassium movement between different compartments, cells or tissues. The first class of functions includes enzyme activation, stabilisation of protein synthesis, neutralisation of negative charges on proteins and maintenance of cytoplasmic pH homeostasis (Marschner, 1995). The optimal potassium concentration for enzyme activation and protein synthesis is in the range of 100 mM (Wyn Jones & Pollard, 1983). Thus, for optimal metabolic activity, cells rely on controlled potassium concentrations of around 100 mM in metabolically active compartments, i.e. the cytoplasm, the nucleus and the stroma of chloroplasts, and the matrix of mitochondria. Other roles of potassium are linked to its high mobility. This is particularly evident where potassium movement is the driving force for osmotic changes – as, for example, in stomatal movement, light-driven and seismonastic movements of organs, or phloem transport. In other cases, potassium movement provides a charge-balancing counter-flux essential for sustaining the movement of other ions. Thus, energy production through $H^+$-ATPases relies on overall $H^+/K^+$ exchange (Tester & Blatt, 1989; Wu et al., 1991) and transport of sugars, amino acids and nitrate is accompanied by $K^+$ fluxes (Marschner, 1995). The most general phenomenon that requires directed movement of potassium is growth. Accumulation of potassium (together with an anion) in plant vacuoles creates the necessary osmotic potential for cell extension. Rapid cell extension relies on high mobility of the used osmoticum and therefore only few other inorganic ions (e.g. $Na^+$) can replace potassium in this role (Reckmann et al., 1990). Once cell growth has come to a halt, maintenance of osmotic potentials can be carried out by less mobile sugars and potassium ions can partly be recovered from vacuoles (Poffenroth et al., 1992; Marschner, 1995).

#### 11.1.1.2 Symptoms of potassium starvation and impact on agriculture

Plants have mechanisms to accumulate potassium from very low external concentrations. And because of its high-mobility, potassium can be quickly reallocated between different compartments and tissues under fluctuating external potassium conditions (see below). Plants therefore grow well over a wide range of external potassium supply (approx. 10 $\mu$M to 10 mM). As for all nutrients, critical concentrations for starvation or toxicity depend on other environmental factors. In particular, stress conditions that are directly linked to potassium availability, such as drought and salinity, narrow the window for potassium sufficiency. Because of the vital role that potassium plays in plant growth and metabolism, potassium-deficient plants show a very general phenotype, which is characterised by reduced growth especially of aerial parts and lateral roots. The various physiological components of potassium deficiency, such as limited cell extension, reduced photosynthesis and impaired regulation of transpiration, can easily be linked to known functions of potassium (see above), but in many cases it is difficult to determine which $K^+$-dependent process is the most crucial in creating deficiency symptoms. For example reduced photosynthesis can be linked to both the impossibility of establishing a $H^+$ gradient and direct inhibition of several $K^+$-dependent photosynthetic enzymes.

Potassium deficiency is one of the most important factors for reduced crop yield in agriculture, and therefore potassium constitutes, together with nitrate and phosphate, the main component of routine mineral fertilisation. Insufficient potassium supply reduces not only the quantity but also the quality of the produce as well as its suitability for processing. For example potassium is important for fruit firmness and texture (Laegreid *et al.*, 1999). Decreased phloem transport in $K^+$-starved plants impacts on sugar content of tubers and other storage organs. Disturbance of sugar metabolism and allocation may also be the reason for increased susceptibility to fungal attack and frost damage of $K^+$-deficient crops (Marschner, 1995).

### 11.1.1.3    Potassium mutants
Forward genetics has so far failed to identify genes involved in potassium nutrition. Screening for mutants altered in potassium transport is difficult because, as pointed out before, plants do not display $K^+$-related symptoms over a wide range of external supply and once symptoms appear they are very general and relate to many basic physiological processes. One possibility to obtain mutants affected in $K^+$ transport is to screen for altered tissue $K^+$ contents. Recent advances in automatic tissue ion extraction and development of methods for simultaneous monitoring of a large number of elements (inductively coupled plasma spectroscopy) have made large-scale screening possible, and first mutants with altered tissue $K^+$ content have now been obtained (Lahner *et al.*, 2003). This approach is particularly promising as it can reveal interactions between different nutrients. Technical advance in the isolation of specific cell types will add further interest to such screens. Mutations in individual $K^+$ transporters (KUP4 and KUP2) have been discovered in morphological screens, e.g. for altered root hair or hypocotyl growth (Rigas *et al.*, 2001; Elumalai *et al.*, 2002). Analysis of these mutants has revealed novel and unexpected functions of $K^+$ transporters and it can be anticipated that forward genetics will continue to reveal specific roles of $K^+$ transporters confined to certain developmental stages, organs or cell types. Reverse genetics has so far been of limited success as transporters are often redundant. Nevertheless, knockout mutants for some $K^+$ transporters (AKT1

and SKOR; Gaymard *et al.*, 1998; Hirsch *et al.*, 1998) have been useful to elucidate their physiological impact. A detailed description of these mutant phenotypes will be given later in the text.

### 11.1.1.4 Potassium homeostasis

Function of potassium in enzyme activation and protein biosynthesis relies on stable high potassium concentrations in the cytoplasm and other metabolically active compartments. Therefore, cellular potassium homeostasis is necessary. First direct evidence for cellular potassium homeostasis was provided by employing ion-selective microelectrodes, which combine voltage-, $K^+$- and $H^+$-sensitivity in a triple-barrelled micropipette. Inclusion of a $H^+$ sensor allowed determination of the compartment in which the electrode was located after impalement, i.e. neutral pH indicated cytoplasmic location, and acid pH vacuolar location. Use of this method (Walker *et al.*, 1996a) could show that in barley root cells cytoplasmic $K^+$ is indeed stabilised around 70 mM, whereas vacuolar concentrations are more flexible and mirror the external supply. More recently it was found, however, that $K^+$ homeostasis is not necessarily effective in all cell types. In barley epidermal leaf cells, cytoplasmic $K^+$ levels as low as 15 mM were measured (Cuin *et al.*, 2003). These cells were still alive but had probably very low metabolic activity. Indeed, low epidermal $K^+$ concentrations reflect selective tissue allocation of $K^+$ within leaves, which allows the plant to protect metabolically active mesophyll cells against potassium deficiency. Such mechanism seems to kick into action when toxic $Na^+$ is present in excess (Cuin *et al.*, 2003). This hypothesis gained further support from studies on *Thellungiella halophila*, a salt-tolerant relative of *Arabidopsis thaliana*. Volkov *et al.* (2004) found that in *T. halophila*, under low-salt conditions the $K^+$ content of epidermal cells was greater than the one of mesophyll cells, whereas in high-salt conditions the opposite was observed. This resulted in relatively high $K^+/Na^+$ ratios in the mesophyll, an essential prerequisite for maintenance of enzyme activity under salt stress (see below).

We can conclude that the need to maintain metabolic activity within individual cells is more essential in some tissues than in others. Thus, cellular $K^+$ homeostasis is subcontracted to tissue $K^+$ homeostasis. The fact that potassium is able to move in both xylem and phloem allows redistribution of $K^+$ between different tissues (Touraine *et al.*, 1988). For example potassium in root tips is supplied from the shoot via the phloem rather than through direct uptake from the soil environment (Pritchard *et al.*, 2000). Tissue homeostasis becomes particularly important if the plants encounter shortage in potassium supply and is apparent in the fact that potassium deficiency symptoms appear first in older leaves. This is in contrast to deficiency symptoms of less mobile nutrients (e.g. $Ca^{2+}$), which deplete first in the growing tissues. Thus, plants are able to reallocate $K^+$ so that young, growing tissues are protected against starvation.

It is evident that both cellular and tissue potassium homeostasis rely on the presence of $K^+$-transport systems in different subcellular membranes and in different cell types, and that these systems must have different affinities and modes of function, depending on prevailing electrochemical potentials and direction of transport (see below). The big challenge in an era of functional genomics is to assign individual genes to different membranes and tissues and to elucidate their specific roles in tissue and cellular potassium homeostasis. This is not an easy task, considering that more than 50 genes potentially encode $K^+$ transporters and thus redundancy occurs (Mäser et al., 2001). A further complication lies in the fact that the sheer presence of a $K^+$ transporter in a specific membrane does not reveal whether this transporter is involved in general nutrition and homeostasis of potassium or is carrying out a specific task linked to a particular metabolic or osmotic process. This can be illustrated if one imagines $K^+$ ions as drivers of trains representing vital biological processes. If shortage of train drivers occurs, drivers may be recruited to certain parts of the plant to optimise their usage. In this case the train driver becomes a passenger. Our task is to identify the trains and to distinguish between those where $K^+$ is the driver and those where $K^+$ is the passenger. It may turn out that the distinction cannot always been made, i.e. that $K^+$ functions and $K^+$-allocation routes are coordinated in a way that the drivers drive themselves to their workplace. The next question is then how this system is controlled and fine-tuned. Although regulation of some $K^+$ transporters (i.e. channels) has been studied in detail, research into perception of potassium status and its signalling at the cellular and whole-plant level is in its early stages. Later sections of the review will deal with these issues.

## 11.1.2   The physiology of salt stress

### 11.1.2.1   The problem with salt
An estimated 7% of land area worldwide and approximately 5% of cultivated land are affected by soil salinity (Szabolcs, 1987; Munns et al., 1999). The problem in agricultural areas is exacerbated by secondary salinisation caused by bad irrigation practices. One third of irrigated land has been estimated to be seriously harmed by salinity; the result for crop yield is devastating, since despite its relatively small area, irrigated land produces approximately one third of the world's food (Munns, 2002). A diverse range of plant species, so-called halophytes, can thrive on high salt but most plants including all major crops are sensitive to NaCl concentrations above 50–100 mM (glycophytes). Salt stress is a complex phenomenon that has at least two components, ion toxicity and water stress, the latter being accompanied by oxidative stress (Moran et al., 1994; Pitman & Läuchli, 2002). At a tissue level, morphological symptoms of severe salt stress are necrosis and shortened lifetime of leaves. At the cellular level, salt stress is accompanied by damage of the cytoskeleton (Aon et al., 2000), leading to disruption of the cell wall/cytoplasm communication (Baluska et al., 2003) and signalling pathways (Samaj et al., 2004). Salt stress often goes hand in hand with nutrient deficiency (especially for $K^+$ and $Ca^{2+}$) caused by competitive inhibition of ion transporters and a general reduction of nutrient uptake due to inhibition of root growth (Maathuis & Amtmann, 1999; Tester & Davenport, 2003). Furthermore, salt stress exacerbates the detrimental effect of other stress factors especially drought. Several excellent and comprehensive reviews on plant salt stress have been published over the last years (Serrano et al., 1999; Hasegawa et al., 2000; Läuchli & Lüttge, 2002; Tester & Davenport, 2003). In this chapter we

will focus on issues related to homeostasis and transport of $Na^+$ and $K^+$. Although chloride toxicity is an important part of salt stress, very little work has been done to characterise its molecular basis.

### 11.1.2.2  Sodium toxicity

The reason for treating sodium stress and potassium deficiency in the same chapter lies in the close interaction between these two ions. $Na^+$ and $K^+$ resemble each other in ion radius and share basic physicochemical properties. Because of this similarity, one of the main reasons for $Na^+$ toxicity lies in the fact that $Na^+$ interferes with vital $K^+$-dependent processes (see above), i.e. $Na^+$ can displace $K^+$ from enzyme-binding sites (Wyn Jones & Pollard, 1983). One example of a proven target of $Na^+$ toxicity in yeast is the HAL2 nucleotidase, which operates in sulphate activation (Murguia et al., 1996). $K^+$ counteracts the inhibition of HAL2 by $Na^+$ (Murguia et al., 1995). We can conclude that it is the $K^+/Na^+$ ratio in the cytoplasm that is critical for stress tolerance rather than the absolute $Na^+$ concentration (Maathuis & Amtmann, 1999). Indeed, elevated $K^+$ (and $Ca^{2+}$) in the soil solution alleviate salt stress. The latter finding can also be explained with competitive inhibition between $K^+$, $Na^+$ and $Ca^{2+}$ uptake, as these ions share common transport pathways (see below). Although $Na^+$ is not a crucial nutrient for most plants, it can be beneficial for plant growth in certain conditions. As potassium, it is very mobile and soluble and is therefore a good osmoticum. It is also a 'cheap' osmoticum as its uptake does not require energy. Hence, when added in low millimolar, non-toxic concentrations, $Na^+$ can promote growth in $K^+$-deficient conditions by alleviating osmotic stress (Maathuis & Amtmann, 1999).

### 11.1.2.3  Sodium mutants

Screening for mutants with modified response to salinity has been based on evaluating growth on saline medium. Screening for salt tolerance in yeast has led to the identification of so-called *HAL* genes (Serrano, 1996), which encode components of signalling pathways, e.g. protein kinases HAL4 and HAL5 (Mulet et al., 1999) and transcription factors HAL6-9 (Mendizabal et al., 1998), or salt toxicity targets such as HAL2 (Murguia et al., 1995). Some *HAL* genes have homologues in plants (e.g. the flavoprotein-like *HAL3*, Espinosa-Ruiz et al., 1999). In *A. thaliana* the opposite approach, i.e. screening for salt oversensitive (SOS) mutants using a root bending assay, has been successful in identifying so-called *SOS* genes (Zhu, 2000; Gong et al., 2001) encoding a plasma membrane $Na^+/H^+$ antiporter (SOS1; Shi et al., 2000), regulatory elements (SOS2 and SOS3; Liu & Zhu, 1998; Zhu et al., 1998) as well as a cell-wall adhesive protein (SOS5; Shi et al., 2003a). Screens in *A. thaliana* seedlings have isolated mutant lines with altered osmotolerance (Saleki et al., 1993; Werner & Finkelstein, 1995). Interestingly, salt tolerance in these mutants is restricted to the seedling stage, which implies that different salt tolerance strategies operate in different developmental stages (Zhu, 2000). Reverse genetics in *A. thaliana* have elucidated the role of individual transporters. Thus, over-expression of NHX1, a vacuolar $Na^+/H^+$ antiporter (Apse et al., 1999), and AVP1, a vacuolar pyrophosphatase (Gaxiola et al., 2001), resulted in increased salt tolerance whereas knockout lines for the ABC transporter MRP5 (Lee *et al.*, 2004) and the $Na^+$ transporter HKT1 (Berthomieu *et al.*, 2003) were hypersensitive to $Na^+$ stress.

### 11.1.2.4  Sodium homeostasis

It is clear that plant cells have to control their cytoplasmic $Na^+$ concentration under saline conditions. In principle, this can be achieved by three strategies: (1) restriction of $Na^+$ uptake (exclusion), (2) active $Na^+$ export and (3) storage of $Na^+$ in intracellular compartments (compartmentation). Each of these strategies has its limitations:

1)  Because plants have to maintain $K^+$ uptake, $Na^+$ exclusion requires highly selective $K^+$ transporters and indeed some $K^+$ uptake channels have extremely high selectivity for $K^+$ over $Na^+$. However, because of the close physicochemical resemblance of the two ions, selectivity has its limitation. For example, a $K^+$ channel with a 50/1 selectivity for $K^+$ over $Na^+$ will still take up equal amounts of $Na^+$ and $K^+$ if the external solution contains concentrations of 100 mM $Na^+$ and 2 mM $K^+$, a realistic scenario in saline soils. Furthermore, $Na^+$ exclusion would prevent the plant from exploiting $Na^+$ as a cheap osmoticum.

2)  Export of $Na^+$ is 'uphill', i.e. against the electrochemical gradient and therefore requires energy. Operating an active $Na^+$ export system would soon exhaust cellular energy pools if it had to counteract unlimited $Na^+$ uptake. It has the additional disadvantage that $Na^+$ will accumulate in the apoplast and further decrease the extracellular water potential.

3)  Accumulation of $Na^+$ in compartments, i.e. the vacuole also requires energy but allows the use of $Na^+$ as an osmoticum while keeping cytoplasmic $Na^+$ concentrations low. The main limitation of this strategy is the volume provided by the vacuolar space. To avoid that the plant runs out of salt storage space, growth has to keep up with $Na^+$ uptake. This strategy therefore works only if the relation between growth and overall $Na^+$ uptake is positive.

It seems that most plants operate a combination of all three strategies: limit the uptake, compartmentalise what is needed for turgor and export the rest. However, depending on conditions, plant species and tissue type, the relative proportions of the three factors will vary (Flowers *et al.*, 1977; Greenway & Munns, 1980). Transport pathways supporting each of the three strategies have been characterised and are described later in the chapter. It is worth pointing out that the same strategies for $Na^+$ management are pursued at the whole-plant level. Thus, $Na^+$ uptake into specific tissues can be restricted. $Na^+$ can be actively extruded from the plant, e.g. by shedding old leaves or through specialised salt glands, and $Na^+$ can be stored in selected tissues, e.g. older leaves. The basic mechanisms for cellular and whole-plant $Na^+$ homeostasis are shown in Fig. 11.1. In summary, adaptation to high-salt conditions requires $Na^+$ homeostasis at the cellular and at the whole-plant level. As in the case of potassium, we can expect a wide range of transporters differing in their selectivity, regulation and localisation to cooperate in this task.

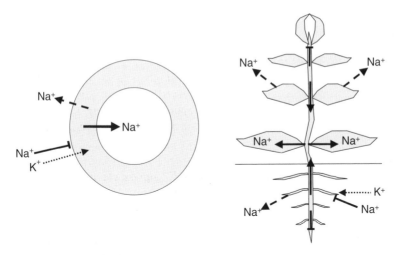

**Fig. 11.1** Strategies for $Na^+$ homeostasis at the cellular level (*left*) and at the whole-plant level (*right*). At the cellular level, low concentrations of $Na^+$ in the cytoplasm can be achieved via $K^+/Na^+$ selectivity of uptake systems (e.g. ion channels; ➡), active export of $Na^+$ by energised transport systems (•••➡) or compartmentation of $Na^+$ into the vacuole (➡). At the whole-plant level, $Na^+$ can be excluded from certain tissues through selective uptake or reabsorption of $Na^+$ from the transpiration stream (➡), exported from the plant through active release from roots into the soil solution, leaf shedding or salt glands (•••➡), or compartmentalised in specific tissues (e.g. older leaves; ➡).

## 11.2   Setting the scene for $K^+$ and $Na^+$ transport

### 11.2.1   Driving forces for $K^+$ and $Na^+$ movement across membranes

The driving force for ion movement across a membrane has two components: a chemical component, i.e. the concentration difference between the two compartments separated by the membrane, and an electrical component, i.e. the difference between the electrical potentials in the two compartments. The prevailing negative potential of the cytoplasm generally facilitates passive uptake of cations such as $K^+$ and $Na^+$. The exact amount of the potential difference between the cytoplasm and its neighbouring compartments determines the concentration at which passive influx of a cation has to be replaced by an active (energy-requiring) uptake system. According to the Nernst equation, a plasma membrane potential of $-180$ mV (a common value in plants) allows passive cation accumulation by a factor of approximately 1000. Assuming a cytoplasmic $K^+$ concentration of 100 mM, this means that $K^+$ uptake from the external medium does not require energy as long as the external $K^+$ concentration is higher than 100 µM. Similar considerations apply to the uptake of $Na^+$ from the external medium. The potential difference between the vacuole and the cytoplasm is much smaller (around 25 mV) and therefore the driving force for accumulation or release of $K^+$ or $Na^+$ into/from the vacuole follows more or less the concentration differences for the two ions between the compartments.

Early experiments measuring unidirectional fluxes of isotope tracers for various nutrients revealed biphasic kinetics of $K^+$ uptake into barley roots (Epstein *et al.*, 1963). Epstein and co-workers concluded that at least two different types of $K^+$ transporters are involved in $K^+$ uptake: a high-affinity system active at external $K^+$ concentrations below 0.5 M, and a low-affinity system mediating $K^+$ uptake at higher external concentrations (Epstein, 1973). Based on determination of root cell membrane potentials and cytoplasmic $K^+$ concentrations, it was later proposed that the high-affinity system involves active $K^+$ co-transport systems coupling $K^+$ movement to downhill flow of protons whereas the low-affinity component is realised by passive flow of $K^+$ through channels (Maathuis & Sanders, 1996a). Subsequent molecular identification of $K^+$ transporters allowed assigning of specific genes to these two components (Maathuis *et al.*, 1997; Rodriguez-Navarro, 2000). However, recent findings suggest that such assignment is simplistic and that a distinct separation between the two components is not always possible. For example it appears that under certain conditions and in certain cell types, membrane potentials can be negative enough to allow channels to play an important role in $K^+$ uptake from very low external concentrations (Hirsch *et al.*, 1998; Brüggemann *et al.*, 1999). There is also evidence that one protein can fulfil both high- and low-affinity transport modes (Fu & Luan, 1998). This means that assignment of individual genes to specific roles in $K^+$ homeostasis cannot be made on the basis of generalised thermodynamic considerations but has to be guided by detailed experiments. These include the determination of compartmental concentrations and membrane potentials under varying environmental conditions and in the specific cell type where the gene is expressed, as well as careful characterisation of mutant phenotypes.

## 11.2.2 *Tissues and membranes involved in $K^+$ and $Na^+$ transport*

On their travel through the plant, $K^+$ and $Na^+$ follow both symplastic and apoplastic routes. The apoplast represents the cell wall continuum while the symplast is formed by the cellular cytoplasm linked through plasmodesmata. The first point of contact between the plant and the surrounding medium is the root epidermis. However, because of the apoplastic continuum of peripheral root tissues, ions can take an apoplastic route to enter the root and to travel through peripheral tissues towards the stele. This apoplastic journey ends at the endodermis where secondary thickening of the cell wall (the Casparian strip) restricts further apoplastic ion movement. Although the endodermis represents the final barrier for apoplastic movement, it appears that much of the apoplast/symplast transfer of $K^+$ already takes place in epidermal and cortical cells as these cell types have to satisfy their own $K^+$ requirements. For example it was shown that inward-rectifying $K^+$ channels in the plasma membrane of root cortex cells increase their activity (Maathuis & Sanders, 1995) or expression levels (Buschmann *et al.*, 2000) in response to low external $K^+$ supply. Uptake of $K^+$ and $Na^+$ into the root symplast can be achieved by channels over a wide range of external concentrations (see above). Both $K^+$-selective inward-rectifying channels and non-selective voltage-independent channels can carry out this task (for review, see Amtmann & Sanders, 1999). If the external supply is very low, active transporters might also be needed for uptake (see above). To maintain cytoplasmic $K^+$/$Na^+$ homeostasis in root cells, the vacuole is used as a flexible storage space (Walker et al., 1996b; Carden et al., 2003). Thus, transporters mediating uptake and release of $K^+$ or $Na^+$ into/from the vacuole have to be present in the tonoplast of root cells. Depending on the prevailing electrochemical gradients across the tonoplast, these will include both passive and active transport systems.

Once the ions have entered the root symplast they cross the endodermis before they are released back into the apoplast for further long-distance transport into aerial parts of the plant via the transpiration stream in the xylem. Symplast/apoplast transfer of ions in the root stele requires a second transmembrane pathway, which may be situated either at the inner plasma membrane of endodermal cells or at the plasma membrane of xylem parenchyma cells. Because the electrochemical potential of the root xylem is unknown, it is not clear whether passive cation uptake into the xylem through channels is thermodynamically possible or whether active, i.e. energy-consuming, transport systems have to be employed. Experimental evidence exists for both types of transporters (see below). Pathways of $K^+$ through different root tissues have been reviewed in detail by Tester and Leigh (2001).

$K^+$ and $Na^+$ travel from roots to shoots with the transpiration stream. During this journey the cations may be reabsorbed from the xylem into adjacent parenchyma cells and this could be an important mechanism for restricting delivery of toxic $Na^+$ to the shoot. Several pathways have been suggested for the uptake of $K^+$ and $Na^+$ from the xylem into leaf cells (Karley et al., 2000b). The first route involves absorption of cations from the xylem into adjacent cells (bundle sheath cells or parenchyma) and thus requires a transmembrane step across the plasma membrane of these cells. Again both inward channels as well as active systems may be employed depending on the electrochemical gradients between the xylem and the cytoplasm of the adjacent cells. Once the cations have entered the leaf symplast they either move symplastically into mesophyll cells or are exported into the mesophyll apoplast and continue to travel apoplastically towards peripheral tissues, i.e. epidermal cells. The degree of symplastic continuity (i.e. the existence of plasmodesmata) between mesophyll and epidermis is unknown but considering that cytoplasmic concentrations of $K^+$ in the mesophyll and the epidermis can differ considerably (Cuin et al., 2003; see above), it seems reasonable to assume that ion uptake into epidermal cells is controlled by a transmembrane step. Indeed, inward-rectifying as well as voltage-independent channels have been characterised in the plasma membrane of epidermal cells (Elzenga & Van Volkenburgh, 1994; Karley et al., 2000a). Differential ion concentrations in mesophyll and epidermis could be explained either by selective export of ions from mesophyll cells (altering apoplastic concentrations around the epidermis) or by differential uptake of ions by mesophyll and epidermis. In this context it is interesting that in barley, $Na^+$ permeable currents are larger in the plasma membrane of epidermal cells than in the plasma membrane of mesophyll cells, whereas $K^+$-selective inward currents have similar characteristics in both cell types (Karley et al., 2000a). Different amounts of ion transport through morphologically similar channel types could be explained by differential regulation. Regulation of $K^+$ channels has been studied in detail for guard cells (Blatt, 2000) but until now no study has addressed the issue of differential regulation of ion transport in mesophyll versus epidermis cells. A second route for ion delivery to the epidermis has been suggested based on the observation that certain solutes selectively accumulate in peripheral vein extensions. This pathway bypasses the mesophyll and involves export of ions from bundle sheath cells of vein extensions directly into the apoplast adjacent to the epidermis for subsequent uptake into epidermal cells (Karley *et al.*, 2000b).

Apoplastic concentrations of $K^+$ in leaves are usually in the low millimolar range, thus allowing passive uptake through ion channels if negative membrane potentials prevail. However, membrane potentials of leaf cells can vary with light (Elzenga *et al.*, 1995; Shabala & Newman, 1999) and it has therefore been suggested that active uptake may be necessary under certain conditions (Shabala, 2003). Genes for both $K^+$ channels and putative active $K^+$ transporters are expressed in leaf tissues as well as flowers (reviewed by Shabala, 2003; Véry & Sentenac, 2003). Important storage capacity of leaves for nutrients is provided by vacuoles. Very little work has been done to characterise vacuolar ion transport in leaf tissues other than guard cells (Allen & Sanders, 1997).

To complete this overview of $K^+$ and $Na^+$ transport through the plant, we have to consider reallocation of ions by the phloem. Uptake of ions into the phloem and their release in sink tissues requires transmembrane passage as there is no symplastic connection between the phloem and the surrounding tissues (except for the root elongation zone; Oparka *et al.*, 1994). Transport of $K^+$ in the phloem is important for $K^+$ nutrition of specific tissues (e.g. the root tip; see above) under normal conditions and becomes particularly important under $K^+$-deficient conditions where leaf $K^+$ stores can be mobilised to supply roots with essential $K^+$. It appears that $K^+$ phloem transport plays a central role in the fine-tuning of $K^+$ fluxes within tissues and the whole plant. Thus, exchange of $K^+$ between xylem and phloem sap controls the amount of $K^+$ that reaches younger leaves (Jeschke & Stelter, 1976) and $K^+$ recirculation via the phloem has been suggested to act as a feedback signal for root $K^+$ uptake and xylem loading (Drew & Saker, 1984; Marschner *et al.*, 1996). Furthermore, phloem transport of $K^+$ impacts on source–sink allocation of other solutes as different rates of $K^+$ uptake and release into/from the phloem create an osmotic gradient, which controls the flow rate of the sap (Mengel & Haeder, 1977; Lang, 1983; Marschner *et al.*, 1996). Recent localisation of the *AKT2/3* gene encoding a $K^+$-selective channel to the phloem (Marten *et al.*, 1999; Lacombe *et al.*, 2000) was an important step forward in assigning specific transporters to this function (see below). Phloem transport of $Na^+$ might be important to avoid over-accumulation of this potentially toxic ion in young leaves, thus supporting reabsorption of $Na^+$ from the xylem in roots or lower parts of the shoot (Wolf *et al.*, 1991). Transporters involved in this function remain to be identified at the molecular level.

Following $K^+$ and $Na^+$ on an imaginary voyage through the plant pictures the immense complexity of transport events involved in $K^+$ and $Na^+$ homeostasis.

We have seen that the ions have to cross membranes at multiple sites and that the energetic and mechanistic requirements for transport vary with the specific electrochemical conditions encountered at every site. Bearing this in mind it is not surprising that approximately 5% of all *A. thaliana* genes encode membrane transporters (Arabidopsis Genome Initiative, 2000).

## 11.3   Functional genomics of $K^+$ and $Na^+$ transport: linking experimental evidence

An inventory of putative cation transporters in *A. thaliana* has revealed 10 gene families with putative role in the transport of $K^+$ or $Na^+$ (Mäser *et al.*, 2001). Some of these families are large (e.g. 13 *KUP/HAK/KT* transporter genes, 20 cyclic-nucleotide-gated channels (CNGCs)) but often only a few genes in the family have been functionally characterised. The picture we have obtained so far from the characterisation of individual family members shows that on one hand most families are heterogeneous with respect to the transported substrate, substrate affinity and regulation and that on the other hand transporter characteristics such as substrate specificity and voltage dependence cross family boundaries. For example $K^+$-selective inward-rectifying channels can be found both in the Shaker family and in the CNGC family, but the Shaker family also includes $K^+$-selective channels with outward-rectifying voltage dependence whereas CNGCs can also be non-selective and voltage-independent (see Chapters 6 and 7 in this volume). Most gene families show diverse tissue expression profiles among their members and the same can be anticipated for environmental conditions and developmental stages. Thus, although a certain degree of functional redundancy within gene families as well as between gene families can be expected, this redundancy might not extend into every environmental condition or developmental stage. We have to accept that characterisation of ion currents *in planta* is not sufficient for gene prediction, and vice versa that sequence homology *per se* is not a good indicator for individual characteristics of transporters and their function *in planta*. Therefore, in most cases one has to go through a time-consuming experimental process to reveal mode of function and physiological role of an individual transporter gene. Fortunately, today we have a large toolbox at hand to assist in this enterprise (Barbier-Brygoo *et al.*, 2001; see also Chapter 1), with technical approaches ranging from physiological experiments *in planta* over mutant analysis to the study of individual genes in heterologous expression systems.

Most of our current knowledge of transporters is based on a combination of (electro)-physiological experiments *in situ*, molecular biology, heterologous expression and genetics (Fig. 11.2). In a perfect case these basic experimental approaches (light-shade background in Fig. 11.2) form a full cycle, providing the proof that a certain gene is linked to a particular function *in vivo*. Steps in this cycle are (i) identification and characterisation of a transporter *in situ* (e.g. measuring currents and fluxes in native membranes), (ii) cloning of a gene (through PCR, cDNA

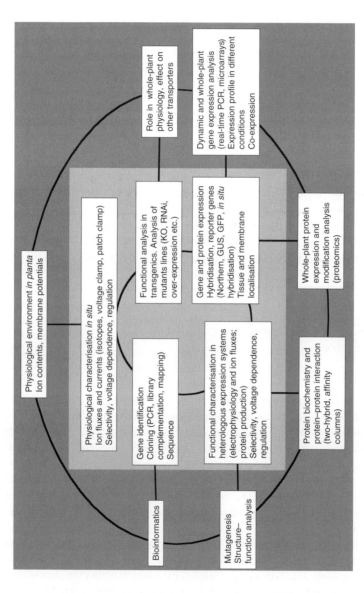

**Fig. 11.2** Experimental techniques for functional characterisation of ion transporters. A set of basic experiments in plants and heterologous expression systems using various experimental techniques leads to the first proof that a particular gene is responsible for a particular physiological phenomenon (light-shade background). Further experiments analysing structural and biochemical properties of the encoded protein as well as upstream and downstream regulatory events provide a higher level of knowledge (dark-shade background) that can be used to refine functional characterisation of the protein.

library complementation or mutation mapping), (iii) functional characterisation in heterologous expression systems (e.g. yeast or *Xenopus* oocytes), (iv) analysis of gene expression in the plant and (v) analysis of transgenic plants and proof that manipulation of the gene affects the parameters measured in (i) (Fig. 11.2). The exact order of experiments is not fixed, e.g. in a forward genetics approach we will start with point (v), and continue with points (ii), (iii), (iv) and (i). Each experimental stage allows access to a hyper-level of knowledge (dark-shade background in Fig. 11.2) and additional information gained transforms the circle into a spiral, i.e. at the end of one cycle we have gained extra knowledge, which can be used to reassess function *in planta*. Unfortunately, in most cases, evidence from at least one experimental approach is lacking owing to technical problems. For example, the transporter might not functionally express in heterologous systems, transgenics might not show a phenotype because of redundancy or the relevant tissue might not be accessible to physiological methods. This leaves us at the moment with a patchy picture for most transporters and the challenge to identify and fill experimental gaps.

In the remainder of this chapter we will gather our current knowledge on individual transporters involved in $K^+$ nutrition or $Na^+$ stress. This knowledge is based on various combinations of the experimental procedures shown in Fig. 11.2. We will point out links that have been established between evidence originating from different experimental approaches as well as technical shortcomings preventing us from fully understanding individual transporter functions.

## 11.4 Functional types of transporters involved in $K^+$ homeostasis and salt stress

### 11.4.1 Transport pathways for $K^+$ and $Na^+$

#### 11.4.1.1 Voltage-dependent channels

Voltage-dependent channels have been characterised in the plasma membrane and tonoplast of many cell types and species using electrophysiology. In whole-cell or whole-vacuole voltage clamp experiments, they can be recognised by their kinetic behaviour. Upon a stepwise change of voltage, the current through voltage-dependent channels will change over a certain period of time (usually in the range of several 100 ms) before reaching a new steady-state level (so-called time-dependent currents). The reason for this is that a change in voltage not only changes the ohmic conductance of the channel but also modulates its open probability, i.e. channels open or close in response to a sudden voltage change. Voltage dependence of the gating can have various shapes: it can be either bell-shaped (as for some Cl-selective channels) or sigmoid. In the latter case, the open probability increases either with negative voltage (so-called 'hyperpolarisation-activated' or 'inward-rectifying' channels) or with positive voltage (so-called 'depolarisation-activated' or 'outward-rectifying' channels). An extreme case of rectification is achieved when the activation threshold of a channel is tied to the equilibrium potential of the permeable ions. For example many $K^+$-selective inward rectifiers (KIRs) open only at membrane voltages, which are more negative than the reversal potential (zero current potential) and thus only pass inward currents. Such valve-like behaviour supports uptake of $K^+$ at hyperpolarised membrane voltages whilst preventing its loss under depolarising conditions and has been interpreted as an important feature of KIRs involved in $K^+$ nutrition (Maathuis et al., 1997). Rectification of $K^+$-selective outward-rectifying currents (KORs) is usually less strong and thus some inward current can pass through KORs (but see Blatt, 1988; Blatt & Gradmann, 1997). A generalised function for plasma membrane KORs based on their rectification properties might lie in the stabilisation of the membrane potential at voltages close to the equilibrium potential of $K^+$, $E_K$ (Maathuis et al., 1997; Hille, 2001). Some KORs play an important role in whole-plant $K^+$ homeostasis as they deliver ions to the xylem for long-distance transport (Roberts & Tester, 1995; Wegner & De Boer, 1997; Gaymard et al., 1998).

KIRs and KORs have been identified in the plasma membrane of all root tissues. The relative function of KIRs in $K^+$ uptake and KORs in long-distance transport is reflected in a polar distribution in maize where KIRs are mostly found in the cortex whereas KORs are more frequent in stelar cells (Roberts & Tester, 1995). However, the opposite distribution was found in A. thaliana (Maathuis et al., 1998). KIRs and KORs are also present in the plasma membrane of leaf mesophyll and epidermis cells (Kourie & Goldsmith, 1992; Spalding et al., 1992; Karley et al., 2000a; Shabala, 2003) and they have specialised functions in guard cells and pulvinar motor cells where inward and outward movement of $K^+$ drives changes in cell volume (Moran et al., 1988; Blatt, 1992, 2000; Blatt & Armstrong, 1993; Moshelion & Moran, 2001).

KIRs are generally very selective for $K^+$ over $Na^+$ ($K^+/Na^+ > 50$) and will therefore not provide an important pathway for $Na^+$ uptake (Amtmann & Sanders, 1999). KORs are usually less selective amongst cations ($K^+/Na^+$ around 10). Since under high-salt conditions the driving force for $Na^+$ is directed inward even at voltages more positive than $E_K$, it has been suggested that $Na^+$ uptake through KORs could occur (Schachtman et al., 1991). However, a role of KORs in $Na^+$ uptake has so far not been demonstrated and is unlikely to make a major contribution to $Na^+$ uptake (Amtmann & Sanders, 1999). Outward-rectifying channels that do not distinguish between $K^+$ and $Na^+$ have been identified in the xylem parenchyma of barley (Wegner & De Boer, 1997). These channels could provide a pathway for $Na^+$ reabsorption from the xylem.

Voltage-dependent channels are also present in the tonoplast of all plant cells so far investigated. So-called SV ('slow vacuolar') channels have slow activation kinetics, are outward rectifying (with respect to the cytoplasm) and allow passage of both $K^+$ and $Na^+$. So-called FV ('fast vacuolar') channels are selective for $K^+$ over other ions. They open very quickly and their open probability increases with both hyperpolarisation and depolarisation (Allen & Sanders, 1997). Depending on membrane potential and ion concentrations, SV and FV channels might be involved in passive uptake or release of $K^+$ and $Na^+$ into/from the vacuole but other functions have also been suggested, e.g. $Ca^{2+}$ signalling (Ward & Schroeder, 1994;

Allen & Sanders, 1995) or providing a 'shunt conductance' for the vacuolar ATPase (Davies & Sanders, 1995).

### 11.4.1.2 Voltage-independent channels

Voltage-independent channels (VICs) do not change their open probability with voltage and are thus visible as instantaneous currents in whole-cell voltage clamp protocols. VICs with selectivity for cations over anions have been characterised in the plasma membrane of root cells of various species including barley, maize, wheat, rye and A. thaliana (White & Lemitiri-Clieh, 1995; Amtmann et al., 1997; Roberts & Tester, 1997; Tyerman & Skerrett, 1999; Maathuis & Sanders, 2001; Demidchik & Tester, 2002). In all these species the channels discriminate only weakly between $K^+$ and $Na^+$ and it was therefore suggested that they represent the main pathway for $Na^+$ uptake into plants (Amtmann & Sanders, 1999). This notion is supported by the finding that currents through VICs are inhibited by external $Ca^{2+}$, which coincides with $Ca^{2+}$ inhibition of $Na^+$ fluxes into plant roots (Tyerman & Skerrett, 1999; Davenport & Tester, 2000). Furthermore, in A. thaliana, it was discovered that VICs are inhibited by cyclic nucleotides and this finding correlated with the observation that addition of cyclic nucleotides alleviated salt-stress symptoms of A. thaliana seedlings (Maathuis & Sanders, 2001). Recent results from T. halophila, a close relative of A. thaliana further strengthened the role of VICs in salt stress. T. halophila is salt-tolerant and when grown in high salt accumulates significantly less $Na^+$ than does A. thaliana (Volkov et al., 2004). Patch clamp analysis of root cells revealed that all major plasma membrane channels had higher selectivity for $K^+$ over $Na^+$ than the respective channels in A. thaliana; this difference was particularly striking for VICs (Volkov et al., 2004).

Non-selective VICs have also been found in leaf cells of pea and barley (Elzenga & Van Volkenburgh, 1994; Karley et al., 2000a). Interestingly, in barley, VIC-mediated currents were larger in epidermal than in mesophyll cells and this coincides with the finding that $Na^+$ is preferentially accumulated in the leaf epidermis. Finally, a $K^+$-selective VIC (VK, 'vacuolar $K^+$', channel) is present in the tonoplast, providing a pathway for $K^+$ uptake/release into/from vacuoles (Ward & Schroeder, 1994; Allen & Sanders, 1996).

### 11.4.1.3 Genes encoding cation-selective channels

The most important gene families of A. thaliana harbouring voltage-dependent channels are the family of Shaker channels (nine genes) and the KCO family (six genes) (for details of structure and molecular function, see Chapter 6 in this volume as well as Véry & Sentenac, 2002). Shaker channels in plants comprise both inward-rectifying (AKT and KAT channels) and outward-rectifying (GORK and SKOR) channels as well as the 'silent channel' AtKC (Véry & Sentenac, 2003). AKT1 and KAT were the first plant channels to be cloned using a yeast $K^+$ complementation screen (Anderson et al., 1992; Sentenac et al., 1992). AKT1 is one example where evidence from many different experimental approaches has converged to determine its role (compare Fig. 11.2). Functional analysis after heterologous expression (Gaymard et al., 1996) and gene expression analysis (Lagarde *et al.*, 1996) indicated that AKT1 is a root expressed $K^+$ inward rectifier with similar characteristics to those of a channel identified by patch clamping root cortex protoplasts (Maathuis & Sanders, 1995). A T-DNA insertion knockout mutant *akt1* was characterised using a variety of techniques ($^{86}$Rb fluxes, membrane potential measurements, patch clamp), which led to the final conclusion that AKT1 is indeed a major player in root $K^+$ uptake (Hirsch *et al.*, 1998). However, some results with respect to the physiological role of AKT1 for $K^+$ nutrition came as a surprise: *akt1* knockout plants did not show a $K^+$-deficient phenotype in the concentration range of typical low-affinity $K^+$ supply (between 100 μM to several mM) but growth was inhibited at external $K^+$ concentrations, which previously had been assigned to active high-affinity uptake (i.e. 10 μM). These findings are remarkable in several aspects: First, they show that membrane potentials are at least temporarily negative enough to allow passive $K^+$ uptake at very low external concentrations (indeed membrane potentials of $-230$ mV were measured in this study). Second, they imply that at high external $K^+$ concentrations (100 μM to 1 mM) a considerable proportion of total $K^+$ uptake (20–50%) is provided by transporters other than AKT1 and that this proportion is sufficient for plant growth. Third, the experiments by Spalding and co-workers suggested that at least one other $K^+$-transport system works in parallel with AKT1 at low $K^+$ concentrations. Detailed analysis of the *akt1* phenotype indicated that this system is blocked by ammonium, and stimulated by $Na^+$ (Spalding *et al.*, 1999). Further insight into AKT1 function was provided by comparing ion currents in *akt1* knockout plants with those in plants that have a knockout mutation in another Shaker gene, *AtKC*. Reintanz *et al.* (2002) showed that AtKC alone did not produce currents in root hairs but modulated currents mediated by AKT1. This modulatory action is probably achieved through hetero-tetramerisation between AKT1 and AtKC subunits (Véry & Sentenac, 2003). The possibility that different members of the Shaker family coaggregate into functional channels (Pilot *et al.*, 2001; Zimmermann *et al.*, 2001) introduces a new level of complexity into the assignment of individual genes to observed currents (Dietrich *et al.*, 2001; Szyroki *et al.*, 2001).

Two members of the plant Shaker family encode outward-rectifying channels, GORK and SKOR. SKOR has been shown to be important for $K^+$ root/shoot allocation (Gaymard *et al.*, 1998). Several lines of evidence point to SKOR-mediating $K^+$ uptake into the root xylem. It is expressed in the root pericycle and stelar parenchyma cells. Knockout plants *skor* have reduced $K^+$ content in the leaves whereas root $K^+$ contents were unchanged. Furthermore, *SKOR* expression is down-regulated by ABA (Gaymard *et al.*, 1998; Pilot *et al.*, 2003). These results coincide with patch clamp experiments, which identified outward-rectifying channels in the plasma membrane of xylem parenchyma cells of barley and maize (Roberts & Tester, 1995; Wegner & De Boer, 1997). In maize these channels were inhibited after application of ABA as well as drought (Roberts, 1998). The combined results suggest that (i) transport of $K^+$ into the root xylem is at least partly channel mediated, and (ii) reduced uptake of $K^+$ into the root xylem is an adaptive feature to drought stress.

Shaker channels might also be involved in phloem transport. Promoter-GUS constructs of the *AKT2/3* gene have reported expression in the phloem and both gene products (AKT3 is an N-terminal truncation of AKT2) have been characterised after heterologous expression in *Xenopus* oocytes (Marten *et al.*, 1999; Lacombe *et al.*, 2000). AKT2/3 is different from other Shaker channels in two aspects. First, it is only weakly rectifying, displaying both instantaneous and time-dependent current components. The weak voltage dependence of AKT2/3 means that this channel allows both inward and outward movement of $K^+$ and therefore theoretically could function in both phloem loading with $K^+$ in source tissues and unloading in sink tissues (Lacombe *et al.*, 2000). Whether this is indeed possible depends on the electrochemical potentials of different parts of the phloem, which remain unknown. Second, AKT2/3 is *inhibited* by external protons whereas other known inward rectifiers are *activated* by apoplastic acidification (Hoth *et al.*, 1997; Geiger *et al.*, 2002). The latter finding agrees with patch clamp data on $K^+$ channels in maize leaves, which showed differential pH regulation in mesophyll and bundle sheath cells (Keunecke & Hansen, 2000). So far no mutant lines for AKT2/3 have been analysed.

It should be pointed out that although some functions of Shaker-type channels have been clearly delimited in *A. thaliana*, this information cannot necessarily be extended into other plant species. Important species differences were found concerning transcriptional regulation of Shaker genes by environmental stress (e.g. for ice plant, Su *et al.*, 2001; rice, Golldack *et al.*, 2003; and wheat, Buschmann *et al.*, 2000). It can therefore be anticipated that integration of individual channels into whole-plant physiology differs between species. Furthermore, many Shaker channels are expressed in more than one cell type (e.g. *AKT1* is expressed not only in roots but also in leaf mesophyll and epidermis) and their expression patterns overlap (e.g. phloem expresses *KAT2* alongside *AKT2/3*). More research is needed at the cellular level to assign individual roles of the same gene in different cell types and of different genes within the same cell type.

The second gene family harbouring voltage-dependent channels is the KCO family. Only one out of six genes in this family has been functionally characterised. KCO1 from *A. thaliana* was identified as an outward-rectifying $K^+$ channel after expression in insect cells (Czempinski *et al.*, 1997). It is expressed in various tissues and is localised in the tonoplast (Czempinski *et al.*, 2002). Analysis of an *A. thaliana kco1* knockout line suggested that KCO1 is a component of vacuolar SV currents (Schönknecht *et al.*, 2002). However, important differences exist between functional features of SV currents and KCO1-mediated currents in insect cells. It is not clear at this stage whether such differences are due to functional modification of the channel in the heterologous system or whether the *kco1* knockout phenotype has not revealed the full functional spectrum of this gene. Any possible involvement of KCO1 in cellular $K^+$ homeostasis remains hypothetical.

Two large families of putatively ligand-gated channels, so-called CNGCs (20 genes in *A. thaliana*; Mäser *et al.*, 2001; Talke *et al.*, 2003) and so-called glutamate receptors (20 members; Davenport, 2002), might be involved in $K^+$ and $Na^+$ transport. AtCNGC1 and AtCNGC2 have been characterised after heterologous expression in *Xenopus* oocytes and COS cells (Leng *et al.*, 1999, 2002; Hua *et al.*, 2003). Both channels evoke voltage-dependent inward currents but whereas CNGC2 is selective for $K^+$ over $Na^+$, CNGC1 is also permeable for $Na^+$. Two important features make it unlikely that CNGC1 is responsible for the $Na^+$ currents observed in *A. thaliana* root protoplasts (Maathuis & Sanders, 2001; see above). First, CNGC1 is an inward rectifier whereas cyclic-nucleotide-dependent $Na^+$ currents in *A. thaliana* root cells are voltage-independent. Second, cyclic nucleotides activate CNGC1 but inhibit root VICs. Another member of the *A. thaliana* CNGC family, CNGC4, is involved in the hypersensitive response in leaves and was identified as a non-selective VIC in *Xenopus* oocytes (Balague *et al.*, 2003). Unfortunately, heterologous expression has proven difficult for many CNGCs. Nevertheless, considering the observed breadth of selectivity and voltage-dependence among the few characterised CNGCs, it is not unlikely that $Na^+$-permeable root VICs are indeed encoded by this gene family. Very few results have so far been obtained for plant glutamate receptors. A detailed description of ligand-gated channels and their possible involvement in salt stress can be found in Chapter 7 of this volume.

### 11.4.1.4 Active transport of $K^+$ and $Na^+$

As argued before, active, i.e. energy-consuming, transport systems for $K^+$ and $Na^+$ must be present in plant cells to account for $K^+$ uptake from very low external concentrations, $Na^+$ extrusion from the cytoplasm and accumulation of both ions in vacuoles. Many such systems are well described in bacteria and yeast. In particular, $Na^+$ pumps, directly energised by ATP, account for extremely high salt tolerance in yeast. The search for $Na^+$-ATPases has been successful for several halotolerant algae (Gimmler, 2000) but never delivered results in higher plants. Today, with an increasing number of plant genome information available we can safely assume that $Na^+$-ATPases are not present in higher plants. Considering the close resemblance between $Na^+$ pumps in yeast and $Ca^{2+}$ pumps in higher plants, this is somewhat astonishing (Garciadeblas *et al.*, 2001; Benito & Rodriguez-Navarro, 2003). By contrast, proton-coupled $Na^+$ transport in both vacuole and plasma membrane has been well established in many higher plant species (Mennen *et al.*, 1990; Wilson & Shannon, 1995; Barkla & Pantoja, 1996). Direct evidence for active $K^+$-transport systems is scarce, but Maathuis and Sanders (1996b) demonstrated proton-coupled high-affinity $K^+$ transport in *A. thaliana* root cells with the use of electrophysiological techniques. Over the last 10 years, a large number of plant genes have been identified either through genome sequencing or through yeast complementation screens, which encode putative active transporters for $K^+$ and $Na^+$. They can be divided into those that resemble proton–antiport systems (*NHX*, *CHX* and *KEA* genes; Mäser *et al.*, 2001) and those that resemble high-affinity $K^+$ transporters in yeast and bacteria (*KUP/HAK/KT* and *HKT* genes). Some of these gene families also include members, which mediate low-affinity transport and therefore do not require energisation. They usually differ from ion channels in topology and lower transport rates ('permeases'). However, a clear distinction between ion channels and permeases is not always possible (Cammack & Schwartz, 1996; Lin *et al.*, 1996) and *HKT* genes might represent one such case (Durell *et al.*, 1999).

### 11.4.1.5    The KUP/HAK/KT family

Plant genes with homology to bacterial $K^+$ uptake permeases (KUP; Schleyer & Bakker, 1993) and fungal high-affinity $K^+$ transporters (HAK; Banuelos *et al.*, 1995) were characterised by several laboratories using various heterologous expression systems (Quintero & Blatt, 1997; Santa-Maria *et al.*, 1997; Fu & Luan, 1998; Kim *et al.*, 1998) and named *KT, HAK, KUP* respectively. The *KUP/HAK/KT* gene family contains 13 members in *A. thaliana* and 17 members in rice, which can be subdivided into four groups on a phylogenetic tree (Rubio *et al.*, 2000; Banuelos *et al.*, 2002). So far, only members of group I and II have been functionally characterised. Group I genes seem to be generally more permeable to $K^+$ than $Na^+$ and include high-affinity systems likely to be energised by $H^+$ symport, although no direct evidence for proton coupling has yet been established. Group II genes mediate low-affinity $K^+$ uptake and are in some cases also permeable for $Na^+$. Dual affinity has been suggested for KUP1 (Fu & Luan, 1998) but because of background low-affinity transport capacity of the heterologous expression system this finding is disputable. KUP/HAK/KT transporters are expressed in many tissues (Rubio *et al.*, 2000). Subcellular localisation to the vacuole was achieved for one gene from rice (Banuelos *et al.*, 2002). Owing to the lack of reverse genetics studies, the role for *KUP/HAK/KT* genes in $K^+$ nutrition or $Na^+$ compartmentation is hitherto unknown. Transcriptional up-regulation upon $K^+$ starvation was reported for *KUP3* in roots of *A. thaliana* seedlings (Kim *et al.*, 1998) and for *HAK5* in leaves of mature plants (Rubio *et al.*, 2000). Recent microarray results from our laboratory (Armengaud *et al.*, in press) showed an increase of *HAK5* transcript level in roots of $K^+$-starved *A. thaliana* seedlings, which was quickly reversed upon resupply of $K^+$ (within 2 h). The same result was obtained in real-time PCR experiments (Ahn *et al.*, 2004). The latter two studies agree that in growing *A. thaliana* seedlings *HAK5* is the only member of the *KUP/HAK/KT* family to be transcriptionally regulated by external $K^+$. A role of HAK5 in high-affinity $K^+$ uptake therefore seems likely. Interestingly, *HAK5* transcripts were also up-regulated in *A. thaliana* cultured cells exposed to high NaCl (B. Wang & A. Amtmann, unpublished results, 2003), indicating a possible role of this gene in $K^+/Na^+$ homeostasis during salt stress. Specific functions of two other family members, *KUP2* and *KUP4*, were revealed in forward genetic screens. Mutant phenotypes related to cell elongation in hypocotyls and root hairs and could not be rescued by additional $K^+$ supply (Rigas *et al.*, 2001; Elumalai *et al.*, 2002). The possibility that some KUPs carry out localised functions in polar cell growth rather than in $K^+$ nutrition is supported by findings from a microarray study that indicated responsiveness of *KUP* genes to $Ca^{2+}$ rather than $K^+$ stress (Maathuis *et al.*, 2003).

### 11.4.1.6    HKT

A transporter with homology to TRK-type $K^+$-transport systems of yeast and bacteria was first cloned from wheat (Schachtman & Schroeder, 1994) in a complementation screen of $K^+$-uptake-deficient yeast. Functional characterisation in *Xenopus* oocytes revealed that TaHKT1 works as a $Na^+/K^+$ symport system, which provides $Na^+$-dependent high-affinity $K^+$ uptake in low $Na^+$ concentrations (Rubio et al., 1995). When $Na^+$ is present in high concentrations it competes with $K^+$ for its binding site and mediates low-affinity $Na^+/Na^+$ symport (Gassman et al., 1996). The function of HKT1 in high-affinity $K^+$ uptake was a matter of debate (Rubio et al., 1995; Gassman et al., 1996). Membrane potential measurements showed that in several aquatic species $K^+$-induced depolarisations were indeed $Na^+$-dependent, but there was no indication for $Na^+$ stimulation of $K^+$ transport in wheat, barley or A. thaliana (Maathuis et al., 1996). Subsequent cloning and functional characterisation of HKT1 homologues from several plant species, including A. thaliana, Oryza sativa, Eucalyptus camaldulensis and Mesembryanthemum crystallinum (Uozumi et al., 2000; Horie et al., 2001; Liu et al., 2001; Su et al., 2003), showed that cation selectivity of HKT1-type transporters varies between species and isoforms ranging from transporters that transport $Na^+$ alone (AtHKT1, OsHKT) to those that are permeable for both $Na^+$ and $K^+$ (EcHKT1 and EcHKT2, OsHKT2, McHKT1). Interestingly, $K^+$ uptake through EcHKT transporters is stimulated by $Na^+$ but also occurs in the absence of $Na^+$ (Liu et al., 2001). In several species, HKT genes are transcriptionally regulated by low $K^+$ (up-regulation in wheat, barley and rice; Wang et al., 1998; Horie et al., 2001) or by high $Na^+$ (down-regulation in rice, transient up-regulation and subsequent down-regulation in ice plant; Horie et al., 2001; Su et al., 2003). Strong support of a role of AtHKT1 in $Na^+$ uptake in A. thaliana came from experiments screening for suppression of the salt-hypersensitive phenotype of sos3 mutants. SOS3 is a cytoplasmic $Ca^{2+}$-binding protein that regulates the plasma membrane $Na^+/H^+$ antiporter SOS1 via a protein kinase SOS2 (SOS signalling pathway; see below). Mutation in HKT1 suppressed both the salt-sensitive and the $K^+$-deficient phenotype of sos3 mutants (Rus et al., 2001). This is a clear indication that AtHKT1 is not a high-affinity $K^+$ transporter, but plays an important role in $Na^+$ uptake under high-salt conditions. An alternative (or complimentary) role of AtHKT1 was proposed recently (Berthomieu et al., 2003) on the basis of another HKT mutation, sas2. Compared to wild-type plants, sas2 mutants displayed increased $Na^+$ sensitivity, which was accompanied by over-accumulation of $Na^+$ in the shoots and under-accumulation in the roots. Localisation of AtHKT1 transcript to phloem tissue in both roots and shoots led to the novel hypothesis that HKT1 is involved in recirculation of $Na^+$ from shoots to roots via the phloem. Whether HKT1 can indeed function in both phloem loading (in shoots) and unloading (in roots) as suggested by the authors still requires proof, as such dual function depends on opposite electrochemical gradients between phloem and adjacent cells in the two tissues (compare role of AKT2/3; see above). The interest of HKT1 protein for ion transport extends into its structural characterisation. A topological model has been proposed which assigns eight transmembrane-spanning domains and four pore regions to the amino acid sequence of AtHKT1 (Kato et al., 2001). This topology agrees with previous models for other members of the TRK superfamily, which suggest that all these proteins have evolved from a simple KcsA $K^+$-channel protein (Durell & Guy, 1999).

## 11.4.1.7 Antiporter genes

More than 50 genes in *A. thaliana* encode proteins with resemblance to bacterial and fungal antiporters coupling active transport of cations to downhill transport of protons (Mäser *et al.*, 2001). Considering the importance of primary proton pumping in plants, it is obvious that proton coupling plays an important function in active transport of $K^+$ and $Na^+$, either for uptake or extrusion of the ions across the plasma membrane or for ion accumulation in the vacuole. On the basis of sequence similarity to bacterial and fungal antiporters, three gene families in *A. thaliana* have putative functions in active transport of $K^+$ and/or $Na^+$: KEA ($K^+$-exchange antiporters), CHX (cation exchangers), both members of the CPA2 family, and NHX ($Na^+$ exchangers), belonging to the CPA1 family of antiporters (Mäser *et al.*, 2001). *KEA*-type genes are related to a plasma-membrane $K^+$-export system in yeast (KHA1; Ramirez *et al.*, 1998) but still await functional characterisation in plants. *CHX*-type genes (26 genes in *A. thaliana*, mostly annotated as $Na^+/H^+$ antiporters) are distantly related to *KEA*- and *NHX*-type genes. No member of this large family has been functionally characterised in plants to date, but recent findings from the laboratory of J. Pritchard and H.J. Newbury (D. Hall *et al.*, unpublished results, 2004) indicate the presence of a CHX-type transporter in the root endodermis. Knockout of this transporter decreased $Na^+$ levels both in the xylem and in the leaves. These results suggest an active, probably proton-driven, mechanism to export $Na^+$ from the endodermis.

Several genes of the *A. thaliana NHX* family have been cloned and functionally characterised in *A. thaliana* mutant lines or after heterologous expression in yeast. *SOS1* (*AtNHX7*) was identified as the locus underlying the salt-oversensitive mutant *sos1* (Wu *et al.*, 1996; Shi *et al.*, 2000) and is closely related to $Na^+/H^+$ antiporters of bacteria and fungi. The encoded protein is predicted to have 12 transmembrane-spanning domains and a long C-terminus, which is located in the cytoplasm where it might interact with various regulators (Shi *et al.*, 2000). SOS1 was localised to the plasma membrane and is strongly expressed in root parenchyma cells at the xylem–symplast boundary (Shi *et al.*, 2002). Plants over-expressing *SOS1* display increased salt tolerance both during seedling development and as mature plants and show reduced $Na^+$-accumulation in shoots (Shi *et al.*, 2003b). A function of SOS1 in both delivery and retrieval of $Na^+$ to/from the xylem has been proposed (Shi *et al.*, 2002; Tester & Davenport, 2003). Thus, together with the outward-rectifier SKOR1 and a CHX-type transporter (see above), SOS1 represents a third putative pathway for cation delivery to the root xylem. To gain further insight into the functional differentiation between these three transporters, subtissue localisation, membrane potential measurements and functional analysis of double and triple knockouts are required. Interestingly, *sos1* mutants are also deficient in high-affinity $K^+$ uptake (Wu *et al.*, 1996; Shi *et al.*, 2000). SOS1 does not transport $K^+$ itself but it has been suggested that it is linked with $K^+$-uptake systems via the SOS regulatory pathway (Zhu *et al.*, 1998; Shi *et al.*, 2000; see below).

Another gene of the *A. thaliana NHX* family, NHX1, is located in the tonoplast and acts in compartmentalising $Na^+$ into the vacuole (Apse *et al.*, 1999).

Experiments measuring pH-dependent acridine orange fluorescence in vacuoles isolated from *NHX1* over-expressing *A. thaliana* plants revealed that NHX mediates
electroneutral $Na^+/H^+$ exchange (Apse *et al.*, 1999). As for SOS1, a topology of
12 transmembrane-spanning domains and a long C-terminus is predicted from the
sequence of NHX1. However, HA tagging of hydrophilic regions in NHX1 showed
that only nine hydrophobic regions span the entire membrane, thus resulting in
a model where the N-terminus is located in the cytoplasm and the C-terminus in
the vacuole (Yamaguchi *et al.*, 2003). Interestingly, truncation of the C-terminus
differentially affected transport rates of $Na^+/H^+$ and $K^+/H^+$, thus pointing to the
possibility that NHX1 is regulated from the vacuolar side. Over-expression of *NHX1*
in *A. thaliana* as well as *Brassica napus* and *Lycopersicon esculentum* increases salt
tolerance of mature plants (Apse *et al.*, 1999; Zhang & Blumwald, 2001; Zhang *et
al.*, 2001). Over-expression of *NHX1* enhances salt accumulation in leaves without
affecting $Na^+$ levels in fruits or lipid composition of seeds, thus indicating that
NHX1 could be a useful tool for engineering salt-tolerant crops. It is surprising
that salt tolerance can be achieved through increased vacuolar compartmentalisation alone, without the necessity to restrict root $Na^+$ uptake or to enhance active
$Na^+$ extrusion. $Na^+$ compartmentalisation is limited by the overall vacuolar volume
supplied by leaf cells and one might therefore assume that salt tolerance in *NHX1*
over-expressing plants is closely related to growth and transpiration rates as well as
development. Such relationships need further investigation in various environments
and agricultural settings before *NHX1* over-expression can be considered a general
cure for salt sensitivity in crops.

Two further members of the *NHX* family (*NHX2* and *NHX5*) were identified on the
basis of sequence similarity to *NHX1* and analysed with respect to $Na^+$ transport and
transcriptional regulation by salt (Yokoi *et al.*, 2002). Like *NHX1*, *NHX2* is strongly
expressed in both roots and shoots and localises to the tonoplast. *NHX5* transcripts
are also present in roots and shoots but at a lower level. All three genes suppress the
$Na^+/Li^+$-sensitive phenotype of a yeast mutant defective in the vacuolar antiporter
ScNHX1. *NHX5* shows significant differences to *NHX1* and *NHX2* with respect to
transcriptional regulation. Whereas *NHX1* and *NHX2* are up-regulated by osmotic
stress ($Na^+$ and sorbitol) in an ABA-dependent fashion, *NHX5* responds only to
ionic stress ($Na^+$) and this response is ABA-independent (Yokoi *et al.*, 2002). In
summary, it appears that NHX-type transporters in *A. thaliana* are involved in active
$Na^+$ extrusion from the cytoplasm but differ in membrane location and regulation.
Most NHX-type antiporters also affect directly or indirectly cellular $K^+$ homeostasis
(Zhu *et al.*, 1998; Apse *et al.*, 2003), an important observation that requires further
investigation.

### 11.4.1.8   Other cation transporters

An unusual cDNA sequence with no homology to any known gene in prokaryotes
or eukaryotes was isolated from wheat in a complementation assay screening for
cDNAs suppressing the $K^+$-deficient phenotype of a yeast mutant deleted in the high-
affinity $K^+$-transport system Trk1,2 (Schachtman *et al.*, 1997). LCT1 (low-affinity

cation transporter) mediates non-selective transport of cations with low affinity. In yeast, the wheat LCT1 can operate the uptake of $K^+$ ($Rb^+$), $Ca^{2+}$, $Na^+$ and heavy metals (Schachtman *et al.*, 1997; Clemens *et al.*, 1998). Expression of LCT1 in a salt-sensitive yeast mutant deleted in the $Na^+$-export pump ENA1 results in increased intracellular $Na^+$ levels and salt hypersensitivity, which can be alleviated by other cations, including $K^+$ and $Ca^{2+}$, through competitive inhibition (Amtmann *et al.*, 2001). *LCT1* transformed yeast was also sensitive to high concentrations of $Ca^{2+}$, thus confirming suggestions that $Na^+$ and $Ca^{2+}$ transport in plants is mediated by the same transporter. In contrast to VICs, $Na^+$ transport through LCT1 in the millimolar range is not affected by micromolar $Ca^{2+}$ concentrations. The main function of LCT1 in plants might be the uptake of $Ca^{2+}$, but under salt stress, LCT1 could be an important component of $Na^+$ uptake. Unfortunately, functional characterisation of LCT1 *in planta* has not yet been carried out and its tissue expression is unknown.

A symport system coupling the transport of $Na^+$ to the transport of myoinositol (MI) has been identified in the halophyte *M. crystallinum*. MI is the precursor for stress-induced metabolites such as pinitol and ononitol (Vernon & Bohnert, 1992) and its synthesis increases in response to salt stress in *M. crystallinum* but not in glycophytes (Ishitani *et al.*, 1996). A family of MI transporters (ITRs) plays important roles in osmoregulation in animals and during mating in yeast. Two *ITR* clones, *MITR1* and *MITR2*, were isolated from a *M. crystallinum* cDNA library based on homology with *A. thaliana* and yeast ITRs (Chauhan *et al.*, 2000). *MITR1* is expressed ubiquitously while expression of *MTR2* is restricted to shoots. Both proteins are localised at the tonoplast. Salt-induced expression of *MITR* genes was found to be coordinated with expression of tonoplast $Na^+/H^+$ antiporters of the NHX family (see above) and the authors suggest a model where MITRs function in the retrieval of $Na^+$ from root vacuoles, thus keeping root vacuolar $Na^+$ low, and NHX transporters mediate accumulation of $Na^+$ in stem and leaf vacuoles (Chauhan *et al.*, 2000).

### 11.4.2    *Providing the driving force for $K^+$ and $Na^+$ transport: proton pumps*

Primary proton pumps in the plasma membrane and the tonoplast of plant cells are major players in $K^+$ homeostasis and salt tolerance. Three types of proton pumps can be distinguished: the plasma membrane $H^+$-ATPase, a P-type pump consisting of a single catalytic subunit; the vacuolar $H^+$-ATPase, a multi-subunit protein complex; and the vacuolar $H^+$-PPase, a single subunit protein that is fuelled by pyrophosphatase rather than ATP (see Chapter 2). All three enzymes actively pump protons out of the cytoplasm (either into the apoplast or into the vacuolar lumen). Resulting proton gradients can be used to drive $H^+$-coupled transport of $K^+$ and $Na^+$, such as high-affinity uptake of $K^+$ across the plasma membrane, active extrusion of $Na^+$ across the plasma membrane or accumulation of $K^+$ or $Na^+$ in the vacuole. Plasma membrane $H^+$-ATPases also establish the inside negative membrane potential, which allows uptake of cation through ion channels. Furthermore, pump and channel activity are sometimes directly linked through membrane potential and apoplastic pH. Thus, the gating of $K^+$ inward-rectifying channels in guard cells and plant roots depends on membrane potential and apoplastic pH with negative voltages and acidic apoplastic pH increasing channel activity (Blatt, 1992; Ilan et al., 1996; Grabov & Blatt, 1997; Amtmann et al., 1999; see below).

Many studies have shown that abundance and activity of plasma membrane and vacuolar proton pumps impact on $K^+$ homeostasis and salt tolerance and vice versa. For example, inhibition of the plasma membrane proton pump decreased $K^+$ uptake in red beet plasma membrane vesicles (Briskin & Gawienowski, 1996). Salt stress has been shown to increase both transcript abundance and activity of plasma membrane and vacuolar proton pumps (for reviews, see Palmgren & Axelsen, 1998; Kluge et al., 2003).

However, a direct proof of function of $H^+$-ATPases in ion homeostasis is difficult to obtain because they are encoded by large gene families. A. thaliana possesses at least 11 genes encoding plasma membrane proton pumps with overlapping expression patterns (Axelsen & Palmgren, 2001) and the subunits of the V-type $H^+$-ATPase are encoded by nearly 30 genes (Sze et al., 2002). Thus, individual gene knockout is problematic. Nevertheless, transgenic plants affected in single isoforms of plasma membrane pumps have recently delivered some results. Thus, co-suppression of tobacco PMA4 led to failures in sugar transport and stomatal opening (Zhao et al., 2000). $K^+$ and nitrogen nutrition was not affected in pma4 plants but further studies in nutrient-deficient conditions are required to study the role of this gene in high-affinity uptake. An effect on $Na^+$ and $K^+$ transport was observed in A. thaliana lines carrying a T-DNA insertion in the AHA4 gene, encoding a plasma membrane proton pump, which is expressed in the root endodermis (Vitart et al., 2001). Although no phenotype was observed under normal conditions, aha4 knockout plants were salt-sensitive and had lower $K^+$ and higher $Na^+$ contents in leaves than did wild-type plants when salt stress was applied. Since the aha4 mutation seems to be semi-dominant, it has been suggested that the observed phenotype is due to production of a dominant negative protein or RNA, which disrupts the activity of other pumps in the root endodermis (Vitart et al., 2001). Clearly, more work is needed to elucidate the functions of individual plasma membrane pumps, in particular analysis of double and triple knockouts should provide further insights.

The vacuolar $H^+$-ATPase is particularly resistant to mutant analysis as it not only comprises 12 subunits but some of these subunits are encoded by several genes and the number of isoforms per gene is species-dependent (Kluge et al., 2003). Interesting results have been obtained monitoring gene expression of all V-type pump genes on macro- or microarrays (Golldack & Dietz, 2001; Maathuis et al., 2003). In general, genes were up-regulated by salt stress and down-regulated by cation deficiency (i.e. $Ca^{2+}$ and $K^+$), thus supporting a function of this pump in vacuolar cation accumulation under salt-stress and nutrient-sufficient conditions. However, the number and subunit-type of the affected genes differed between species and stresses. It appears that regulation of the vacuolar $H^+$-ATPase is a complex process that involves differential transcriptional and post-transcriptional regulation as well as differential turnover of individual subunits.

By contrast, a role of the vacuolar PPase in salt stress has been well established at the molecular level. This enzyme is a single protein but three genes exist in the *A. thaliana* genome, *AVP1*-3, of which only *AVP1* has been studied. Over-expression of *AVP1* in a salt-sensitive yeast strain and in *A. thaliana* resulted in increased salt tolerance (Gaxiola *et al.*, 1999, 2001). *AVP1* over-expressing plants were also resistant against drought stress, and vacuolar membrane vesicles isolated from the transgenic plants had higher cation uptake capacity than did the wild type. In yeast, AVP1-mediated salt tolerance was dependent on the presence of the vacuolar $Na^+/H^+$ antiporter ScNHX1 as well as the vacuolar chloride channel GEF1 (Gaxiola *et al.*, 1999). The combined results show that AVP1 establishes a proton gradient driving cation/$H^+$ antiport at the tonoplast, but relies on the existence of a pathway that allows charge-balancing flow of anions. Other *AVP* genes have not yet been studied in transgenics. Interestingly, the *AVP2* transcript is down-regulated during $K^+$ and $Ca^{2+}$ deficiency but did not change during salt stress (Maathuis *et al.*, 2003). This gene might therefore have a specific role in nutrient homeostasis.

### 11.4.3   Other transporters involved in $K^+$ homeostasis and salt stress

#### 11.4.3.1   ABC transporters
The large gene family of ABC transporters (more than 100 genes in *A. thaliana*; Sanchez-Fernandez *et al.*, 2001) is mostly implicated in the detoxification of heavy metals and organic toxins, e.g. herbicides (Rea *et al.*, 1998), by plants. However, there is increasing evidence that some ABC transporters are involved in plant responses to salinity and nutrient deficiency. Several ABC transporters have been shown to be transcriptionally regulated by ion stress. In particular, two members of the *PDR* gene family (*PDR7* and *PDR8*) showed opposite transcriptional responses to high salinity and $K^+$ deficiency (Maathuis *et al.*, 2003), confirming previous findings on regulation of PDR-type ABC transporters by environmental stress and hormones (Smart & Fleming, 1996). A putative function of ABC transporters in ion homeostasis, either directly through transport of inorganic cations or indirectly through transport of signalling molecules, is an intriguing possibility, which requires further investigation through mutant analysis and heterologous expression studies. First results have just emerged: a knockout mutation of *AtMRP5* produces a salt but not mannitol hypersensitive phenotype with low tissue $K^+/Na^+$ ratios (Lee *et al.*, 2004). Furthermore, there is circumstantial evidence that *AtMRP5* acts as a sulfony-lurea receptor, similar to animal SURs, which upon binding of sulfonylurea activates a $K^+$ channel (Gaedeke *et al.*, 2001; Lee *et al.*, 2004). Previously, $K^+$-channel modulation by sulfonylurea had been measured in guard cells (Leonhardt *et al.*, 1997) and AtMRP5 is indeed necessary for guard cell functioning (Gaedeke *et al.*, 2001).

More experiments are needed to elucidate whether (and which) $K^+$ channels are targeted by AtMRP5 and whether sulfonylurea acts as a signalling molecule in plant ion homeostasis.

### 11.4.3.2  Aquaporins

$K^+$ and $Na^+$ accumulation in vacuoles is vital for maintaining osmotic pressure during growth and salt stress. To carry out osmotic adjustment, plants have to take up water. Although phospholipid bilayers are permeable for water, it has been shown that membrane intrinsic proteins, so-called water channels or aquaporins, considerably increase water flux across membranes (Chrispeels *et al.*, 2001; see Chapter 8). The gene family of aquaporins is large (50 members in *A. thaliana*) and individual genes vary in tissue expression, membrane localisation and substrate specificity. Some aquaporins are involved in the transport of small organic ions such as urea. It appears that transcriptional regulation of aquaporins is linked to the external ion concentration. A microarray study by Maathuis *et al.* (2003) found that $K^+$ and $Ca^{2+}$ deficiency as well as salt stress strongly affected the expression of many aquaporins. Most genes of this family were down-regulated by ion deficiency ($K^+$ or $Ca^{2+}$) and up-regulated by ion excess ($Na^+$). In the case of $Na^+$ stress a response in aquaporins was expected as this treatment affected the water potential of the growth solution. By contrast the response of aquaporins to nutrient deficiency is remarkable, as these treatments were not accompanied by changes in water potential. One plausible explanation for the observation is that water channels 'mastermind' the overall velocity of ion flow through the root apoplast. If cellular water permeability is high, the ratio of apoplastic/symplastic flow is low. As a result, apoplastic solute flow is slow and ions 'hang around' for some time, therefore facilitating their uptake into the root symplast. If cellular water permeability is low, the ratio of apoplastic/symplastic flow is high, apoplastic solute flow is fast and ions are quickly moved towards the endodermis for subsequent uptake into the xylem and long-distance transport to the leaves. Thus, up- and down-regulation of aquaporins could direct overall solute flow through the plant. This hypothesis still needs testing, but it would explain the surprising sensitivity of aquaporins to $Ca^{2+}$, which is not a major osmoticum but is mainly transported in the apoplast (White, 2001). The transcriptional response of aquaporins to salt stress is plausible, as water has to be taken up by cells to adjust their osmotic potential. Interestingly, Maathuis *et al.* (2003) found that the expression profile of aquaporin genes over the first 24 h of stress mirrored the changes in plant water potential after salt stress, i.e. an initial down-regulation followed by subsequent up-regulation. This time-course was slightly delayed in tonoplast-localised aquaporins compared to plasma-membrane-localised water channels. In summary, it appears that transcriptional regulation of aquaporins is an important factor in plant ion homeostasis and future research in this field should provide new insight into whole-plant kinetics of ion transport.

A schematic overview summarising current knowledge of location, substrates and transport direction of characterised transporters is shown in Fig. 11.3.

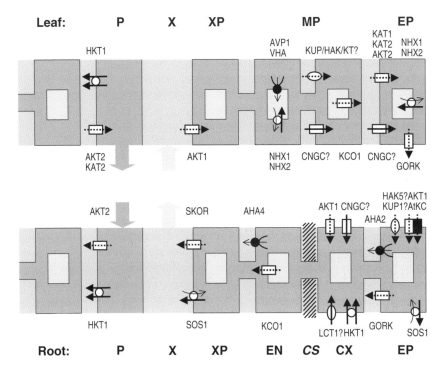

**Fig. 11.3** Overview of individual transporters involved in K$^+$ and Na$^+$ homeostasis. Channels are shown as boxes, pumps as filled circles, co-transporters as open circles, and transport systems of unknown mechanism as ellipses. Direction of ion transport is given by arrows. Selective K$^+$ transport is indicated with ⋯▶ arrows, and Na$^+$ transport or non-selective cation transport with ➔ arrows. H$^+$ transport is shown with → arrows. The following cell types are shown: P: phloem, X: xylem, XP: xylem parenchyma, EN: root endodermis, CS: Casparian strip, CX: root cortex, MP: leaf mesophyll, EP: epidermis (including guard cells and root hairs). Membranes are represented by single black lines; cytoplasmic space is indicated by dark grey; extracellular or vacuolar space by light grey. Only transporters for which genes have been identified are shown.

## 11.5   Regulation and integration of K$^+$ and Na$^+$ transport

In the previous sections we have shown that a large number of functionally diverse transport proteins are involved in the transport of K$^+$ and Na$^+$, either directly or in a supportive role. As the list of functionally characterised transporters increases, so does our wish to understand how these transporters are regulated and integrated in whole-plant ion homeostasis.

Regulatory events involved in plant responses to changes in K$^+$ supply and to salt stress can be divided into three fundamental steps: (i) perception of the external ion concentration and creation of a primary signal, (ii) signal transduction at the cellular level and (iii) signal transduction at the whole-plant level. Research into the perception of ionic stress is still in its infancy but progress has been made during recent years in identifying components of intracellular signalling events implicated in plant responses to salt stress. Limited knowledge is also available for the role of plant hormones in whole-plant signalling of ion stress. In this section we will concentrate on regulatory pathways linking external $K^+$ and $Na^+$ concentrations with ion transporters.

### 11.5.1   Perception of $K^+$ and $Na^+$

How plants sense external ion concentrations is a fundamental question that remains unsolved to date. Not only do we not know which are the receptors, we don't even know what the stimulus is. Are the ions themselves perceived or are physical or chemical factors accompanying changes in ion concentrations the main stimulus? For example, high $Na^+$ concentrations in the apoplast increase the ionic strength and thus change surface potentials; $Na^+$ replaces plasma-membrane-bound $Ca^{2+}$, resulting in decreased mechanical stability of the plasma membrane and salt-induced changes in the osmotic potential exert pressure on the membrane, which can be perceived by stretch-activated channels (Ramahaleo *et al.*, 1996). In contrast to salt stress, which occurs at high millimolar $Na^+$ concentrations, $K^+$ fluctuations in the soil are in the low millimolar or micromolar range. Therefore, $K^+$ stress will not cause major changes in ionic strength or water potential and is less likely to have profound effects on the physical properties of the plasma membrane.

The most likely candidates for receptors of ion stress are integral membrane proteins in the plasma membrane that change their biochemical or biophysical properties in response to $K^+$ or $Na^+$. Recognition of the stress at the receptor might either directly affect ion transport at the plasma membrane or produce a primary cytoplasmic signal that triggers further downstream responses. Good candidates for receptors are ion transporters themselves. In particular, voltage-dependent $K^+$ channels have been shown to change their gating properties and/or conductance in response to external $K^+$ (Blatt, 1988; Schroeder *et al.*, 1994; Blatt & Gradmann, 1997; Maathuis & Sanders, 1997), resulting in a change in net uptake or release of $K^+$ as well as in a change of the membrane potential. In the simplest model of a regulatory loop, $K^+$-dependent channels allow direct adaptation of $K^+$ transport to apoplastic $K^+$ concentrations. For example, some inward-rectifying $K^+$ channels shift their activation potential with $E_K$, thus preventing $K^+$ efflux through this channel type (Maathuis *et al.*, 1997). In a somewhat larger but still plasma-membrane-delimited regulatory loop, the plasma membrane proton pump takes the role of the primary receptor (Amtmann *et al.*, 1999). Indeed, direct activation of pump activity by $K^+$ and $Na^+$ has been shown (Serrano, 1989; Hall & Williams, 1991; Ayala *et al.*, 1996) and would result in hyperpolarisation of the membrane potential and acidification of the apoplast, both of which will directly affect $K^+$ and $Na^+$ transport through ion channels and proton-coupled co-transporters. Circumstantial evidence supports the individual steps of such regulatory loop:

1) Activation of proton pumps by salt stress is positively correlated with salt tolerance (e.g. Niu *et al.*, 1993; Golldack & Dietz, 2001), in particular the effect is stronger in salt-tolerant species than in salt-sensitive species (Kefu *et al.*, 2003).

2) The cell wall pH is indeed clamped by proton pump activity (as shown by measurements of apoplastic pH after application of fusicoccin; Felle, 1998; Amtmann *et al.*, 1999).

3) Apoplastic pH regulates root $K^+$ channels (Zimmermann *et al.*, 1998) and the $K_m$ of this effect is in the physiological range of cell wall pH. In particular, apoplastic acidification has opposite effects on $K^+$-selective channels (up-regulated) and non-selective channels (inhibited) in roots (Amtmann *et al.*, 1999; Demidchik & Tester, 2002) and the combined effect should increase the overall $K^+/Na^+$ selectivity of channel-mediated cation uptake.

To prove that this chain of events is really occurring *in vivo*, one would have to simultaneously measure apoplastic pH and cation currents in response to salt application, which has not yet been achieved. In addition to proton pumps, other proteins might regulate the membrane potential and thus affect $Na^+$ and $K^+$ uptake ratios. A small hydrophobic membrane protein PMP3 was found to have such function in yeast and several homologues of PMP3 have been identified in plants (Navarre & Goffeau, 2000).

Although plasma-membrane-delimited regulatory events can make an important contribution to ion stress adaptation, it is clear that intracellular downstream events have to occur to include intracellular compartments (i.e. the vacuole) as well as transcriptional events in cellular ion homeostasis. In this case, receptors in the plasma membrane have to produce a cytoplasmic signal. In the simplest case the signal consists in changes of cytoplasmic concentrations of $K^+$ and $Na^+$, which are perceived by ion transporters either in the plasma membrane or in endosomal membrane. For example, in *Aster* sp. a rise of cytoplasmic $Na^+$ results in deactivation of inward-rectifying $K^+$ channels in the guard cell plasma membrane. This effect, which appears to be mediated by cytoplasmic $Ca^{2+}$, was observed in a halophytic species (*A. tripolium*) but not in its glycophytic relative *A. amellus* and might explain differences in stomatal behaviour of the two species during salt stress (Véry *et al.*, 1998). The adaptation to varying external $K^+$ is unlikely to be mediated by cytoplasmic $K^+$ since cytoplasmic $K^+$ concentrations have been shown to remain stable over a wide range of external $K^+$ concentrations (Walker *et al.*, 1996a). Other plasma membrane transporters such as voltage-dependent plasma membrane $Ca^{2+}$ channels provide a possible link between membrane potential and the cytoplasmic $Ca^{2+}$ concentration. Cytoplasmic signals could also be produced by non-transporter-type receptors such as G-protein-coupled receptors (Hooley, 1998) or membrane-anchored cell wall proteins (Showalter, 2001). Research into ion stress perception has still a long way to go, especially with respect to $K^+$.

### 11.5.2 Intracellular signalling of cation stress

#### 11.5.2.1 Cytoplasmic Ca²⁺, kinases and phosphatases

Cytoplasmic $Ca^{2+}$ is a ubiquitous second messenger in plant cells, which plays a central role in many responses to environmental stimuli. An increase of cytoplasmic $Ca^{2+}$ has been measured after salt application (Knight *et al.*, 1997). Whether cytoplasmic $Ca^{2+}$ also changes in response to external $K^+$ is not known. An important target of cytoplasmic $Ca^{2+}$ signals is $Ca^{2+}$-dependent protein phosphatases such as calcineurin. Several calcineurin-related genes were identified on microarrays as being responsive to $K^+$ stress (Armengaud et al., in press) and we therefore predict that $Ca^{2+}$ signalling does plays an important role in cellular $K^+$ homeostasis. Calcineurin is a salt-tolerance determinant in yeast where it induces expression of the plasma membrane $Na^+$-ATPase ENA1 via a zinc finger transcription factor CRZ1/HAL8 (Serrano et al., 1999). Interestingly, in yeast, Calcineurin also regulates $K^+$ uptake during salt stress, probably through phosphorylation of TRK1, a high-affinity $K^+$-uptake system (Mendoza et al., 1994). In a screen for salt oversensitivity in A. thaliana, Liu and Zhu (1998) identified a Calcineurin-related protein SOS3, which is myristoylated and contains three EF hands for $Ca^{2+}$ binding (Ishitani et al., 2000). Unlike Calcineurin, SOS3 is not a phophatase but interacts physically with the serine/threonine protein kinase SOS2 (Halfter et al., 2000). The SOS2/SOS3 protein kinase complex is a central element of salt signalling in A. thaliana and controls ion homeostasis probably through transcriptional or post-transcriptional regulation of ion transporters (Xiong et al., 2002). Indeed, salt-induced up-regulation of the plasma membrane $Na^+/H^+$ antiporter SOS1 is reduced in sos2 and sos3 knockout mutants (Shi et al., 2000). The vacuolar $Na^+/H^+$ antiporter NHX1 and the vacuolar $Ca^{2+}/H^+$ antiporter CAX1 are also targets of SOS2 regulation, but in these cases SOS2 acts independently of SOS3 (Cheng et al., 2004; Qiu et al., 2004). Thus, it is likely that the SOS pathway coordinates plasma membrane and vacuolar $Na^+$ transport and links $Na^+$ and $Ca^{2+}$ homeostasis. The former is reflected in the fact that NHX1 is up-regulated in sos1 mutants (Yokoi et al., 2002). Furthermore, the SOS signalling pathway might also play a role in $K^+$ homeostasis. For example, sos1 mutants are not only salt hypersensitive but also deficient in $K^+$ uptake, although SOS1 does not transport $K^+$ (Shi et al., 2000). The fact that both SOS2 and SOS3 are members of large gene families opens the exciting possibility that other members of these families also interact with ion transporters, thus forming specific signalling triplets consisting each in a $Ca^{2+}$-binding protein, a kinase and a target transporter (Cheng et al., 2004).

Interestingly, SOS2 interacts with ABI2, a protein phosphatase that acts downstream of ABA, suggesting that the SOS module is connected with ABA signalling pathways (Xiong & Zhu, 2002b). Several other protein phosphatases as well as MAP kinases have been implicated in plant salt tolerance (Xiong & Zhu, 2002b) and $K^+$ transport (Véry & Sentenac, 2003). For example, the A. thaliana phosphatase AtPP2CA was found to interact with the phloem-expressed $K^+$-channel AKT2 (Vranova et al., 2001; Cherel et al., 2002) and changes both current and rectification properties of this channel. As in the case of SOS2, AKT2 regulation might be linked to ABA as AtPP2CA is up-regulated by ABA (Cherel et al., 2002).

### 11.5.2.2 Cyclic nucleotides
Shaker-type channels and channels of the CNGC family contain potential cyclic-nucleotide binding regions in their C-terminal sequences. A link between cyclic

nucleotide signalling and salt tolerance has been suggested by Maathuis and Sanders (2001), who found that addition of membrane permeable cyclic nucleotides increased salt tolerance in *A. thaliana* seedlings. These findings were in accordance with the observation that a $Na^+$-permeable voltage-independent channel in the root cortex of *A. thaliana* was inhibited by cyclic nucleotides. Animal CNGCs are activated rather than inhibited by cyclic nucleotides and at least two members of the *A. thaliana* CNGC family are also activated by cyclic nucleotides when expressed in heterologous systems (Leng *et al.*, 2002). At this stage it is not clear whether the observed effect on salt tolerance by cyclic nucleotides is indeed due to direct channel regulation by cyclic nucleotides. Detailed characterisation of all members of the CNGC family in *A. thaliana* mutant lines is required to prove a role of CNGCs in salt sensitivity (Talke *et al.*, 2003). For detailed description of cyclic-nucleotide-gated channels, see Chapter 7 of this volume.

### 11.5.2.3    *Other regulators of ion transport*

In animal cells, oxydoreductases (so-called beta subunits) interacts with Shaker-type $K^+$ channels and modulate their transport characteristics. Several homologues of beta subunits have been identified in plant genomes, and it has been confirmed *in vitro* that they can interact with plant Shaker channels (Tang *et al.*, 1996). However to date, target channels and the exact mode of action have not been characterised *in vivo*.

Regulation of the proton pump by 14-3-3 proteins is well established and has been studied at the molecular level (e.g. Fuglsang *et al.*, 2003). Thus, 14-3-3 proteins could regulate ion transport via the pump activity. Furthermore, there is evidence for direct regulation of $K^+$ channels by 14-3-3 proteins both at the plasma membrane (Saalbach *et al.*, 1997; Booij *et al.*, 1999) and at the tonoplast (van den Wijngaard *et al.*, 2001; De Boer, 2002). 14-3-3 proteins target phosphorylated proteins and therefore phosphatases/kinases must be part of putative signalling pathways involving 14-3-3 proteins.

It should be pointed out that although many cytoplasmic regulatory elements have been found to affect $Na^+$ and $K^+$ transporters, in none of the cases it is known whether and how this regulation is linked to the perception of ionic stress at the plasma membrane. To date, the SOS module is by far the best studied regulatory pathway where a known *early event* in the response to salt stress (a rise in cytoplasmic $Ca^{2+}$) has been linked to the *modulation of ion transporters* (SOS1) via *cytoplasmic signalling components* including a $Ca^{2+}$-binding protein (SOS3) and a kinase (SOS2).

### 11.5.3    *Hormonal control of ion homeostasis*

In order to achieve ion homeostasis at the whole-plant level, cellular signals need to be integrated between different tissues. To prevent damage of particularly sensitive tissues one would expect that plants do not rely on individual cells in each tissue to react to potentially harmful ion conditions in the apoplast but instead have a regulatory system in place that transmits information on ion contents between different tissues. For example, we found that many shoot genes show significant expression changes within 2–6 h of $K^+$ resupply after starvation, although no significant change in shoot $K^+$ concentration was recorded during this period of time (Armengaud *et al.*, in press). How do plants register and integrate tissue ion status? We can assume that phytohormones have an important function in this process.

### 11.5.3.1  Abscisic acid

The plant hormone abscisic acid (ABA) is involved in responses to many abiotic stresses. Application of salt increases the level of endogenous ABA, and application of exogenous ABA can increase salt tolerance (Xiong & Zhu, 2002a). Several transcription factors have been identified as participating in ABA-dependent transcriptional regulation of target genes (Shinozaki & Yamaguchi-Shinozaki, 2000), many of which have been implicated in salt tolerance (Rock, 2000). In *M. crystallinum*, the vacuolar ATPase is up-regulated by ABA (Barkla *et al.*, 1999) and in *A. thaliana* salt-induced proline accumulation can be triggered by ABA through up-regulation of *AtPCS1* (Savouré *et al.*, 1997). ABA signalling is also an important component of plant responses to drought stress and is directly linked to both $K^+$-channel activity in the guard cell plasma membrane (Blatt, 1990; Blatt & Armstrong, 1993; Armstrong *et al.*, 1995; Zhu *et al.*, 2002) and abundance of outward-rectifying $K^+$ channels in the plasma membrane of root stelar cells (Gaymard *et al.*, 1998; Roberts, 1998). A role of ABA in plant responses to external $K^+$ has not been established, i.e. it is not known whether endogenous ABA levels change in response to $K^+$-deprivation or resupply. Pilot *et al.* (2003) investigated the transcriptional response of all Shaker-type $K^+$ channels to various environmental factors and hormones and found that although several channels were regulated by ABA and salt stress, none of them showed transcriptional regulation by $K^+$ deprivation. We have recently carried out a comprehensive microarray analysis of regulation of more than 1000 putative membrane transporters by ABA and external $K^+$ (C. Blake, P. Armengaud & A. Amtmann, unpublished results, 2004). Whilst more than 200 transporters were regulated by ABA, only a small number of transporters were affected by external $K^+$ supply. Approximately half of the $K^+$-dependent transporter genes were also regulated by ABA, indicating that probably both ABA-dependent and ABA-independent signalling pathways operate in $K^+$ signalling. Interestingly, $K^+$ deprivation modulated the response of several transporters to ABA, including aquaporins and V-type ATPases (C. Blake, P. Armengaud & A. Amtmann, unpublished results, 2004).

### 11.5.3.2  Jasmonic acid and polyamines

New evidence from our laboratory indicates that another plant hormone, jasmonic acid (JA), plays a central role in plant adaptation to $K^+$ deficiency (Armengaud *et al.*, in press). JA is a well-characterised signalling component of plant responses to pathogen attack and wounding (Turner *et al.*, 2002). Endogenous levels of JA also increase in response to several abiotic stresses including salinity. In a genome-wide microarray study, we found that a significant number of JA biosynthesis genes and known JA-responsive genes were up-regulated in $K^+$-starved *A. thaliana* seedlings and quickly down-regulated after resupply of $K^+$ in the growth medium (Armengaud, *et al.*, in press). Several JA biosynthesis and JA responsive genes have also been reported to be up-regulated by sulphur starvation (Hirai *et al.*, 2003; Niki-forova *et al.*, 2003), suggesting that JA might play a general role in plant adaptation to nutrient deficiency. Although the physiological role of JA in $K^+$ stress remains to be elucidated, these findings provide us for the first time with tools to study the involvement of known signalling pathways in $K^+$ nutrition. In particular, it should be interesting to study nutrient management in JA mutants (Moons *et al.*, 1997) and to investigate potential cross-talk between pathogen and nutrient signalling.

JA itself and metabolites induced by JA have been implicated in the regulation of ion channels. For example, $K^+$ currents in guard cells are modulated by methyl jasmonate (Evans, 2003). This effect might be mediated by cytoplasmic alkalinisa-tion (Gehring *et al.*, 1997) or cytoplasmic $Ca^{2+}$ (Leon *et al.*, 1998; Kenton *et al.*, 1999). Interaction of polyamines with ion channels are well documented in animals (Williams, 1997). Polyamine production is JA-dependent and our microarray study showed that several genes involved in polyamine biosynthesis were up-regulated by $K^+$ starvation (e.g. *ADC2* encoding arginine decarboxylase; cf. Watson & Malmberg, 1996). Elevated levels of the polyamine putrescine have long been known to be a typical feature of $K^+$-deficient plants (Coleman & Richards, 1956; Young & Galston, 1984) whereas the polyamines spermidine and spermine appear to have a protective role in salt-stressed plants (e.g. rice; Chattopadhayay *et al.*, 2002). The latter effect could partly be ascribed to reduced uptake of $Na^+$. Patch clamp experi-ments on barley mesophyll cells and *V. faba* guard cells revealed spermidine-induced blockage of the vacuolar FV channel and the plasma membrane $K^+$ inward rectifier respectively (Brüggemann *et al.*, 1998; Liu *et al.*, 2000). Whether putrescine has an effect on plant ion channels has not been studied to date. It should be particularly interesting to investigate whether putrescine has an effect on ion channels involved in $K^+$ redistribution under $K^+$-deficient conditions.

## 11.6 Future prospects

### 11.6.1 Technologies

Mutant characterisation has been proven to be a useful tool to study the role of trans-porters in $K^+$ nutrition and $Na^+$ stress *in planta* and we can expect that this trend will continue. A screen for increased salt tolerance of 30 000 *A. thaliana* activation tag lines has recently been accomplished and several salt-tolerant mutant lines have been identified (G. Price & P. Dominy, unpublished results, 2004). These are now being characterised fully; some of the mutants show disruptions in novel positive regulators of salt tolerance in plants, whilst others carry disruptions in components of the ABA-sensitive and ABA-insensitive signalling pathways. Mapping of lines with altered tissue ion contents (Lahner *et al.*, 2003), in particular $K^+$, is also eagerly awaited as it might reveal first genes with function in perception and signalling of $K^+$. It appears that the perception of external $K^+$ supply and tissue $K^+$ status is a particularly understudied area of research. Microarray technology allows us to make a fresh start here. In our laboratory we have recently completed a comprehensive analysis of early transcriptional responses to $K^+$ using *A. thaliana* full-genome arrays and we have identified a large number of genes that responded quickly and selectively to a change in external $K^+$ supply (Armengaud *et al.*, in press). The identified genes can now be used as markers of $K^+$ perception and exploited for identification of upstream elements, for example in expression screens of mutagenised promoter-reporter lines.

Microarray analysis has already been useful for pointing out co-regulation of transporters with respect to different environmental stresses (Maathuis *et al.*, 2003). As the number of experimental conditions evaluated on microarrays increases, clusters of co-regulated transporters will emerge. Although gene expression is only one of many levels at which transporter activity is controlled, expression clusters will provide useful frameworks of regulatory networks, which subsequently can be complemented with other players identified, for example, in two-hybrid screens. In the early days of microarray technology, there was a general worry that transporters might be too weakly expressed and regulated to be detected on microarrays. However, this concern was not justified. An oligonucleotide array containing 50mer probes for more that 1000 putative membrane transporters showed significant signals for 90% of probes on the array although only root tissue was analysed (Maathuis *et al.*, 2003). Unfortunately, genome-wide analysis of membrane transporters at the protein level is still problematic due to difficulties to extract proteins from their membrane environment (Adessi *et al.*, 1997; Barbier-Brygoo *et al.*, 2001). However, progress has been made to develop proteomics protocols for membrane proteins (Seigneurin-Berny *et al.*, 1999; Santoni *et al.*, 2000). The same holds for structural analysis of membrane proteins, with an increasing number of crystal structures now available (Ben-Shem *et al.*, 2003; Jiang *et al.*, 2003).

## 11.6.2   Model plants

The majority of evidence collected in this review derives from *A. thaliana* because of its particular suitability for molecular and genetics methods. Progress has been made to establish model crop species such as rice, barley and tomato. On the environmental side, halophytic model systems are required to move from studying stress responses of salt-sensitive species towards investigating successful long-term survival strategies of salt-tolerant plants. *M. crystallinum* has been established as a halophytic model (Bohnert & Cushman, 2000) and we have presented here several recent findings on transporters in this species. *T. halophila* (salt cress) has recently emerged as a useful halophytic model. *T. halophila* is a close relative of *A. thaliana* and therefore allows direct comparison between a glycophyte and a halophyte (Bressan *et al.*, 2001). It appears that salt tolerance in *T. halophila* is to a large part due to $K^+/Na^+$ management and first results point to voltage-independent root ion channels as a crucial determinant of differential $K^+/Na^+$ homeostasis in the two species (Volkov *et al.*, 2004). The identification of genes encoding this type of channel is one of the big challenges that lie ahead. The fact that *A. thaliana* and *T. halophila* are both accessible to transformation techniques opens the exciting possibility to exchange genes between the two species and thus study glycophyte transporters in a halophytic background and vice versa.

## 11.7 Concluding remarks

It is clear that the combination of different experimental approaches has been extremely useful for investigating the role of ion transporters in nutrition and abiotic stress. Traditional approaches to study ion transport, such as measurement of tissue solute content, isotope fluxes, membrane potential and electric currents, have obtained new value in the analysis of mutant lines and heterologous expression systems, and automisation of these techniques has created new tools for mutant screening and high-throughput analysis (e.g. Lahner *et al.*, 2003). In many laboratories, electrophysiologists, biochemists, molecular biologists and geneticists work hand in hand to elucidate the function of a transporter and usually one researcher works with several different techniques. This is not only beneficial for the long-term future of scientific training but also allows questions to be approached from many different angles. Considering the complexity of plant ion transport with respect to functional diversification of single genes and networking of their regulation, it is clear that a truly multidisciplinary programme is the only way to make progress in understanding the integration of ion transport in response to nutrient deficiency and salinity.

In this review we have tried (i) to familiarise the reader with the physiological and electrochemical environment in which ion transporters operate and (ii) to integrate the vast amount of data available on $K^+$ and $Na^+$ transporters by (iii) assigning each transporter a place and a role within the response of the whole plant to $K^+$ and $Na^+$ stress. In this attempt we have not only been hampered by our limited mental capacities but also by the fact that the existing evidence – although impressive – is still very patchy. In many cases, patchiness results from the lack of knowledge concerning a few very basic parameters such as membrane localisation of the transporters, membrane potential and ion concentrations of the expressing cell type or primary signals. Considering how difficult and time-consuming it is, for example, to unequivocally link mutation and phenotype in a knockout line or to fully characterise a single transporter by heterologous expression, it is clear that painting the picture of ion transport in $K^+$ nutrition and salinity has to be a collective effort. Unfortunately, within this collective effort some tasks are more rewarding than others and therefore certain awkward and low-profile jobs remain unattended (e.g. measurement of membrane potentials). Critical analysis of missing links is therefore crucial and we have tried to make the reader aware of shortcomings in the existing experimental evidence. During the writing of this review we also became aware of a potential pitfall in trying to make 'sense' from data that is not yet ready to reveal the truth. The demand of several journals to assign function to a gene rather than merely collect descriptive data is intended to force scientists to produce 'meaningful' data, but unfortunately also forces them to oversimplify the interpretation of their findings. Clear-cut functional labels provided in papers come handy when writing reviews but carry the danger to produce a heritage of pseudo-knowledge. In future, one could imagine to start the review process from the entire set of accumulated raw data and use computerised network analysis to extract biological meaning in a completely unbiased approach. This requires storage of raw data in well-organised databases. Current efforts to build databases of gene expression data, protein location and metabolic pathways (e.g. TAIR) are pointing in the right direction. Such a data pool could not only be questioned from different angles but also be constantly reanalysed and reinterpreted as we progress in our research.

# References

Adessi, C., Miege, C., Albrieux, C. & Rabilloud, T. (1997) Two-dimensional electrophoresis of membrane proteins: a current challenge for immobilized pH gradients. *Electrophoresis*, **18**, 127–135.

Ahn, S.J., Shin, R. & Schachtman, D.P. (2004) Expression of KT/KUP genes in *Arabidopsis* and the role of root hairs in $K^+$ uptake. *Plant Physiol.*, **134**, 1135–1145.

Allen, G.J. & Sanders, D. (1995) Calcineurin, a type 2B protein phosphatase, modulates the $Ca^{2+}$-permeable slow vacuolar ion channel of stomatal guard cells. *Plant Cell*, **7**, 1473–1483.

Allen, G.J. & Sanders, D. (1996) Control of ionic currents in guard cell vacuoles by cytosolic and luminal calcium. *Plant J.*, **10**, 1055–1069.

Allen, G.J. & Sanders, D. (1997) Vacuolar ion channels of higher plants. *Adv. Bot. Res.*, **25**, 217–252.

Amtmann, A., Fischer, M., Marsh, E.L., Stefanovic, A., Sanders, D. & Schachtman, D.P. (2001) The wheat cDNA LCT1 generates hypersensitivity to sodium in a salt-sensitive yeast strain. *Plant Physiol.*, **126**, 1061–1071.

Amtmann, A., Jelitto, T.C. & Sanders, D. (1999) $K^+$-selective inward-rectifying channels and apoplastic pH in barley roots. *Plant Physiol.*, **120**, 331–338.

Amtmann, A., Laurie, S., Leigh, R. & Sanders, D. (1997) Multiple inward channels provide flexibility in $Na^+/K^+$ discrimination at the plasma membrane of barley suspension culture cells. *J. Exp. Bot.*, **48**, 481–497.

Amtmann, A. & Sanders, D. (1999) Mechanisms of $Na^+$ uptake by plant cells. *Adv. Bot. Res.*, **29**, 75–112.

Anderson, J.A., Huprikar, S.S., Kochian, L.V., Lucas, W.J. & Gaber, R.F. (1992) Functional expression of a probable *Arabidopsis thaliana* potassium channel in *Saccharomyces cerevisiae*. *Proc. Natl. Acad. Sci. U.S.A.*, **89**, 3736–3740.

Aon, M.A., Cortassa, S., Gomez-Casati, D.F. & Iglesias, A.A. (2000) Effects of stress on cellular infrastructure and metabolic organization in plant cells. *Int. Rev. Cytol.*, **194**, 239–273.

Apse, M.P., Aharon, G.S., Snedden, W.A. & Blumwald, E. (1999) Salt tolerance conferred by overexpression of a vacuolar $Na^+/H^+$ antiport in *Arabidopsis*. *Science*, **285**, 1256–1258.

Apse, M.P., Sottosanto, J.B. & Blumwald, E. (2003) Vacuolar cation/$H^+$ exchange, ion homeostasis, and leaf development are altered in a T-DNA insertional mutant of AtNHX1, the *Arabidopsis* vacuolar $Na^+/H^+$ antiporter. *Plant J.*, **36**, 229–239.

Arabidopsis Genome Initiative, T. (2000) Analysis of the genome sequence of the flowering plant *Arabidopsis thaliana*. *Nature*, **408**, 796–815.

Armengaud, P., Breitling, R. and Amtmann, A. (in press) The potassium-dependent transcriptome of *Arabidopsis thaliana* reveals a prominent role of jasmonic acid in nutrient signaling. *Plant Physiol.*

Armstrong, F., Leung, J., Grabov, A., Brearley, J., Giraudat, J. & Blatt, M.R. (1995) Sensitivity to abscisic acid of guard-cell K$^+$ channels is suppressed by abi1-1, a mutant *Arabidopsis* gene encoding a putative protein phosphatase. *Proc. Natl. Acad. Sci. U.S.A.*, **92**, 9520–9524.

Axelsen, K.B. & Palmgren, M.G. (2001) Inventory of the superfamily of P-type ion pumps in *Arabidopsis*. *Plant Physiol.*, **126**, 696–706.

Ayala, F., Oleary, J.W. & Schumaker, K.S. (1996) Increased vacuolar and plasma membrane H$^+$-ATPase activities in *Salicornia bigelovii* Torr in response to NaCl. *J. Exp. Bot.*, **47**, 25–32.

Balague, C., Lin, B., Alcon, C., *et al.* (2003) HLM1, an essential signaling component in the hypersensitive response, is a member of the cyclic nucleotide-gated channel ion channel family. *Plant Cell*, **15**, 365–379.

Baluska, F., Samaj, J., Wojtaszek, P., Volkmann, D. & Menzel, D. (2003) Cytoskeleton-plasma membrane-cell wall continuum in plants. Emerging links revisited. *Plant Physiol.*, **133**, 482–491.

Banuelos, M.A., Garciadeblas, B., Cubero, B. & Rodriguez-Navarro, A. (2002) Inventory and functional characterization of the HAK potassium transporters of rice. *Plant Physiol.*, **130**, 784–795.

Banuelos, M.A., Klein, R.D., Alexander-Bowman, S.J. & Rodriguez-Navarro, A. (1995) A potassium transporter of the yeast *Schwanniomyces occidentalis* homologous to the Kup system of *Escherichia coli* has a high concentrative capacity. *EMBO J.*, **14**, 3021–3027.

Barbier-Brygoo, H., Gaymard, F., Rolland, N. & Joyard, J. (2001) Strategies to identify transport systems in plants. *Trends Plant Sci.*, **6**, 577–585.

Barkla, B.J. & Pantoja, O. (1996) Physiology of ion transport across the tonoplast of higher plants. *Annu. Rev. Plant Phys. Plant Mol. Biol.*, **47**, 159–184.

Barkla, B.J., Vera-Estrella, R., Maldonado-Gama, M. & Pantoja, O. (1999) Abscisic acid induction of vacuolar H$^+$-ATPase activity in *Mesembryanthemum crystallinum* is developmentally regulated. *Plant Physiol.*, **120**, 811–820.

Benito, B. & Rodriguez-Navarro, A. (2003) ) Molecular cloning and characterization of a sodium-pump ATPase of the moss *Physcomitrella patens*. *Plant J.*, **36**, 382–389.

Ben-Shem, A., Frolow, F. & Nelson, N. (2003) Crystal structure of plant photosystem I. *Nature*, **426**, 630–635.

Berthomieu, P., Conejero, G., Nublat, A., *et al.* (2003) Functional analysis of AtHKT1 in *Arabidopsis* shows that Na$^+$ recirculation by the phloem is crucial for salt tolerance. *EMBO J.*, **22**, 2004–2014.

Blatt, M.R. (1988) Potassium-dependent, bipolar gating of K$^+$ channels in guard cells. *J. Membr. Biol.*, **102**, 235–246.

Blatt, M.R. (1990) Potassium channel currents in intact stomatal guard cells: rapid enhancement by abscisic acid. *Planta*, **180**, 445–455.

Blatt, M.R. (1992) K$^+$ channels of stomatal guard-cells – characteristics of the inward rectifier and its control by pH. *J. Gen. Physiol.*, **99**, 615–644.

Blatt, M.R. (2000) Cellular signaling and volume control in stomatal movements in plants. *Annu. Rev. Cell Dev. Biol.*, **16**, 221–241.

Blatt, M.R. & Armstrong, F. (1993) K$^+$ channels of stomatal guard-cells – abscisic-acid-evoked control of the outward rectifier mediated by cytoplasmic pH. *Planta*, **191**, 330–341.

Bohnert, H.J. & Cushman, J.C. (2000) The ice plant cometh: lessons in abiotic stress tolerance. *J. Plant Growth Regul.*, **19**, 334–346.

Blatt, M.R. & Gradmann, D. (1997) K$^+$-sensitive gating of the K$^+$ outward rectifier in *Vicia* guard cells. *J. Membr. Biol.*, **158**, 241–256.

Booij, P.P., Roberts, M.R., Vogelzang, S.A., Kraayenhof, R. & De Boer, A.H. (1999) 14-3-3 proteins double the number of outward-rectifying K$^+$ channels available for activation in tomato cells. *Plant J.*, **20**, 673–683.

Bressan, R.A., Zhang, C., Zhang, H., Hasegawa, P.M., Bohnert, H.J. & Zhu, J.K. (2001) Learning from the *Arabidopsis* experience. The next gene search paradigm. *Plant Physiol.*, **127**, 1354–1360.

Briskin, D.P. & Gawienowski, M.C. (1996) Role of the plasma membrane H$^+$-ATPase in K$^+$ transport. *Plant Physiol.*, **111**, 1199–1207.

Brüggemann, L., Dietrich, P., Becker, D., Dreyer, I.I., Palme, K. & Hedrich, R. (1999) Channel-mediated high-affinity K$^+$ uptake into guard cells from *Arabidopsis. Proc. Natl. Acad. Sci. U.S.A.*, **96**, 3298–3302.

Brüggemann, L.I., Pottosin, I.I. & Schonknecht, G. (1998) Cytoplasmic polyamines block the fast-activating vacuolar cation channel. *Plant J.*, **16**, 101–105.

Buschmann, P.H., Vaidyanathan, R., Gassmann, W. & Schroeder, J.I. (2000) Enhancement of Na$^+$ uptake currents, time-dependent inward-rectifying K$^+$ channel currents, and K$^+$ channel transcripts by K$^+$ starvation in wheat root cells. *Plant Physiol.*, **122**, 1387–1397.

Cammack, J.N. & Schwartz, E.A. (1996) Channel behavior in a gamma-aminobutyrate transporter. *Proc. Natl. Acad. Sci. U.S.A.*, **93**, 723–727.

Carden, D.E., Walker, D.J., Flowers, T.J. & Miller, A.J. (2003) Single-cell measurements of the contributions of cytosolic Na$^+$ and K$^+$ to salt tolerance. *Plant Physiol.*, **131**, 676–683.

Chattopadhayay, M.K., Tiwari, B.S., Chattopadhyay, G., Bose, A., Sengupta, D.N. & Ghosh, B. (2002) Protective role of exogenous polyamines on salinity-stressed rice (*Oryza sativa*) plants. *Physiol. Plant.*, **116**, 192–199.

Chauhan, S., Forsthoefel, N., Ran, Y., Quigley, F., Nelson, D.E. & Bohnert, H.J. (2000) Na$^+$/myo-inositol symporters and Na$^+$/H$^+$-antiport in *Mesembryanthemum crystallinum. Plant J.*, **24**, 511–522.

Cheng, N.H., Pittman, J.K., Zhu, J.K. & Hirschi, K.D. (2004) The protein kinase SOS2 activates the *Arabidopsis* H$^+$/Ca$^{2+}$ antiporter CAX1 to integrate calcium transport and salt tolerance. *J. Biol. Chem.*, **279**, 2922–2926.

Cherel, I., Michard, E., Platet, N., *et al.* (2002) Physical and functional interaction of the Arabidopsis K$^+$ channel AKT2 and phosphatase AtPP2CA. *Plant Cell*, **14**, 1133–1146.

Chrispeels, M.J., Morillon, R., Maurel, C., Gerbeau, P., Kjellbom, P. & Johansson, I. (2001) Aquaporins of plants: structure, function, regulation, and role in plant water relations. *Aquaporins*, **51**, 277–334.

Clemens, S., Antosiewicz, D.M., Ward, J.M., Schachtman, D.P. & Schroeder, J.I. (1998) The plant cDNA LCT1 mediates the uptake of calcium and cadmium in yeast. *Proc. Natl. Acad. Sci. U.S.A.*, **95**, 12043–12048.

Coleman, R.G. & Richards, F.J. (1956) Physiological studies in plant nutrition, XVIII: Some aspects of nitrogen metabolism in barley and others plants in relation to potassium deficiency. *Ann. Bot.*, **20**, 393–409.

Cuin, T.A., Miller, A.J., Laurie, S.A. & Leigh, R.A. (2003) Potassium activities in cell compartments of salt-grown barley leaves. *J. Exp. Bot.*, **54**, 657–661.

Czempinski, K., Frachisse, J.M., Maurel, C., Barbier-Brygoo, H. & Mueller-Roeber, B. (2002) Vacuolar membrane localization of the *Arabidopsis* 'two-pore' K$^+$ channel KCO1. *Plant J.*, **29**, 809–820.

Czempinski, K., Zimmermann, S., Ehrhardt, T. & Muller-Rober, B. (1997) New structure and function in plant K$^+$ channels: KCO1, an outward rectifier with a steep Ca$^{2+}$ dependency. *EMBO J.*, **16**, 2565–2575.

Davenport, R. (2002) Glutamate receptors in plants. *Ann. Bot.*, **90**, 549–557.

Davenport, R.J. & Tester, M. (2000) A weakly voltage-dependent, nonselective cation channel mediates toxic sodium influx in wheat. *Plant Physiol.*, **122**, 823–834.

Davies, J.M. & Sanders, D. (1995) ATP, pH and Mg$^{2+}$ modulate a cation current in *Beta vulgaris* vacuoles: a possible shunt conductance for the vacuolar H$^+$-ATPase. *J. Membr. Biol.*, **145**, 75–86.

De Boer, A.H. (2002) Plant 14-3-3 proteins assist ion channels and pumps. *Biochem. Soc. Trans.*, **30**, 416–421.

Demidchik, V. & Tester, M. (2002) Sodium fluxes through nonselective cation channels in the plasma membrane of protoplasts from *Arabidopsis* roots. *Plant Physiol.*, **128**, 379–387.

Dietrich, P., Sanders, D. & Hedrich, R. (2001) The role of ion channels in light-dependent stomatal opening. *J. Exp. Bot.*, **52**, 1959–1967.

Drew, M.C. & Saker, L.R. (1984) Uptake and long-distance transport of phosphate, potassium and chloride in relation to internal ion concentrations in barley. Evidence for a non-allosteric regulation. *Planta*, **160**, 500–507.

Durell, S.R. & Guy, H.R. (1999) Structural models of the KtrB, TrkH, and Trk1,2 symporters based on the structure of the KcsA K$^+$ channel. *Biophys J.*, **77**, 789–807.

Durell, S.R., Hao, Y., Nakamura, T., Bakker, E.P. & Guy, H.R. (1999) Evolutionary relationship between K$^+$ channels and symporters. *Biophys. J.*, **77**, 775–788.

Elumalai, R.P., Nagpal, P. & Reed, J.W. (2002) A mutation in the Arabidopsis KT2/KUP2 potassium transporter gene affects shoot cell expansion. *Plant Cell*, **14**, 119–131.

Elzenga, J.T.M., Prins, H.B.A. & Van Volkenburgh, E. (1995) Light-induced membrane-potential changes of epidermal and mesophyll-cells in growing leaves of *Pisum sativum*. *Planta*, **197**, 127–134.

Elzenga, J.T.M. & Van Volkenburgh, E. (1994) Characterization of ion channels in the plasma-membrane of epidermal-cells of expanding pea (*Pisum sativum Arg*) Leaves. *J. Membr. Biol.*, **137**, 227–235.

Epstein, I. (1973) Mechanisms of ion transport through plant cell membranes. *Crit. Rev. Cytol.*, **34**, 123–168.

Epstein, I., Rains, D.W. & Elzam, O.E. (1963) Resolution of dual mechanisms of potassium absorption by barley roots. *Proc. Natl. Acad. Sci. U.S.A.*, **49**, 684–692.

Espinosa-Ruiz, A., Belles, J.M., Serrano, R. & Culianez-MacIa, F.A. (1999) *Arabidopsis thaliana* AtHAL3: a flavoprotein related to salt and osmotic tolerance and plant growth. *Plant J.*, **20**, 529–539.

Evans, N.H. (2003) Modulation of guard cell plasma membrane potassium currents by methyl jasmonate. *Plant Physiol.*, **131**, 8–11.

Felle, H.H. (1998) The apoplastic pH of the Zea mays root cortex as measured with pH-sensitive micro-electrodes: aspects of regulation. *J. Exp. Bot.*, **49**, 987–995.

Flowers, T.J., Troke, P.F. & Yeo, A. (1977) The mechanism of salt tolerance in halophytes. *Annu. Rev. Plant Phys. Plant Mol. Biol.*, **28**, 89–121.

Fu, H.H. & Luan, S. (1998) AtKuP1: a dual-affinity K$^+$ transporter from *Arabidopsis*. *Plant Cell*, **10**, 63–73.

Fuglsang, A.T., Borch, J., Bych, K., Jahn, T.P., Roepstorff, P. & Palmgren, M.G. (2003) The binding site for regulatory 14-3-3 protein in plant plasma membrane H$^+$-ATPase: involvement of a region promoting phosphorylation-independent interaction in addition to the phosphorylation-dependent C-terminal end. *J. Biol. Chem.*, **278**, 42266–42272.

Gaedeke, N., Klein, M., Kolukisaoglu, U., *et al.* (2001) The *Arabidopsis thaliana* ABC transporter AtMRP5 controls root development and stomata movement. *EMBO J.*, **20**, 1875–1887.

Garciadeblas, B., Benito, B. & Rodriguez-Navarro, A. (2001) Plant cells express several stress calcium ATPases but apparently no sodium ATPase. *Plant Soil*, **235**, 181–192.

Gassman, W., Rubio, F. & Schroeder, J.I. (1996) Alkali cation selectivity of the wheat root high-affinity potassium transporter HKT1. *Plant J.*, **10**, 869–852.

Gaxiola, R.A., Li, J., Undurraga, S., *et al.* (2001) Drought- and salt-tolerant plants result from overex-pression of the AVP1 H$^+$-pump. *Proc. Natl. Acad. Sci. U.S.A.*, **98**, 11444–11449.

Gaxiola, R.A., Rao, R., Sherman, A., Grisafi, P., Alper, S.L. & Fink, G.R. (1999) The *Arabidopsis thaliana* proton transporters, AtNhx1 and Avp1, can function in cation detoxification in yeast. *Proc. Natl. Acad. Sci. U.S.A.*, **96**, 1480–1485.

Gaymard, F., Cerutti, M., Horeau, C., *et al.* (1996) The baculovirus/insect cell system as an alternative to *Xenopus* oocytes. First characterization of the AKT1 K$^+$ channel from *Arabidopsis thaliana*. *J. Biol. Chem.*, **271**, 22863–22870.

Gaymard, F., Pilot, G., Lacombe, B., *et al.* (1998) Identification and disruption of a plant shaker-like outward channel involved in K$^+$ release into the xylem sap. *Cell*, **94**, 647–655.

Gehring, C.A., Irving, H.R., McConchie, R. & Parish, R.W. (1997) Jasmonates induce intracellular alkalinization and closure of Paphiopedilum guard cells. *Ann. Bot.*, **80**, 485–489.

Geiger, D., Becker, D., Lacombe, B. & Hedrich, R. (2002) Outer pore residues control the H$^+$ and K$^+$ sensitivity of the *Arabidopsis* potassium channel AKT3. *Plant Cell*, **14**, 1859–1868.

Gimmler, H. (2000) Primary sodium plasma membrane ATPases in salt-tolerant algae: facts and fictions. *J. Exp. Bot.*, **51**, 1171–1178.

Golldack, D. & Dietz, K.J. (2001) Salt-induced expression of the vacuolar $H^+$-ATPase in the common ice plant is developmentally controlled and tissue specific. *Plant Physiol.*, **125**, 1643–1654.

Golldack, D., Quigley, F., Michalowski, C.B., Kamasani, U.R. & Bohnert, H.J. (2003) Salinity stress-tolerant and -sensitive rice (*Oryza sativa* L.) regulate AKT1-type potassium channel transcripts differently. *Plant Mol. Biol.*, **51**, 71–81.

Gong, Z., Koiwa, H., Cushman, M.A., *et al.* (2001) Genes that are uniquely stress regulated in salt overly sensitive (sos) mutants. *Plant Physiol.*, **126**, 363–375.

Grabov, A. & Blatt, M.R. (1997) Parallel control of the inward-rectifier $K^+$ channel by cytosolic free $Ca^{2+}$ and pH in *Vicia* guard cells. *Planta*, **201**, 84–95.

Greenway, H. & Munns, R.A. (1980) Mechanisms of salt tolerance in non-halophytes. *Annu. Rev. Plant Phys. Plant Mol. Biol.*, **31**, 149–190.

Halfter, U., Ishitani, M. & Zhu, J.K. (2000) The *Arabidopsis* SOS2 protein kinase physically interacts with and is activated by the calcium-binding protein SOS3. *Proc. Natl. Acad. Sci. U.S.A.*, **97**, 3735–3740.

Hall, J.L. & Williams, L.E. (1991) Properties and functions of proton pumps in higher-plants. *Pestic. Sci.*, **32**, 339–351.

Hasegawa, M., Bressan, R., Zhu, J.K. & Bohnert, H.J. (2000) plant cellular and molecular responses to high salinity. *Annu. Rev. Plant Phys. Plant Mol. Biol.*, **51**, 463–499.

Hille, B. (2001) *Ion Channel of Excitable Membranes*, Sinauer Associates, Sunderland, MA.

Hirai, M.Y., Fujiwara, T., Awazuhara, M., Kimura, T., Noji, M. & Saito, K. (2003) Global expression profiling of sulfur-starved *Arabidopsis* by DNA macroarray reveals the role of *O*-acetyl-l-serine as a general regulator of gene expression in response to sulfur nutrition. *Plant J.*, **33**, 651–663.

Hirsch, R.E., Lewis, B.D., Spalding, E.P. & Sussman, M.R. (1998) A role for the AKT1 potassium channel in plant nutrition. *Science*, **280**, 918–921.

Hooley, R. (1998) Plant hormone perception and action: a role for G-protein signal transduction? *Philos. Trans. R. Soc. Lond. B Biol. Sci.*, **353**, 1425–1430.

Horie, T., Yoshida, K., Nakayama, H., Yamada, K., Oiki, S. & Shinmyo, A. (2001) Two types of HKT transporters with different properties of $Na^+$ and $K^+$ transport in *Oryza sativa*. *Plant J.*, **27**, 129–138.

Hoth, S., Dreyer, I., Dietrich, P., Becker, D., MullerRober, B. & Hedrich, R. (1997) Molecular basis of plant-specific acid activation of $K^+$ uptake channels. *Proc. Natl. Acad. Sci. U.S.A.*, **94**, 4806–4810.

Hua, B.G., Mercier, R.W., Leng, Q. & Berkowitz, G.A. (2003) Plants do it differently. A new basis for potassium/sodium selectivity in the pore of an ion channel. *Plant Physiol.*, **132**, 1353–1361.

Ilan, N., Schwartz, A. & Moran, N. (1996) External protons enhance the activity of the hyperpolarization-activated $K^+$ channels in guard cell protoplasts of *Vicia faba*. *J. Membr. Biol.*, **154**, 169–181.

Ishitani, M., Liu, J., Halfter, U., Kim, C.S., Shi, W. & Zhu, J.K. (2000) SOS3 function in plant salt tolerance requires N-myristoylation and calcium binding. *Plant Cell*, **12**, 1667–1678.

Ishitani, M., Majumder, A.L., Bornhouser, A., Michalowski, C.B., Jensen, R.G. & Bohnert, H.J. (1996) Coordinate transcriptional induction of myo-inositol metabolism during environmental stress. *Plant J.*, **9**, 537–548.

Jeschke, W.D. & Stelter, W. (1976) Measurment of longitudinal ion profiles in single root of *Hordeum* and *Atriplex* by use of flameless atomic absorption spectroscopy. *Planta*, **128**, 107–112.

Jiang, Y., Lee, A., Chen, J., *et al.* (2003) X-ray structure of a voltage-dependent $K^+$ channel. *Nature*, **423**, 33–41.

Karley, A.J., Leigh, R.A. & Sanders, D. (2000a) Differential ion accumulation and ion fluxes in the mesophyll and epidermis of barley. *Plant Physiol.*, **122**, 835–844.

Karley, A.J., Leigh, R.A. & Sanders, D. (2000b) Where do all the ions go? The cellular basis of differential ion accumulation in leaf cells. *Trends Plant Sci.*, **5**, 465–470.

Kato, Y., Sakaguchi, M., Mori, Y., *et al.* (2001) Evidence in support of a four transmembrane-pore-transmembrane topology model for the *Arabidopsis thaliana* $Na^+/K^+$ translocating AtHKT1 protein, a member of the superfamily of $K^+$ transporters. *Proc. Natl. Acad. Sci. U.S.A.*, **98**, 6488–6493.

Kefu, Z., Hai, F., San, Z. & Jie, S. (2003) Study on the salt and drought tolerance of *Suaeda salsa* and *Kalanchoe claigremontiana* under iso-osmotic salt and water stress. *Plant Sci.*, **165**, 837–844.

Kenton, P., Mur, L.A.J. & Draper, J. (1999) A requirement for calcium and protein phosphatase in the jasmonate-induced increase in tobacco leaf acid phosphatase specific activity. *J. Exp. Bot.*, **50**, 1331–1341.

Keunecke, M. & Hansen, U.P. (2000) Different pH-dependences of $K^+$ channel activity in bundle sheath and mesophyll cells of maize leaves. *Planta*, **210**, 792–800.

Kim, E.J., Kwak, J.M., Uozumi, N. & Schroeder, J.I. (1998) AtKUP1: an *Arabidopsis* gene encoding high-affinity potassium transport activity. *Plant Cell*, **10**, 51–62.

Kluge, C., Lahr, J., Hanitzsch, M., Bolte, S., Golldack, D. & Dietz, K.J. (2003) New insight into the structure and regulation of the plant vacuolar $H^+$-ATPase. *J. Bioenerg. Biomembr.*, **35**, 377–388.

Knight, H., Trewavas, A.J. & Knight, M.R. (1997) Calcium signalling in *Arabidopsis thaliana* responding to drought and salinity. *Plant J.*, **12**, 1067–1078.

Kourie, J. & Goldsmith, M.H.M. (1992) $K^+$ channels are responsible for an inwardly rectifying current in the plasma-membrane of mesophyll protoplasts of *Avena sativa*. *Plant Physiol.*, **98**, 1087–1097.

Lacombe, B., Pilot, G., Michard, E., Gaymard, F., Sentenac, H. & Thibaud, J.B. (2000) A shaker-like $K^+$ channel with weak rectification is expressed in both source and sink phloem tissues of *Arabidopsis*. *Plant Cell*, **12**, 837–851.

Laegreid, M., Bockman, O.C. & Kaarstad, O. (1999) *Agriculture, Fertilizers and the Environement*, CABI, Oxon, UK.

Lagarde, D., Basset, M., Lepetit, M., *et al.* (1996) Tissue-specific expression of *Arabidopsis AKT1* gene is consistent with a role in $K^+$ nutrition. *Plant J.*, **9**, 195–203.

Lahner, B., Gong, J., Mahmoudian, M., *et al.* (2003) Genomic scale profiling of nutrient and trace elements in *Arabidopsis thaliana*. *Nat. Biotech.*, **21**, 1215–1221.

Lang, A. (1983) Tugor-related translocation. *Plant Cell Env.*, **6**, 683–689.

Läuchli, A. & Lüttge, U. (2002) *Salinity: Environment–Plants–Molecules*, Kluwer Academic, Dordrecht, The Netherlands.

Lee, E.K., Kwon, M., Ko, J.H., *et al.* (2004) Binding of sulfonylurea by AtMRP5, an *Arabidopsis* multidrug resistance-related protein that functions in salt tolerance. *Plant Physiol.*, **134**, 528–538.

Leigh, R.A. & Jones, R.G.W. (1984) A hypothesis relating critical potassium concentrations for growth to the distribution and functions of this ion in the plant-cell. *New Phytol.*, **97**, 1–13.

Leng, Q., Mercier, R.W., Hua, B.G., Fromm, H. & Berkowitz, G.A. (2002) Electrophysiological analysis of cloned cyclic nucleotide-gated ion channels. *Plant Physiol.*, **128**, 400–410.

Leng, Q., Mercier, R.W., Yao, W. & Berkowitz, G.A. (1999) Cloning and first functional characterization of a plant cyclic nucleotide-gated cation channel. *Plant Physiol.*, **121**, 753–761.

Leon, J., Rojo, E., Titarenko, E. & Sanchez-Serrano, J.J. (1998) Jasmonic acid-dependent and -independent wound signal transduction pathways are differentially regulated by $Ca^{2+}$/calmodulin in *Arabidopsis thaliana*. *Mol. Gen. Genet.*, **258**, 412–419.

Leonhardt, N., Marin, E., Vavasseur, A. & Forestier, C. (1997) Evidence for the existence of a sulfonylurea-receptor-like protein in plants: modulation of stomatal movements and guard cell potassium channels by sulfonylureas and potassium channel openers. *Proc. Natl. Acad. Sci. U.S.A.*, **94**, 14156–14161.

Lin, F., Lester, H.A. & Mager, S. (1996) Single-channel currents produced by the serotonin transporter and analysis of a mutation affecting ion permeation. *Biophys J.*, **71**, 3126–3135.

Liu, J. & Zhu, J.K. (1998) A calcium sensor homolog required for plant salt tolerance. *Science*, **280**, 1943–1945.

Liu, K., Fu, H.H., Bei, Q.X. & Luan, S. (2000) Inward potassium channel in guard cells as a target for polyamine regulation of stomatal movements. *Plant Physiol.*, **124**, 1315–1325.

Liu, W., Fairbairn, D.J., Reid, R.J. & Schachtman, D.P. (2001) Characterization of two HKT1 homologues from *Eucalyptus camaldulensis* that display intrinsic osmosensing capability. *Plant Physiol.*, **127**, 283–294.

Maathuis, F. & Sanders, D. (1996a) Mechanisms of potassium absorption by higher plant roots. *Physiol. Plant.*, **96**, 158–168.

Maathuis, F. & Sanders, D. (1997) Regulation of $K^+$ absorption in plant root cells by external $K^+$: interplay of different $K^+$ transporters. *J. Exp. Bot.*, **48**, 451–458.

Maathuis, F., Verlin, D., Smith, F.A., Sanders, D., Fernandez, J.A. & Walker, N.A. (1996) The physiological relevance of $Na^+$-coupled $K^+$-transport. *Plant Physiol.*, **112**, 1609–1616.

Maathuis, F.J., Filatov, V., Herzyk, P., *et al.* (2003) Transcriptome analysis of root transporters reveals participation of multiple gene families in the response to cation stress. *Plant J.*, **35**, 675–692.

Maathuis, F.J., Ichida, A.M., Sanders, D. & Schroeder, J.I. (1997) Roles of higher plant $K^+$ channels. *Plant Physiol.*, **114**, 1141–1149.

Maathuis, F.J., May, S.T., Graham, N.S., *et al.* (1998) Cell marking in *Arabidopsis thaliana* and its application to patch-clamp studies. *Plant J.*, **15**, 843–851.

Maathuis, F.J. & Sanders, D. (1995) Contrasting roles in ion transport of two $K^+$-channel types in root cells of *Arabidopsis thaliana*. *Planta*, **197**, 456–464.

Maathuis, F.J. & Sanders, D. (1996b) Characterization of csi52, a $Cs^+$ resistant mutant of *Arabidopsis thaliana* altered in $K^+$ transport. *Plant J.*, **10**, 579–589.

Maathuis, F.J. & Sanders, D. (2001) Sodium uptake in *Arabidopsis* roots is regulated by cyclic nucleotides. *Plant Physiol.*, **127**, 1617–1625.

Maathuis, F.J.M. & Amtmann, A. (1999) $K^+$ nutrition and $Na^+$ toxicity: the basis of cellular $K^+/Na^+$ ratios. *Ann. Bot.*, **84**, 123–133.

Marschner, H. (1995) *Mineral Nutrition of Higher Plants*, Academic Press, London.

Marschner, H., Kirkby, E.A. & Cakmak, I. (1996) Effect of mineral nutritional status on shoot-root partitioning of photoassimilates and cycling of mineral nutrients. *J. Exp. Bot.*, **47**, 1255–1265.

Marten, I., Hoth, S., Deeken, R., *et al.* (1999) AKT3, a phloem-localized $K^+$ channel, is blocked by protons. *Proc. Natl. Acad. Sci. U.S.A.*, **96**, 7581–7586.

Mäser, P., Thomine, S., Schroeder, J.I., *et al.* (2001) Phylogenetic relationships within cation transporter families of *Arabidopsis*. *Plant Physiol.*, **126**, 1646–1667.

Mendizabal, I., Rios, G., Mulet, J.M., Serrano, R. & de Larrinoa, I.F. (1998) Yeast putative transcription factors involved in salt tolerance. *FEBS Lett.*, **425**, 323–328.

Mendoza, I., Rubio, F., Rodriguez-Navarro, A. & Pardo, J.M. (1994) The protein phosphatase calcineurin is essential for NaCl tolerance of *Saccharomyces cerevisiae*. *J. Biol. Chem.*, **269**, 8792–8796.

Mengel, K. & Haeder, H.E. (1977) Effect of potassium supply on the rate of phloem sap exudation and the composition of phloem sap of *Ricinus communis*.*Plant Physiol.*, **59**, 282–284.

Mennen, H., Jacoby, B. & Marschner, H. (1990) Is sodium proton antiporter ubiquitous in plant cells? *J. Plant Phys.*, **137**, 180–183.

Moons, A., Prinsen, E., Bauw, G. & Van Montagu, M. (1997) Antagonistic effects of abscisic acid and jasmonates on salt stress-inducible transcripts in rice roots. *Plant Cell*, **9**, 2243–2259.

Moran, J.F., Becana, M., Iturbe-Ormaetxe, I., Frechilla, S., Klucas, R.V. & Aparicio-Tejo, P. (1994) Drought induces oxidative stress in pea plants. *Planta*, **194**, 346–352.

Moran, N., Ehrenstein, G., Iwasa, K., Mischke, C., Bare, C. & Satter, R.L. (1988) Potassium channels in motor cells of *Samanea saman* – a patch-clamp study. *Plant Physiol.*, **88**, 643–648.

Moshelion, M. & Moran, N. (2001) Potassium-efflux channels in extensor and flexor cells of the motor organ of *Samanea saman* are not identical. Effects of cytosolic calcium. *Plant Physiol.*, **125**, 1142–1150.

Mulet, J.M., Leube, M.P., Kron, S.J., Rios, G., Fink, G.R. & Serrano, R. (1999) A novel mechanism of ion homeostasis and salt tolerance in yeast: the Hal4 and Hal5 protein kinases modulate the Trk1–Trk2 potassium transporter. *Mol. Cell Biol.*, **19**, 3328–3337.

Munns, R. (2002) Comparative physiology of salt and water stress. *Plant Cell Env.*, **25**, 239–250.

Munns, R., Cramer, G.R. & Ball, M.C. (1999) Interactions between rising CO2, soil salinity and plant growth, in *Carbon Dioxide and Environmental Stress* (eds, W. Luo & H.A. Mooney), Academic Press, London, pp. 139–167.

Murguia, J.R., Belles, J.M. & Serrano, R. (1995) A salt-sensitive $3'(2'),5'$-bisphosphate nucleotidase involved in sulfate activation. *Science*, **267**, 232–234.

Murguia, J.R., Belles, J.M. & Serrano, R. (1996) The yeast HAL2 nucleotidase is an in vivo target of salt toxicity. *J. Biol. Chem.*, **271**, 29029–29033.

Navarre, C. & Goffeau, A. (2000) Membrane hyperpolarization and salt sensitivity induced by deletion of PMP3, a highly conserved small protein of yeast plasma membrane. *EMBO J.*, **19**, 2515–2524.

Nikiforova, V., Freitag, J., Kempa, S., Adamik, M., Hesse, H. & Hoefgen, R. (2003) Transcriptome analysis of sulfur depletion in *Arabidopsis thaliana*: interlacing of biosynthetic pathways provides response specificity. *Plant J.*, **33**, 633–650.

Niu, X., Narasimhan, M.L., Salzman, R.A., Bressan, R.A. & Hasegawa, P.M. (1993) NaCl regulation of plasma membrane $H^+$-ATPase gene expression in a glycophyte and a halophyte. *Plant Physiol.*, **103**, 713–718.

Oparka, K.J., Duckett, C.M., Prior, D.A.M. & Fischer, D.B. (1994) Real-time imaging of phloem unloading in the root tip of *Arabidopsis*. *Plant J.*, **6**, 759–766.

Palmgren, M.G. & Axelsen, K.B. (1998) Evolution of P-type ATPases. *Biochim. Biophys. Acta*, **1365**, 37–45.

Pilot, G., Gaymard, F., Mouline, K., Cherel, I. & Sentenac, H. (2003) ) Regulated expression of *Arabidopsis* shaker $K^+$ channel genes involved in $K^+$ uptake and distribution in the plant. *Plant Mol. Biol.*, **51**, 773–787.

Pilot, G., Lacombe, B., Gaymard, F., *et al.* (2001) Guard cell inward $K^+$ channel activity in *Arabidopsis* involves expression of the twin channel subunits KAT1 and KAT2. *J. Biol. Chem.*, **276**, 3215–3221.

Pitman, M.G. & Läuchli, A. (2002) Global impact of salinity and agricultural ecosystems, in *Salinity: Environment–Plants–Molecules* (eds A. Läuchli & U. Lüttge), Kluwer Academic Publishers, Dordrecht, The Netherlands, pp. 3–20.

Poffenroth, M., Green, D.B. & Tallman, G. (1992) Sugar concentration in guard cells of *Vicia faba* illuminated with red or blue light. *Plant Physiol.*, **98**, 1460–1471.

Pritchard, J., Winch, S.K. & Gould, N. (2000) Phloem water relations and root growth. *Aust. J. Plant Phys.*, **27**, 539–548.

Qiu, Q.S., Guo, Y., Quintero, F.J., Pardo, J.M., Schumaker, K.S. & Zhu, J.K. (2004) Regulation of vacuolar $Na^+/H^+$ exchange in *Arabidopsis thaliana* by the salt-overly-sensitive (SOS) pathway. *J. Biol. Chem.*, **279**, 207–215.

Quintero, F.J. & Blatt, M.R. (1997) A new family of $K^+$ transporters from *Arabidopsis* that are conserved across phyla. *FEBS Lett.*, **415**, 206–211.

Ramahaleo, T., Alexandre, J. & Lassalles, J.P. (1996) Stretch activated channels in plant cells. A new model for osmoelastic coupling. *Plant Physiol. Biochem.*, **34**, 327–334.

Ramirez, J., Ramirez, O., Saldana, C., Coria, R. & Pena, A. (1998) A *Saccharomyces cerevisiae* mutant lacking a $K^+/H^+$ exchanger. *J. Bacteriol.*, **180**, 5860–5865.

Rea, P.A., Li, Z.S., Lu, Y.P., Drozdowicz, Y.M. & Martinoia, E. (1998) From vacuolar GS-X pumps to multispecific ABC transporters. *Annu. Rev. Plant Phys. Plant Mol. Biol.*, **49**, 727–760.

Reckmann, U., Scheibe, R. & Raschke, K. (1990) Rubisco activity in guard cells compared with the solute requirement for stomatal opening. *Plant Physiol.*, **92**, 246–253.

Reintanz, B., Szyroki, A., Ivashikina, N., *et al.* (2002) AtKC1, a silent *Arabidopsis* potassium channel alpha-subunit modulates root hair $K^+$ influx. *Proc. Natl. Acad. Sci. U.S.A.*, **99**, 4079–4084.

Rigas, S., Debrosses, G., Haralampidis, K., *et al.* (2001) TRH1 encodes a potassium transporter required for tip growth in *Arabidopsis* root hairs. *Plant Cell*, **13**, 139–151.

Roberts, S.K. (1998) Regulation of $K^+$ channels in maize roots by water stress and abscisic acid. *Plant Physiol.*, **116**, 145–153.

Roberts, S.K. & Tester, M. (1995) Inward and outward $K^+$-selective currents in the plasma membrane of protoplasts from maize root cortex and stele. *Plant J.*, **8**, 811–825.

Roberts, S.K. & Tester, M. (1997) A patch clamp study of $Na^+$ transport in maize roots. *J. Exp. Bot.*, **48**, 431–440.

Rock, C.D. (2000) Pathways to abscisic acid-regulated gene expression. *New Phytol.*, **148**, 357–396.

Rodriguez-Navarro, A. (2000) Potassium transport in fungi and plants. *Biochim. Biophys. Acta*, **1469**, 1–30.

Rubio, F., Gassmann, W. & Schroeder, J.I. (1995) Sodium-driven potassium uptake by the plant potassium transporter Hkt1 and mutations conferring salt tolerance. *Science*, **270**, 1660–1663.

Rubio, F., Santa-Maria, G.E. & Rodriguez-Navarro, A. (2000) Cloning of *Arabidopsis* and barley cDNAs encoding HAK potassium transporters in root and shoot cells. *Physiol. Plant.*, **109**, 34–43.

Rus, A., Yokoi, S., Sharkhuu, A., *et al.* (2001) AtHKT1 is a salt tolerance determinant that controls Na$^+$ entry into plant roots. *Proc. Natl. Acad. Sci. U.S.A.*, **98**, 14150–14155.

Saalbach, G., Schwerdel, M., Natura, G., Buschmann, P., Christov, V. & Dahse, I. (1997) Over-expression of plant 14-3-3 proteins in tobacco: enhancement of the plasmalemma K$^+$ conductance of mesophyll cells. *FEBS Lett.*, **413**, 294–298.

Saleki, R., Young, P.G. & Lefebvre, D.D. (1993) Mutants of *Arabidopsis thaliana* capable of germination under saline conditions. *Plant Physiol.*, **101**, 839–845.

Samaj, J., Baluska, F. & Hirt, H. (2004) From signal to cell polarity: mitogen-activated protein kinases as sensors and effectors of cytoskeleton dynamicity. *J. Exp. Bot.*, **55**, 189–198.

Sanchez-Fernandez, R., Davies, T.G., Coleman, J.O. & Rea, P.A. (2001) The *Arabidopsis thaliana* ABC protein superfamily, a complete inventory. *J. Biol. Chem.*, **276**, 30231–30244.

Santa-Maria, G.E., Rubio, F., Dubcovsky, J. & Rodriguez-Navarro, A. (1997) The HAK1 gene of barley is a member of a large gene family and encodes a high-affinity potassium transporter. *Plant Cell*, **9**, 2281–2289.

Santoni, V., Molloy, M. & Rabilloud, T. (2000) Membrane proteins and proteomics: un amour impossible? *Electrophoresis*, **21**, 1054–1070.

Savouré, A., Hua, X.J., Bertauche, N., Van Montagu, M. & Verbruggen, N. (1997) Abscisic acid-independent and abscisic acid-dependent regulation of proline biosynthesis following cold and osmotic stresses in *Arabidopsis thaliana*. *Mol. Gen. Genet.*, **254**, 104–109.

Schachtman, D.P., Kumar, R., Schroeder, J.I. & Marsh, E.L. (1997) Molecular and functional characterization of a novel low-affinity cation transporter (LCT1) in higher plants. *Proc. Natl. Acad. Sci. U.S.A.*, **94**, 11079–11084.

Schachtman, D.P. & Schroeder, J.I. (1994) Structure and transport mechanism of a high-affinity potassium uptake transporter from higher plants. *Nature*, **370**, 655–658.

Schachtman, D.P., Tyerman, S.D. & Terry, B.R. (1991) The K$^+$/Na$^+$ selectivity of a cation channel in the plasma membrane of root cells does not differ in salt tolerant and salt sensitive wheat species. *Plant Physiol.*, **97**, 598–605.

Schleyer, M. & Bakker, E.P. (1993) Nucleotide sequence and 3′-end deletion studies indicate that the K$^+$-uptake protein kup from *Escherichia coli* is composed of a hydrophobic core linked to a large and partially essential hydrophilic C terminus. *J. Bacteriol.*, **175**, 6925–6931.

Schönknecht, G., Spoormaker, P., Steinmeyer, R., *et al.* (2002) KCO1 is a component of the slow-vacuolar (SV) ion channel. *FEBS Lett.*, **511**, 28–32.

Schroeder, J.I., Ward, J.M. & Gassmann, W. (1994) Perspectives on the physiology and structure of inward-rectifying K$^+$ channels in higher plants: biophysical implications for K$^+$ uptake. *Annu. Rev. Biophys. Biomol. Struct.*, **23**, 441–471.

Seigneurin-Berny, D., Rolland, N., Garin, J. & Joyard, J. (1999) Technical advance – differential extraction of hydrophobic proteins from chloroplast envelope membranes: a subcellular-specific proteomic approach to identify rare intrinsic membrane proteins. *Plant J.*, **19**, 217–228.

Sentenac, H., Bonneaud, N., Minet, M., *et al.* (1992) Cloning and expression in yeast of a plant potassium-ion transport-system. *Science*, **256**, 663–665.

Serrano, R. (1989) Structure and fonction of plasma membrane ATPase. *Annu. Rev. Plant Phys. Plant Mol. Biol.*, **40**, 61–94.

Serrano, R. (1996) Salt tolerance in plants and microorganisms: toxicity targets and defense responses. *Int. Rev. Cytol.*, **165**, 1–52.

Serrano, R., Mulet, J.M., Rios, G., *et al.* (1999) A glimpse of the mechanisms of ion homeostasis during salt stress. *J. Exp. Bot.*, **50**, 1023–1036.

Shabala, S. (2003) Regulation of potassium transport in leaves: from molecular to tissue level. *Ann. Bot.*, **92**, 627–634.

Shabala, S. & Newman, I.I. (1999) Light-induced changes in hydrogen, calcium, potassium, and chloride ion fluxes and concentrations from the mesophyll and epidermal tissues of bean leaves. Understanding the ionic basis of light-induced bioelectrogenesis. *Plant Physiol.*, **119**, 1115–1124.

Shi, H., Ishitani, M., Kim, C. & Zhu, J.K. (2000) The *Arabidopsis thaliana* salt tolerance gene SOS1 encodes a putative $Na^+/H^+$ antiporter. *Proc. Natl. Acad. Sci. U.S.A.*, **97**, 6896–6901.

Shi, H., Kim, Y., Guo, Y., Stevenson, B. & Zhu, J.-K. (2003a) The *Arabidopsis* SOS5 locus encodes a putative cell surface adhesion protein and is required for normal cell expansion. *Plant Cell*, **15**, 19–32.

Shi, H., Lee, B.H., Wu, S.J. & Zhu, J.K. (2003b) Overexpression of a plasma membrane $Na^+/H^+$ antiporter gene improves salt tolerance in *Arabidopsis thaliana*. *Nat. Biotech.*, **21**, 81–85.

Shi, H., Quintero, F.J., Pardo, J.M. & Zhu, J.K. (2002) The putative plasma membrane $Na^+/H^+$ antiporter SOS1 controls long-distance $Na^{(+)}$ transport in plants. *Plant Cell*, **14**, 465–477.

Shinozaki, K. & Yamaguchi-Shinozaki, K. (2000) Molecular responses to dehydration and low temperature: differences and cross-talk between two stress signaling pathways. *Curr. Opin. Plant Biol.*, **3**, 217–223.

Showalter, A.M. (2001) Arabinogalactan-proteins: structure, expression and function. *Cell Mol. Life Sci.*, **58**, 1399–1417.

Smart, C.C. & Fleming, A.J. (1996) Hormonal and environmental regulation of a plant PDR5-like ABC transporter. *J. Biol. Chem.*, **271**, 19351–19357.

Spalding, E.P., Hirsch, R.E., Lewis, D.R., Qi, Z., Sussman, M.R. & Lewis, B.D. (1999) Potassium uptake supporting plant growth in the absence of AKT1 channel activity: inhibition by ammonium and stimulation by sodium. *J. Gen. Physiol.*, **113**, 909–918.

Spalding, E.P., Slayman, C.L., Goldsmith, M.H.M., Gradmann, D. & Bertl, A. (1992) Ion channel in *Arabidopsis* plasma membrane. Transport characteristics and involvement in light-induced voltage changes. *Plant Physiol.*, **99**, 96–102.

Su, H., Balderas, E., Vera-Estrella, R., *et al.* (2003) Expression of the cation transporter McHKT1 in a halophyte. *Plant Mol. Biol.*, **52**, 967–980.

Su, H., Golldack, D., Katsuhara, M., Zhao, C. & Bohnert, H.J. (2001) Expression and stress-dependent induction of potassium channel transcripts in the common ice plant. *Plant Physiol.*, **125**, 604–614.

Szabolcs, I. (1987) The global problems of salt-affected soils. *Acta Agron. Hung.*, **36**, 159–172.

Sze, H., Schumacher, K., Muller, M.L., Padmanaban, S. & Taiz, L. (2002) A simple nomenclature for a complex proton pump: VHA genes encode the vacuolar $H^+$-ATPase. *Trends Plant Sci.*, **7**, 157–161.

Szyroki, A., Ivashikina, N., Dietrich, P., *et al.* (2001) KAT1 is not essential for stomatal opening. *Proc. Natl. Acad. Sci. U.S.A.*, **98**, 2917–2921.

Talke, I.N., Blaudez, D., Maathuis, F.J. & Sanders, D. (2003) CNGCs: prime targets of plant cyclic nucleotide signalling? *Trends Plant Sci.*, **8**, 286–293.

Tang, H., Vasconcelos, A.C. & Berkowitz, G.A. (1996) Physical association of KAB1 with plant $K^+$ channel alpha subunits. *Plant Cell*, **8**, 1545–1553.

Tester, M. & Blatt, M.R. (1989) Direct measurement of $K^+$ channels in thylakoid membranes by incorporation of vesicles into planar lipid bilayers. *Plant Physiol.*, **91**, 249–252.

Tester, M. & Davenport, R. (2003) $Na^+$ tolerance and $Na^+$ transport in higher plants. *Ann. Bot.*, **91**, 503–527.

Tester, M. & Leigh, R.A. (2001) Partitioning of nutrient transport processes in roots. *J. Exp. Bot.*, **52**, 445–457.

Touraine, B., Grignon, N. & Grignon, C. (1988) Charge balance in NO3-fed soybean – estimation of $K^+$ and carboxylate recirculation. *Plant Physiol.*, **88**, 605–612.

Turner, J.G., Ellis, C. & Devoto, A. (2002) The jasmonate signal pathway. *Plant Cell*, **14** (Suppl.), S153–S164.

Tyerman, S.D. & Skerrett, I.M. (1999) Root ion channels and salinity. *Sci. Horticult.*, **78**, 175–235.

Uozumi, N., Kim, E.J., Rubio, F., *et al.* (2000) The *Arabidopsis* HKT1 gene homolog mediates inward $Na^+$ currents in *Xenopus laevis* oocytes and $Na^+$ uptake in *Saccharomyces cerevisiae*. *Plant Physiol.*, **122**, 1249–1259.

van den Wijngaard, P.W., Bunney, T.D., Roobeek, I., Schonknecht, G. & De Boer, A.H. (2001) Slow vacuolar channels from barley mesophyll cells are regulated by 14-3-3 proteins. *FEBS Lett.*, **488**, 100–104.

Vernon, D.M. & Bohnert, H.J. (1992) A novel methyl transferase induced by osmotic stress in the facultative halophyte *Mesembryanthemum crystallinum*. *EMBO J.*, **11**, 2077–2085.

Véry, A.A., Robinson, M.F., Mansfield, T.A. & Sanders, D. (1998) Guard cell cation channel are involved in $Na^+$-induced stomatal closure in a halophyte. *Plant J.*, **14**, 509–521.

Véry, A.A. & Sentenac, H. (2002) Cation channels in the *Arabidopsis* plasma membrane. *Trends Plant Sci.*, **7**, 168–175.

Véry, A.A. & Sentenac, H. (2003) Molecular mechanisms and regulation of $K^+$ transport in higher plants. *Annu. Rev. Plant Biol.*, **54**, 575–603.

Vitart, V., Baxter, I., Doerner, P. & Harper, J.F. (2001) Evidence for a role in growth and salt resistance of a plasma membrane $H^+$-ATPase in the root endodermis. *Plant J.*, **27**, 191–201.

Volkov, V., Wang, B., Dominy, P.J., Fricke, W. & Amtmann, A. (2004) *Thellungiella halophila*, a salt-tolerant relative of *Arabidopsis thaliana*, possesses effective mechanisms to discriminate between potassium and sodium. *Plant Cell Env.*, **27**, 1–14.

Vranova, E., Tahtiharju, S., Sriprang, R., *et al.* (2001) The AKT3 potassium channel protein interacts with the AtPP2CA protein phosphatase 2C. *J. Exp. Bot.*, **52**, 181–182.

Walker, D.J., Leigh, R.A. & Miller, A.J. (1996a) Potassium homeostasis in vacuolate plant cells. *Proc. Natl. Acad. Sci. U.S.A.*, **93**, 10510–10514.

Walker, N.A., Sanders, D. & Maathuis, F.J. (1996b) High-affinity potassium uptake in plants. *Science*, **273**, 977–979.

Wang, T.B., Gassmann, W., Rubio, F., Schroeder, J.I. & Glass, A.D. (1998) Rapid up-regulation of HKT1, a high-affinity potassium transporter gene, in roots of barley and wheat following withdrawal of potassium. *Plant Physiol.*, **118**, 651–659.

Ward, J.M. & Schroeder, J.I. (1994) Calcium-activated $K^+$ channels and calcium-induced calcium release by slow vacuolar ion channels in guard cell vacuoles implicated in the control of stomatal closure. *Plant Cell*, **6**, 669–683.

Watson, M.B. & Malmberg, R.L. (1996) Regulation of *Arabidopsis thaliana* (L.) Heynh arginine decarboxylase by potassium deficiency stress. *Plant Physiol.*, **111**, 1077–1083.

Wegner, L.H. & De Boer, A.H. (1997) Properties of two outward-rectifying channels in root xylem parenchyma cells suggest a role in $K^+$ homeostasis and long-distance signaling. *Plant Physiol.*, **115**, 1707–1719.

Werner, J. & Finkelstein, R.R. (1995) *Arabidopsis* mutants with reduced response to NaCl and osmotic stress. *Physiol. Plant.*, **93**, 659–666.

White, P.J. (2001) The pathways of calcium movement to the xylem. *J. Exp. Bot.*, **52**, 891–899.

White, P.J. & Lemitiri-Clieh, F. (1995) Potassium current across the plasma membrane of protoplast derived from rye roots: a patch clamp study. *J. Exp. Bot.*, **46**, 497–511.

Williams, K. (1997) Interactions of polyamines with ion channels. *Biochem. J.*, **325** (Pt 2), 289–297.

Wilson, C. & Shannon, M.C. (1995) Salt-induced $Na^+/H^+$ antiport in root plasma-membrane of a glycophytic and halophytic species of tomato. *Plant Sci.*, **107**, 147–157.

Wolf, O., Munns, R., Tonnet, M.L. & Jeschke, W.D. (1991) The role of the stem in the partitioning of $Na^+$ and $K^+$ in salt tolerant barley. *J. Exp. Bot.*, **42**, 697–704.

Wu, S.J., Ding, L. & Zhu, J.K. (1996) SOS1, a genetic locus essential for salt tolerance and potassium acquisition. *Plant Cell*, **8**, 617–627.

Wu, W., Peters, J. & Berkowitz, G.A. (1991) Surface charge-mediated effects of $Mg^{2+}$ and $K^+$ flux across the chloroplast envelope are associated with regulation of stroma pH and photosynthesis. *Plant Physiol.*, **97**, 580–587.

Wyn Jones, R.J. & Pollard, A. (1983) Proteins, enzymes and inorganic ions, in *Encyclopedia of Plant Physiology* (eds A. Lauchli & A. Pirson), Springer, Berlin, pp. 528–562.

Xiong, L., Schumaker, K.S. & Zhu, J.K. (2002) Cell signaling during cold, drought, and salt stress. *Plant Cell*, **14** (Suppl.), S165–S183.

Xiong, L. & Zhu, J.K. (2002a) Molecular and genetic aspects of plant responses to osmotic stress. *Plant Cell Env.*, **25**, 131–139.

Xiong, L. & Zhu, J.K. (2002b) Salt stress signal transduction in plants, in *Plant Signal Transduction* (eds D. Scheel & C. Wasternack), Oxford University Press, Oxford, UK, pp. 165–197.

Yamaguchi, T., Apse, M.P., Shi, H. & Blumwald, E. (2003) Topological analysis of a plant vacuolar $Na^+/H^+$ antiporter reveals a luminal C terminus that regulates antiporter cation selectivity. *Proc. Natl. Acad. Sci. U.S.A*, **100**, 12510–12515.

Yokoi, S., Quintero, F.J., Cubero, B., *et al.* (2002) Differential expression and function of *Arabidopsis thaliana* NHX $Na^+/H^+$ antiporters in the salt stress response. *Plant J.*, **30**, 529–539.

Young, N.D. & Galston, A.W. (1984) Physiological control of arginine decarboxylase activity in $K^+$-deficient oat shoots. *Plant Physiol.*, **76**, 331–335.

Zhang, H.X. & Blumwald, E. (2001) Transgenic salt-tolerant tomato plants accumulate salt in foliage but not in fruit. *Nat. Biotech.*, **19**, 765–768.

Zhang, H.X., Hodson, J.N., Williams, J.P. & Blumwald, E. (2001) Engineering salt-tolerant *Brassica* plants: characterization of yield and seed oil quality in transgenic plants with increased vacuolar sodium accumulation. *Proc. Natl. Acad. Sci. U.S.A.*, **98**, 12832–12836.

Zhao, R., Dielen, V., Kinet, J.M. & Boutry, M. (2000) Cosuppression of a plasma membrane $H^+$-ATPase isoform impairs sucrose translocation, stomatal opening, plant growth, and male fertility. *Plant Cell*, **12**, 535–546.

Zhu, J., Gong, Z., Zhang, C., *et al.* (2002) OSM1/SYP61: a syntaxin protein in Arabidopsis controls abscisic acid-mediated and non-abscisic acid-mediated responses to abiotic stress. *Plant Cell*, **14**, 3009–3028.

Zhu, J.K. (2000) Genetic analysis of plant salt tolerance using *Arabidopsis*. *Plant Physiol.*, **124**, 941–948.

Zhu, J.K., Liu, J. & Xiong, L. (1998) Genetic analysis of salt tolerance in arabidopsis. Evidence for a critical role of potassium nutrition. *Plant Cell*, **10**, 1181–1191.

Zimmermann, S., Hartje, S., Ehrhardt, T., Plesch, G. & Mueller-Roeber, B. (2001) The $K^+$ channel SKT1 is co-expressed with KST1 in potato guard cells – both channels can co-assemble via their conserved K-T domains. *Plant J.*, **28**, 517–527.

Zimmermann, S., Talke, I., Ehrhardt, T., Nast, G. & Muller-Rober, B. (1998) Characterization of SKT1, an inwardly rectifying potassium channel from potato, by heterologous expression in insect cells. *Plant Physiol.*, **116**, 879–890.

# 12    Membrane transport and soil bioremediation

Susan Rosser and Peter Dominy

## 12.1    Introduction

The supply of an adequate, safe diet for everyone is perhaps the major challenge for the twenty-first century, if political, economic and social stability is to be achieved. Over the next 100 years, food production will have to be at least doubled. How best can this be achieved?

One strategy that will be adopted to some extent is the appropriation of unfarmed land for cultivation. Unfortunately, the best fertile regions currently not exploited are the temperate and tropical rain forests, and there is now strong public resistance to clearing these biomes. Approximately half of the world's surface area is considered to be unsuitable for agriculture. In some of these regions, this is attributable to metrological (too hot or cold, too wet or dry, etc.) or geographical (too steep, too rocky, etc.) factors. In other regions it is due to the activities (free concentrations) of primary (naturally occurring) pollutants, such as $Al^{3+}$, $Cd^+$, $Na^+$, etc., which suppress primary production. In addition, there are regions where human activities have caused levels of persistent 'secondary pollutants' in the soil – principally organic compounds and heavy metals – thus seriously impairing primary production. There is now much interest in cleaning up these regions to enable safe crop production using conventional practices and crops and in developing novel crop lines that perform well in contaminated soils and yield safe products.

In this chapter, we assess the prospects for manipulating plant transport processes for the improvement of plant production in soils contaminated with organic compounds and heavy metals. The task is complicated for a number of reasons. One problem is that the range of organic and inorganic pollutants is very diverse and therefore bioremediation strategies are likely to be quite varied. At this stage, we are able to identify only a few specific processes that are likely to be of general relevance. A second problem, as mentioned above, relates to strategies for improvement. One involves biostabilisation of the unmodified pollutant in the rhizosphere, so that their activities are so low that they are no longer toxic or taken up into the shoot. The other, more ambitious strategy involves the mobilisation of the pollutant in the soil, and its subsequent detoxification (modification) or sequestration in living cells (plant or microbe). A third problem is that, while a great deal is known about transport mechanisms in model plant (e.g. *Arabidopsis*) and bacterial (e.g. *Escherichia coli*) systems, very little is known of the commensurate mechanisms in organisms surviving in contaminated soils; therefore, much of what follows is founded on extrapolation from sensitive model organisms. Clearly, plants and microbes that survive in contaminated soils may possess unique mechanisms of which we know little.

Before proceeding, it is worth considering why so much interest is now focused on phytostabilisation and phytoremediation. As a result of the process of photosynthesis, plants possess substrates with a broad range of redox potentials. Bacteria can achieve a similar redox span, but rarely in the same cell; anaerobiosis shifts cell redox status to more reducing potentials than do aerobically grown cells. Fungi, which are essentially aerobes, have a more limited redox metabolism, and microbial growth in soils is often slow because of the supply of carbon skeletons for energy and anabolic processes. Plants, as phototrophs, are not limited in this way. However, prokaryotes show a far greater diversity of metabolic processes than do eukaryotes, and therefore there is great interest in expressing bacterial enzymes in plants.

## 12.2  Phytostabilisation

The formation of ligands between the pollutant and microbe- or plant-derived exudates in the rhizosphere decreases pollutant activity (but not its concentration) per unit volume of soil/root matrix and is clearly bio- or phytostabilisation. The uptake and intracellular complexing of a pollutant in a microbe or root cell also decreases its activity (but not concentration) per unit volume of soil/root matrix and could also be classed as bio- or phytostabilisation. However, the ambiguity arises when some of the intracellular pollutant in the rhizosphere is metabolised (detoxified) or relocated to the shoot; by most definitions, this is phytoremediation. In what follows, we have subdivided our discussions into phytostabilisation and phytoremediation. It is important to realise that some component processes (e.g. uptake) may be utilised by both strategies.

### 12.2.1  Root exudation

In addition to their well-characterised roles in the provision of mechanical support, water and nutrient uptake, roots have the ability to synthesise and secrete an astonishingly diverse array of compounds that can modify soils (Flores *et al.*, 1999). Plants perceive and respond to their environment via the zone of soil immediately surrounding plant roots (rhizosphere). It is in this zone that plants exchange organic and inorganic substrates between the root and soil, leading to changes in the biochemical and physical properties of the rhizosphere. Roots receive 30–60% of all photosynthetically fixed carbon, and release 5–21% of that carbon into the rhizosphere via root exudates (Salt *et al.*, 1998). Exudates consist primarily of low-molecular-weight compounds such as amino acids, organic acids, sugars, phenolics and an array of other secondary metabolites (Jones, 1998). Roots also release high-molecular-weight compounds mainly in the form of mucilage (high-molecular-weight polysaccharide) and enzymes (Knee *et al.*, 2001).

Root exudates are known to play a vital role in a variety of rhizosphere processes, including nutrient acquisition (Romheld, 1990; Zhang *et al.*, 1991; Hopkins *et al.*,

1998), plant–microbe interactions (Janczarek *et al.*, 1997; Scheidemann & Wetzel, 1997), regulation of plant growth (Einhellig *et al.*, 1993) and determination of microbial community structure in the plant rhizosphere (Grayston & Campbell, 1996; Yang & Crowley, 2000). All such processes potentially influence the effectiveness of phytoremediation but components of root exudates are also thought to interact directly with both organic and inorganic soil contaminants, altering their bioavailability, motility and transport in the soil. Despite the fact that root exudation is a process constituting a major carbon cost to plants, studies fully defining the composition of exudates from a wide range of plant species have yet to be undertaken. In addition, the mechanisms and regulation of root secretion have only recently begun to be investigated.

Organic anions are commonly detected in the rhizosphere as a result of root exudation and they are thought to play a vital role in the acquisition of nutrients and in defence against inorganic ion stresses. Low-molecular-weight, non-amino organic acid anions such as citrate, malate, fumarate and malonate are intermediates in the tricarboxylic acid cycle and their concentrations tend to be strictly regulated. Their concentrations in the cytosol are relatively stable but in the vacuole they can vary by one or two orders of magnitude depending on metabolic activity and nutrient availability (Gerhardt *et al.*, 1987).

Organic anion exudation aids plants in three ways – first by increasing availability of nutrients, second by reducing the activities (free concentration) of toxic cations in the rhizosphere and finally by reducing the accumulation of potentially toxic metabolites (e.g. lactic acid, resulting from anaerobic respiration under anoxic conditions) in the cytoplasm. The first two processes rely on the capacity of organic anions to bind cations such as $Fe^{3+}$, $Cu^{2+}$, $Zn^{2+}$ and $Al^{3+}$. The stability of the ligand–metal complex is determined by the number of carboxyl groups and their relative arrangements with other carboxyl or hydroxyl moieties (Hue *et al.*, 1986; Tam & McColl, 1990; Bolan *et al.*, 1994). In general, the tricarboxylates (citrate$^{3+}$) chelate these cations more strongly than dicarboxylates (malate$^{2-}$, oxalate$^{2-}$, malonate$^{2-}$), which in turn are stronger chelaters than the relatively weak monocarboxylates (acetate$^-$). Organic anions have also been shown to chelate heavy metals (Poulsen & Hansen, 2000), and the artificial increase of soil anion concentration has been demonstrated to increase the accumulation of heavy metals like $Cr^{3+}$, $Cd^{2+}$ or $U^-$ in leaves (Cieslinski *et al.*, 1998; Srivastava *et al.*, 1999).

Plant-derived organic ligands may enhance the sorption of metals to the solid soil phase by a process resulting in the formation of metal–ligand complexes, which have higher affinities for the solid soil phase. Organic ligands can also increase metal sorption to soil surfaces by binding to the soil surfaces and increasing the negative electrostatic potential on the soil matrix or by directly binding the aqueous metal cation (Haas & Horowitz, 1986; Naidu & Harter, 1998). The crystallisation of some metal (hydr)oxides can be inhibited by organic ligands because of an increased negative surface charge of the (hydr)oxides, increasing the capacity for metal adsorption (Xue & Huang, 1995). The situation is however a complex one since organic ligands may also have the reverse effect of stimulating metal solubilisation (Elliott & Denneny, 1982; Mench & Martin, 1991; Naidu & Harter, 1998). The presence of positively charged organic ligands can increase metal solubility by competing with metal ions for cation-adsorption sites or by reducing the negative electrostatic potential on surfaces (Chubin & Street, 1981). Organic ligands could also dissolve metal precipitates and dissolve the minerals adsorbing the metal (Jauregui & Reisenauer, 1982; Pohlman & McColl, 1986; Dinkelaker *et al.*, 1989). It should also be noted that although less well characterised, many of the described processes also have an impact on the bioavailability of organic pollutants.

The best-characterised organic anion effect on metal availability is their interaction with aluminium. Although not a target for phytoremediation, aluminium toxicity has been identified as a major limiting factor of plant productivity on acidic soils, which comprise up to 50% of the world's potentially arable lands (Vonuexkull & Mutert, 1995). When the soil pH drops below 4, Al is solubilised into the toxic trivalent cation $Al^{3+}$. However, some plants have evolved Al-tolerance mechanisms, enabling them to grow in Al-toxic, acidic soil environments (for review, see Kochian, 1995; Ma *et al.*, 2003). A well-characterised mechanism of Al resistance is the specific release of organic acid anions, such as malate, citrate and oxalate, from the roots of Al-resistant plants (Delhaize *et al.*, 1993; Pellet *et al.*, 1995; Ma *et al.*, 1997). Organic acids have also been proposed to be involved in the chelation and sequestration of a number of metal ions, including cadmium, zinc and nickel. However, this has yet to be demonstrated conclusively. Organic acids are thought to detoxify $Al^{3+}$ by complexing these cations in the cytosol or at the root–soil interface. Overexpression of a citrate synthase gene from *Pseudomonas aeruginosa* in *Nicotiana tabacum* and *Carica papaya* resulted in increased citrate production and increased growth in the presence of $Al^{3+}$ (de la Fuente *et al.*, 1997). Two distinct patterns of organic acid release have been identified on the basis of the timing of secretion (Ma, 2000b). In pattern I, there is no apparent delay between Al exposure and the onset of organic acid release. For example, in wheat and buck wheat, the secretion of malate or oxalate respectively was detectable within 15–30 min after Al exposure (Delhaize *et al.*, 1993; Zheng *et al.*, 1998). In contrast, in pattern II, organic acid secretion is delayed for several hours post $Al^{3+}$ exposure. For example, in *Cassia tora*, maximal citrate efflux occurs 4 h after Al exposure (Ma *et al.*, 1997), and in rye, citrate and malate efflux increases steadily with time over 10 h (Pellet *et al.*, 1995; Pineros & Kochian, 2001). The rapidity of the response of pattern I suggests that Al activates a pre-existing transporter on the plasma membrane and that induction of new protein synthesis is not required (Ma, 2000a; Ma *et al.*, 2001). Anion channels are membrane-bound transporter proteins that allow the passive flow of anions down their electrochemical gradient. The large electrical potential difference across the plasma membrane of plant cells (inside negative) ensures that, under most circumstances, anions move from the cytosol to the apoplasm (see Chapters 2 and 7). Examples of such channels are the Al-responsive malate-permeable channels in the plasma membrane of wheat root cells (Ryan *et al.*, 1997; Zhang *et al.*, 2001).

The delay observed in pattern II-type secretion appears to indicate the requirement for gene transcription and new or increased synthesis of specific proteins. Such induced proteins could be involved in organic acid synthesis or in the transport of organic acid anions, but the exact mechanisms have yet to be defined. Recently, in a screen of activation-tagged *Arabidopsis thaliana* lines, a cadmium-tolerant mutant was identified that overexpressed a putative organic acid transporter (C. Thompson & P. Dominy, unpublished results, 2004); the functional characterisation of this transporter has not yet been completed. Other metal ions can also activate organic anion efflux from roots. Copper ($Cu^{2+}$) tolerance in *A. thaliana* is associated with a rapid release of citrate (Murphy *et al.*, 1999), and enhanced lead ($Pb^{2+}$) tolerance in certain rice varieties is correlated with oxalate secretion (Yang *et al.*, 2000).

In addition to their role in metal bioavailability, root exudates have also been shown to enhance availability and bioremediation of organic contaminants. A study investigating hybrid poplar degradation of atrazine found that the addition of root exudates resulted in an enhanced mineralisation of 23% compared with acetate-amended controls (Burken & Schnoor, 1996). Increased pyrene mineralisation was observed in an alfalfa rhizosphere soil (Reilley *et al.*, 1996).

It is well known that the longer organic pollutants are in the environment the less bioavailable they become (Hatzinger & Alexander, 1995; Alexander, 2000). This decrease in bioavailability with aging has been observed for many persistent organic pollutants (POPs) including poly-alkylated hydrocarbons (PAHs; Hatzinger & Alexander, 1995), poly-chlorinated biphenyls (PCBs; Carroll *et al.*, 1994) and DDT (Robertson & Alexander, 1998) in both laboratory- and field-weathered soil. However, recently it has also been observed that some plants can uptake POPs in spite of the effects of aging (Mattina *et al.*, 2000; White, 2001). It has been shown that this ability to take up weathered POPs by some plants is the result of differences in the patterns of root exudation by specific plants (White, 2001; White & Kottler, 2002). It is likely that the exudates in the process of chelating metal micronutrients partially destroy the local inorganic soil matrix, resulting in the release of the soil organic matter and with it the contaminants previously sequestered within (White, 2002).

Because aluminium toxicity is a function of soil acidity, a resistance mechanism involving increased alkalinisation of the rhizosphere has been proposed. An *Arabidopsis* mutant with increased aluminium tolerance was isolated recently and was shown to exhibit aluminium-induced increase in rhizosphere pH at the root tip (Degenhardt *et al.*, 1998). It is likely that similar processes have an influence on plant resistance to other metals and toxic compounds but to date these have not been characterised. The ability of strawberry clover roots to change rhizosphere pH and take up cadmium was examined using culture tubes containing nutrient agar, a moderate level of cadmium and a pH indicator dye. The results provided evidence for a negative correlation between rhizosphere pH and cadmium uptake.

### 12.2.2   Enrichment of microbial degraders

It is well known that microbial biomass and metabolism increase rapidly around plant roots (Mawdsley & Burns, 1994; Wiehe *et al.*, 1994; Bowers *et al.*, 1996). This microbial activity is stimulated by the release of sugars from roots (Farrar & Jones, 2000) but organic anion exudates can also support microbial growth (Jones, 1998).

This organic anion utilisation is likely to decrease their effectiveness in dissolving minerals and detoxifying metals. However, laboratory studies have suggested that organic anions complexed with metals are more resistant to microbial uptake and degradation (Boudot, 1992; Braum & Helmke, 1995). Enhanced microbial populations and activity in the rhizosphere have also been shown to be beneficial in the biodegradation of organic pollutants. Soils planted with alfalfa and alpine bluegrass (*Poa alpina* L.) were shown to be significantly enriched with microbial populations that could degrade a range or organic contaminants including phenanthrene (Nichols *et al.*, 1997). Microorganisms capable of utilising chlorobenzoate either as a carbon source or as a co-metabolite were enriched by 40 times in soil planted with ryegrass when compared with bulk soil (Haby & Crowley, 1996) and there is a selective enrichment of phenanthrene degraders in soil planted with slender oat (*Avena barbata* Pott ex Link) compared with that of unplanted controls (Miya & Firestone, 2000). Miya and Firestone went on to show that it was the root exudates which significantly enhanced phenanthrene degradation in rhizosphere soils, either by increasing phenanthrene bioavailability and/or increasing microbial population size and activity (Miya & Firestone, 2001). Using a most probable number method, the exudate-amended soils were shown to maintain a larger phenanthrene degrader population that control soils (Miya & Firestone, 2001).

### 12.2.3 Enhancement of microbial biodegradation activity

Many plant-derived chemicals stimulate microbial degradation of xenobiotics (Donnelly *et al.*, 1994; Fletcher & Hegde, 1995; Haby & Crowley, 1996; Miya & Firestone, 2001; Isidorov & Jdanova, 2002). One of the best examples is the induction and enhancement of microbial degradation of the polycyclic aromatic hydrocarbon (PAH) naphthalene by the plant-derived compound salicylate, which is responsible for the induction of plant systemic acquired resistance (Yen & Gunsalus, 1982; Van der Meer *et al.*, 1992). PAH pollution is the result of the industrial use of fossil fuels. Salicylate is a metabolic intermediate in the bacterial degradation of naphthalene, which interacts with a transcriptional activator protein (nahR) promoting a conformational change inducing the transcription of the naphthalene degradation pathway genes. Salicylate has also been demonstrated to enhance the removal of other PAHs, such as fluoranthene, pyrene, benz[*a*]anthracene, chrysene and benz[*a*]pyrene (Chen & Aitken, 1999). Researchers have begun to investigate the possibility of enhancing PAH degradation by amending PAH-contaminated soil with salicylate in an effort to induce PAH degradation by indigenous soil microorganisms (Ogunseitan *et al.*, 1991; Colbert *et al.*, 1993). Given the huge number of plant-derived PAH analogues, it should come as no surprise that the rhizosphere has repeatedly been shown to contain very diverse populations of PAH-degrading microorganisms (Daane *et al.*, 2001).

In addition to its role in stimulating PAH degradation, recent evidence also suggested a link between salicylate and the removal of the highly persistent xenobiotic polychlorinated biphenyls (Singer *et al.*, 2000, 2001; Master & Mohn, 2001).

Under aerobic conditions, the remediation of highly chlorinated compounds such as polychlorinated biphenyls (PCBs) relies on co-metabolism to provide the microorganisms with carbon and energy during the degradation process. Co-metabolism is the process by which an enzyme produced for the metabolism of a primary substrate (co-metabolite) is, fortuitously, capable of degrading a secondary substrate from which it gains no energy or nutritional advantages. PCB-degrading bacteria can readily be isolated from the soil environment using biphenyl as a co-metabolite (Brunner et al., 1985). However, biphenyl cannot be applied to PCB-contaminated sites because it is toxic and can itself cause environmental damage. Therefore, there is a need to identify alternative PCB degradation co-metabolites or inducers to enhance in situ PCB bioremediation. The possibility that flavinoids could enhance PCB degradation was investigated by Donnelly et al. (1994), who demonstrated that naringin (flavinoid found in grapefruit), myricetin (flavinoid found in berries, herbs and pollen) and coumarin induced PCB degradation in strains of Ralstonia eutropha, Burkholdaria cepacia and Corynebacterium sp. The authors suggested the potential of using a network of plant roots as a natural injection system for delivery of PCB-degradation-enhancing flavinoids to microorganisms present in PCB-contaminated soil over long time periods.

It has been proposed that plant terpenes might be the natural substrates for enzymes involved in PCB oxidation (Focht, 1995). Several terpinoid compounds, including carvone, cumene, carvacol, limonene, thymol and p-cymene, were screened for their ability to induce PCB degradation in Arthrobacter sp. Strain B1B (Gilbert & Crowley, 1997). Many of the terpinoids investigated are found in pine needles, citrus, juniper, thyme, oregano, spearmint and other aromatic plants (Gilbert & Crowley, 1997; Hernandez et al., 1997). The best result was obtained with S-carvone, which induced the biotransformation of 62% of the commercial PCB mixture Aroclor 1242. The degradation of each of the terpinoids was not tested, so it was not possible to conclude if the induction was co-metabolic, growth-linked or via an alternative mechanism. Researchers have also found that terpenes can also induce the degradation of other xenobiotic compounds such as toluene, phenol and trichloroethene often through novel pathways (Dabrock et al., 1992; Kim et al., 2003).

It is easy to understand how a plant secondary metabolite with a similar carbon skeleton could induce enzymes responsible for xenobiotc degradation. However, there are many examples of the biotransformation of a pollutant in the presence of a plant secondary metabolite that has very little structural similarity. One likely explanation for such circumstances is the induction of broad-specificity enzymes such as the cytochrome P450s. Cytochrome P450 mono-oxygenases (hydroxylases) catalyse the insertion of oxygen into a wide variety of different substrates, including natural steroids, fatty acids and xenobiotics. They are ubiquitous, being found in animals, plants, insects and microorganisms, and are the subject of a great deal of research because of their importance in drug development, pollutant degradation and metabolism (Guengerich, 2001). The broad specificity P450 enzymes in microorganisms have the potential to be induced by non-toxic plant secondary metabolites resulting in the biotransformation of target compounds. In addition, plant cytochrome P450s are involved in the biosynthesis of secondary metabolites including flavanoids, alkaloids and hormones, but also in the detoxification of herbicides and other xenobiotics.

Plant production of many secondary metabolites has been linked with stress. Thus under stress conditions, P450 enzymes can be linked to enhanced xenobiotic degradation in the rhizosphere via the production of degradation-inducing secondary metabolites and also to the plant's own internal pollutant detoxification mechanisms. Recent research has demonstrated the potential to generate plants that overproduce secondary metabolites (Verpoorte & Memelink, 2002). It is therefore possible to envisage a system in which transgenic plants could overproduce terpenes in contaminated soil as a response to toxic stress, resulting in a stimulation of rhizosphere populations to degrade the pollutant. The use of plant secondary metabolites that can act as inducers to stimulate microbial degradation of pollutants has enormous advantages. First, the plant secondary metabolites are often biologically effective at very low concentrations and are inherently 'environmentally friendly'. In addition, they offer the opportunity to develop sustainable systems, allowing the continuous input of inducers of biodegradation to the rhizosphere over long time periods. By introducing plants to a contaminated site that produces appropriate secondary metabolites, a readily available and cheap barrier could be established for long-term site protection or remediation.

### 12.2.4 Mechanisms of exudation

It is known that malate, citrate and other anions accumulate to high concentrations in the vacuole. However, exudation of these compounds must involve their transport from the cytosol to the external medium. In the cytosol, where pH is approximately neutral, most of the acids are fully dissociated anions, with approximately 80% of citric acid occurring as citrate$^{3-}$ anion and 99% of malic acid in the malate$^{2-}$ form. There is a large electrical potential difference across the plasma membrane of the majority of plant cells (inside negative), resulting in an electrochemical gradient favouring the passive diffusion of organic anions out of the cell via membrane channels.

Plants produce tens of thousands of secondary metabolites, many of which are potentially cytotoxic, preventing their accumulation in the cytoplasm. The suggestion that phytochemicals are transported from site of synthesis to storage sites or for exocytosis in vesicles or specialist organelles is gaining credibility (Grotewold & Lin, 2000). The isoquinoline alkaloid biosynthetic pathway relies on vesicles to traffic intermediates from one subcellular compartment to another, preventing their diffusion into the cytosol (Facchini, 2001). Phenylpropanoids and flavinoids are frequently found in root exudates and are thought to be synthesised on the cytoplasmic surface of the endoplasmic reticulum (ER) (Winkel-Shirley, 2001), for example, the catechin flavanoids, which are known to be cytotoxic and have antimicrobial properties, are secreted by the roots of knapweed plants (Bais et al., 2002). Although the mechanisms responsible for their transport from the ER to the plasma membrane have not been defined, it has been postulated that it is via ER-derived vesicles fusing to the cell membrane releasing their cargo (Walker *et al.*, 2003). In addition, ER-derived vesicles containing green autofluorescent compounds have been observed in maize cells expressing the P regulator of 3-deoxy flavinoid biosynthesis (Grotewold *et al.*, 1998). The vesicles fuse to form large fluorescent bodies that migrate to the surface of the cell where they fuse to the cell membrane, releasing the fluorescent compound (Grotewold *et al.*, 1998). The use of such autofluorescent compounds or the fluorescent β-carbolines found in the root exudates of *Oxalis tuberosa* should aid the investigation of the mechanisms of phytochemical secretion (Bais *et al.*, 2002).

Not a great deal is known about the release of ions via exocytosis, but estimates of vesicular volumes and production rates suggest their potential in providing a significant pathway for ion exudation (Kronestedt-Robards & Robards, 1991) and there is growing interest in the role of exocytosis in solute transport (MacRobbie, 1999).

To date, the vast majority of characterised plant ABC (ATP-binding cassette) transporters have been found in the vacuolar membrane and are thought to be responsible for the sequestration of cytotoxins (Theodoulou, 2000; Sanchez-Fernandez *et al.*, 2001; Martinoia *et al.*, 2002). Very little is currently known about plasma membrane ABC transporters, but the *Arabidopsis* AtPGP1 auxin transporter has been localised to the plasma membrane and shown to be involved in cell elongation by pumping auxin into adjacent cells (Sidler *et al.*, 1998; Noh *et al.*, 2001). Recently, Jasinski *et al.* (2001) identified a plasma membrane ABC transporter (NpABC1) from *Nicotiana plumbaginifolia* by treating cells with a range of secondary metabolites. The addition of sclareolide, an antifungal diterpene, resulted in the expression of NpABC1 (Jasinski *et al.*, 2001). This work suggests that plasma membrane ABC transporters are involved in the secretion of secondary metabolites but further work is required to ascertain the identity and role of plasma membrane ABC transporters involved in root exudation of specific compounds.

## 12.3  Phytoextraction

Over the last decade there have been major advances in our understanding of how plants acquire nutrient ions from the rhizosphere. Many transporters have been cloned and characterised, and reviews of these processes are presented in this book. However, although some advances have been made recently, much less is known of the transport processes associated with the uptake and efflux of toxic metals (e.g. $Zn^{2+}$, $Cd^{2+}$, $Ni^{2+}$, $Pb^{2+}$, etc.). Virtually nothing is known of how plants acquire, excrete or translocate organic pollutants or their derivatives; it is tacitly assumed that ABC transporters perform this task but there is very scant evidence to support this contention. For this reason, this section will focus only on the transport of metals.

Some studies on metal transport have utilised model plant systems (e.g. *Arabidopsis* and tobacco) and progress is now accelerating. Others have focused on naturally tolerant plants (e.g. true metallophytes such as plants from the genus *Thlaspi* or

*Alyssum,* or from *Arabidopsis halleri* – all from the *Brassicaceae*), as it is argued that these plants possess fundamental traits that are essential for tolerance, traits that are simply not found in the model plant systems.

### 12.3.1    Uptake of heavy metals from the rhizosphere

Several excellent review articles have appeared recently on this topic and so here we provide only a cursory summary and include a few new observations. For a more detailed discussion, readers are referred elsewhere, e.g. Lasat, 2002; McGrath *et al.*, 2002; McGrath & Zhao, 2003.

It is generally assumed that the uptake of non-nutrient ions occurs through trans-porters for nutrient ions. The kinetics for uptake of toxic ions into roots has not been studied extensively, the only exception being $Zn^{2+}/Cd^{2+}$ uptake. Several physiolog-ical studies have been conducted on a range of plants that express different levels of tolerance to these cations, and the general conclusion is that uptake is a carrier-mediated, saturable (Michaelis–Menten-like) process (Cutler & Rains, 1974; Lasat *et al.*, 1996; Hart *et al.*, 1998). Figure 12.1 summarises the mechanisms implicated in heavy metal transport.

Recently, heterologous expression of a cDNA from the metallophyte *Thlaspi caerulescens* was shown to complement a $Cd^{2+}$-sensitive yeast mutant. The cloned sequence, *TcZNT1*, showed good homology to the ZIP/IRT family of cation transporters, and was highly expressed in the plasma membrane of root cells in *T. caerulescens*, but not in the non-accumulator *Thlaspi arvense* (Pence *et al.*, 2000). Other members of the ZIP/IRT family have been implicated in heavy metal uptake in plants. In *Arabidopsis*, the iron transporters IRT1/2 have been shown to transport other cations, including $Zn^{2+}$, $Cd^{2+}$ and $Mn^{2+}$ (Eide *et al.*, 1996; Korshunova *et al.*, 1999; Guerinot, 2000).

Further, heterologous expression of *Arabidopsis* NRAMP transporters in transport-deficient *Saccharomyces cerevisiae* mutants restored $Fe^{2+}$ and $Mn^{2+}$ up-take, but also increased sensitivity to $Cd^{2+}$ and intracellular $Cd^{2+}$ levels, implicating a plasma membrane localisation *in planta* (Thomine *et al.*, 2003). Some of these transporters (AtNRAMP1/3/4) are expressed in root tissues and may, therefore, con-tribute to $Zn^{2+}$ and $Cd^{2+}$ uptake (Curie *et al.*, 2000; Thomine *et al.*, 2000; Thomine *et al.*, 2003). A more detailed description of the role of NRAMP transporters in the uptake of toxic metals in metallophytes and non-metallophytes is required.

An electrophysiological study has suggested that AtCNGC1, a cyclic nucleotide-gated ion channel (see Chapter 8), may be permeable to $Pb^{2+}$ and represents a major pathway for $Pb^{2+}$ uptake in *Arabidopsis* roots (Sunkar *et al.*, 2000). Further, in a recent screen of *Arabidopsis* activation-tagged lines, a $Cd^{2+}$-tolerant mutant was identified, which carried a knockout in a related gene (C. Thompson & P. Dominy, unpublished results, 2004).

In conclusion, the uptake of organic and inorganic pollutants from the rhizosphere into plant roots is poorly understood. Because of the membrane potential across the plant cell membrane, the uptake of heavy metal cations does not require an active mechanism, and to date, most classes of passive cation transporters have been implicated in this role. Therefore, the prospects for engineering plants more tolerant of heavy metals by reducing uptake (i.e. a gene knockout approach) are not clear, as nutrient ion acquisition may be compromised. All plants take up heavy metals to some extent but there are insufficient data at present to establish whether the kinetics of uptake by transport mechanisms in hyperaccumulators are significantly different from non-accumulators. At this stage, therefore, it is unclear whether the key physiological strategies required for hyperaccumulation are associated with uptake mechanisms as well as those dealing with the intracellular level of metals. Very little indeed is known of the substrates or the mechanisms by which modified or unmodified organic pollutants are taken up into plants.

### 12.3.2   Formation and transport of intracellular chelates

Relatively low levels of pollutants are toxic in the cytoplasm and it is not surprising, therefore, that tolerant plants have developed a battery of mechanisms for rapidly reducing their activities in root cells. In this regard, sequestration appears to be a vital strategy for survival, either by removal to endomembrane compartments or by complexing to chelating agents.

Phytochelatins, in particular, and metallothionines are small cysteine-rich peptides that are known to be expressed at high levels in tolerant plants and which donate SH ligands for complexing free cations or cation radicals in the cytoplasm. The pH of the cytoplasm ($>7.5$) probably precludes the formation of complexes with other ligand donors (e.g. malate, citrate, histidine). It is now well established that in yeast, complexed heavy metals are removed from the cytoplasm and transported into the vacuole. In *S. cerevisiae*, the cytosolic tripeptide glutathione ($\gamma$-ECG) complexes Cd by forming ligands through the SH group of the internal cysteine; the resulting complex, Cd($\gamma$-ECG)$_2$, is then transported into the central vacuole by the ScYCF1 ABC transporter (Li *et al.*, 1997). In contrast, in *Schizosaccharomyces pombe*, phytochelatins (PCs) are synthesised from glutathione and these also provide ligands for complexing $Cd^{2+}$ to form low-molecular-weight PCs. These are subsequently transported into the vacuole by the SpHMT1 ABC transporter (Kreutz *et al.*, 1996); for an excellent review see Cobbett & Goldsbrough (2002). Both of these processes are dependent on ATP but not the establishment of a proton electrochemical gradient (pmf); details of the precise mechanisms are unclear. A similar PC-mediated mechanism is believed to operate in higher plants, although a direct role for a glutathione-mediated system operating in some plants cannot be ruled out. It has been shown in tobacco protoplasts (Vogeli-Lange & Wagner, 1990) and in oat roots (Salt & Rauser, 1995) that PC complexes are transported into vacuoles in an ATP-dependent process that does not require a pmf. Several ABC transporters from the MRP subfamily have been shown to sequester glutathione, glucuronide and malonylated conjugates into vacuoles (Martinoia *et al.*, 2002), but it is not clear which members of this diverse family of transporters are involved in GSH- or PC-conjugate transport into the vacuoles of model plants or metallophytes (Cobbett & Goldsbrough, 2002).

There is evidence that toxic cations can be transported out of the cytoplasm into the apoplast by CDF transporters and by PDR ABC transporters, or into endomembrane/vacuolar compartments by CDF and CAX/MHX transporters. Several mammalian homologues of plant CDFs have been well characterised (ZnT1, ZnT3), and these transport $Zn^{2+}$ from the cytoplasm to outside the cell (Palmiter & Findley, 1995; Wenzel *et al.*, 1997), or into endomembrane compartments (ZnT2) (Palmiter *et al.*, 1996). In addition, two *S. cerevisiae* homologues (ZRC1 and COT1) transport Cd/Zn and Co (respectively) into the vacuole (Li & Kaplan, 1998). It is tempting to speculate that plant homologues of these CDF transporters play a similar role in removing free, uncomplexed heavy metals from the cytoplasm.

Yeast and plants appear to sequester $Ca^{2+}$ in their vacuoles operating $Ca^+/H^+$ exchangers (antiporters). Overexpressions of these Ca exchanger (CAX) sequences in plants result not only in higher levels of endogenous $Ca^{2+}$ but also of $Cd^{2+}$, suggesting CAX transporters are involved in the sequestration of divalent cations in endo- (CAX1) and vacuolar (CAX2) membranes (Hirschi, 2001). Recently, a protein has been identified in *Arabidopsis* (AtMHX1), which is a $H^+$-coupled antiporter that transports $Zn^{2+}$ and $Mn^{2+}$ into plant vacuoles (Shaul *et al.*, 1999). Whether any of the other members of the burgeoning members of plant cation/$H^+$ antiporters ($\sim$60 in *Arabidopsis*) perform similar tasks remains to be determined. Presumably, a comparable number of cation/$H^+$ antiporters are found in metallophytes and are also involved in toxic cation transport out of the cytoplasm.

The newly divided, undifferentiated cells in the root meristem of intact roots exposed to high levels of soil pollutants are faced with a very different set of problems. Cells in this region are avacuolate, and are not afforded the protection of the endodermis enjoyed by some more mature cells. They will be exposed to a significant amount of 'undiluted' soil solution as it is pulled by the transpiration stream through the intercellular spaces and up into the shoot. Undifferentiated cells in this region will therefore be exposed to relatively high levels of pollutants and have no means of protecting themselves by vacuolar sequestration. Presumably, these undifferentiated cells may adopt any one or a combination of the following strategies: decreased pollutant uptake at the plasma membrane, enhanced pollutant efflux at the plasma membrane, increased capacity for metabolic detoxification/cytosolic sequestration as inert organic complexes. Candidate transporters that may be involved in pollutant uptake have been outlined above and include NRAMPs ($Cd^{2+}/Fe^{2+}/Zn^{2+}$), ZIP/IRT ($Zn^{2+}/Cd^{2+}/Fe^{2+}/Mn^{2+}$) and CNGC ($Pb^{2+}/Cd^{2+}$); it is conceivable that the activities of these uptake mechanisms are impaired in undifferentiated cells. Enhanced efflux of toxic metals may be achieved by increased activity of cation diffusion facilitators (CDFs) such as the metal transport proteins or MTPs (Williams *et al.*, 2000) and ABC transporters, particularly members of the PDR subfamily (Theodoulou, 2000; Sanchez-Fernandez *et al.*, 2001; Martinoia *et al.*, 2002).

Intracellular metabolism of organic xenobiotics by plants occurs in four phases. In Phase I, an active functional group is introduced or revealed, for example by the action of esterases, amidases or cytochrome P450s. In Phase II, this active group is conjugated to a water-soluble moiety, such as glutathione, glucose or malonate via the respective transferase enzymes. Certain xenobiotics are inherently reactive and therefore susceptible to conjugation without prior activation. Although glucosyl transferases and malonyl transferases with activities towards herbicides have been documented in plants, the best-studied group of Phase II enzymes are the glutathione transferases (GST), which catalyse the conjugation of the tripeptide, glutathione or an analogue to an electrophilic site in an activated xenobiotic (Marrs, 1996; Edwards *et al.*, 2000). Conjugates are removed from the cytoplasm in Phase III by the action of energy-dependent transporters. It has been proposed, therefore, that conjugation 'tags' xenobiotic metabolites for sequestration or further degradation (Sandermann, 1994). In plants, glutathione conjugates are transported into the vacuole by the action of multidrug resistance-associated proteins (MRP), members of the ATP-binding cassette (ABC) transporter family (reviewed by Rea, 1999; Theodoulou, 2000). Within the vacuole, further metabolism occurs, constituting the poorly characterised Phase IV, and the xenobiotic metabolites may be deposited as bound residues in the cell wall (Sandermann, 1994).

The ABC superfamily consists of a large group of related proteins whose members mediate a wide range of transport processes in organisms ranging from bacteria to man (Higgins, 1992). Common to these proteins is the presence of one or two copies each of two core structural elements: a hydrophobic transmembrane-spanning (TMS) domain consisting of multiple membrane-spanning segments, and a cytosolically orientated highly conserved ABC. This molecular architecture allows the proteins to transport substrates across biological membranes, usually against a concentration gradient, via the binding and hydrolysis of ATP. The transmembrane domains are thought to form a chamber and to determine the substrate specificity of the transporter (Egner *et al.*, 2000; Chang & Roth, 2001). ABC transporters have been extensively studied in bacterial and mammalian systems and are associated with the acquisition of multiple drug resistance by pathogens as well as the occurrence of multidrug-resistant tumours (Higgins, 1992), but plant ABC transporters have only recently begun to be studied extensively (reviewed by Theodoulou, 2000). The sequencing of the *A. thaliana* genome has led to the identification of 129 putative members of the ABC transporter family, including 22 members of the multidrug resistance (MDR) subfamily, 15 members of the multidrug resistance related proteins (MRP) and 12 relatives of the yeast pleiotropic drug resistance (PDR) protein. However, the potential substrates for the vast majority of these transporters remain totally unknown and the subcellular localisation has only been substantiated for one member each of the MDR-like (Sidler *et al.*, 1998) and PDR-like (Jasinski *et al.*, 2001) families. The most progress has been made in the characterisation of the MRP-like proteins in plants for which a function in the transport of glutathione conjugates, glucuronides and glucuronide conjugates, and chloropyll catabolites into the vacuole has been shown in a heterologous yeast expression system (Lu *et al.*, 1997; Lu *et al.*, 1998; Tommasini *et al.*, 1998; Liu *et al.*, 2001).

Martinoia *et al.* (1993) demonstrated that intact vacuoles isolated from barley mesophyll mediated MgATP-dependent accumulation of glutathione conjugates (GS). The model substrate, *N*-ethylmaleimide-GS (NEM-GS), and a glutathione conjugate of the herbicide, metolachlor (metolachlor-GS), were investigated. The transport of NEM-GS and metolachlor-GS was shown to be sensitive to vanadate (phosphoryl transition state inhibitor, which inhibits transporters forming an acylphosphate during catalytic cycles and strongly inhibits GS-conjugate uptake), but unaffected by inhibitors of the vacuolar $H^+$-ATPase, and chemicals that collapse the tonoplast proton gradient, indicating that uptake of glutathione conjugates into the vacuole is mediated by a specific ATPase, and not by a secondary active process. Oxidised glutathione (GSSG) was also shown to be a substrate for this transporter, whereas reduced glutathione (GSH) was not (Tommasini et al., 1993). It has also been shown that MgATP-energised transport of the model conjugate dinitrophenol-GS (DNP-GS) by vacuolar membrane vesicles from Arabidopsis, beet, maize and mung bean (Li et al., 1995). Some further evidence has been provided that GS-X transport is part of an integrated detoxification pathway. The cereal safener cloquintocet mexyl doubled the vacuolar transport activity for both glutathione and glucoside conjugates of herbicides in barley; glutathione S-transferase (GST) activity was also increased by the treatment (Gaillard et al., 1994). Similarly, Li et al. (1995) found that pretreatment of mung bean seedlings with the model GST substrate, 1-chloro-2,4-dinitrobenzene, increased the activity of DNP-GS transport in tonoplast vesicles (Li et al., 1995). In both studies, application of the exogenous compound increased the $V_{max}$ but did not affect the $K_m$, suggesting that the higher activity was due to the increased expression of the transporter and not the altered affinity for substrate (Gaillard et al., 1994; Li et al., 1995). See Fig. 12.1 for a summary.

The identification and characterisation of new vacuolar pumps is of vital importance if plants are to be engineered for detoxification purposes. To date, research into herbicide resistance and xenobiotic detoxification has (a) involved the isolation of mutants or the engineering of plants in which the target for xenobiotic action is no longer sensitive, (b) involved the generation of mutants with elevated cellular levels of glutathionine (GSH) or with increased glutathione-S-transferase activities or (c) involved the application of chemical agents (safeners) that elevate GSH and/or glutathione-S-transferase levels or activities. However, these mechanisms have serious limitations. First, the use of mutated target gene products is limited in its application to those xenobiotics that directly interact with the target in question. Second, technologies based on elevated cellular GSH levels or increased glutathione-S-transferase catalytic efficiencies are limited by the capacity of cells to subsequently metabolise and/or sequester the conjugates generated. The success of all of these detoxification processes depends on transport of GSH conjugates into the vacuole and thus are reliant on the activity of the vacuolar GS-X pumps that tend to have broad substrate specificity. Since the plant vacuole frequently constitutes 40–90% of total intracellular volume and the GS-X pumps mediate the uptake of xenobiotics into this compartment, the use of engineered vacuolar pumps has great potential for the promotion of hyperaccumulation.

### 12.3.3 Transport to the shoot

Most plants show a strong preference to sequester heavy metals in their roots (typical shoot/root ratios are <0.1 when expressed on wt cation/dry wt basis). However, a few species of higher plants, approximately 400, are 'hyperaccumulators' that can concentrate very high levels of heavy metals in their shoots ($\sim$10% $Zn^{2+}/Mn^{2+}$, $\sim$1% $Co^{2+}$, $Cu^{2+}$, $Ni^{2+}$, $As^{2+}$ and $Se^{2+}$, and $\sim$0.1% $Cd^{2+}$), expressed on a tissue dry-weight basis. These hyperaccumulators can concentrate metals in their shoots to levels that are 100–1000 times greater than those found in non-accumulators, and can show shoot/root ratios of >1 (Brown *et al.*, 1995; Shen *et al.*, 1997; Zhao *et al.*, 2000).

Clearly, hyperaccumulators have adopted strategies for coping with high levels of heavy metals in the root, but also for their export from the root to the shoot, and their sequestration in the aerial parts of the plant. In general, there appears to be a good correlation between transpiration rate and shoot heavy metal levels. Therefore, the transfer of heavy metals to the shoot is believed to occur through the xylem (McGrath *et al.*, 2002). It is this delivery and loading of the pollutant onto the xylem that is believed to be the key physiological strategy that confers the hyperaccumulator trait. This may be achieved by a symplastic or apoplastic pathway. The symplastic route would require the cell-to-cell centripetal diffusion through plasmodesmata of the metal, probably in its conjugated form, from its site of uptake to the xylem parenchyma. As the levels of conjugated metals in the xylem sap are generally found to be low, presumably the metal is deconjugated and excreted into the xylem transpiration stream. Perhaps the more likely strategy is the apoplastic movement of metals to the xylem in the transpiration stream where the low pH facilitates ligation to free organic chelators. The root cells of hyperaccumulators may not, therefore, directly load metals onto the xylem, but excrete excess levels into the apoplast where they are swept by the transpiration stream up into the shoot.

The 1b subfamily of P-type ATPases are known to transport cations across membranes. There are eight members of this subfamily now recognised in *Arabidopsis* (AtHMA1-6 and AtRAN1 and AtPAA1), all of which have eight putative membrane-spanning domains, the sixth of which contains the Cys-Pro-Cys (CPC) putative metal translocation motif (Williams *et al.*, 2000; Axelsen & Palmgren, 2001; Cobbett *et al.*, 2003). AtRAN1 is reported to be localised to the Golgi membrane and involved in $Cu^{2+}$ transport, but the location of AtPMA1-6 and AtPAA1 is unclear (Hirayama *et al.*, 1999). Recently, however, an elegant reverse genetics approach was used to show that AtHMA2 and AtHMA4 are involved in $Zn^{2+}$ ($Cd^{2+}$) transport to the shoot. Subsequent GUS reporter gene experiments have shown that AtHMA2 and AtMHA4 are localised to the vasculature, basal anther cells and siliques (Hussain *et al.*, 2004). It appears that in *Arabidopsis*, type 1b (CPC) ATPases are involved in Zn (Cd) transport between parenchyma cells and the vascular tissues. Differences between the activities of homologues in hyperaccumulators and non-accumulators may account for the high levels of metals found in the tissues of the former. Comparative studies on the role of these transporters in hyperaccumulators and non-accumulators are required to confirm this possibility.

Nickel hyperaccumulators from the genus *Alyssum* appear to secrete histidine into the xylem, which complexes with the free cation, presumably to reduce phytotoxicity. Other $Ni^{2+}$ hyperaccumulators also show high levels of histidine in their transpiration stream, but no comparable increases have been detected in the $Zn^{2+}/Cd^{2+}$ hyperaccumulator *A. halleri*, the $Mn^{2+}$ hyperaccumulator *Grevillea exul* (Smith *et al.*, 1999) or the $Ni^{2+}$ hyperaccumulator *Thlaspi goesingense* (Persans *et al.*, 1999). The mechanisms by which histidine or other possible organic ligands (citrate, malate, etc.) are secreted into the xylem are not known.

### 12.3.4 Distribution and compartmentation in the shoot

In hyperaccumulators, most of the heavy metals entering the roots must be unloaded from the xylem in the shoot and subsequently distributed to tissues and cellular compartments where they exert a minimal effect on metabolism. It is well established that the epidermal cells and trichomes of leaves act as sinks for excess levels of nutrient and toxic ions. It is not surprising, therefore, that for most metallophytes studied, $Zn^{2+}$ (Vazquez *et al.*, 1994; Kupper *et al.*, 1999; Frey *et al.*, 2000) and $Ni^{2+}$ (Kramer *et al.*, 1997; Psaras *et al.*, 2000; Kupper *et al.*, 2001) accumulate in epidermal tissues, although there is some conjecture on the distribution between cell types (McGrath *et al.*, 2002). For example, the $Zn^{2+}/Cd^{2+}$ hyperaccumulator *A. halleri* does not accumulate metals in epidermal cells although elevated levels were found in the trichomes (Kupper *et al.*, 2000).

If the general principle is accepted that hyperaccumulators transfer excess levels of metals from the vasculature to the epidermis, the question arises 'how is this achieved?' One possibility is that the metal arriving in the shoot xylem (presumably conjugated to an organic acid, see above) moves from the xylem directly into the xylem parenchyma apoplast and eventually by a combination of mass flow and diffusion to the apoplast of epidermal cells. Presumably, an energy-dependent process is then required to take up the metal into an epidermal or trichome cell. Alternatively, metals transported in the xylem to the shoot may be taken up by the xylem parenchyma and transported symplastically to the mesophyll cell/epidermal cell interface, but as plasmodesmatal connections between these tissues are rare, the metals must be unloaded into the apoplast and then taken up by an energy-dependent process. In either case, the epidermal cells that accumulate the metals must express proteins on their cell and endomembranes for transporting metals. This may be achieved by the action of passive transporters (e.g. CNGCs, CDF, ZIP/IRT, etc.), although to maintain a suitable membrane potential, a counter ion would also have to be imported. Alternatively, import may be achieved by a $H^+$-coupled mechanism (e.g. NRAMP, CAX/MHX), or by primary active transporters such as metal-transporting CPx-type ATPases (Williams *et al.*, 2000) and ABC transporters (Theodoulou, 2000; Sanchez-Fernandez *et al.*, 2001; Martinoia *et al.*, 2002). See Fig. 12.1 for a summary.

## 12.4  Discussion

It is not surprising, given the chemically diverse range of pollutants found in the environment and the narrow spectrum of tolerance expressed by plants growing in

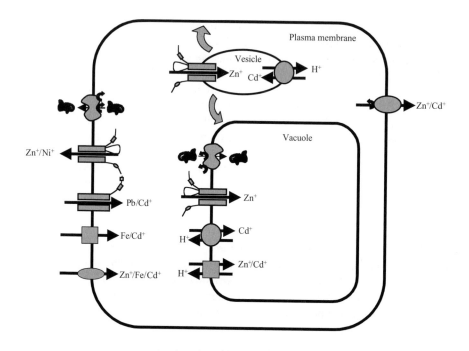

## Key

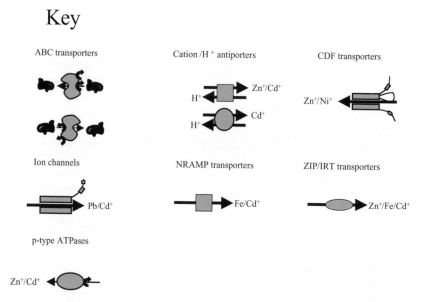

**Fig. 12.1** Summary of mechanisms involved in toxic metal ion transport in plants. This cartoon has been compiled from studies on a range of metallophytes and non-metallophytes. In some cases, functional characterisation has been performed *in planta*, and in others metal transport has been confirmed by heterologous expression in yeast (see text for details). The examples given are as follows. (*Continued*)

contaminated soils, that progress on identifying key traits and their corresponding genes that confer tolerance has been slow. Fortunately, many of the plants that are naturally tolerant of high levels of metals (*Thlaspi* spp. and *A. halleri* – $Zn^{2+}/Cd^{2+}$; *Alyssum* spp. – $Ni^{2+}$) are *Brassicas* and are closely related to *A. thaliana*, and the impressive array of experimental resources available for this model plant will greatly accelerate our understanding of how metallophytes cope with metal ion stress. It is conceivable that *A. thaliana* itself possesses all of the necessary genetic informa- tion to respond as a true metallophyte, even a hyperaccumulator, but that through evolution, it has lost its ability to express the appropriate genes in the appropriate tissues at the required time. It is now clear that the notion that evolution has honed plants to be highly efficient, aggressive competitors is an oversimplification. For example, *A. thaliana* possesses three transcription factors (CBF1, CBF2 and CBF3) that are involved in activating cold acclimation responses to prevent frost damage. Ectopic expression of one, CBF3, has no apparent cost to the plant in non-freezing conditions but allows survival at temperatures ~20°C lower than does CBF1, the transcription factor that controls many of the complex acclimation responses in WT lines (Zarka *et al.*, 2003; C. Somerville, personal communication, 2004); in this particular case, *Arabidopsis* possesses the genetic potential to allow it to survive much better at low temperatures, but it does not do so. It is conceivable that the

---

**Fig. 12.1** (continued)

*ABC transporters*: AtPDR (plasma membrane, upper), AtMRP and SpHMT1 (tonoplast, lower). These 'full-size' ABC transporters have two nucleotide folds (NBF) and two membrane-spanning domains (MSD); the pleiotropic drug resistance (PDR) family contains the NBF-MSD-NBF-MSD arrangement, while the multidrug resistance-related family contains the arrangement of MSD-NBF-MSD-NBF.

*Ion channels*: AtCNGC2. Cyclic nucleotide-gated ion channels are related to the Shaker family of cation channels, but are characterised by C-terminal calmodulin and cyclic nucleotide-binding domains.

*Cation/$H^+$ antiporters*: AtCAX1 (endomembrane), AtCAX2 (vacuole) and AtMHX1 (vacuole). Mem- bers have 10–14 putative MSDs. Involved in the transport of the micronutrients $Ca^{2+}$, $Mg^{2+}$ and $Mn^{2+}$, but also transport $Cd^{2+}$.

*NRAMP transporters*: AtNRAMP3. Natural resistance-associated macrophage protein. Believed to be involved in $Fe^{2+}$ and $Mn^{2+}$ transport, but also appear to be permeable to $Cd^{2+}$.

*CDF transporters* (cation diffusion facilitator effluxers): AtMTP1 (ZAT), TgMTP1-3, ScCOT1, ScZRC1 (all vacuolar influx) and RnZnT1 (plasma membrane). Characterised by six putative MSDs, N-terminal signature, C-terminal cation efflux domain (Pfam 01545) and a cytoplasmic His-rich domain (all shown). Can transport $Zn^{2+}$ ($Cd^{2+}$), possibly $Co^{2+}$ and $Ni^{2+}$.

*ZIP/IRT transporters*: AtIRT1, AtIRT2, AtZIP1-3. Members have eight putative MSDs with a His-rich cytoplasmic domain between MSD III and IV. These proteins normally transport the micronutrients $Fe^{3+}$ and $Zn^{2+}$.

*p-type ATPases*, AtMHA2, AtMHA4. This group of transporters are characterised by eight putative MSDs, the sixth possessing a CPC motif thought to be involved in metal translocation. Some members are involved in $Zn^{2+}$ ($Cd^{2+}$) transport.

Transporters with sites for ATP hydrolysis are indicated by small, dark curved arrows. A role for vesicular trafficking is shown but has not been confirmed experimentally. At, *Arabidopsis thaliana*; Sp, *Schizosac- charomyces pombe*; Tg, *Thlaspi goesingense*; Sc, *Saccharomyces cerevisiae*; Rn, *Rattus norvegicus*. For more details, see text.

manipulation of a single locus in the *Arabidopsis* genome could convert the plant into a true metallophyte, but how can such loci be identified?

Conventional genetic approaches almost always result in gene knockouts (loss-of-function mutations) that usually give rise to hypersensitivity; in screens, the mutants of interest die. The major exception to this rule is the deletion of negative regulators that suppress the expression or activity of desirable traits. A more useful strategy might be activation tagging (Weigel *et al.*, 2000) as this approach will generate both loss-of-function (gene knockout) and gain-of-function (gene activation) mutations. In addition, gene activation circumvents the problem of gene redundancy, which is often encountered with knockout mutants. This approach should, therefore, identify mutants that are more tolerant (survive) against a background of the sensitive (WT) phenotype (which die). In a recent screen of $\sim$60 000 activation-tagged lines for $Cd^{2+}$ tolerance, 16 mutants were identified and 9 have been partially characterised. Amongst the gene activation mutants (overexpressing lines) are a COBRA protein with reported PC synthase-like activity, a cytochrome P450, an organic acid transporter and a zinc RING finger protein, all members of gene families that have been implicated in heavy metal tolerance in plants. Amongst those mutants that carry gene knockouts are an autophagy-related protein and a cyclic-nucleotide-gated ion channel, again members of gene families that have been implicated in stress and heavy metal tolerance (C. Thompson & P. Dominy, unpublished results, 2004). This experimental strategy appears to identify sensible candidate genes for manipulation in *A. thaliana*. Further screens of activation-tagged lines may reveal other $Cd^{2+}$, metal and organic pollutant tolerant lines. Further, if the transformation efficiency can be improved, a similar strategy should be possible in metallophytes (e.g. *Thlaspi* spp. and *Alyssum* spp.).

At this stage, very little can be concluded on the role of transport processes in the biostabilisation or bioremediation of organic pollutants as very little is known. Activation tagging in *A. thaliana* may offer a way forward; similar resources are currently being assembled in rice (Jeong *et al.*, 2002) and maize (Marsch-Martinez *et al.*, 2002). Recently, Drake *et al.* (2002) postulated the use of antibody expressing transgenic plants to neutralise bioactive molecules in the rhizosphere or to accumulate and concentrate such molecules in leaves. Suggested target molecules would be any pollutant for which it is possible to generate a monoclonal antibody but particularly a range of pesticides and several suspected endocrine-disrupting chemical families (such as phthalates, alkylphenols, PCBs, dioxins and both natural and synthetic steroid estrogens). In order to test the feasibility of such an approach, Drake *et al.* (2002) generated a model system of two transgenic plant lines expressing a murine monoclonal antibody. In the first line, functional antibody was rhizosecreted and shown to bind with antigen in the surrounding media. In the second line, a transmembrane sequence retained the antibody in the plasma membrane. Antigen added to the media was transported through the plant and sequestered in leaves, with the antibody as an immune complex.

In conclusion, rapid advances have been made recently in our understanding of how *A. thaliana* copes with metal ion stress, and new experimental tools developed in this model plant system are likely to accelerate progress. Many metallophytes are close relatives of *A. thaliana*, and this fortuitous occurrence will also help enormously with understanding how metallophytes respond to metal ion stress. In some cases, it might be that simply blocking the major pathway for metal acquisition (using gene knockout technologies) will allow better growth and reduce the concentration of metal in the above ground parts. Manipulation of root exudates may also help stabilise contaminated soils and result in produce with lower levels of toxins. The prospects for phytoextraction of metals from contaminated soils are not good, at least in the short term. Metal hyperaccumulators can attain impressive concentrations of metals in their shoots, but their slow growth and low biomass make their use less attractive (McGrath *et al.*, 2002). It remains to be seen if biotechnology can be used to engineer faster growth in hyperaccumulators, or to convert faster growing plants into hyperaccumulators.

On initial examination, the phytoremediation of soils contaminated with organic pollutants is, by comparison, a far more complex problem. The chemical diversity of organic pollutants present in contaminated soils is orders of magnitude greater than the diversity of metal pollutants. However, plants with their extraordinary range of secondary metabolites have become adept at dealing with a huge array of potentially cytotoxic natural products. Unlike metal remediation where the only option is hyperaccumulation, phytoremediation of organic pollutants has the potential to result in complete mineralisation of the contaminant. In order to fully harness these detoxification mechanisms for phytoremediation, we must first understand them fully and deduce their physiological roles. More research in this area is vital if phytoremediation is going to be a viable, realistic alternative to current expensive and environmentally destructive remediation processes.

# References

Alexander, M. (2000) Aging, bioavailability, and overestimation of risk from environmental pollutants, *Environ. Sci. Technol.*, **34** (20), 4259–4265.

Axelsen, K.B. & Palmgren, M.G. (2001) Inventory of the superfamily of P-type ion pumps in *Arabidopsis*, *Plant Physiol.*, **126** (2), 696–706.

Bais, H.P., Walker, T.S., Stermitz, F.R., Hufbauer, R.A. & Vivanco, J.M. (2002) Enantiomeric-dependent phytotoxic and antimicrobial activity of (+/−)-catechin. A rhizosecreted racemic mixture from spotted knapweed, *Plant Physiol.*, **128** (4), 1173–1179.

Bolan, N.S., Naidu, R., Mahimairaja, S. & Baskaran, S. (1994) Influence of low-molecular-weight organic-acids on the solubilization of phosphates, *Biol. Fertil. Soils*, **18** (4), 311–319.

Boudot, J.P. (1992) Relative efficiency of complexed aluminum, noncrystalline Al hydroxide, allophane and imogolite in retarding the biodegradation of citric-acid, *Geoderma*, **52** (1/2), 29–39.

Bowers, J.H., Nameth, S.T., Riedel, R.M. & Rowe, R.C. (1996) Infection and colonization of potato roots by *Verticillium dahliae* as affected by *Pratylenchus penetrans* and *P. crenatus*, *Phytopathology*, **86** (6), 614–621.

Braum, S.M. & Helmke, P.A. (1995) White lupin utilizes soil-phosphorus that is unavailable to soybean, *Plant Soil*, **176** (1), 95–100.

Brown, S.L., Chaney, R.L., Angel, J.S. & Baker, A.J.M. (1995) Zinc and cadmium uptake by the hyperaccumulator *Thlaspi caerulescens* grown in nutrient solution, *Soil Sci. Soc. Am. J.*, **59**, 125–133.

Brunner, W., Sutherland, F.H. & Focht, D.D. (1985) Enhanced biodegradation of polychlorinated-biphenyls in soil by analog enrichment and bacterial inoculation, *J. Environ. Qual.*, **14** (3), 324–328.

Burken, J.G. & Schnoor, J.L. (1996) Phytoremediation: plant uptake of atrazine and role of root exudates, *J. Environ. Eng.-ASCE*, **122** (11), 958–963.

Carroll, K.M., Harkness, M.R., Bracco, A.A. & Balcarcel, R.R. (1994) Application of a permeant polymer diffusional model to the desorption of polychlorinated-biphenyls from Hudson River sediments, *Environ. Sci. Technol.*, **28** (2), 253–258.

Chang, G. & Roth, C.B. (2001) Structure of MsbA from *E. coli*: a homolog of the multidrug resistance ATP binding cassette (ABC) transporters, *Science*, **293** (5536), 1793–1800.

Chen, S.H. & Aitken, M.D. (1999) Salicylate stimulates the degradation of high molecular weight polycyclic aromatic hydrocarbons by *Pseudomonas saccharophila* P15, *Environ. Sci. Technol.*, **33** (3), 435–439.

Chubin, R.G. & Street, J.J. (1981) Adsorption of cadmium on soil constituents in the presence of complexing ligands, *J. Environ. Qual.*, **10** (2), 225–228.

Cieslinski, G., Van Rees, K.C.J., Szmigielska, A.M., Krishnamurti, G.S.R. & Huang, P.M. (1998) Low-molecular-weight organic acids in rhizosphere soils of durum wheat and their effect on cadmium bioaccumulation, *Plant Soil*, **203** (1), 109–117.

Cobbett, C. & Goldsbrough, P. (2002) Phytochelatins and metallothionines: roles in heavy metal detoxification and homeostasis, *Annu. Rev. Plant Biol.*, **53**, 159–182.

Cobbett, C.S., Hussain, D. & Haydon, M.J. (2003) Structural and functional relationships between type 1(B) heavy metal-transporting P-type ATPases in *Arabidopsis*, *New Phytol.*, **159** (2), 315–321.

Colbert, S.F., Schroth, M.N., Weinhold, A.R. & Hendson, M. (1993) Enhancement of population-densities of *Pseudomonas putida* Ppg7 in agricultural ecosystems by selective feeding with the carbon source salicylate, *Appl. Environ. Microbiol.*, **59** (7), 2064–2070.

Curie, C., Alonso, J.M., Le Jean, M., Ecker, R. & Briat, J.-F. (2000) Involvement of NRAMP1 from *Arabidopsis thaliana* in iron transport, *Biochem. J.*, **347**, 749–755.

Cutler, J.M. & Rains, D.W. (1974) Characterization of cadmium uptake by plant tissue, *Plant Physiol.*, **54**, 67–71.

Daane, L.L., Harjono, I., Zylstra, G.J. & Haggblom, M.M. (2001) Isolation and characterization of polycyclic aromatic hydrocarbon-degrading bacteria associated with the rhizosphere of salt marsh plants, *Appl. Environ. Microbiol.*, **67** (6), 2683–2691.

Dabrock, B., Riedel, J., Bertram, J. & Gottschalk, G. (1992) Isopropylbenzene (cumene) – a new substrate for the isolation of trichloroethene-degrading bacteria, *Arch. Microbiol.*, **158** (1), 9–13.

de la Fuente, J.M., Ramirez-Rodriguez, V., Caberera-Ponce, J.L. & Herrera-Estrella, L. (1997) Aluminum tolerance in transgenic plants by alteration of citrate synthesis, *Science*, **276**, 1566–1568.

Degenhardt, J., Larsen, P.B., Howell, S.H. & Kochian, L.V. (1998) Aluminum resistance in the *Arabidopsis* mutant alr-104 is caused by an aluminum-induced increase in rhizosphere pH, *Plant Physiol.*, **117** (1), 19–27.

Delhaize, E., Craig, S., Beaton, C.D., Bennet, R.J., Jagadish, V.C. & Randall, P.J. (1993) Aluminum tolerance in wheat (*Triticum aestivum* L), 1: uptake and distribution of aluminum in root apices, *Plant Physiol.*, **103** (3), 685–693.

Dinkelaker, B., Romheld, V. & Marschner, H. (1989) Citric-acid excretion and precipitation of calcium citrate in the rhizosphere of white lupin (*Lupinus albus* L.), *Plant Cell Environ.*, **12** (3), 285–292.

Donnelly, P.K., Hegde, R.S. & Fletcher, J.S. (1994) Growth of Pcb-degrading bacteria on compounds from photosynthetic plants, *Chemosphere*, **28** (5), 981–988.

Drake, P.M.W., Chargelegue, D., Vine, N.D., Van Dolleweerd, C.J., Obregon, P. & Ma, J.K.C. (2002) Transgenic plants expressing antibodies: a model for phytoremediation, *FASEB J.*, **16** (14), 1855–1860.

Edwards, R., Dixon, D.P. & Walbot, V. (2000) Plant glutathione *S*-transferases: enzymes with multiple functions in sickness and in health, *Trends Plant Sci.*, **5** (5), 193–198.

Egner, R., Bauer, B.E. & Kuchler, K. (2000) The transmembrane domain 10 of the yeast Pdr5p ABC antifungal efflux pump determines both substrate specificity and inhibitor susceptibility, *Mol. Microbiol.*, **35** (5), 1255–1263.

Eide, D., Broderius, M., Fett, J. & Guerinot, M.L. (1996) A novel iron-regulated metal transporter from plants identified by functional expression in yeast, *Proc. Natl. Acad. Sci. U.S.A.*, **93**, 5624–5628.

Einhellig, F.A., Rasmussen, J.A., Hejl, A.M. & Souza, I.F. (1993) Effects of root exudate sorgoleone on photosynthesis, *J. Chem. Ecol.*, **19** (2), 369–375.

Elliott, H.A. & Denneny, C.M. (1982) Soil adsorption of cadmium from solutions containing organic-ligands, *J. Environ. Qual.*, **11** (4), 658–663.

Facchini, P.J. (2001) Alkaloid biosynthesis in plants: biochemistry, cell biology, molecular regulation, and metabolic engineering applications, *Annu. Rev. Plant Physiol. Plant Mol. Biol.*, **52**, 29–66.

Farrar, J.F. & Jones, D.L. (2000) The control of carbon acquisition by roots, *New Phytol.*, **147** (1), 43–53.

Fletcher, J.S. & Hegde, R.S. (1995) Release of phenols by perennial plant-roots and their potential importance in bioremediation, *Chemosphere*, **31** (4), 3009–3016.

Flores, H.E., Vivanco, J.M. & Loyola-Vargas, V.M. (1999) 'Radicle' biochemistry: the biology of root-specific metabolism, *Trends Plant Sci.*, **4** (6), 220–226.

Focht, D.D. (1995) Strategies for the improvement of aerobic metabolism of polychlorinated-biphenyls, *Curr. Opin. Biotechnol.*, **6** (3), 341–346.

Frey, B., Keller, C., Zeirold, K. & Schulin, R. (2000) Distribution of zinc in functionally different leaf epidermis cells of the hyperaccumulator *Thlaspi caerulescens*, *Plant Cell Environ.*, **23**, 675–687.

Gaillard, C., Dufaud, A., Tommasini, R., Kreuz, K., Amrhein, N. & Martinoia, E. (1994) A herbicide antidote (safener) induces the activity of both the herbicide detoxifying enzyme and of a vacuolar transporter for the detoxified herbicide, *FEBS Lett.*, **352** (2), 219–221.

Gerhardt, R., Stitt, M. & Heldt, H.W. (1987) Subcellular metabolite levels in spinach leaves – regulation of sucrose synthesis during diurnal alterations in photosynthetic partitioning, *Plant Physiol.*, **83** (2), 399–407.

Gilbert, E.S. & Crowley, D.E. (1997) Plant compounds that induce polychlorinated biphenyl biodegradation by *Arthrobacter* sp. strain B1B, *Appl. Environ. Microbiol.*, **63** (5), 1933–1938.

Grayston, S.J. & Campbell, C.D. (1996) Functional biodiversity of microbial communities in the rhizospheres of hybrid larch (*Larix eurolepis*) and Sitka spruce (*Picea sitchensis*), *Tree Physiol.*, **16** (11/12), 1031–1038.

Grotewold, E., Chamberlin, M., Snook, M., *et al.* (1998) Engineering secondary metabolism in maize cells by ectopic expression of transcription factors, *Plant Cell*, **10** (5), 721–740.

Grotewold, E. & Lin, Y.K. (2000) Trafficking of plant secondary metabolites, *Mol. Biol. Cell*, **11**, 2584.

Guengerich, F.P. (2001) Common and uncommon cytochrome P450 reactions related to metabolism and chemical toxicity, *Chem. Res. Toxicol.*, **14** (6), 611–650.

Guerinot, M.L. (2000) The ZIP family of metal transporters, *Biochim. Biophys. Acta*, **1465**, 190–198.

Haas, C.N. & Horowitz, N.D. (1986) Adsorption of cadmium to kaolinite in the presence of organic material, *Water Air Soil Pollut.*, **27** (1/2), 131–140.

Haby, P.A. & Crowley, D.E. (1996) Biodegradation of 3-chlorobenzoate as affected by rhizodeposition and selected carbon substrates, *J. Environ. Qual.*, **25** (2), 304–310.

Hart, J.J., Welch, R.M., Norvell, W.A., Sullivan, L.A. & Kochian, L.V. (1998) Characterization of cadmium binding, uptake, and translocation in intact seedlings of bread and durum wheat cultivars, *Plant Physiol.*, **116** (4), 1413–1420.

Hatzinger, P.B. & Alexander, M. (1995) Effect of aging of chemicals in soil on their biodegradability and extractability, *Environ. Sci. Technol.*, **29** (2), 537–545.

Hernandez, B.S., Koh, S.C., Chial, M. & Focht, D.D. (1997) Terpene-utilizing isolates and their relevance to enhanced biotransformation of polychlorinated biphenyls in soil, *Biodegradation*, **8** (3), 153–158.

Higgins, C.F. (1992) ABC transporters – from microorganisms to man, *Annu. Rev. Cell Biol.*, **8**, 67–113.

Hirayama, T., Kieber, J.J., Hirayama, N., *et al.* (1999) Responsive-to-antagonist1, a Menkes/Wilson disease-related copper transporter, is required for ethylene signaling in *Arabidopsis*, *Cell*, **97** (3), 383–393.

Hirschi, K.D. (2001) Vacuolar $H^+$/$Ca^{2+}$ transporters: who's directing the traffic? *Trend Plant Sci.*, **6**, 100–104.

Hopkins, B.G., Whitney, D.A., Lamond, R.E. & Jolley, V.D. (1998) Phytosiderophore release by sorghum, wheat, and corn under zinc deficiency, *J. Plant Nutr.*, **21** (12), 2623–2637.

Hue, N.V., Craddock, G.R. & Adams, F. (1986) Effect of organic-acids on aluminum toxicity in subsoils, *Soil Sci. Soc. Am. J.*, **50** (1), 28–34.

Hussain, D., Haydon, M.J. Wang, Y., *et al.*, (2004) P-type ATPase heavy metal transporters with roles in essential zinc homeostasis in *Arabidosis*, *Plant Cell* **16**, 1327–1339.

Isidorov, V. & Jdanova, M. (2002) Volatile organic compounds from leaves litter, *Chemosphere*, **48** (9), 975–979.

Janczarek, M., UrbanikSypniewska, U. & Skorupska, A. (1997) Effect of authentic flavonoids and the exudate of clover roots on growth rate and inducing ability of nod genes of *Rhizobium leguminosarum* bv trifolii, *Microbiol. Res.*, **152** (1), 93–98.

Jasinski, M., Stukkens, Y., Degand, H., Purnelle, B., Marchand-Brynaert, J. & Boutry, M. (2001) A plant plasma membrane ATP binding cassette-type transporter is involved in antifungal terpenoid secretion, *Plant Cell*, **13** (5), 1095–1107.

Jauregui, M.A. & Reisenauer, H.M. (1982) Dissolution of oxides of manganese and iron by root exudate components, *Soil Sci. Soc. Am. J.*, **46** (2), 314–317.

Jeong, D.H., An, S., Kang, H.G., *et al.* (2002), T-DNA insertional mutagenesis for activation tagging in rice, *Plant Physiol.*, **130** (4), 1636–1644.

Jones, D.L. (1998) Organic acids in the rhizosphere – a critical review, *Plant Soil*, **205** (1), 25–44.

Kim, B.H., Oh, E.T., So, J.S., Ahn, Y. & Koh, S.C. (2003) Plant terpene-induced expression of multiple aromatic ring hydroxylation oxygenase genes in *Rhodococcus* sp strain T104, *J. Microbiol.*, **41** (4), 349–352.

Knee, E.M., Gong, F.C., Gao, M.S., *et al.* (2001), Root mucilage from pea and its utilization by rhizosphere bacteria as a sole carbon source, *Mol. Plant-Microbe Interact.*, **14** (6), 775–784.

Kochian, L.V. (1995) Cellular mechanisms of aluminum toxicity and resistance in plants, *Annu. Rev. Plant Physiol. Plant Mol. Biol.*, **46**, 237–260.

Korshunova, Y.O., Eide, D., Clark, W.G., Guerinot, M.L. & Pakrasi, H.B. (1999) The IRT1 protein from *Arabidopsis thaliana* is a metal transporter with a broad substrate range, *Plant Mol. Biol.*, **40**, 37–44.

Kramer, U., Grime, G.W., Smith, J.A.C., Hawes, C.R. & Baker, A.J.M. (1997) Micro-PIXE as a technique for studying nickel localization in leaves of the hyperaccumulator plant *Alyssum lesbiacum*, *Nucl. Instr. Meth. Phys. Res.*, **130**, 346–350.

Kreutz, K., Tommasini, R. & Martinoia, E. (1996) Old enzymes for a new job: herbicide detoxification in plants, *Plant Physiol.*, **111**, 349–353.

Kronestedt-Robards, E. & Robards, A.W. (1991) Exocytosis in gland cells, in *Endocytosis, Exocytosis and Vesicle Traffic in Plants* (eds C.R. Hawes, J.O.D. Coleman & D.E. Evans), Cambridge University Press, Cambridge, UK, pp. 199–232.

Kupper, H., Lombi, E., Zhao, F.J. & McGrath, S.P. (2000) Cellular compartmentation of cadmium and zinc in relation to other elements in the hyperaccumulator *Arabidopsis halleri*, *Planta*, **212**, 75–84.

Kupper, H., Lombi, E., Zhao, F.J., Wieshammer, G. & McGrath, S.P. (2001) Cellular compartmentation of nickel in the hyperaccumulators *Alussum lesbiacum*, *Alyssum bortoilonii* and *Thlaspi goesingense*, *J. Exp. Bot.*, **52**, 2291–2300.

Kupper, H., Zhao, F.J. & McGrath, S.P. (1999) Cellular compartmentation of zinc in leaves of the hyperaccumulator *Thlaspi caerulescens*, *Plant Physiol.*, **119**, 305–311.

Lasat, M.M. (2002) Phytoextraction of toxic metals: a review of biological mechanisms, *J. Environ. Qual.*, **31** (1), 109–120.

Lasat, M.M., Baker, A. & Kochian, L.V. (1996) Physiological characterization of root $Zn^{2+}$ absorption and translocation to shoots in Zn hyperaccumulator and nonaccumulator species of *Thlaspi*, *Plant Physiol.*, **112** (4), 1715–1722.

Li, L. & Kaplan, J. (1998) Defects in the yeast high affinity iron transport system result in increased metal sensitivity because of an increased expression of transporters with a broad transition metal specificity, *J. Biol. Chem.*, **273**, 22181–22187.

Li, Z.-S., Lu, Y.-P., Zhen, R.-G., Szczypka, M., Thiele, D.J. & Rea, P.A. (1997) A new pathway for vacuolar cadmium sequestration in *Saccharomyces cerevisiae*: YCF1-catalysed transport of *bis*(glutathionato) cadmium., *Proc. Natl. Acad. Sci. U.S.A.*, **94**, 42–47.

Li, Z.S., Zhao, Y. & Rea, P.A. (1995) Magnesium adenosine 5′-triphosphate-energized transport of glutathione-S-conjugates by plant vacuolar membrane-vesicles, *Plant Physiol.*, **107** (4), 1257–1268.

Liu, G.S., Sanchez-Fernandez, R., Li, Z.S. & Rea, P.A. (2001) Enhanced multispecificity of *Arabidopsis* vacuolar multidrug resistance-associated protein-type ATP-binding cassette transporter, AtMRP2, *J. Biol. Chem.*, **276** (12), 8648–8656.

Lu, Y.P., Li, Z.S., Drozdowicz, Y.M., Hortensteiner, S., Martinoia, E. & Rea, P.A. (1998) AtMRP2, an *Arabidopsis* ATP binding cassette transporter able to transport glutathione S-conjugates and chlorophyll catabolites: functional comparisons with AtMRP1, *Plant Cell*, **10** (2), 267–282.

Lu, Y.P., Li, Z.S. & Rea, P.A. (1997) AtMRP1 gene of *Arabidopsis* encodes a glutathione S-conjugate pump: isolation and functional definition of a plant ATP-binding cassette transporter gene, *Proc. Natl. Acad. Sci. U.S.A.*, **94** (15), 8243–8248.

Ma, J.F. (2000a) Role of organic acids in detoxification of aluminum in higher plants, *Plant Cell Physiol.*, **41** (4), 383–390.

Ma, J.F. (2000b) Role of organic acids in detoxification of aluminum in higher plants, *Plant Cell Physiol.*, **41** (4), 383–390.

Ma, J.F., Ryan, P.R. & Delhaize, E. (2001) Aluminium tolerance in plants and the complexing role of organic acids, *Trends Plant Sci.*, **6** (6), 273–278.

Ma, J.F., Ueno, H., Ueno, D., Rombola, A.D. & Iwashita, T. (2003) Characterization of phytosiderophore secretion under Fe deficiency stress in *Festuca rubra*, *Plant Soil*, **256** (1), 131–137.

Ma, J.F., Zheng, S.J. & Matsumoto, H. (1997) Specific secretion of citric acid induced by Al stress in *Cassia tora* L., *Plant Cell Physiol.*, **38** (9), 1019–1025.

MacRobbie, E.A.C. (1999) Vesicle trafficking: a role in trans-tonoplast ion movements?, *J. Exp. Bot.*, **50**, 925–934.

Marrs, K.A. (1996) The functions and regulation of glutathione S-transferases in plants, *Annu. Rev. Plant Physiol. Plant Mol. Biol.*, **47**, 127–158.

Marsch-Martinez, N., Greco, R., Van Arkel, G., Herrera-Estrella, L. & Pereira, A. (2002) Activation tagging using the En-I maize transposon system in *Arabidopsis*, *Plant Physiol.*, **129** (4), 1544–1556.

Martinoia, E., Grill, E., Tommasini, R., Kreuz, K. & Amrhein, N. (1993) Atp-dependent glutathione S-conjugate export pump in the vacuolar membrane of plants, *Nature*, **364** (6434), 247–249.

Martinoia, E., Klein, M., Geisler, M., *et al.* (2002), Multifunctionality of plant ABC transporters – more than just detoxifiers, *Planta*, **214**, 345–355.

Master, E.R. & Mohn, W.W. (2001) Induction of bphA, encoding biphenyl dioxygenase, in two polychlorinated biphenyl-degrading bacteria, psychrotolerant *Pseudomonas* strain Cam-1 and mesophilic *Burkholderia* strain LB400, *Appl. Environ. Microbiol.*, **67** (6), 2669–2676.

Mattina, M.J.I., Iannucci-Berger, W. & Dykas, L. (2000) Chlordane uptake and its translocation in food crops, *J. Agric. Food Chem.*, **48** (5), 1909–1915.

Mawdsley, J.L. & Burns, R.G. (1994) Inoculation of plants with a *Flavobacterium* species results in altered rhizosphere enzyme-activities, *Soil Biol. Biochem.*, **26** (7), 871–882.

McGrath, S.P. & Zhao, F.J. (2003) Phytoextraction of metals and metalloids from contaminated soils, *Curr. Opin. Biotechnol.*, **14**, 277–282.

McGrath, S.P., Zhao, F.J. & Lombi, E. (2002) Phytoremediation of metals, metalloids, and radionuclides, *Adv. Agronomy*, **75**, 1–56.

Mench, M. & Martin, E. (1991) Mobilization of cadmium and other metals from 2 soils by root exudates of *Zea mays* L, *Nicotiana tabacum*-L and *Nicotiana rustica* L, *Plant Soil*, **132** (2), 187–196.

Miya, R.K. & Firestone, M.K. (2000) Phenanthrene-degrader community dynamics in rhizosphere soil from a common annual grass, *J. Environ. Qual.*, **29** (2), 584–592.

Miya, R.K. & Firestone, M.K. (2001) Enhanced phenanthrene biodegradation in soil by slender oat root exudates and root debris, *J. Environ. Qual.*, **30** (6), 1911–1918.

Murphy, A.S., Eisenger, W.R., Schaff, J.E., Kochian, L.V. & Taiz, L. (1999) Early copper-induced leakage of $K^+$ from *Arabidopsis* seedlings is mediated by ion channels and coupled to citrate efflux, *Plant Physiol.*, **121** (4), 1375–1382.

Naidu, R. & Harter, R.D. (1998) Effect of different organic ligands on cadmium sorption by and extractability from soils, *Soil Sci. Soc. Am. J.*, **62** (3), 644–650.

Nichols, T.D., Wolf, D.C., Rogers, H.B., Beyrouty, C.A. & Reynolds, C.M. (1997) Rhizosphere microbial populations in contaminated soils, *Water Air Soil Pollut.*, **95** (1–4), 165–178.

Noh, B., Murphy, A.S. & Spalding, E.P. (2001) Multidrug resistance-like genes of *Arabidopsis* required for auxin transport and auxin-mediated development, *Plant Cell*, **13** (11), 2441–2454.

Ogunseitan, O.A., Delgado, I.L., Tsai, Y.L. & Olson, B.H. (1991) Effect of 2-hydroxybenzoate on the maintenance of naphthalene-degrading pseudomonads in seeded and unseeded soil, *Appl. Environ. Microbiol.*, **57** (10), 2873–2879.

Palmiter, R.D., Cole, T.B. & Findley, S.D. (1996) ZnT-2, a mammalian protein that confers resistance by to zinc by facilitating vesicular sequestration, *EMBO J.*, **15**, 1784–1791.

Palmiter, R.D. & Findley, S.D. (1995) Cloning and functional characterization of a mammalian zinc transporter that confers resistance to zinc, *EMBO J.*, **14**, 639–649.

Pellet, D.M., Grunes, D.L. & Kochian, L.V. (1995) Organic-acid exudation as an aluminum-tolerance mechanism in maize (*Zea mays* L.), *Planta*, **196** (4), 788–795.

Pence, N.S., Larsen, P.B., Ebbs, S.D., *et al.* (2000) The molecular physiology of heavy metal transport in the Zn/Cd hyperaccumulator *Thlaspi caerulescens*, *Proc. Natl. Acad. Sci. U.S.A.*, **97** (9), 4956–4960.

Persans, M.W., Yan, X.G., Patnoe, J., Kramer, U. & Salt, D.E. (1999) Molecular dissection of the role of histidine in nickel hyperaccumulation in *Thlaspi goesingense* (Halacsy), *Plant Physiol.*, **121**, 1117–1126.

Pineros, M.A. & Kochian, L.V. (2001) A patch-clamp study on the physiology of aluminum toxicity and aluminum tolerance in maize. Identification and characterization of $Al^{3+}$-induced anion channels, *Plant Physiol.*, **125** (1), 292–305.

Pohlman, A.A. & McColl, J.G. (1986) Kinetics of metal dissolution from forest soils by soluble organic-acids, *J. Environ. Qual.*, **15** (1), 86–92.

Poulsen, I.F. & Hansen, H.C.B. (2000) Soil sorption of nickel in presence of citrate or arginine, *Water Air Soil Pollut.*, **120** (3/4), 249–259.

Psaras, G.K., Constantinidis, T., Costopoulos, B. & Manetas, Y. (2000) Relative abundance in the leaf epidermis of eight hyperaccumulators: evidence that the metal is excluded from both guard cell and trichomes, *Ann. Bot.*, **86**, 73–78.

Rea, P.A. (1999) MRP subfamily ABC transporters from plants and yeast, *J. Exp. Bot.*, **50**, (Special Issue), 895–913.

Reilley, K.A., Banks, M.K. & Schwab, A.P. (1996) Dissipation of polycyclic aromatic hydrocarbons in the rhizosphere, *J. Environ. Qual.*, **25** (2), 212–219.

Robertson, B.K. & Alexander, M. (1998) Sequestration of DDT and dieldrin in soil: disappearance of acute toxicity but not the compounds, *Environ. Toxicol. Chem.*, **17** (6), 1034–1038.

Romheld, V. (1990) The soil-root interface in relation to mineral-nutrition, *Symbiosis*, **9** (1–3), 19–27.

Ryan, P.R., Skerrett, M., Findlay, G.P., Delhaize, E. & Tyerman, S.D. (1997) Aluminum activates an anion channel in the apical cells of wheat roots, *Proc. Natl. Acad. Sci. U.S.A.*, **94** (12), 6547–6552.

Salt, D.E. & Rauser, W.E. (1995) MgATP-dependent transport of phytochelatins across the tonoplast of oat roots, *Plant Physiol.*, **107**, 1293–1301.

Salt, D.E., Smith, R.D. & Raskin, I. (1998) Phytoremediation, *Annu. Rev. Plant Physiol. Plant Mol. Biol.*, **49**, 643–668.

Sanchez-Fernandez, R., Davis, T.G.E., Coleman, J.O.D. & Rea, P.A. (2001) The *Arabidopsis thaliana* ABC protein superfamily; a complete inventory, *J. Biol. Chem.*, **276**, 30231–30244.

Sandermann, H. (1994) Higher-plant metabolism of xenobiotics – the Green Liver concept, *Pharmacogenetics*, **4** (5), 225–241.

Scheidemann, P. & Wetzel, A. (1997) Identification and characterization of flavonoids in the root exudate of *Robinia pseudoacacia*, *Trees-Struct. Funct.*, **11** (5), 316–321.

Shaul, O., Hilgemann, D.W., de-Almeida-Engler, J., Van Montague, M.V., Inze, D. & Galili, G. (1999) Cloning and characterization of a novel $Mg^{2+}/H^+$ exchanger, *EMBO J.*, **18**, 3973–3980.

Shen, Z.G., Zhao, F.J. & McGrath, S.P. (1997) Uptake and transport of zinc in the hyperaccumulator *Thlaspi caerulescens* and non-hyperaccumulator *Thlaspi ochroleucum*, *Plant Cell Environ.*, **20**, 898–906.

Sidler, M., Hassa, P., Hasan, S., Ringli, C. & Dudler, R. (1998) Involvement of an ABC transporter in a developmental pathway regulating hypocotyl cell elongation in the light, *Plant Cell*, **10** (10), 1623–1636.

Singer, A.C., Gilbert, E.S., Luepromchai, E. & Crowley, D.E. (2000) Bioremediation of polychlorinated biphenyl-contaminated soil using carvone and surfactant-grown bacteria, *Appl. Microbiol. Biotechnol.*, **54**(6), 838–843.

Singer, A.C., Jury, W., Luepromchai, E., Yahng, C.S. & Crowley, D.E. (2001) Contribution of earthworms to PCB bioremediation, *Soil Biol. Biochem.*, 33 (6), 765–776.

Smith, J.A.C., Harper, F.A., Leighton, R.S., Thompson, I.P., Vaughan, D.J. & Baker, A.J.M. Comparative analysis of metal uptake, transport and sequestration in hyperaccumulator plants, in *Proceedings of the 5th International Conference on the Biogeochemistry of Trace Elements* (eds W.W. Wenzel, D.C. Adriano, B. Alloway, H.E. Doner & C. Keller), Vienna, pp. 22–32.

Srivastava, S., Prakash, S. & Srivastava, M.M. (1999) Studies on mobilization of chromium with reference to its plant availability – role of organic acids, *Biometals*, **12** (3), 201–207.

Sunkar, R., Kaplan, B., Bouche, N., *et al.* (2000) Expression of a truncated tobacco *NtCBP4* channel in transgenic plants and disruption of the homologous *Arabidopsis CNGC1* gene confers $Pb^{2+}$ tolerance, *Plant J.*, **24**, 533–542.

Tam, S.C. & McColl, J.G. (1990) Aluminum-binding and calcium-binding affinities of some organic-ligands in acidic conditions, *J. Environ. Qual.*, **19** (3), 514–520.

Theodoulou, F.L. (2000) Plant ABC transporters, *Biochim. Biophys. Acta*, **1465**, 79–103.

Thomine, S., Lelievre, F., Debarbieux, E., Schroeder, J.I. & Barbier-Brygoo, H. (2003) AtNRAMP3, a multispecific vacuolar metal transporter involved in plant responses to iron deficiency, *Plant J.*, **34** (5), 685–695.

Thomine, S., Wang, R., Ward, J.M., Crawford, N.M. & Schroeder, J.I. (2000) Cadmium and iron transport by members of plant metal transporter family in *Arabidopsis* with homology to *Nramp* genes., *Proc. Natl. Acad. Sci. U.S.A.*, **97**, 4991–4996.

Tommasini, R., Martinoia, E., Grill, E., Dietz, K.J. & Amrhein, N. (1993) Transport of oxidized glutathione into barley vacuoles – evidence for the involvement of the glutathione-*S*-conjugate ATPase, *Z. Naturforsch., C: J. Biosci.*, **48** (11/12), 867–871.

Tommasini, R., Vogt, E., Fromenteau, M., *et al.* (1998), An ABC-transporter of *Arabidopsis thaliana* has both glutathione-conjugate and chlorophyll catabolite transport activity, *Plant J.*, **13** (6), 773–780.

Van der Meer, J.R., Devos, W.M., Harayama, S. & Zehnder, A.J.B. (1992) Molecular mechanisms of genetic adaptation to xenobiotic compounds, *Microbiol. Rev.*, **56** (4), 677–694.

Vazquez, M.D., Poschenrieder, C., Barcelo, J., Baker, A.J.M., Hatton, P. & Cope, G.H. (1994) Compartmentation of zinc in roots and leaves of the zinc hyperaccumulator *Thlaspi caerulescens*, *Bot. Acta*, **107**, 243–250.

Verpoorte, R. & Memelink, J. (2002) Engineering secondary metabolite production in plants, *Curr. Opin. Biotechnol.*, **13** (2), 181–187.

Vogeli-Lange, R. & Wagner, G.J. (1990) Sub-cellular localization of cadmium and cadmium-binding peptides in tobacco leaves. Implications of a transport function for cadmium-binding peptides, *Plant Physiol.*, **92**, 1086–1093.

Vonuexkull, H.R. & Mutert, E. (1995) Global extent, development and economic-impact of acid soils, *Plant Soil*, **171** (1), 1–15.

Walker, T.S., Bais, H.P., Grotewold, E. & Vivanco, J.M. (2003) Root exudation and rhizosphere biology, *Plant Physiol.*, **132** (1), 44–51.

Weigel, D., Ahn, J.H., Blazquez, M.A., *et al.* (2000) Activation tagging in *Arabidopsis*, *Plant Physiol.*, **122** (4), 1003–1014.

Wenzel, H.J., Cole, T.B., Born, D.E., Shwartzkroin, P.A. & Palmiter, R.D. (1997) Ultrastructural localization of zinc transporter-3 (ZnT-3) to synaptic vesicle membranes within mossy fibre boutons in the hippocampus of mouse and monkey, *Proc. Natl. Acad. Sci. U.S.A.*, **94**, 12676–12681.

White, J.C. (2001) Plant-facilitated mobilization and translocation of weathered 2,2-bis(*p*-chlorophenyl)-1,1-dichloroethylene (*p*,*p'*-DDE) from an agricultural soil, *Environ. Toxicol. Chem.*, **20** (9), 2047–2052.

White, J.C. (2002) Differential bioavailability of field-weathered $p,p'$-DDE to plants of the *Cucurbita* and *Cucumis* genera, *Chemosphere*, **49** (2), 143–152.

White, J.C. & Kottler, B.D. (2002) Citrate-mediated increase in the uptake of weathered 2,2-bis($p$-chlorophenyl)1,1-dichloroethylene residues by plants, *Environ. Toxicol. Chem.*, **21** (3), 550–556.

Wiehe, W., Hechtbuchholz, C. & Hoflich, G. (1994) Electron-microscopic investigations on root colonization of *Lupinus albus* and *Pisum sativum* with 2 associative plant-growth promoting rhizobacteria, *Pseudomonas fluorescens* and *Rhizobium leguminosarum* Bv trifolii, *Symbiosis*, **17** (1), 15–31.

Williams, L.E., Pittman, J.L. & Hall, J.L. (2000) Emerging mechanisms for heavy metal transport in plants, *Biochim. Biophys. Acta*, **1465**, 104–126.

Winkel-Shirley, B. (2001) Flavonoid biosynthesis. A colorful model for genetics, biochemistry, cell biology, and biotechnology, *Plant Physiol.*, **126** (2), 485–493.

Xue, J. & Huang, P.M. (1995) Zinc adsorption-desorption on short-range ordered iron-oxide as influenced by citric-acid during its formation, *Geoderma*, **64** (3/4), 343–356.

Yang, C.H. & Crowley, D.E. (2000) Rhizosphere microbial community structure in relation to root location and plant iron nutritional status, *Appl. Environ. Microbiol.*, **66** (1), 345–351.

Yang, Y.Y., Jung, J.Y., Song, W.Y., Suh, H.S. & Lee, Y. (2000) Identification of rice varieties with high tolerance or sensitivity to lead and characterization of the mechanism of tolerance, *Plant Physiol.*, **124** (3), 1019–1026.

Yen, K.M. & Gunsalus, I.C. (1982) Plasmid gene organization – naphthalene salicylate oxidation, *Proc. Natl. Acad. Sci. U.S.A.-Biol. Sci.*, **79** (3), 874–878.

Zarka, D.G., Vogel, J.T., Cook, D. & Thomashow, M.F. (2003) Cold induction of *Arabidopsis* CBF genes involves multiple ICE (inducer of CBF expression) promoter elements and a cold-regulatory circuit that is desensitized by low temperature, *Plant Physiol.*, **133** (2), 910–918.

Zhang, F.S., Romheld, V. & Marschner, H. (1991) Release of zinc mobilizing root exudates in different plant – species as affected by zinc nutritional-status, *J. Plant Nutr.*, **14** (7), 675–686.

Zhao, F.J., Lombi, E., Breedon, T. & McGrath, S.P. (2000) Zinc hyperaccumulation and cellular distribution in *Arabidopsis halleri*, *Plant Cell Environ.*, **23**, 507–514.

Zheng, S.J., Ma, J.F. & Matsumoto, H. (1998) Continuous secretion of organic acids is related to aluminium resistance during relatively long-term exposure to aluminium stress, *Physiol. Plantarum*, **103** (2), 209–214.

# Index